INTRODUCTION TO
ELECTRICAL ENGINEERING

McGRAW-HILL SERIES IN ELECTRICAL AND COMPUTER ENGINEERING

Senior Consulting Editor

Stephen W. Director, Carnegie Mellon University

CIRCUITS AND SYSTEMS
COMMUNICATIONS AND SIGNAL PROCESSING
COMPUTER ENGINEERING
CONTROL THEORY
ELECTROMAGNETICS
ELECTRONICS AND ELECTRONIC CIRCUITS
INTRODUCTORY
POWER AND ENERGY
RADAR AND ANTENNAS
VLSI

Previous Consulting Editors

Ronald N. Bracewell, Colin Cherry, James F. Gibbons, Willis W. Harman,
Hubert Heffner, Edward W. Herold, John G. Linvill, Simon Ramo,
Ronald A. Rohrer, Anthony E. Siegman, Charles Susskind,
Frederick E. Terman, John G. Truxal, Ernst Weber, and John R. Whinnery

INTRODUCTORY

Senior Consulting Editor

Stephen W. Director, Carnegie Mellon University

Belove, Schachter, and Schilling
DIGITAL AND ANALOG SYSTEMS, CIRCUITS AND DEVICES
Fitzgerald, Higginbotham, and Grabel
BASIC ELECTRICAL ENGINEERING
Paul, Nasar, and Unnewehr
INTRODUCTION TO ELECTRICAL ENGINEERING
Peebles and Giuma
PRINCIPLES OF ELECTRICAL ENGINEERING
Piel and Truxal
TECHNOLOGY: HANDLE WITH CARE

INTRODUCTION TO ELECTRICAL ENGINEERING

Second Edition

C. R. Paul
S. A. Nasar

Department of Electrical Engineering
University of Kentucky

L. E. Unnewehr

Lewis Consulting Service
Michigan City, Indiana

McGraw-Hill, Inc.

New York   St. Louis   San Francisco
Auckland   Bogotá   Caracas
Lisbon   London   Madrid
Mexico   Milan   Montreal
New Delhi   Paris   San Juan
Singapore   Sydney   Tokyo   Toronto

INTRODUCTION TO ELECTRICAL ENGINEERING

 This book is printed on recycled paper containing a minimum of 50% total recycled fiber with 10% postconsumer de-inked fiber.

234567890 DOC DOC 9098765432

ISBN 0-07-011322-X

This book was set in Times Roman by Techset Composition Limited
The editors were Anne T. Brown, K. Kimbell, and Jack Maisel.
The cover was designed by Carla Bauer.
R. R. Donnelley & Sons Company was printer and binder.

Library of Congress Cataloging-in-Publication Data

Paul, Clayton R.
     Introduction to electrical engineering / C. R. Paul, S. A. Nasar,
L. E. Unnewehr. — 2nd ed.
         p.   cm.
     Includes index.
     ISBN 0-07-011322-X
     1. Electric engineering.   I. Nasar, S. A.   II. Unnewehr, L. E.,
     (date)   III. Title.
TK146.P345   1992   91-39610
621.3—dc20

CONTENTS

Preface to the Second Edition   xi

Preface to the First Edition   xiii

CHAPTER 1 Introduction   1

PART 1   LINEAR ELECTRIC CIRCUITS

CHAPTER 2 Circuit Elements and Laws   9

2.1	Charge and Electric Forces	11
2.2	Voltage	13
2.3	Current and Magnetic Forces	16
2.4	Lumped-Circuit Elements	18
2.5	Kirchhoff's Voltage and Current Laws	21
2.6	The Resistor	27
2.7	Voltage and Current Sources	31
2.8	Signal Waveforms	34
2.9	Analysis of Simple Circuits	40
	Problems	48

CHAPTER 3   Analytical Techniques for Resistive Circuits   57

3.1 Series and Parallel Connections of Resistors   57
3.2 Voltage and Current Division   65
3.3 Superposition   69
3.4 Thévenin and Norton Equivalent Circuits   76
3.5 Node Voltage and Mesh Current Analysis   81
3.6 Maximum Power Transfer   94
         Problems   96

CHAPTER 4   Energy Storage Elements   104

4.1 The Capacitor   104
4.2 The Inductor   113
4.3 Continuity of Capacitor Voltages and Inductor Currents 119
4.4 Transformers and Coupled Coils   119
4.5 Circuit Differential Equations for Circuits Containing Energy
         Storage Elements   128
4.6 Response of the Energy Storage Elements to DC Sources   131
         Problems   134

CHAPTER 5   AC Circuits   145

5.1 Sinusoidal (AC) Sources   146
5.2 Response of Circuits to AC Sources   149
5.3 Complex Numbers and Complex Algebra   151
5.4 Use of the Complex Exponential Source   159
5.5 The Phasor Circuit   164
5.6 Average Power, Reactive Power, and Power Factor   172
5.7 Frequency Response of Circuits   185
         Problems   201

CHAPTER 6   Transients   210

6.1 The RL Circuit   212
6.2 The RC Circuit   222
6.3 Writing the Solution by Inspection for Circuits Containing
         One Energy Storage Element   228
6.4 Circuits Involving Two Energy Storage Elements   235
         Problems   259

PART 2 ELECTRONIC CIRCUITS

CHAPTER 7   Diodes   267

    7.1   The Semiconductor Diode   267
    7.2   Load Lines   275
    7.3   Piecewise-Linear Approximation and Small-Signal Analysis   278
    7.4   The Ideal Diode   283
    7.5   Rectifiers   288
    7.6   The Limiter   298
    7.7   Zener Diodes   298
         Problems   306

CHAPTER 8   Transistors and Amplifiers   308

    8.1   Principles of Linear Amplifiers   308
    8.2   The Junction Field-Effect Transistor   311
    8.3   The JFET Amplifier   316
    8.4   Small-Signal Analysis of the JFET Amplifier   327
    8.5   Frequency Response of the JFET Amplifier   338
    8.6   The Bipolar Junction Transistor   342
    8.7   The BJT Amplifier   347
    8.8   Small-Signal Analysis of the BJT Amplifier   356
    8.9   Frequency Response of the BJT Amplifier   370
    8.10 Integrated Circuits   373
         Problems   384

CHAPTER 9   The Operational Amplifier   389

    9.1   The Ideal Operational Amplifier   389
    9.2   The Practical Operational Amplifier   395
    9.3   Inductorless (Active) Filters   399
    9.4   Analog Computers   405
         Problems   408

CHAPTER 10   Digital Electronic Circuits   411

    10.1 Diode Gates   412
    10.2 The BJT Switch   417
    10.3 Logic Families   425
    10.4 Complementary Metal-Oxide-Semiconductor (CMOS) Gates   431
    10.5 Multivibrators   437
    10.6 A Practical Note   448
         Problems   449

PART 3   ELECTRIC MACHINES AND TRANSFORMERS

CHAPTER 11   Polyphase Circuits   **455**

11.1   Three-Phase Systems   455
11.2   Three-Phase Connections   457
11.3   Power in Three-Phase Systems   460
11.4   Wye-Delta Equivalence   464
    Problems   467
    References   468

CHAPTER 12   Transformers   **469**

12.1   Magnetic Circuits   469
12.2   Core Material   471
12.3   Analogy With DC Resistive Circuits   472
12.4   Core Losses   473
12.5   Magnetic and Electric Circuits of a Transformer   476
12.6   Transformer Construction   477
12.7   Principle of Operation of a Transformer   477
12.8   Voltage, Current and Impedance Transformations   479
12.9   Non-ideal Transformer and its Equivalent Circuit   481
12.10 Transformer Polarity   486
12.11 Autotransformers   488
    Problems   489
    References   491

CHAPTER 13   DC Machines   **492**

13.1   The Faraday Disk and Faraday's Law   493
13.2   The Heteropolar, or Conventional, DC Machine   495
13.3   Constructional Details   499
13.4   Classification according to Forms of Excitation   502
13.5   Performance Equations   503
13.6   Effects of Saturation—Buildup of Voltage in a Shunt Generator   509
13.7   Generator Characteristics   510
13.8   Motor Characteristics   512
13.9   Starting of DC Motors   513
13.10 Speed Control of DC Motors   515
13.11 Losses and Efficiency   523
    Problems   525
    References   528

CHAPTER 14   Synchronous Machines   **529**

    14.1   Some Construction Details   530
    14.2   Magnetomotive Forces (MMFs) and Fluxes due to Armature
           and Field Windings   533
    14.3   The Synchronous Speed   536
    14.4   Synchronous Generator Operation   537
    14.5   Performance of a Round-Rotor Synchronous Generator   538
    14.6   Synchronous Motor Operation   542
    14.7   Performance of a Round-Rotor Synchronous Motor   543
           Problems   546
           References   547

CHAPTER 15   Induction Machines   **548**

    15.1   Operation of a Three-Phase Induction Motor   550
    15.2   Slip   550
    15.3   Development of Equivalent Circuits   552
    15.4   Performance Calculations   555
    15.5   Performance Criteria of Induction Motors   560
    15.6   Speed and Torque Control of Induction Motors   561
    15.7   Starting of Induction Motors   568
           Problems   571

CHAPTER 16   Small AC Motors   **573**

    16.1   Single-Phase Induction Motors   573
    16.2   Small Synchronous Motors   580
    16.3   AC Commutator Motors   582
    16.4   Two-Phase Motors   584
    16.5   Stepper Motors   586
           Problems   588

PART 4   CONTROL AND INSTRUMENTATION

CHAPTER 17   Feedback Control Systems   **591**

    17.1   Basic Concepts of Control   591
    17.2   Some Elementary Feedback Systems   605
    17.3   Linear Control Systems Analysis   611
    17.4   Types of Feedback Control Systems   617
           Problems   624
           References   629

CHAPTER 18   Electrical Instrumentation   **630**

    18.1   Some Basic Concepts of Electrical Measurements   630
    18.2   Measurement of Voltage, Current, and Power   634
    18.3   Measurement of Impedance   646
    18.4   Oscilloscopes   648
           Problems   651
           References   652

CHAPTER 19   Digital Logic Circuits   **653**

    19.1   Binary Bits   653
    19.2   Introduction to Boolean Algebra   656
    19.3   Implementation of Boolean Logic   661
    19.4   Sequential Logic   671
    19.5   Hardware Implementation of Logic Circuits   679
    19.6   IC Logic Families   680
           Problems   689
           References   690

CHAPTER 20   Digital Systems   **691**

    20.1   Pulse Modulation and Sampline   695
    20.2   Signal Conversion   705
    20.3   Data Acquisition   713
    20.4   Arithmetic Digital Systems   716
    20.5   Types of Integrated Circuits   729
    20.6   Microprocessors   739
    20.7   Microcomputers   746
           Problems   749
           References   750

APPENDIX A   Unit Conversion   **751**

APPENDIX B   Table of Constants   **752**

APPENDIX C   Fourier Series   **753**

APPENDIX D   Laplace Transforms   **754**

APPENDIX E   Solution of Simultaneous Equations   **756**

           Answers to Problems   **763**

           Index   **767**

PREFACE TO THE SECOND EDITION

In this second edition of the book, a major portion of Chapter 1 has been rewritten. Other significant revisions in the second edition are as follows. In Part 1, Chapter 2 has been expanded by adding examples of the solution of some simple circuits (single loop and single node pair). Chapter 3 now includes examples of writing node voltage equations for circuits containing voltage sources. Examples of writing mesh current equations for circuits containing current sources are also included. The material on Gauss elimination and Cramer's rule has been moved to Appendix E. Chapter 4 is essentially unchanged, although additional material on mutual inductance has been included. Additional material on the transfer function has been added to Chapter 5. The differential operator method has been removed from Chapter 6.

The primary change to Part II, "Electronic Circuits," is that Chapters 8 and 9 have been combined. The small-signal analysis of amplifiers is now covered immediately after discussions of the BJT and FET elements and their biasing in Chapter 8. Chapters 10 and 11 of the first edition now become Chapter 9 and 10, respectively.

In Part III, the chapter on electric power systems has been omitted. In its place, a short chapter on small ac motors (Chapter 16) has been added. The chapter on control of electric motors has been deleted. The pertinent material, however, has been included in the respective chapters on dc and ac machines. Finally, throughout this part, more solved examples and chapter problems have been added.

In Part IV of the book, in Chapter 17, a discussion of the classification of feedback control systems by several methods is presented. Steady state error response is defined and illustrated. Proportional and integral control are defined theoretically and by numerous examples and illustrations. The emphasis of the chapter has been slanted more toward industrial control applications and hardware. Several analytical methods, including Routh criterion and root-locus, have been eliminated. Several chapter problems have also been added. In Chapter 18, a section on digital oscilloscopes has been added, including a new photograph of one of the latest models of digital scopes. Several photos and associated discussions of older instruments have been eliminated. Almost no change has been made in Chapter 19, except for the elimination of excess wordage. This is a theoretical and analytical presentation which did not require updating. Chapter 20 has been updated considerably, and several photos and discussions of outmoded digital circuits have been eliminated. The section on memory chips has been expanded and the latest characteristics of memory chips have been defined. A discussion of IC types for the nonsemiconductor specialist has been added, including descriptions of programmable devices (PLDs), programmable ROMs, and semicustom chips. Some simple analysis methods used in evaluating chips are presented, including means of estimating chip power loss and chip size.

McGraw-Hill and the authors would like to thank the following reviewers for their many helpful comments and suggestions: Anthony England, University of Michigan; John A. Fleming, Texas A&M University; Virginia D. McWhorter, Virginia Polytechnic Institute and State University; D. P. Malone, State University of New York at Buffalo; Bahram Mirshab, Lawrence Institute of Technology; Frank W. Smith, University of Tennesse at Chattanooga; and Harold Smith, University of Toronto.

C. R. Paul
S. A. Nasar
L. E. Unnewehr

PREFACE TO THE FIRST EDITION

The field of electrical engineering has developed at a dramatic rate in the last 30 years. Solid-state electronics virtually shouldered vacuum-tube electronics aside after the second world war, electrical engineering courses in automatic control soon became commonplace, and within the past 15 years integrated-circuit technology has dynamically expanded the electrical engineering curriculum. These topics have come to be integral parts of the discipline. Nevertheless, very few of such older, more traditional subjects as electric machinery, power, and instrumentation have been deemphasized. Even though their treatment may have been compacted, they continue to form an important part of the curriculum.

This text provides a survey of all topics integral to the electrical engineering discipline. Seeking to characterize the field of electrical engineering, one invariably finds four broad, central topics—electric circuits, electronics, electric machines and power systems, and systems control and instrumentation. The text has been divided along those lines.

Part 1—Linear Electric Circuits—provides the basic circuit-analysis techniques and concepts used throughout the subsequent topics. The basic circuit elements and the concepts of voltage, current, ideal sources, resistors, signal waveforms and Kirchhoff's laws are introduced in Chap. 2. This forms the basis for resistive-circuit analysis, presented in Chap. 3. We have chosen to delay the discussion of energy storage elements—the capacitor and the inductor—until Chap. 4. Circuit-analysis techniques can be illustrated with resistive circuits without the additional complication of differential equations.

We have chosen to discuss sinusoidal steady-state analysis (Chap. 5) prior to general waveform excitation (Chap. 6). We believe that this ordering of the material is easier for the student to assimilate. First-order RL and RC circuits, along with RLC circuits, are discussed for general excitation in Chap. 6. Simple circuits are used to illustrate the concepts of transient response. More general circuits are handled using the differential operator in the last section of Chap. 6.

Part 2—Electronic Circuits—covers the traditional material concerning the analysis of discrete electronic devices in Chaps. 7 (Diodes), 8 (Transistors and Amplifiers), and 9 (Small-Signal Analysis of Amplifiers). The operational amplifier and its uses are covered in Chap. 10. Chapter 11 provides a discussion of the use of these discrete elements to form digital circuits. The systems application of these digital circuits is reserved for discussion in Part 4.

Part 3—Electric Power and Machines—covers three-phase circuits (Chap. 12), magnetic circuits and transformers (Chap. 13), and electric machines (Chaps. 14, 15, 16). The control of electric motors and certain important topics pertinent to power systems are presented in Chaps. 17 and 18, respectively

Part 4—Control and Instrumentation—brings together the techniques and concepts of the previous three parts in the design of systems to accomplish specific tasks. The important topics of instrumentation and transducers are covered in Chap. 20. This part closes with a discussion of the predominant method of modern system implementation—digital techniques—in Chaps. 21 and 22. Chapter 22 also contains material on data acquisition and coding.

The appendixes cover unit conversion (A), physical constants (B), Fourier series (C), and Laplace transforms (D).

The text is suitable for a one-year (two semesters or three quarters) introductory course offered to non-EE majors. We suggest coverage of Part 1 (Linear Electric Circuits) and Part 2 (Electronic Circuits) in the first half year, with Part 3 (Electric Power and Machines) and Part 4 (Control and Instrumentation) covered in the second half year.

The level of the material requires the student to have completed the basic college-level courses in algebra, trigonometry, and elementary calculus, as well as basic physics. A knowledge of differential equations will be helpful but not mandatory, since we develop those solution techniques within the text when they naturally arise.

We have endeavored in this text to provide the concepts, analytical methods, and topics basic to the field of electrical engineering. The depth of coverage has consciously been kept at the minimum level for each subject—yet at the same time we have chosen not merely to give equations and rules alone, without logical development. Rote memorization provides no long-term value to the student. We believe that students will not only be the more attracted to a subject and retain more from it when we give them simple, logical developments of the principles, but that they will also be the better able to apply these principles to future situations.

We would like to express our thanks for the many useful comments and suggestions provided by colleagues who reviewed this text during the course of its development, especially to Robert Cromwell; H. Hayre, University of Houston; Martin Kaliski, Northeastern University; Martin Kaplan, Drexel University; Douglas Miron, South Dakota State University; Louis Roemer, University of Akron; Lee Rosenthal; Martha Sloan, Michigan Technological University; William Swanson, San Jose State University; and R. P. Misra, New Jersey Institute of Technology.

C. R. Paul
S. A. Nasar
L. E. Unnewehr

INTRODUCTION TO ELECTRICAL ENGINEERING

Introduction

Electrical engineering is concerned with the design and analysis of electric circuits, systems, and devices. In the early history of this branch of engineering, some distinction was made between the terms "electrical" and "electronic," the former term being associated with the relatively high current applications in power and machinery and the latter term with radio and communications circuits. For many years, this distinction has become increasingly blurred, and today the general term "electrical" is used to describe all circuits involving electron conduction and convection and their associated electromagnetic fields. In this text, the terms "electrical" and "electronic" are used more from historical convention than as descriptors of generic difference, and the basic theory developed here is applicable to all electric circuits when proper limits and restrictions are satisfied. The broader area known as electrical engineering also encompasses many additional classes of electrical phenomena, such as superconductivity, electrical conduction in human tissues and organs, evaluation of extraterrestrial signals, solar-electrical energy conversion, permanent magnet systems, and many others. And the major concepts of electrical engineering are endemic to computer science and technology, which affect almost every aspect of life in industrial societies today.

The purpose of this text is to present the basic theory and practice of electrical engineering to students with many different types of backgrounds and interests. While some of the material to be presented may be of value in working with the common electric circuits found in daily life, such as household wiring systems, automotive signals and controls, computer arithmetic etc., the majority of the material is presented to give the reader an understanding of the theory of electric circuits and systems for use in a wide range of applications. A knowledge of these basic concepts is vital to most engineers, scientists, and technologists today. The necessity of having some understanding of electrical principles will only increase in coming years. For example, the development of alternative energy sources will be a major concern of industrial societies in the near future, and this will involve the solution of problems in many disciplines. The development of solar-electrical conversion systems, such as photovoltaic converters, will require an understanding of the electrical, optical, thermal, and materials systems involved; the fluidization of coal, the largest energy source in many industrial countries, and the development of safe nuclear power are both essential for future energy needs and will likewise involve many engineering and science disciplines, including electric power, controls, and computer technology. Improved manufacturing processes present another engineering challenge for coming years to increase productivity and to reduce the monotony of the assembly line. This will require new and more, economical control systems, improved electric motor characteristics, and more sophisticated computer supervisory systems.

Energy conservation and protecting the environment are continuing challenges involving all engineering disciplines. A major area for energy savings is the prosaic electric motor. Approximately 60 percent of all electric energy used in industrialized countries is used in electric motors. Therefore, even small improvements in the average efficiency of electric motors are significant. Motor efficiency can be increased through the use of improved materials (materials engineer), improved heat transfer (mechanical engineer), more sophisticated control (control and computer engineer), and by cleverer motor designs and configurations (electrical engineer). Personal transportation is another area where small improvements in efficiency can make large savings in total energy and reduced dependence on oil imports. There is little doubt that engineers in coming years will be continuing the impressive work already in progress in this area: increased electronic engine control, more efficient auxiliary systems, wider use of alternative fuels and energy sources including battery-operated vehicles, etc. Other environmental problems that must be addressed in the near future are nuclear waste disposal; improved scrubbers for collecting particles and emissions from industrial processes; the recently stated concern of the effects of electromagnetic radiation upon humans from electric power lines, computer displays, microwave ovens, etc.; protection of the atmosphere, and so on. And closely related to these more obvious environmental issues in the growing field of medical electronics which is involved not only in sensing environmental conditions hazardous to humans and other species, but also in sensing and observing (and sometimes helping to correct) many human and mammalian internal processes, diseases, and dysfunctions.

The achievements of electrical instrumentation and communications have been spectacular in recent decades. We have all witnessed almost routine communications between earth stations and many types of space vehicles and their occupants; we have observed a Voyager vehicle transmitting signals from beyond the Solar System after a voyage of over 9 years; we have seen that long-cherished dream of "one world" become a reality from the communications standpoint and watch on the nightly TV news such diverse news stories as a scientific project in Antarctica, the effects of food shortages in Ethiopia, fighting in Afghanistan, the "song" of a whale off the shores of Alaska, etc., all unthinkable only a few decades ago. It might be asked whether any engineering challenges remain in this vast technology of communications and related computer and instrumentation technology; and, of course, the answer is a resounding yes. While the density of transistors per silicon chip may be approaching physical limits, there is the potential of "high-temperature" super-conducting materials just around the corner as well as other relatively new materials such as gallium arsenide. And the potential for improved computer codes, computer networking, and computer architecture seems almost unlimited. And many unpredictable developments certainly await; after all, the integrated circuit was developed only recently—probably within the lifetime of most readers of this text.

This text is only an "introduction" to the exciting field of electrical engineering, but the authors hope that it will help inspire the reader to apply the basic principles presented here to the many challenges described above. The text is divided into four major areas which are summarized as follows:

Electric Circuits This part begins by defining the basic electrical parameters and their units. Kirchhoff's and other basic laws of circuits are defined and illustrated. The electrical circuit elements—resistor, inductor, and capacitor—are defined and discussed. The circuits containing these elements are generally considered linear in the mathematical sense; examples of nonlinearity are considered. Mutual inductance and coupled circuits are presented and the theory of these circuits is developed. The principles of superposition and Norton and Thévenin equivalent circuits are developed. Power, reactive power, and phasor diagrams are defined and illustrated. Many methods of circuit analysis are presented which are of value in the study of most types of problems in steady-state dc and ac circuits and in transient analysis.

Engineers must acquire a knowledge of electric circuit analysis; in other words, a proficiency in carrying out a mathematical study of various types of interconnections of electrical elements having paths for the flow of electric current. On the surface it may appear that such a skill is useful only to electrical engineers. However, as shown in the following chapters, electric circuit analysis is a prerequisite for the study of the major areas of electrical engineering. Electric circuit analysis not only aids in pursuing further studies in electrical engineering but also broadens our scope and helps us to communicate with others involved in a common project (since an engineer invariably becomes a member of a team consisting of other engineers and managers). For a coordinated group effort it is important that various members of the group meaningfully communicate with each other. The concepts pertaining to electric circuits extend to problems such as traffic flow, socioeconomic systems, control and instrumentation, fluid flow, etc.

Electronic Circuits As noted previously, the term *electronic* is a historical carryover from the days of early radio and TV developments. The term was used to distinguish circuits of very low current levels and which also contained "electronic devices"—in those days, vacuum tubes. The term is used in this text only in the latter sense, although vacuum tubes have been almost totally supplanted by semiconductor devices. However, the level of electric current is no longer meaningful, since power semiconductor devices (and some vacuum tubes) operate at thousands of amperes. The purpose of this section is first to define these semiconductor devices, present their characteristics, discuss their temperature and power limitations, and then to illustrate their use in communication, control, computer, and power circuits. Conduction in semiconductors is defined and discussed. In control, computer, and power applications, semiconductors are used in an on-off mode, and this requires some modification of the circuit analysis techniques. Often called the *switching mode*, this mode also results in voltage and current waveforms considerably different from those used in the common dc and sine wave ac analysis. However, under appropriate restrictions, the basic circuit laws are applicable.

The range of applications of electronic circuits is enormous and is probably limited only by our imagination. Our goal in this major area is to develop analytical techniques for quantitative studies of some basic electronic circuits. Relying on the methods developed in the study of electric circuits, we study the different types of diodes and transistors and then use these electronic components to obtain electronic circuits which perform such functions as amplifications, waveshaping, and switching.

A thorough understanding of the operation and analysis of electronic circuits is essential to the engineer and is of great value in choosing and applying electronic circuits to perform the desired task.

Electric Machines and Transformers   Harnessing and utilizing energy has always been a key factor in improving the quality of life. The use of electrical energy has aided our ability to develop socially and economically and live with physical comfort. The major area of electrical engineering concerned with the generation, transmission, distribution, and utilization of electrical energy comes under the heading of "electric power and machines," a portion of which is covered in Part 3 of this book. The subject matter contained in this part encompasses a wide variety of disciplines of engineering. The importance of electric machines in almost every aspect of life can hardly be overemphasized. The number of electric motors in the average U.S. residence today is probably a minimum of 10 and can easily exceed 50. There are at least five electric machines on even the most spartan compact automobile, and this number is increasing steadily as emission and fuel economy systems are added. An aircraft contains many more electric machines. More people travel by means of electrical propulsion each day—in elevators, escalators, and horizontal people movers —than by any other mode of propulsion. Many electric machines have been on the moon and play an important role in most aerospace systems. Of course, electric machines are involved in every industrial and manufacturing process of a technological society. Electrical blackouts constantly remind us of our almost total dependence on electric power. Therefore, an understanding of the principles of energy conversion, electric machines, and power systems is important for all who wish to extend the usefulness of electrical technology in order to ameliorate the problems of energy, pollution, and poverty that presently face humanity.

Power technology is associated with the origins of electrical engineering back in the mid-nineteenth century which largely made possible the industrial society that we know today. It is associated with such famous names as Edison, Tesla, Steinmetz, Brush, Siemens, and Kelvin and was originally oriented toward producing electrical illumination. It is generally considered to be a "mature" technology, but many exciting things are happening in this old technology today. The many needs for alternative energy sources and improved motor efficiency have already been discussed. There are many exciting developments underway in power semiconductors, permanent magnets, and insulation that will have profound effects upon future power transmission and utilization systems.

Control and Instrumentation The fourth major area of electrical engineering of interest to all engineers is control and instrumentation. Control systems influence our everyday lives just as much as some of the other areas of electrical engineering. For example, household appliances, such as clothes washers and dryers, manufacturing and processing plants, and navigation and guidance systems all utilize the concepts of analysis and design of control systems. Therefore, in general, a control system is an interconnection of components—electrical, mechanical, hydraulic, thermal, etc.—to obtain a desired function in an efficient and accurate fashion. The control engineer is involved in the control of industrial processes and systems. The concepts of control engineering are not limited to any particular branch of engineering. Hence, a basic understanding of control theory is very useful to every engineer involved in the understanding of dynamics of various types of systems.

Measurements of quantities such as voltage, current, temperature, and strain are of utmost importance to the engineer. Measurements tell us much about the universe, the environment, the economy, and ourselves. The basis of all scientific principles and engineering concepts rests, ultimately, on their verification by tests and measurements. Therefore, accurate measurement techniques in everyday engineering problems—and knowing the limitations of these techniques—is of great value to the engineer.

Some of the chapters in Part 4 will be out of date by the time they are read owing to the rapid changes in computer circuit technology, but that is the price of trying to describe the present state of such a fast-changing technology.

We feel that engineers who acquire a basic knowledge of linear electric circuits, electronic circuits, electric machines, and control and instrumentation will have a well-rounded background and be better suited to join a team effort.

Linear
Electric
Circuits

2

Circuit Elements and Laws

In the chapters of Part 1 we study *electric circuits*. An electric circuit is an idealized, mathematical *model* of some physical circuit. The ideal *circuit elements* are the resistor, inductor, capacitor, and the voltage and current sources. Each of these circuit elements is also an ideal mathematical model of some physical circuit element. For example, the resistor circuit element shown in Fig. 2.1*a* is a mathematical approximation of a physical resistor that can be purchased in any electronics store. Figure 2.1*b* and *c* illustrates the difference between the ideal resistor model and an actual resistor. The important electric circuit parameters that we will be interested in obtaining are the *voltage v* and *current i* associated with each element of the circuit. The voltage *v* and current *i* associated with the ideal resistor model and an actual resistor are plotted in Fig. 2.1. The relation between the ideal resistor model's voltage and current is a straight line and can be expressed by Ohm's law as $v = Ri$, where R is the slope of the *v-i* characteristic. This is said to be a *linear* relation, and the ideal resistor model is said to be a linear circuit element. An ideal resistor is assumed to satisfy this linear relation no matter what values the voltage and current assume. For example, a 1-ohm (1-Ω) resistor assumes that a 1-milliampere (1-mA) current through it will produce a 1-millivolt (1-mV) voltage across its terminals. Also a 10^6-ampere (10^6-A) current through it will produce a 10^6-volt (V) voltage across its terminals. Typical currents in household wiring are on the order of 20 A. An actual 1-Ω resistor that we may purchase at an electronics store would not be expected to carry 10^6 A of current without melting! Therefore, the linear resistor model approximates an actual resistor only over some limited range of current and/or voltage at its terminals. This simple example illustrates the essential differences between an *ideal element model* and the *physical element* it is to represent. Nevertheless, this ideal circuit model allows us to *predict*, mathematically, the behavior of the actual circuit element which is the essential purpose of these models.

Actual electric circuits that perform some useful function, such as TVs, radios, digital computers, electronic typewriters, etc., are interconnections of actual circuit

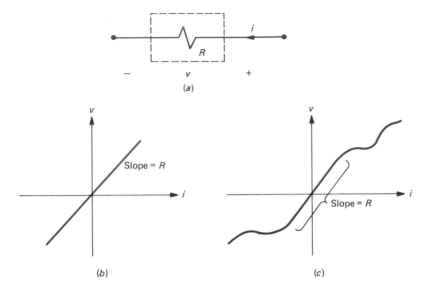

FIGURE 2.1
Ideal vs. actual
resistor circuit
elements.

elements. The particular way in which these circuit elements are interconnected to form an electric circuit determines the particular function of the physical circuit. We will model that physical circuit by replacing the physical circuit elements with their ideal mathematical models. This will allow us to mathematically predict the behavior of this physical circuit without ever having to construct it. This model therefore provides insights, mathematically, into how to *design* a physical electric circuit to perform some designated task.

Figure 2.2 illustrates the modeling process for the case of a phonograph–amplifier–speaker system. The phonograph cartridge converts variations in the grooves of the record to small electric signals via a phonograph cartridge. These electric signals are too small to drive a loudspeaker so that an amplifier is inserted between the cartridge and the speaker. The amplifier enlarges the small signal and passes it to the speaker where the electrical signal variations are converted to corresponding air pressure variations. The resulting model using ideal circuit elements is shown in Fig. 2.2b. Some of these ideal elements of the model represent physical elements in the actual system, and some cannot be directly related to physical elements of the system. For example, the speaker consists of turns of wire wound around a ferromagnetic core which is attached to a paper cone. As current from the amplifier passes through this coil, a magnetic field is generated which pulls the core and cone in and out, thereby generating air pressure waves that correspond to the phonograph musical signal. This suggests the inductance L_{speaker} in the speaker model. All wires also have resistance which suggests the resistance R_{speaker}. The amplifier is evidently composed of many more such elements than are shown in its model. The model represents a condensation of all the effects of the actual amplifier which relate the input voltage, $V_{\text{in}}(t)$, and the output voltage, $V_{\text{out}}(t)$. The model of the phonograph cartridge consists of a resistor $R_{\text{cartridge}}$ and a voltage source $V_{\text{cartridge}}(t)$. Neither of these two ideal elements correspond to physical elements internal to the cartridge. The model nevertheless gives an accurate representation of the output voltage of the

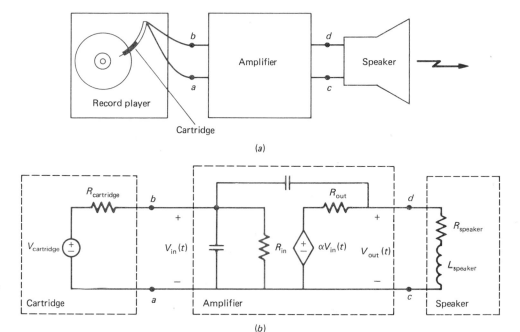

FIGURE 2.2
Modeling a
physical electric
circuit using ideal
electric circuit ele-
ments; (*a*) actual
circuit; (*b*) circuit
ideal model.

cartridge which is the input voltage to the amplifier, $V_{in}(t)$. Thus given the element values in the model and the time variation of $V_{cartridge}(t)$, we can predict the voltage at the input to the speaker, $V_{out}(t)$.

These notions represent the key aspects and benefits of a mathematical circuit model. This text is devoted to the construction and analysis of mathematical models which represent physical, electrical systems. The reader should strive to keep this important goal in mind.

2.1 CHARGE AND ELECTRIC FORCES

Electric charge and its movement are the most basic items of interest in electrical engineering. We will not be concerned with the charge on a microscopic basis such as electrons but will be more concerned with large-scale (relatively speaking) charge movement.

The basic component of charge is the electron, which carries a negative charge of 1.6×10^{-19} coulomb (C), which is the unit of charge. One coulomb of charge therefore represents a tremendous number of electrons. The nuclei of atoms contain an equal amount of positive charge in the heavier protons.

Charges of the same sign tend to repel each other, and charges of opposite sign tend to be attracted together. Thus, charges exert forces on each other. It is this electric force that we are interested in utilizing and controlling. For example, consider two charges Q_1 and Q_2 which are separated a distance r as shown in Fig. 2.3. The force F exerted on one charge by the other varies inversely with the square of the

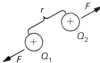

FIGURE 2.3
Illustration of
Coulomb's law.

separation between them and directly as the strengths of the charges according to Coulomb's law

$$F = k\frac{Q_1 Q_2}{r^2} \quad \text{N} \tag{2.1}$$

In the International System of Units (SI), the unit of force is the newton (N), where $1 \text{ N} = 1 \text{ (kg} \cdot \text{m)/s}^2$. In free space the proportionality constant k is 9×10^9 when F is in newtons, Q_1 and Q_2 are in coulombs, and r is in meters (m). The reader should verify that the force between two 1-C charges separated a distance of 1 m in free space is approximately 1 million tons.

Example 2.1   As an example of the use of Coulomb's law, determine the force exerted by a charge $Q_1 = 2 \ \mu C$ on a charge $q = 3 \ \mu C$ when the charges are separated a distance of 2 m in free space, as shown in Fig. 2.4a. Add another charge $Q_2 = -2 \ \mu C$ to this system 1 m above q, as shown in Fig. 2.4b, and determine the force exerted on q.

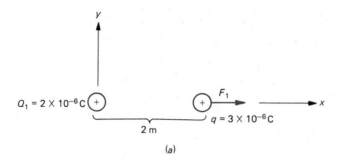

(a)

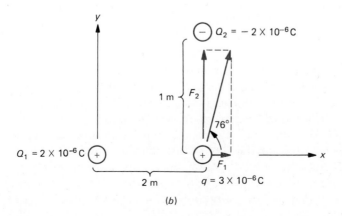

FIGURE 2.4
Example 2.1

(b)

Solution   The rectangular coordinate system shown defines the locations of the charges and the direction of the resulting force. The force Q_1 exerts on q is in the positive x direction and is given by

$$F_1 = 9 \times 10^9 \frac{Q_1 q}{2^2}$$

$$= 1.35 \times 10^{-2} \, \text{N}$$

Adding Q_2, as shown in Fig. 2.4b, gives another contribution to the force on q and is directed in the positive y direction since Q_2 and q are of opposite sign:

$$F_2 = 9 \times 10^9 \frac{Q_2 q}{r^2}$$

$$= 5.40 \times 10^{-2} \, \text{N}$$

The vector combination of F_1 and F_2 gives the resultant force exerted on q:

$$F = \sqrt{F_1^2 + F_2^2}$$

$$= 5.57 \times 10^{-2} \, \text{N}$$

This force is directed upward at an angle of 76° to the positive x axis, as shown.

2.2 VOLTAGE

Since charges exert forces on other charges, energy must be expended in moving a charge in the vicinity of other charges. The unit of energy is the joule (J), where one joule is the energy expended in the application of one newton of force in moving an object through a distance of one meter (1 J = 1 N · m).

For example, consider Fig. 2.5. Moving a charge q from point a to point b in the presence of some other charge Q may require a net expenditure of energy. The force of Q may oppose the movement of q over certain portions of the route, while over other portions this force may be in a direction so as to aid the movement of q. Thus, it reasonably follows that if we move the charge from point a to point b and return it to point a, the net expenditure of energy will be zero. This result is explained by the fact that charge Q has a type of force field around it which tends to repel charges of like sign and attract charges of unlike sign.

The action of the electric force field is similar to that of the gravitational force field. If we raise an object above the surface of the earth, the object acquires a certain amount of potential energy which is released once the object falls back to earth. We characterize the potential of an electric force field for doing work by a similar concept—voltage. If we move a charge q from point a to point b and energy w_{ba} is

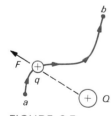

FIGURE 2.5
Voltage (potential difference) between two points in terms of the work required to move a charge between those points.

expended, we say that the voltage v—or potential difference—between these two points is the work required per unit charge:

$$v_{ba} = \frac{w_{ba}}{q} \qquad \text{V} \tag{2.2}$$

Note that, knowing the voltage between two points, we can easily compute the energy required to move some other charge $\hat{Q}$ between these two points by multiplying the voltage by the charge: $v_{ba}\hat{Q}$. Therefore, knowing the voltage between two points is a convenient way of characterizing the force field. The unit of voltage is, from Eq. (2.2), joules per coulomb. This combination of units is referred to in condensed form as volts, where 1 V = 1 J/C.

Knowing the direction (polarity) of this voltage, or electric potential, is obviously as important as knowing its magnitude, since this will determine whether energy is to be expended by us or by the force field when a charge is moved between the two points. Let us reconsider the gravitational potential example. If we lift an object to a point above the surface of the earth, energy is expended by us in doing so, and we say that points above the earth's surface are at a higher gravitational potential. In the same fashion, we describe point b as being at a higher electric potential than point a if net energy must be expended by us in moving a positive charge from a to b. Thus, v_{ba} in Eq. (2.2) is a "positive" number. This is denoted as shown in Fig. 2.6. We use $+$ and $-$ signs to designate which point is at the *assumed* higher potential (the $+$ point). (We use the word "assumed" because the guess may be incorrect, in which case v_{ba} will simply be a negative number as $v_{ba} = -2$ V.) It is important to understand that we need not be concerned with designating the correct point of higher potential, since this can easily be reversed by using a negative sign with the numerical value of the voltage. When we use the double subscript notation as v_{ba}, the first terminal is at the higher assumed potential.

The definition of voltage in (2.2) shows that the amount of charge q that is moved is unimportant since *the voltage between two points is the work per unit positive charge required to move that charge between the two points*. If the work required to move a 2-μC charge from point a to point b turns out to be 10 μJ, then the voltage between these points with point b at the assumed higher potential is

$$v_{ba} = \frac{10\ \mu\text{J}}{2\ \mu\text{C}}$$
$$= 5\ \text{V}$$

Equivalently we could say that the voltage between the two points with point a at the assumed higher potential is

$$v_{ab} = -5\ \text{V}$$

It is important to realize that the *path taken* to move q from point a to point b in order to compute the work w_{ba} and subsequently the voltage between the two points v_{ba} as in Eq. (2.2) is *unimportant*. Figure 2.7 illustrates this for three paths

+   • b

v_{ba}

−   • a

FIGURE 2.6
Designation of voltage between two points.

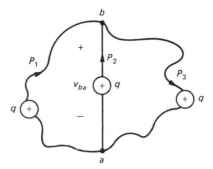

FIGURE 2.7

P_1, P_2, P_3. The work required to move q from a to b along paths P_1 and P_2 is the same, w_{ba}. The work required to move q from b to a along path P_3 is the negative of this: $w_{ab} = -w_{ba}$. This independence of path is sensible to expect. Suppose we move q from a to b along P_1 and return to point a along P_3. The net energy expenditure should be zero because we have returned to the starting point. Thus $w_{ba}|_{P_1} = w_{ba}|_{P_3}$.

Example 2.2

Consider computing the voltage between two points, point a located at $x = 1$ m and point b located at $x = 5$ m, due to a negative charge $Q = -5 \, \mu C$ located at $x = 6$ m, as shown in Fig. 2.8.

Solution   Let us move a positive unit charge q from a to b, compute the energy expended w_{ba}, and then compute the voltage according to Eq. (2.2). The force exerted on q is evidently in the positive x direction because of the opposite signs of the two charges and is given by

$$F = 9 \times 10^9 \frac{Qq}{r^2}$$

$$-\frac{45 \times 10^3}{(6 - x)^2}$$

The energy expended *by us* in moving q *from* point a *to* point b to give v_{ba} is

$$w_{ba} = -\int_{x=1\,m}^{x=5\,m} F \, dx = -\int_{x=1\,m}^{x=5\,m} \frac{45 \times 10^3}{(6 - x)^2} \, dx$$

$$= -36 \times 10^3 \, J$$

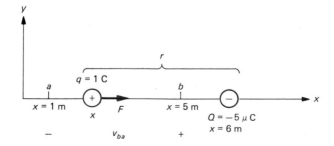

FIGURE 2.8
Example 2.2.

Therefore, the voltage is

$$v_{ba} = \frac{w_{ba}}{q}$$

$$= -36 \times 10^3 \text{ V}$$

Observe the negative sign in front of the integral sign and in the result. This is sensible to expect because the force exerted on q is in the direction of movement and so will require no energy expenditure on our part. Thus the energy expended (by us) in moving q from a to b is clearly going to be negative. Alternatively, the voltage between the two points with point a at the assumed higher voltage is

$$v_{ab} = -v_{ba}$$

$$= 36 \times 10^3 \text{ V}$$

and point a is actually at the higher voltage.

2.3   CURRENT AND MAGNETIC FORCES

Current is the rate of movement of electric charge. For example, consider Fig. 2.9a in which we have shown charge moving along a cylinder. A certain amount of positive charge Q_R^+ and negative charge Q_R^- are moving to the right across some cross section of the cylinder, and similar quantities Q_L^+ and Q_L^- are moving to the left. The *net* positive charge moving to the right is

$$Q = Q_R^+ - Q_R^- - Q_L^+ + Q_L^-$$

since negative charge moving to the right is equivalent to positive charge moving to the left and vice versa. If we observe the charge crossing the area in a certain time

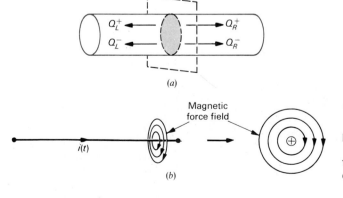

(a)

(b)

FIGURE 2.9
Electric current;
(a) net movement of positive charge;
(b) magnetic force field around a current-carrying wire.

interval Δt, *the current i directed to the right is the rate of movement of net positive charge to the right per unit of time*:

$$i = \frac{Q}{\Delta t} \quad A \tag{2.3}$$

The unit of current is the ampere, where a 1-A current is the movement of 1 C of net positive charge past a point in 1 second (s): $1\ A = 1\ C/s$.

The rate of charge movement may not be constant but may vary with time. To determine the current at a specific time, we reduce the time of observation to a very small interval about that time and obtain

$$i(t) = \frac{dQ}{dt} \tag{2.4}$$

which gives the current as a function of time t.

In most metallic conductors, such as wires, current is exclusively the movement of free electrons in the wire. Since electrons are negative, the charges are thus moving in a direction opposite to the direction of the current designation. The net positive charge movement is nevertheless in the direction we designate for i.

It is important to realize that current is, according to the definition given in Eq. (2.4), the *rate of movement of charge*. Moving a small amount of charge past a point in a short time may give the same value of current as the movement of a large amount of charge past this point in a longer time. For example, Fig. 2.10 shows a plot of

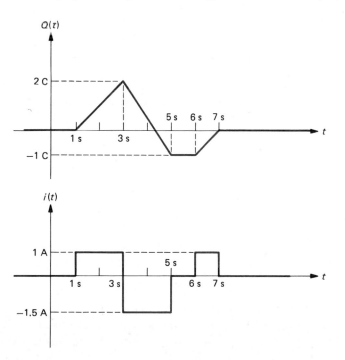

FIGURE 2.10
Relation between current and net positive charge flowing past a point.

the net charge moving to the right past a point versus time and the associated value of current. Over the period of 1 s $\leq t \leq$ 3 s the net positive charge moved is increasing at a rate of 1 C/s. Thus the current over this time interval is the *slope* of this curve according to Eq. (2.4) and is 1 A. Over the time interval 3 s $\leq t \leq$ 5 s, the slope of the net charge transfer charge is -1.5 C/s so that the current over this time interval is -1.5 A. Over the time interval 5 s $\leq t \leq$ 6 s, the net charge transfer is constant, so the current is zero over this time interval. Inverting Eq. (2.4) shows that the net charge transferred to the right at a particular time is the net area under the current curve from the beginning of time to the present:

$$Q(t) = \int_{-\infty}^{t} i(\tau) \, d\tau \tag{2.5}$$

Electric currents produce forces, as do electric charges. The forces produced by electric currents are referred to as *magnetic forces*, and they possess many of the properties of ordinary bar magnets. A current-carrying wire, for example, has a magnetic force field around it just as an electric charge has an electric force field around it. This magnetic force field appears as concentric circles around the wire; at a constant radius r from the wire, the magnetic force is the same (see Fig. 2.9b). The electric force field of a positive charge, on the other hand, always points away from the charge, and vice versa for a negative charge. If we place an ordinary bar magnet of a compass in the vicinity of the wire, the needle of the compass will align with the direction of the magnetic force field.

2.4   LUMPED-CIRCUIT ELEMENTS

We have seen that electric charges and currents both possess force fields and therefore have the ability to do work. We have also seen that these force fields are distributed throughout space. In our study of electric circuits we consider various electric circuit elements which utilize these force fields. In each of these elements we consider the force fields to be confined primarily to some small region around the element. Thus, although the effects of each element are distributed throughout space, we lump them into boxes, as shown in Fig. 2.11. We refer to these as *lumped elements*. We will associate a voltage $v(t)$ between the terminals of the element with a current $i(t)$ entering one terminal of the element (and leaving the other terminal). These quantities will, in general, be functions of time t. In electric circuits the important quantities of interest are the voltages and currents of the circuit elements. If we can determine these in a circuit, we will have "analyzed" the circuit completely. This will be our primary goal.

If we label the element current and element voltage as shown in Fig. 2.11, where the current is *assumed* to enter the terminal of *assumed* higher voltage, then the element is said to be labeled with the *passive sign convention*. The word "passive" means that the element is assumed to absorb power. From Eqs. (2.2) and (2.4) the

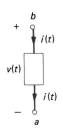

FIGURE 2.11
The passive sign convention for labeling the voltage and current of a circuit element.

power absorbed by the element (or delivered to the element) is

$$p(t) = \frac{dw_{ba}}{dt}$$

$$= \frac{dw_{ba}}{dq}\frac{dq}{dt}$$

$$= v(t)i(t) \qquad \text{W} \qquad\qquad\qquad (2.6)$$

The unit of power is the watt (W), where $1\text{ W} = 1\text{ J/s}$. Thus, the *instantaneous power delivered to (or absorbed by) an element at some time t is the product of the voltage and current associated with that element at that time when the voltage and current are labeled with the passive sign convention.*

As a physical example of the concept of power or energy being delivered to an element, consider the element as being a discharged automobile battery, as shown in Fig. 2.12. To charge this battery, we apply another charged battery to this discharged battery, being careful to connect the two positive terminals and the two negative terminals. Charge is transferred from the positive terminal of the charged battery to the positive terminal of the discharged battery. (In actuality, the electrons are transferred from the negative terminal of the charged battery to the negative terminal of the discharged battery, but the result is the same.) Current i is directed in the direction of net positive charge flow. Observe in Fig. 2.12 that the power being delivered *to* the discharged battery is vi, and v and i are labeled with the passive sign convention *with respect to the terminals of that battery*. The power delivered *by* the charged battery is vi, since these are labeled opposite to the passive sign convention *with respect to the terminals of that battery*.

If either v or i is truly negative (opposite in direction from the assumed direction), then negative power is delivered to the element or, equivalently, absorbed by the element. This is equivalent to saying that the element is delivering power. Several examples are shown in Fig. 2.13. Note that it is not important which terminal is which. It is only important to determine whether the element is labeled with the passive sign convention—the current enters the terminal of higher voltage. For example, in Fig. 2.13*a*

$$P_{\text{absorbed}} = (-v)(i)$$

$$= (-3)(2)$$

$$= -6\text{ W}$$

$$P_{\text{delivered}} = 6\text{ W}$$

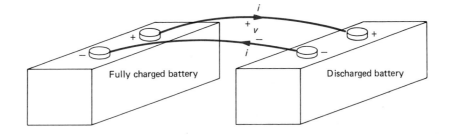

FIGURE 2.12
Illustration of the passive sign convention.

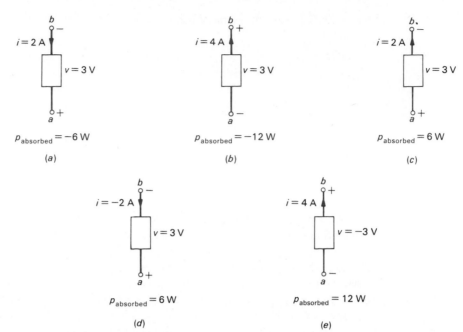

FIGURE 2.13
Illustrations of the
power absorbed or
delivered by an
element.

and the element is delivering 6 W of power P (at that time). In Fig. 2.13b the element is absorbing

$$P_{\text{absorbed}} = (v)(-i)$$
$$= (3)(-4)$$
$$= -12 \text{ W}$$
$$P_{\text{delivered}} = 12 \text{ W}$$

whereas in Fig. 2.13c

$$P_{\text{absorbed}} = (-v)(-i)$$
$$= vi$$
$$= 6 \text{ W}$$
$$P_{\text{delivered}} = -6 \text{ W}$$

Some of the voltages or currents may have negative values. In these cases it is simpler first to write the power expression in terms of the symbols v and i and then to substitute numerical values. For example, in Fig. 2.13e

$$P_{\text{absorbed}} = (v)(-i)$$
$$= (-3)(-4)$$
$$= 12 \text{ W}$$
$$P_{\text{delivered}} = -12 \text{ W}$$

2.5   KIRCHHOFF'S VOLTAGE AND CURRENT LAWS

In later sections we examine the various circuit elements—resistors, inductors, capacitors, and ideal voltage and current sources—which relate the voltage and current of the element. When these elements are interconnected to form an electric circuit, the interconnections establish relationships between the element voltages and currents. These relationships are known as Kirchhoff's laws.

The first law we consider is *Kirchhoff's current law*, abbreviated KCL. The terminals of the elements are referred to as *nodes*. In an electric circuit, these nodes are connected in some fashion. For example, consider the circuit in Fig. 2.14. This circuit consists of five elements. The elements constitute *branches* of the circuit. Kirchhoff's current law dictates that the sum of the currents entering a node must equal zero:

$$\sum_{\text{entering a node}} i = 0 \tag{2.7}$$

Note that a current i which is leaving a node is equivalent to a current $-i$ entering that node. In Fig. 2.14 KCL gives

$$\text{node } a: \quad -i_1 - i_5 + i_4 = 0$$

$$\text{node } b: \quad i_1 - i_2 = 0$$

$$\text{node } c: \quad i_2 + i_5 - i_3 = 0$$

$$\text{node } d: \quad i_3 - i_4 = 0$$

Note that each of these equations may be multiplied by a minus sign so that KCL could have been stated as requiring the sum of the currents leaving a node to be zero:

$$\sum_{\text{leaving a node}} i = 0 \tag{2.8}$$

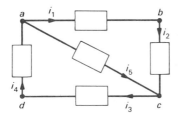

FIGURE 2.14   Example illustrating Kirchhoff's current law (KCL).

As another example illustrating KCL, consider the circuit shown in Fig. 2.15. Summing the currents entering each node gives

node a: $\quad -i_1 - i_2 + i_3 = 0$

node b: $\quad i_1 + i_5 - i_6 = 0$

node c: $\quad i_2 + i_4 - i_5 = 0$

node d: $\quad i_6 - i_3 - i_4 = 0$

If instead of choosing to sum currents entering each node we chose to sum currents leaving each node, we would obtain the negative of these equations, which are identical to these since the results equal zero.

KCL must be satisfied at every node of a circuit and at every instant of time. *KCL is equivalent to the requirement that nodes cannot accumulate or store charge—whatever charge enters a node must immediately leave that node.* For example, if $Q_{in} = Q_1 + Q_2 + Q_3$ denotes the total net positive charge entering a node at some time and $Q_{in} = 0$, then

$$i_{in} = i_1 + i_2 + i_3$$

$$= \frac{dQ_1}{dt} + \frac{dQ_2}{dt} + \frac{dQ_3}{dt}$$

$$= \frac{dQ_{in}}{dt}$$

$$= 0$$

Kirchhoff's voltage law (KVL) provides relationships between the element voltages in a circuit. Kirchhoff's current law similarly related the element currents in a circuit. Before stating KVL, we need to define the terms "voltage drop," "voltage rise," and "circuit loop." Consider Fig. 2.16. Imagine moving a charge q through the element toward the terminal of assumed higher voltage as in Fig. 2.16a. We say that the movement has been in the direction of a *voltage rise* of v. Conversely,

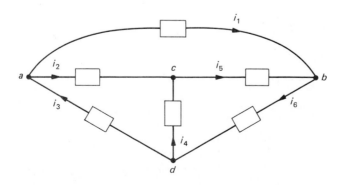

FIGURE 2.15 Example illustrating Kirchhoff's current law (KCL).

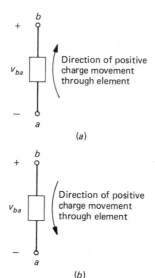

FIGURE 2.16
Definition of
(*a*) voltage rise
and (*b*) voltage
drop.

if the charge movement had been through the element and toward the terminal of assumed lower voltage ($-$), as in Fig. 2.16*b*, we would say that the charge movement has been through a *voltage drop* of *v*. A *circuit loop* is a closed path formed by selecting a starting node, proceeding through the circuits in that path, and returning to that starting node without going through a circuit element more than once. Consider the circuit of Fig. 2.17*a*. The loops consist of the following circuit elements, shown in Fig. 2.17*b*:

L_1:   *AEB*
L_2:   *AFDB*
L_3:   *AEDC*
L_4:   *AFC*
L_5:   *BEFC*
L_6:   *BDC*
L_7:   *EFD*

There are a total of seven possible loops for this circuit.

Kirchhoff's voltage law (KVL) states that the algebraic sum of the voltages encountered in traversing each loop of a circuit must be zero

$$\sum_{L_i} v = 0 \tag{2.9}$$

for all loops of the circuit. The "algebraic sum" means that if we proceed through an element in the direction of a voltage rise, then the element voltage is entered as a positive quantity in the sum. Similarly, if we proceed through an element in the

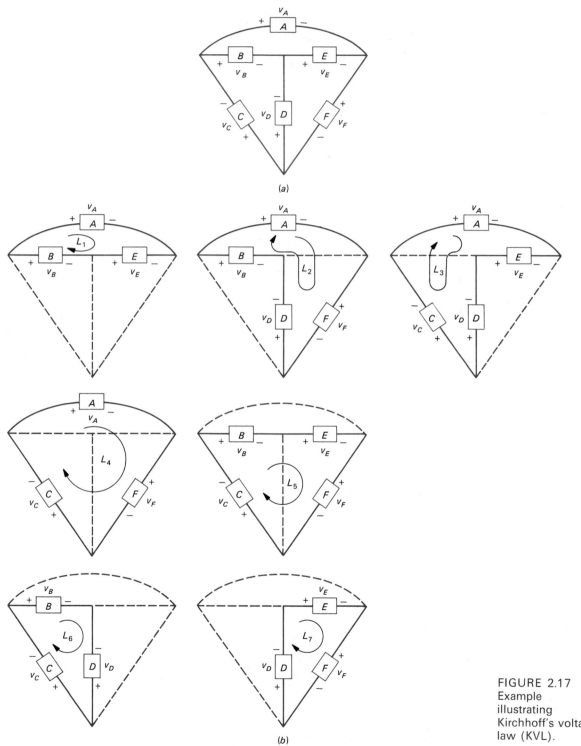

FIGURE 2.17
Example
illustrating
Kirchhoff's voltage
law (KVL).

direction of a voltage drop, the element voltage is entered as a negative quantity in the sum. A voltage rise v is equivalent to a voltage drop $-v$. Thus KVL could be stated in several forms:

$$\sum_{L_i} \text{Voltage rises} = 0 \tag{2.10a}$$

$$\sum_{L_i} \text{Voltage drops} = 0 \tag{2.10b}$$

$$\sum_{L_i} \text{Voltage rises} = \sum_{L_i} \text{voltage drops} \tag{2.10c}$$

The starting node of any circuit loop and the direction of traversal are arbitrary. For example, applying KVL to the loops of the circuit of Fig. 2.17 gives

$$
\begin{aligned}
L_1: & \quad -v_A + v_E + v_B = 0 \\
L_2: & \quad -v_A - v_F - v_D + v_B = 0 \\
L_3: & \quad v_D + v_E = v_A + v_C \\
L_4: & \quad v_A + v_F + v_C = 0 \\
L_5: & \quad v_B + v_E + v_C + v_F = 0 \\
L_6: & \quad v_B + v_C = v_D \\
L_7: & \quad v_D + v_E + v_F = 0
\end{aligned}
$$

We have used a mixture of the three equivalent forms in Eq. (2.10).

KVL is simply a statement of the fact that moving charge around a path and returning to the starting point should require no net expenditure of energy. For example, consider Fig. 2.18. Suppose we move a charge q around the loop through the elements in the direction shown and return to the starting node (node a). Proceeding from node a to node b through element A requires that we expend $3q$ J of energy, since we move in a direction of voltage rise. Moving q from node b to node c through element B requires that we expend $-2q$ J of energy because we move q in the direction of a voltage drop of 2 V. Continuing around the loop, we have the following net expenditures of energy:

Element	Energy Expenditure
A	$3q$ J
B	$-2q$ J
C	$4q$ J
D	$3q$ J
E	$-8q$ J
Net	0 J

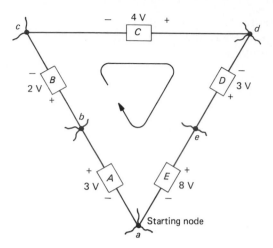

FIGURE 2.18   Example showing that KVL is equivalent to conservation of energy in the movement of charge around a closed loop.

The net expenditure should be zero since we have returned to the starting node. If we factor the charge q, move out of the sum of the net expenditures of energy as

$$(3 \text{ V} - 2 \text{ V} + 4 \text{ V} + 3 \text{ V} - 8 \text{ V})q = 0$$

we arrive at KVL for the loop.

Example 2.3   As an application of these results, consider the circuit shown in Fig. 2.19. Some of the element voltages and currents are known and are shown. Determine the remaining voltages

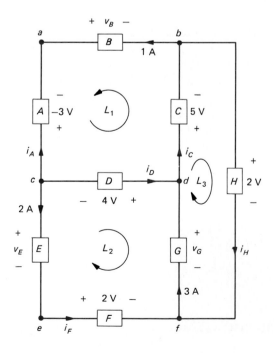

FIGURE 2.19
Example 2.3.

and currents. Show that conservation of power is satisfied for this circuit; i.e., the sum of the powers delivered to each element is zero.

Solution   First let us obtain the element voltages. Applying KVL gives

$$L_1: \quad v_B + (-3) + 4 - 5 = 0$$
$$L_2: \quad v_E + 4 - v_G + 2 = 0$$
$$L_3: \quad v_G - 5 - 2 = 0$$

Solving these gives

$$v_G = 7 \text{ V} \qquad v_E = 1 \text{ V} \qquad v_B = 4 \text{ V}$$

Applying KCL gives

$$\text{node } a: \quad i_A = -1 \text{ A}$$
$$\text{node } e: \quad i_F = 2 \text{ A}$$
$$\text{node } c: \quad i_D = -i_A - 2 = -1 \text{ A}$$
$$\text{node } d: \quad i_C = 3 + i_D = 2 \text{ A}$$
$$\text{node } b: \quad i_H = i_C - 1 = 1 \text{ A}$$

The powers delivered to each element are

Element	Power Delivered to Element
A	$(-3)i_A = 3$ W
B	$-v_B 1 = -4$ W
C	$5i_C = 10$ W
D	$-4i_D = 4$ W
E	$v_E 2 = 2$ W
F	$2i_F = 4$ W
G	$-v_G 3 = -21$ W
H	$2i_H = 2$ W
	$\sum = 0$

2.6 THE RESISTOR

We now begin an investigation of the specific types of circuit elements which we will use to model physical circuits. These are ideal models that are used to construct circuit models which closely represent physical electric circuits.

For many conducting materials, the voltage across a block of the material is linearly related to the current passing through it. This result is credited to the German physicist Georg Simon Ohm, and it bears his name: *Ohm's law*. The symbol for a linear resistor is given in Fig. 2.20 along with the graphical relationship between the element voltage and current. Note that the graph is a straight line with slope R passing through the origin. If the graph were not a straight line or did not pass through the origin, the resistor would be a nonlinear one. Nonlinear resistors are considered in Part 2. For the linear resistor, we may dispense with the graph and write Ohm's law as

$$v(t) = Ri(t) \tag{2.11}$$

The slope R of the graph is called the *resistance*, and its units are ohms (Ω), where $1\ \Omega = 1$ V/A. It is also somewhat common to write Eq. (2.11) as

$$i(t) = \frac{1}{R}\,v(t) \tag{2.12}$$

$$= Gv(t)$$

where

$$G = \frac{1}{R} \tag{2.13}$$

is called the *conductance*, whose units are siemens (S), the reciprocal of ohms. In the past, the unit and symbol of conductance were mhos ($\mho$), mhos in a certain sense being the reciprocal of ohms (Ω). The internationally accepted units, however, are siemens (S).

Note that Ohm's law in Eq. (2.11) *presumes* that the element has been labeled with the passive sign convention. If this is not the case, a minus sign will be needed in Eq. (2.11). Several examples are shown in Fig. 2.21; they should be studied carefully. Compare each of the labels in Fig. 2.21 to that of Fig. 2.20. In Fig. 2.21a, $i_x = i$ but $v_x = -v$. Thus

$$-v_x = Ri_x$$

or

$$v_x = -Ri_x$$

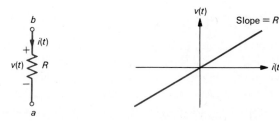

FIGURE 2.20
The ideal linear resistor and Ohm's law.

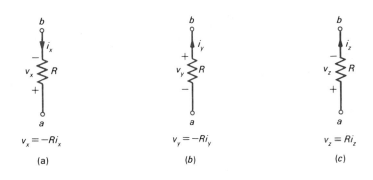

FIGURE 2.21
Ohm's law for
resistors with
various
voltage-current
labeling.

In Fig. 2.21*b*, $v_y = v$ but $i_y = -i$. Thus

$$v_y = R(-i_y)$$
$$= -Ri_y$$

In Fig. 2.21*c*, $v_z = -v$ and $i_z = -i$. Thus

$$-v_z = R(-i_z)$$

or

$$v_z = Ri_z$$

This last result shows that for a linear resistor it doesn't matter which terminal is which, as long as the element voltage and current have been labeled with the passive sign convention with respect to that element.

For many types of conducting materials, the resistance of a block of material whose length *l* and cross-sectional area *A* are known can be calculated from

$$R = \rho \frac{l}{A}$$

where ρ is the resistivity of the material in ohm-meters ($\Omega \cdot$ m). Values for representative types of conducting materials are given in Table 2.1.

TABLE 2.1 Resistivity of Common Conductors	
Material	ρ, $\Omega \cdot$ m
Copper	1.72×10^{-8}
Silver	1.62×10^{-8}
Gold	2.44×10^{-8}
Aluminum	2.62×10^{-8}
Brass	6.67×10^{-8}
Stainless steel	9.09×10^{-7}
Nichrome	1.0×10^{-6}
Sea water	0.25

Example 2.4

Determine the resistance of a 100-foot (100-ft) length of No. 14 gauge copper wire (commonly used in extension cords). The diameter of No. 14 gauge wire is 64.1 mils, where 1 mil = 0.001 inch (in). If this wire is attached to a 120-V residential outlet and used to power a pair of hedge clippers which draw a 1-A current, determine the voltage at the terminals of the hedge clippers.

Solution   The cross-sectional area of the wire is πr^2, where r is the wire radius. We first convert the wire radius in mils (32.05 mils) to meters:

$$32.05 \text{ mils} \times \frac{1 \text{ in}}{1000 \text{ mils}} \times \frac{2.54 \text{ cm}}{1 \text{ in}} \times \frac{1 \text{ m}}{100 \text{ cm}} = 8.14 \times 10^{-4} \text{ m}$$

The resistivity of copper from Table 2.1 is $1.72 \times 10^{-8} \, \Omega \cdot \text{m}$. Thus, the total resistance of one wire in the cord is

$$R = \frac{1.72 \times 10^{-8} \times 30.48}{\pi (8.14 \times 10^{-4})^2}$$

$$= 0.252 \, \Omega$$

where we have converted the length of the cord from feet to meters:

$$100 \text{ ft} \times \frac{12 \text{ in}}{1 \text{ ft}} \times \frac{2.54 \text{ cm}}{1 \text{ in}} \times \frac{1 \text{ m}}{100 \text{ cm}} = 30.48 \text{ m}$$

Each of the two wires of the cord has $0.252 \, \Omega \times 1 \text{ A} = 0.252 \text{ V}$ dropped across it. Thus the available voltage at the terminals of the hedge clippers is

$$120 - 0.252 - 0.252 = 119.5 \text{ V}$$

The linear resistor always absorbs power. This can be seen by substituting Ohm's law into the power equation:

$$P_{\text{absorbed}}(t) = v(t)i(t)$$

$$= Ri^2(t)$$

$$= \frac{v^2(t)}{R} \tag{2.14}$$

A true negative resistor would be one in which the element voltage and current are labeled with the passive sign convention *but* in which $v(t) = -Ri(t)$. In this case the power absorbed would be $-Ri^2(t)$, or, in other words, the resistor would deliver power. The power dissipated in the extension cord of Example 2.4 is $i^2R = (1 \text{ A})^2 \times 0.504 \, \Omega = \frac{1}{2}$ watt (W).

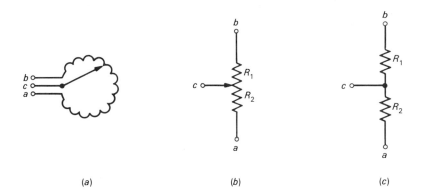

FIGURE 2.22
A variable
resistor—the
potentiometer.

(a)   (b)   (c)

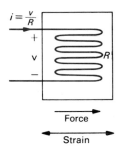

FIGURE 2.23
Illustration of a
resistance
application in
transducers—a
strain gauge.

There are many ways of constructing actual resistors. Carbon resistors are constructed by forming a block of carbon into a cylindrical shape and attaching leads. Wire-wound resistors are constructed of a long length of wire wound on a central axis to conserve space. In either type of resistor, the terminal voltage and current obey Ohm's law. There are also variable resistors known as *potentiometers*; these are usually constructed as a wire-wound resistor with a movable tap which slides along the surface of the wires, as shown in Fig. 2.22. The position of the center tap determines the portion of the total resistance between the tap and each end of the resistor.

Many useful types of elements use this principle of resistance. One common such device is the strain gauge shown in Fig. 2.23. This device is constructed of a long section of wire woven on a flexible backing. When the backing is attached to a material and a force is applied to that material, the wire is stretched, which reduces the wire's cross section and increases its total resistance. This increase in resistance is a measure of the elongation of the wire and hence of the strain in the material.

2.7 VOLTAGE AND CURRENT SOURCES

The voltage and the current associated with a linear resistor are related through Ohm's law. Neither the voltage nor the current is explicitly determined. However, if we should happen to know one quantity, the other is determined by Ohm's law. The ideal voltage and current sources are our next set of ideal element models, and these explicitly specify either the voltage (voltage source) or the current (current source) of the element.

Figure 2.24a shows the symbol and terminal relation for the *ideal voltage source*. The value of this source $v_s(t)$, is given as a function of time t, as an equation, a tabular listing of values, or a graph. A direct-current (dc) source has a constant voltage: $v_s(t) = V$. Observe that the terminal voltage of this element is constrained (forced) to equal $v_s(t)$:

$$v(t) = v_s(t) \tag{2.15}$$

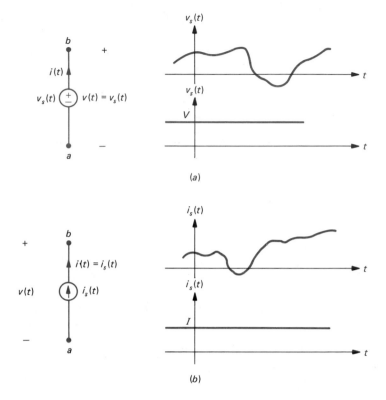

FIGURE 2.24
The ideal voltage
and current
sources:
(*a*) voltage source;
(*b*) current
source.

The $\pm$ polarity symbols within the circle must agree with the $\pm$ terminal voltage polarity, or else a negative sign is required in (2.15). Although the terminal voltage of the ideal voltage source is forced to equal the specified value $v_s(t)$, the terminal current i is not yet determined. The current will be determined once the remaining circuit is attached to the terminals of this source.

The *ideal current source* is shown in Fig. 2.24*b*. As opposed to the ideal voltage source, the ideal current source constrains (forces) the element current to be equal to the specified current:

$$i(t) = i_s(t) \tag{2.16}$$

The current arrow within the circle must agree with the chosen direction of the element current, or else a negative sign is required in (2.16). Although the current through the element is constrained to be the specified function, the voltage across the element terminals is not yet determined. It will be determined once a circuit is attached to the two terminals of the element.

Recall that the ideal (positive) resistor always absorbs power. It can never be a source of energy because the power delivered to it is $p(t) = Ri^2(t) = v^2(t)/R$ which is positive regardless of the sign of the current or the voltage. The ideal sources in Fig. 2.24, on the other hand, have the possibility of being able to deliver energy to any circuit that may be attached to its terminals. For example, the power delivered by the ideal voltage source (to any circuit attached to its terminals) is

$$p(t)_{\text{delivered}} = v_s(t)i(t) \tag{2.17}$$

If $v_s(t)$ and the current $i(t)$ happen to have the same sign at some time, then (2.17) is positive and power is being delivered by the voltage source (to the external circuit). Similarly, the power delivered by the ideal current source (to any circuit attached to its terminals) in Fig. 2.24b is given by

$$p(t)_{\text{delivered}} = v(t)i_s(t) \tag{2.18}$$

If $v(t)$ and $i_s(t)$ happen to have the same sign at some time, then (2.18) is positive and power is being delivered by the current source (to the external circuit).

These two sources are *ideal* in the sense that the voltage or the current of the element is fixed to be a known function regardless of the value of the other element variable. For example, the terminal voltage of the ideal voltage is $v(t) = v_s(t)$ regardless of the value of current which happens to be flowing through the element. It is obviously not realistic to expect to be able to construct such an ideal element. However, actual sources such as batteries can be modeled or approximated using these ideal sources and the ideal linear resistor. An example is shown in Fig. 2.25. Batteries used to start automobiles have a terminal voltage on the order of 12 V as long as no current is being drawn from their terminals. When the battery is used to start an automobile, a large current (on the order of 40 A) is drawn from the battery by the starter. During this event, the terminal voltage of the battery drops from 12 to around 9 V. The voltage-current characteristic shown in Fig. 2.25a illustrates the variation of terminal voltage with terminal current. We can approximate this nonideal source by modeling it as an ideal 12-V dc voltage source and a resistor R_s, as shown in Fig. 2.25b. To show this, write KVL around the loop containing the source and the resistance. The voltage across R_s is $R_s i(t)$ according to Ohm's law. Therefore, KVL gives

$$v = 12 - R_s i \tag{2.19}$$

This is plotted as a straight line in the associated figure. The intersection of the curve with the v axis is (2.19) evaluated for $i = 0$ or 12 V. Obviously the slope of the line is $-R_s$. Although the true voltage-current characteristic of the battery is not a straight line, this *model* provides a simple and reasonable approximation of the true characteristic. This is our first example of the ability to model actual electric circuits and devices by using ideal circuit elements.

Figure 2.25c shows an equally satisfactory model using an ideal current source. To show that the two models are equivalent, write KCL at the top node. The current directed down through R_s is, by Ohm's law, v/R_s. Therefore KCL gives

$$i = \frac{12 \text{ V}}{R_s} - \frac{v}{R_s} \tag{2.20}$$

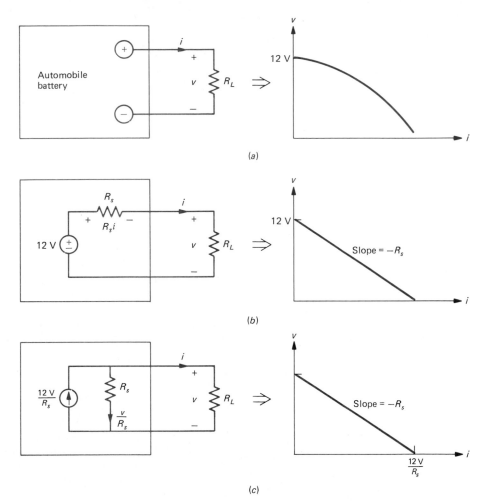

FIGURE 2.25
Representation of
actual sources
using ideal circuit
elements.

Rearranging (2.20) gives (2.19). The equivalency between using an ideal voltage source and resistor to model an actual source or using an ideal current source and a resistor to model the same source is referred to as a *source transformation*. This is used on numerous occasions throughout this text and should be committed to memory.

2.8 SIGNAL WAVEFORMS

The ideal voltage and current sources provide signals which are shaped or otherwise altered by the circuit to which they are attached. We now consider how to describe, or characterize, these signals.

Perhaps the simplest type of waveform is the constant, or dc, waveform shown in Fig. 2.26b. In this section, we denote the waveform as a function of time t—as

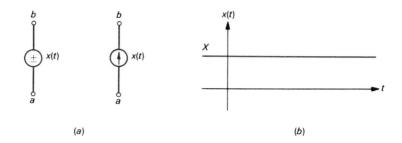

FIGURE 2.26
Constant or (dc)
(*a*) sources and
(*b*) waveforms.

(*a*)   (*b*)

$x(t)$—regardless of whether it is produced by a voltage or a current source. Characterizing this waveform is quite simple (the graph is unnecessary):

$$x(t) = X \tag{2.21}$$

These waveforms are referred to as dc since they characterize the direct current supplied by a constant voltage source or battery.

We will often be interested in waveforms which are not constant in time. Generally, we have no choice but to describe these waveforms graphically. However, there is an important class of time-varying waveform which is quite easy to describe—periodic waveforms. A periodic waveform is one which repeats itself over intervals of time of length T. In other words,

$$x(t) = x(t \pm nT) \tag{2.22}$$

where $n = 1, 2, 3, \ldots$. The time for repetition of the waveform T is called the *period* of the waveform. An example of a periodic waveform is shown in Fig. 2.27. Every point on the waveform is identical nT seconds later or earlier in time. Obviously, for a waveform to be periodic, it must continue indefinitely in time. Note that the dc waveform in Fig. 2.26*b* may be considered to be periodic with an infinite period. The frequency f of a periodic waveform is the reciprocal of its period:

$$f = \frac{1}{T} \quad \text{Hz} \tag{2.23}$$

The unit of frequency is the hertz (Hz), where one hertz represents a repetition of the waveform in one second. Since we only need to describe the waveform over

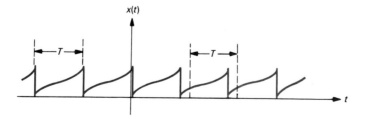

FIGURE 2.27
A general periodic
waveform (signal).

one period, periodic waveforms are simpler to characterize than general waveforms which are not periodic.

The periodic waveform to which we devote considerable attention is the sinusoid shown in Fig. 2.28. The waveform is described by

$$x(t) = A \sin (\omega t + \phi) \tag{2.24}$$

The amplitude of the sinusoid is A, and ϕ is some phase offset. The radian frequency of the wave is $\omega = 2\pi f = 2\pi/T$. If the phase offset is zero, the wave is classified as a sine wave, and if $\phi = 90°$, the wave is classified as a cosine wave, or $A \sin (\omega t + 90°) = A \cos \omega t$.

These sinusoidal waveforms have a practical utility, and we concentrate our study of electric circuits on those containing ideal voltage and current sources which are either dc or sinusoidal. Electric power is generated, disributed, and utilized as sinusoidal voltage or current at a frequency of 60 Hz (50 Hz in Europe). Sinusoidal waveforms are commonly classified as ac waveforms since they are characteristic of the alternating current produced by electric power generators. Radio and television transmission makes use of sinusoids which have much higher frequencies than 60 Hz. For example, standard AM transmission utilizes frequencies in the range of 550 kHz (550×10^3 Hz) to 1600 kHz (1.6×10^6 Hz). These radio transmissions utilize a single frequency in this range which is called a *carrier*. The music or other information to be transmitted is used to vary the amplitude of this carrier (amplitude modulation) so that a receiver tuned to the carrier frequency may extract these variations and hence the information. In FM, or frequency modulation, the information is used to vary the carrier frequency.

Periodic waveforms may also be classified according to their average value. The average value (AV) of a periodic waveform is the net positive area under the curve for one period divided by the period:

$$x_{AV} = \frac{1}{T} \int_0^T x(t) \, dt \tag{2.25}$$

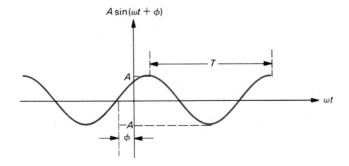

FIGURE 2.28
A sinusoidal
waveform (signal).

The effective, or root-mean-square (RMS), value is the square root of the average of $x^2(t)$:

$$x_{\text{RMS}} = \sqrt{\frac{1}{T} \int_0^T x^2(t)\, dt} \tag{2.26}$$

This definition of the RMS, or effective, value of a periodic waveform arises for the following reason. Consider a resistor R through which a periodic current $i_R(t)$ passes. The instantaneous power dissipated in the resistor is

$$\begin{aligned} p_R(t) &= v_R(t)i_R(t) \\ &= Ri_R^2(t) \end{aligned} \tag{2.27}$$

Since $i_R(t)$ is periodic, $p_R(t)$ will be periodic with the same period. We may compute an average power dissipated in the resistor (converted to heat) as

$$\begin{aligned} P_{\text{AV}} &= \frac{1}{T} \int_0^T p_R(t)\, dt \\ &= R\left[\frac{1}{T} \int_0^T i_R^2(t)\, dt\right] \end{aligned} \tag{2.28}$$

But this may be written as

$$P_{\text{AV}} = RI_{\text{RMS}}^2 \tag{2.29}$$

where, comparing Eqs. (2.28) and (2.29), we have

$$I_{\text{RMS}} = \sqrt{\frac{1}{T} \int_0^T i_R^2(t)\, dt} \tag{2.30}$$

For a particular periodic current waveform $i_R(t)$, it is somewhat inconvenient to compute Eq. (2.28). However, if we know the RMS value of the waveform, we may easily compute the average power dissipated in a resistance via Eq. (2.29) as though the waveform were dc with a value of I_{RMS}. Similarly, we may compute

$$P_{\text{AV}} = \frac{V_{\text{RMS}}^2}{R} \tag{2.31}$$

where

$$V_{\text{RMS}} = \sqrt{\frac{1}{T} \int_0^t v_R^2(t)\, dt} \tag{2.32}$$

Example 2.5   Compute the power dissipated in a 5-Ω resistor which has a current with a waveform given in Fig. 2.29 passing through it.

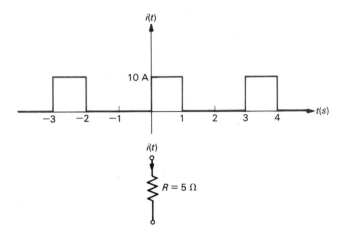

FIGURE 2.29
Example 2.5.

Solution   The period is 3 s, and the RMS value is

$$I_{\text{RMS}} = \sqrt{\frac{1}{3} \int_0^3 i^2(t)\, dt}$$

$$= \sqrt{\frac{1}{3} \int_0^1 10^2\, dt}$$

$$= \sqrt{\frac{100}{3}}$$

$$= 5.77\ \text{A}$$

Thus

$$P_{\text{AV}} = 5\ \Omega \times (5.77)^2$$

$$= 166.7\ \text{W}$$

For the special case of a dc waveform shown in Fig. 2.26, the average and rms values are

$$x_{\text{AV}} = x_{\text{RMS}}$$

$$= X \tag{2.33}$$

For the sinusoid

$$x(t) = A \sin \omega t \tag{2.34}$$

the average value is zero

$$x_{AV} = 0$$

but the RMS value is

$$x_{RMS} = \sqrt{\frac{1}{T} \int_0^T A^2 \sin^2\left(\frac{2\pi}{T} t\right) dt}$$

$$= \frac{A}{\sqrt{2}}$$

$$\doteq 0.707 \, A \tag{2.35}$$

where we have used the trigonometric identity $\sin^2 \theta = \frac{1}{2}(1 - \cos \theta)$. The same result would be obtained for the cosine wave $x(t) = A \cos \omega t$ or for any general sinusoid $x(t) = A \sin (\omega t + \phi)$. Thus, if a sinusoidal current of amplitude A passes through a resistor R, the average power dissipated in that resistor is

$$P_{AV} = RI^2_{RMS}$$

$$= \frac{1}{2} A^2 R \tag{2.36}$$

Other common types of waveforms are the exponential waveforms shown in Fig. 2.30. The exponential waveform shown in Fig. 2.30a is of the form

$$x(t) = Ae^{-t/\tau} \tag{2.37}$$

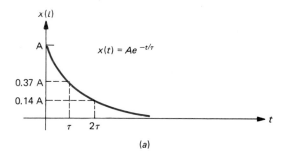

(a)

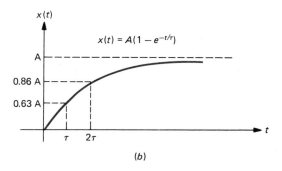

(b)

FIGURE 2.30
Exponential
waveforms.

where τ is referred to as the *time constant*. At $t = 0$, $x(0) = A$. As $t \to \infty$, $x(t) \to 0$. After a time of one time constant has elapsed, the value of the waveform is reduced to 37 percent of its peak. The exponential waveform shown in Fig. 2.30b is characterized by

$$x(t) = A(1 - e^{-t/\tau}) \tag{2.38}$$

This waveform asymptotically approaches a value of A as $t \to \infty$. After a time of one time constant has elapsed, the value of the waveform has risen to 63 percent of its final value.

2.9   ANALYSIS OF SIMPLE CIRCUITS

The interconnection of these ideal circuit elements gives a specific model. Remember that the intent in constructing such an interconnection is to obtain a mathematical model which can be used to predict the behavior of a physical electric circuit or device and to design it to perform a specified task. Therefore, our first goal is to learn to analyze any particular interconnection of these ideal elements.

In the next chapter we will learn several useful techniques that will help us to analyze any particular interconnection. However, numerous simple circuits can be easily analyzed by applying KVL, KCL, and the element relations. For example, consider the circuit shown in Fig. 2.31. The task is to determine the currents through the elements, i_s, i_x, i_y, and the voltages v_x and v_y across the resistors. It is important to remember that the only tools we may employ to determine these currents and voltages are KVL, KCL, and the element relations such as Ohm's law. Note that the lower line in the circuit can be considered to be one node since no elements are connected between its endpoints. Applying KCL at the three nodes shows that the three currents are equal—a point that should become evident to the reader. Thus we can designate one current i flowing around the loop. This has exhausted KCL. Next we apply KVL around the loop. As an intermediate step, let us use Ohm's law to label the voltage across each resistor as shown in terms of the current i. (Be careful to observe the passive sign convention.) Applying KVL gives

$$10 = 3i + 2i$$

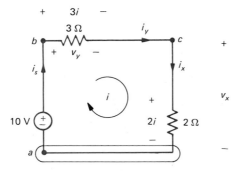

FIGURE 2.31

or

$$i = 2 \text{ A}$$

Once i has been determined, the remaining determination of the element voltages and currents is simple:

$$i_s = i = 2 \text{ A}$$
$$i_y = i = 2 \text{ A}$$
$$i_x = i = 2 \text{ A}$$
$$v_y = 3i = 6 \text{ V}$$
$$v_x = 2i = 4 \text{ V}$$

The power delivered by the ideal voltage source is

$$P_{\text{source}} = 10i_s = 20 \text{ W}$$

The powers absorbed by the two resistors are

$$P_{3\Omega} = 3i^2 \qquad P_{2\Omega} = 2i^2$$
$$= 12 \text{ W} \qquad\quad = 8 \text{ W}$$

Observe that the power delivered by the source is consumed by the resistors. The circuit shown in Fig. 2.31 is said to be a *single-loop circuit*.

A similar example is shown in Fig. 2.32. The task is again to determine the element voltages and currents such that they satisfy KVL, KCL, and the element relations for the specific interconnection. Observe that we may consider the endpoints of all three elements to be single nodes. This type of circuit is referred to as a *single-node-pair circuit* since there is only a single pair of nodes in the entire circuit. Applying KVL around the loops shows, as should begin to be obvious, that the element voltages are equal, which we designated as $v = v_s = v_x = v_y$. Labeling the currents through the two resistors in terms of this common voltage v, using Ohm's

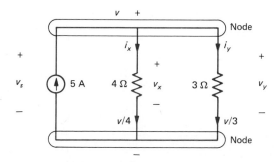

FIGURE 2.32

law, gives the element currents as shown. The only constraint that is left to apply is KCL. Applying KCL at the upper node yields

$$5 = \frac{v}{4} + \frac{v}{3}$$

which gives

$$v = \frac{60}{7} \text{ V}$$

These examples are quite simple, yet the general analytical procedures that we employed are applicable to more complicated circuits. The circuits shown in Fig. 2.31 or Fig. 2.32 are special types, known as single-loop or single-node-pair circuits. The circuit of Fig. 2.31 is a single-loop circuit, whereas the circuit of Fig. 2.32 is a single-node-pair circuit. The general procedure for analyzing these types of circuits is as follows:

1   Use KVL or KCL to define certain voltages or current that are equal in terms of some common designation.

2   Label the resistor voltages or currents in terms of that common variable, using Ohm's law.

3   Use the remaining law, KVL or KCL, to write an equation in terms of that common variable.

As an example, consider the single-node-pair circuit shown in Fig. 2.33. KVL shows that we may define a common voltage v across all the elements. Ohm's law is used to express the currents through the resistors (while being careful to observe the passive sign convention) in terms of this common voltage. Applying KCL at the upper node gives

$$3 - \frac{v}{2} - 5 - \frac{v}{3} - \frac{v}{1} = 0$$

Solving gives

$$v = -\frac{12}{11} \text{ V}$$

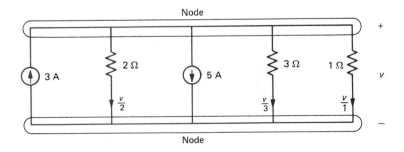

FIGURE 2.33

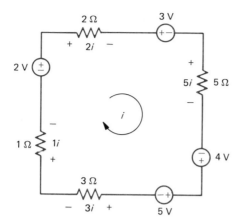

FIGURE 2.34

Note the polarity of the resulting voltage. This polarity was not evident from the start, but a careful application of KVL, KCL, and Ohm's law shows the proper polarity, and it did not matter that we made an incorrect first "guess" about this polarity.

As an example of a single-loop circuit, consider Fig. 2.34. KCL shows that a common current i flowing around the loop may be defined. Labeling the voltages across each resistor, using Ohm's law (being careful to observe the passive sign convention), and applying KVL give

$$-i + 2 - 2i - 3 - 5i + 4 - 5 - 3i = 0$$

Solving gives

$$i = -\tfrac{2}{11} \text{ A}$$

Once again, it did not matter that we made an incorrect guess about the direction of current i.

We can frequently convert some more complex circuits to either single-loop or single-node-pair circuits which we can now analyze. The key is to use the source transformation discussed in Sec. 2.7 and shown in Fig. 2.35. It is very important that we observe the proper polarities of the source. Given the voltage source–resistor

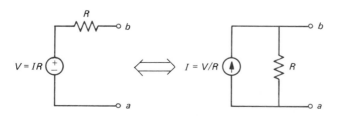

FIGURE 2.35

circuit, we can replace it with the current source–resistor circuit where the value of the current source is

$$I = \frac{V}{R}$$

Similarly, given the current source–resistor circuit, we can replace it with the voltage source–resistor circuit where the value of the voltage source is

$$V = IR$$

The voltage source–resistor circuit is said to be a *series connection* of the two elements. The current source–resistor circuit is said to be a *parallel connection* of the two elements. A *series connection* of circuit elements is one in which the *currents* through the elements must be equal. A *parallel connection* of circuit elements is one in which the *voltages* across the elements must be equal.

Now let us apply these source transformations to reduce a circuit to either a single-loop or a single-node-pair circuit. Consider the circuit shown in Fig. 2.36*a*. Converting the series voltage source–resistor elements to parallel current source–resistor elements gives the circuit shown in Fig. 2.36*b*. Observe that the node *c* has disappeared in this transformation. Defining the common voltage *v*, writing the currents through the resistors, using Ohm's law, in terms of *v*, and applying KCL gives

$$2 = \frac{v}{4} + \frac{v}{2} + \frac{v}{3}$$

Solving this equation gives

$$v = \tfrac{24}{13} \text{ V}$$

Observe in the original circuit in Fig. 2.36*a* that the currents through the 2- and 3-Ω resistors are the same as in Fig. 2.36*b*. However, the current through the 4-Ω resistor in Fig. 2.36*a* is not the same as that through the 4-Ω resistor in Fig.

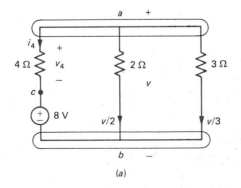

(a)

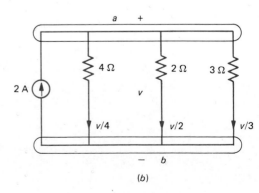

(b)

FIGURE 2.36

2.36b. We can still find the current through and voltage across the 4-Ω resistor, i_4 and v_4, in Fig. 2.36a since the common node voltage v is the same in both circuits. Using KVL in Fig. 2.36a gives

$$v_4 = v - 8$$
$$= -\tfrac{80}{13} \text{ V}$$

Thus the current through the resistor is

$$i_4 = \frac{v_4}{4}$$
$$= -\tfrac{20}{13} \text{ A}$$

This observation of what is equivalent between two circuits is important to understand. Note in Fig. 2.36a that $i_4 = -v/2 - v/3$.

As another example, consider the circuit shown in Fig. 2.37a. The parallel resistor–current source combination can be converted to a series resistor–voltage source combination, shown in Fig. 2.37b. Defining a current I to circulate around this loop, labeling the resistor voltages in terms of this current, and applying KVL gives

$$20 = 3I + 2I + 4I$$

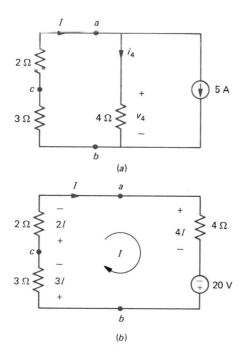

FIGURE 2.37

or

$$I = \tfrac{20}{9} \text{ A}$$

What is the value of the current i_4 in the original circuit? The current entering node a is the same in both circuits. Applying KCL at node a in Fig. 2.37a gives

$$i_4 = I - 5$$
$$= -\tfrac{25}{9} \text{ A}$$

Thus the voltage is

$$v_4 = 4i_4$$
$$= -\tfrac{100}{9} \text{ V}$$

Let us check the conservation of power for this circuit. The powers consumed by the resistors are

$$P_{4\Omega} = 4i_4^2$$
$$= \tfrac{2500}{81} \text{ W}$$
$$P_{2\Omega} = 2I^2$$
$$= \tfrac{800}{81} \text{ W}$$
$$P_{3\Omega} = 3I^2$$
$$= \tfrac{1200}{81} \text{ W}$$

The power delivered by the 5-A current source is

$$P_{5\text{A}} = -v_4(5)$$
$$= \tfrac{500}{9} \text{ W}$$

Observe the negative sign in this equation. Why is it necessary? Thus

$$P_{5\text{A}} = P_{4\Omega} + P_{2\Omega} + P_{3\Omega}$$

and power is conserved in this circuit.

These examples have illustrated some general analytical techniques which may be used to analyze a large class of circuits. In the next chapter, we discover some important, additional analytical techniques which will greatly enlarge the class of circuits that we can analyze. It is important to realize that the fundamental laws which all voltages and currents must satisfy are KCL at all nodes, KVL around all circuit loops, and the element relations (Ohm's law). Frequently it is a simple matter to directly apply these to a circuit and avoid using source transformations. In doing so it is imperative that we define as few currents and/or voltages to be solved for as possible and write all other element voltages and currents in terms of these few unknowns. Figure 2.38a illustrates this technique. We will define only one unknown: the current I through the 2-Ω resistor. Next apply KCL at node a to give the current through the 3-Ω resistor *in terms of I*. Then apply Ohm's law to give the resistor voltages, again *in terms of I*, taking care to observe the passive sign convention. Once KCL and Ohm's law have been exhausted in this manner, we only have KVL left to apply. Applying KVL around the loop containing the 10-V source, the 2-Ω resistor and the 3-Ω resistor gives

$$10 = -2I + 3(5 - I)$$

Solving gives

$$I = 1 \text{ A}$$

Observe that we do not know the voltage across the 5-A current source v_s (it is not necessarily zero) but it can be determined as the voltage across the 3-Ω resistor by applying KVL, $v_s = 3(5 - I) = 12$ V.

Usually there are many ways of analyzing a circuit. The parallel combination of the 3-Ω resistor and 5-A current source in Fig. 2.38a can be converted to an equivalent series combination of a 3-Ω resistor and a 15-V voltage source as shown in Fig. 2.38b. Observe that I is the same in both circuits and can be obtained by applying KVL around the single-loop circuit of Fig. 2.38b:

$$10 + 2I + 3I = 15$$

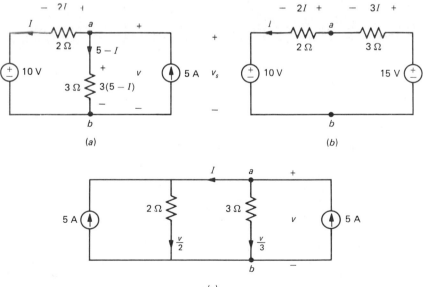

FIGURE 2.38

or

$$I = 1 \text{ A}$$

as before.

Alternatively, the series combination of the 10-V voltage source and 2-Ω resistor in Fig. 2.38a can be converted to the parallel combination of a 5-A current source and a 2-Ω resistor, as shown in Fig. 2.38c. Note the location of I. This is a single-node-pair circuit. Defining the common voltage v and writing the currents through the resistors in terms of this voltage, using Ohm's law, give, by applying KCL at node a,

$$5 - \frac{v}{2} - \frac{v}{3} + 5 = 0$$

or

$$v = 12 \text{ V}$$

This voltage is the same as across the nodes a and b in the original circuit. Current I can be found from Fig. 2.38c by as applying KCL at node a:

$$I = 5 - \frac{v}{3}$$

$$= 1 \text{ A}$$

or from Fig. 2.38a, writing KVL around the loop,

$$10 + 2I = v$$

giving

$$I = 1 \text{ A}$$

PROBLEMS

2.1 Show that the force exerted on a 1-C charge by another 1-C charge when the charges are separated by 1 m is approximately 1 million tons.

2.2 Two charges are held fixed in a two-dimensional coordinate system. One charge is negative, $Q_1 = -2 \times 10^{-9}$ C, and is placed at $x = 0$ m, $y = 2$ m. The other charge is positive, $Q_2 = 3 \times 10^{-9}$ C, and is placed at $x = 3$ m, $y = 0$ m. A positive charge, $q = 10^{-9}$ C, is placed at the origin, $x = 0$ m, $y = 0$ m. Determine the magnitude of the total force exerted on q and the direction of this resultant in terms of the angle measured from the x axis.

2.3 Two charges of equal magnitude, 5×10^{-9} C but opposite sign, are separated by a distance of 10 m. Determine the net force exerted on a positive charge, $q = 2 \times 10^{-9}$ C, that is placed midway between the two charges.

2.4 A charge Q is held fixed and a charge q is to be moved radially away from it from point a, located at a distance R_1 from Q, to point b, located a distance R_2 from Q. Determine an equation for the work required to move q from a to b. Evaluate this for $Q = 2 \times 10^{-6}$ C, $q = 1 \times 10^{-6}$ C, $R_1 = 2$ m, and $R_2 = 5$ m. Does the sign of your answer make sense? Repeat for $Q = -5 \times 10^{-6}$ C, $q = 1 \times 10^{-6}$ C, $R_1 = 2$ m, and $R_2 = 10$ m.

2.5 Two positive charges $Q_1 = 2 \times 10^{-9}$ C and $Q_2 = 5 \times 10^{-9}$ C, are separated by a distance of 10 m. Determine the work required to move a positive charge, $q = 1 \times 10^{-6}$ C, along a straight line between Q_1 and Q_2 from a distance of 1 m away from Q_1 to a distance of 1 m away from Q_2.

2.6 Two charges, $Q_1 = 3 \times 10^{-6}$ C and $Q_2 = -2 \times 10^{-6}$ C, are held fixed and separated by a distance of 6 m. Determine the voltage v_{ba} between two points along a line between the two charges. Point a is 1 m from Q_1 and point b is 2 m from Q_2.

2.7 Two charges are held fixed in a rectangular coordinate system. Charge $Q_1 = 2 \times 10^{-9}$ C is located at $x = 0$ m, $y = 0$ m and charge $Q_2 = 3 \times 10^{-9}$ C is located at $x = 5$ m, $y = 0$ m. Determine the voltage v_{ba} between point a at $x = 1$ m, $y = 0$ m and point b at $x = 3$ m, $y = 0$ m.

2.8 The current through an element is sketched as a function of time in Fig. P2.8. Determine the net positive charge transferred through the element at $t = -1$ s, 0, 1 s, 2 s, 3 s, and 4 s in the direction of the current.

2.9 The net positive charge passing a point to the right is sketched in Fig. P2.9. Determine the current directed to the right at $t = -1.5$ s, -0.5 s, 0.5 s, 1.5 s, 2.5 s, 3.5 s.

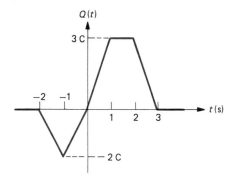

FIGURE P2.9

2.10 The current passing through an element is sketched in Fig. P2.8. Sketch the net positive charge transferred in the direction of the current as a function of time.

2.11 The charge passing a point to the right is sketched in Fig. P2.9. Sketch the current directed to the right as a function of time.

2.12 Automobile storage batteries are rated in terms of their terminal voltage (12 V) and their ampere-hour (Ah) capacities. For a typical 12-V battery having a 115-Ah capacity, determine the length of time this battery will light a 6-W bulb. (Assume that the battery voltage is constant at 12 V, even though this is not true.) Determine the total energy stored in the battery before it is connected to the bulb. Determine the total amount of charge that has passed through the wires that connect the battery and the bulb.

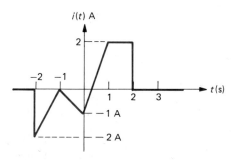

FIGURE P2.8

2.13 Determine the power delivered to (absorbed by) the elements shown in Fig. P2.13.

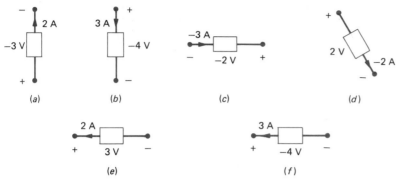

FIGURE P2.13

2.14 Determine currents i_x, i_y, i_z, i_w in the circuit of Fig. P2.14.

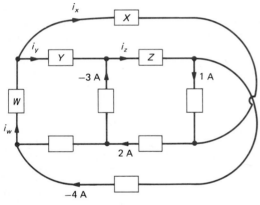

FIGURE P2.14

2.15 Determine currents i_x, i_y, and i_z in the circuit of Fig. P2.15.

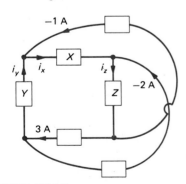

FIGURE P2.15

2.16 Determine currents i_1, i_2, i_3, and i_4 in the circuit of Fig. P2.16.

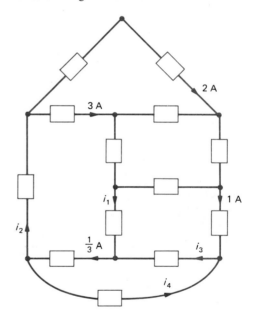

FIGURE P2.16

2.17 Determine currents i_1 and i_2 in the circuit of Fig. P2.17.

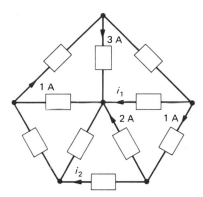

FIGURE P2.17

2.18 Determine currents i_x and i_y in the circuit of Fig. P2.18.

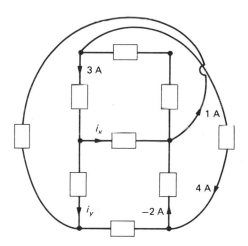

FIGURE P2.18

2.19 Determine voltages v_x, v_y, and v_{ba} in the circuit of Fig. P2.19.

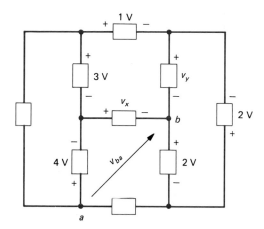

FIGURE P2.19

2.20 Determine voltages v_x, v_y, v_z, and v_w in the circuit of Fig. P2.20.

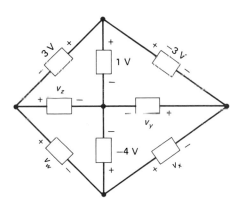

FIGURE P2.20

2.21 Determine voltages v_x, v_y, and v_z in the circuit of Fig. P2.21. Can you determine voltages v_w and v_q?

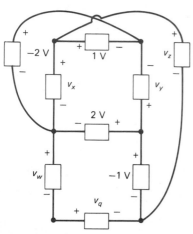

FIGURE P2.21

2.22 Determine voltages v_x, v_y, and v_z in the circuit of Fig. P2.22:

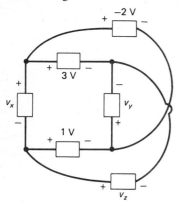

FIGURE P2.22

2.23 Determine voltages v_x, v_y, and v_z in the circuit of Fig. P2.23.

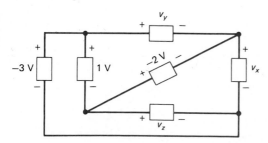

FIGURE P2.23

2.24 Determine voltages v_x, v_y, v_z, and v_w in the circuit of Fig. P2.24.

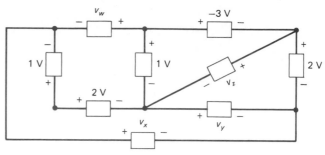

FIGURE P2.24

2.25 Determine voltage v in the circuit of Fig. 2.25.

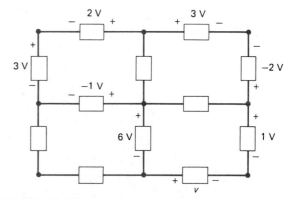

FIGURE P2.25

2.26 Determine voltage v and current i in the circuit of Fig. P2.26. Determine the power delivered to element A. Solve for all other currents and voltages and check conservation of power.

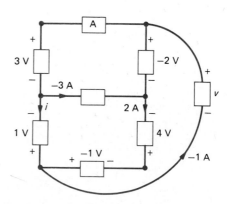

FIGURE P2.26

2.27 For the circuit shown in Fig. P2.27, determine the powers delivered to all four boxes. Check conservation of power for this circuit.

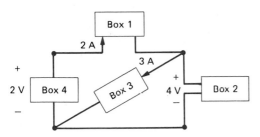

FIGURE P2.27

2.28 A 3-Ω resistor has the periodic current waveform shown in Fig. P2.28a passing through it. Compute the average power dissipated in the resistor. Repeat for the waveform shown in Fig. P2.28b.

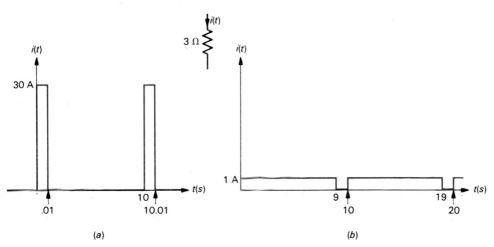

(a)   (b)

FIGURE P2.28

2.29 Determine the voltage v_x and current i_x in the circuit of Fig. P2.29. Show that conservation of power holds for this circuit.

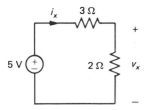

FIGURE P2.29

2.30 Determine the voltage v_x and current i_x in the circuit of Fig. P2.30. Show that conservation of power holds for this circuit.

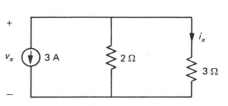

FIGURE P2.30

2.31   Determine the voltage v_x and current i_x in the circuit of Fig. P2.31. Show that conservation of power holds for this circuit.

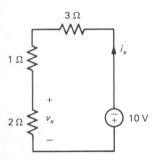

FIGURE P2.31

2.32   Determine the voltage v_x and current i_x in the circuit of Fig. P2.32. Show that conservation of power holds for this circuit.

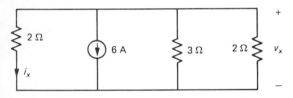

FIGURE P2.32

2.33   Determine the voltage v_x and current i_x in the circuit of Fig. P2.33. Show that conservation of power holds for this circuit.

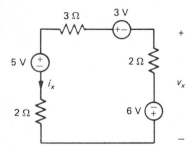

FIGURE P2.33

2.34 Determine the voltage v_x and current i_x in the circuit of Fig. P2.34. Show that conservation of power holds for this circuit.

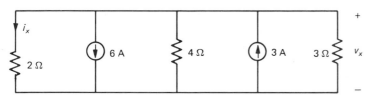

FIGURE P2.34

2.35 Determine the voltage v_x and current i_x in the circuit of Fig. P2.35. Show that conservation of power holds for this circuit.

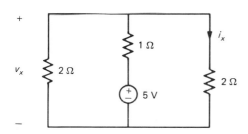

FIGURE P2.35

2.36 Determine the voltage v_x and current i_x in the circuit of Fig. P2.36. Show that conservation of power holds for this circuit,

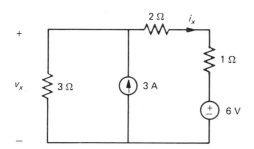

FIGURE P2.36

2.37 Determine the voltage v_x and current i_x in the circuit of Fig. P2.37. Show that conservation of power holds for this circuit.

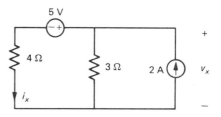

FIGURE P2.37

2.38 Determine the voltage v_x and current i_x in the circuit of Fig. P2.38. Show that conservation of power holds for this circuit.

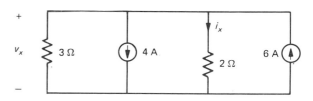

FIGURE P2.38

2.39 Determine the voltage v_x and current i_x in the circuit of Fig. P2.39. Show that conservation of power holds for this circuit.

2.40 Determine the voltage v_x and current i_x in the circuit of Fig. P2.40. Show that conservation of power holds for this circuit.

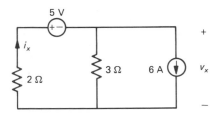

FIGURE P2.39

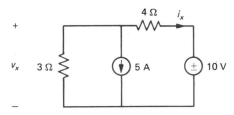

FIGURE P2.40

Analysis Techniques for Resistive Circuits

In Chap. 2 we studied various types of electric circuit elements—the resistor and the ideal voltage and current sources—and the laws which govern their voltages and currents when they are interconnected to form an electric circuit—KVL and KCL. Our ultimate objective in the analysis of an electric circuit is to determine some of or all the branch (element) voltages and currents. These branch voltages and currents are the result of the excitation of the circuit by its ideal sources. The particular interconnections of these circuit elements determine the values of these branch voltages and currents. At the end of Chap. 2 we studied some techniques for the analysis of simple circuits. There are numerous other circuits which cannot be analyzed with those techniques in a simple fashion.

In this chapter we study various additional techniques for solving a circuit for its branch voltages and currents. We concentrate on dc sources, although the techniques are equally valid for other time-varying sources. These techniques are quite general and very important. Although we use them to analyze only resistive circuits—circuits containing sources and resistors—we will soon extend their use to circuits containing the remaining circuit elements—the capacitor and inductor.

3.1 SERIES AND PARALLEL CONNECTIONS OF RESISTORS

We frequently have occasion to consider what are known as *series* and *parallel* connections of circuit elements. These notions were briefly considered in Chap. 2. It is important that we review these ideas since many errors made in the analysis of circuits are a result of not understanding these concepts.

A *series* connection of three elements is shown in Fig. 3.1a. The three elements are joined at common nodes b and c. Elements A and B are said to be connected in series because current i_A, leaving element A and entering node b, *must*, by KCL,

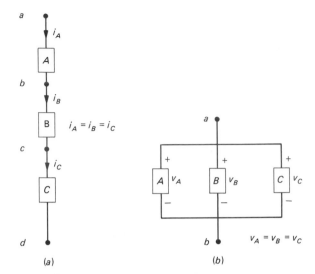

FIGURE 3.1
(*a*) Series and
(*b*) parallel
connections of
circuit elements.

equal current i_B, which enters element *B* and leaves node *b*, and therefore $i_A = i_B$. This is true because there are no other elements attached to that node. Similarly, elements *B* and *C* are said to be connected in series since KCL shows that $i_B = i_C$.

A *parallel* connection of three elements is shown in Fig. 3.1*b*. The three elements are said to be connected in parallel because, by KVL, the voltages across these elements *must* necessarily be equal, that is, $v_A = v_B = v_C$.

Equivalently, we may say that *two elements are connected in series if they are connected at a common node to which no other circuit elements are attached.* Thus KCL dictates that the *currents* through the two elements be identical. Similarly, we may say that *two elements are connected in parallel if the elements are connected at both sets of terminals.* In this case, KVL dictates that the element *voltages* be identical.

Some common misconceptions and improper classification of typical connections are shown in Fig. 3.2. Note in Fig. 3.2*a* that elements *B* and *C* are in parallel because we must have, by KVL, $v_B = v_C$. However, elements *A* and *B* are *not*

FIGURE 3.2
Improper
classification of
series and parallel
connections. (*a*) *B*
and *C* are in
parallel, but *A* and
B are *not* in series.
(*b*) *B* and *C* are in
series, but *A* and
B are *not* in
parallel.

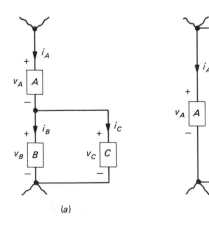

in series, nor are elements A and C, since $i_A \neq i_B$ and $i_A \neq i_C$. Similarly, in Fig. 3.2b elements B and C are in series because KCL requires that $i_B = i_C$. However, elements A and B are *not* in parallel, nor are elements A and C, since $v_A \neq v_B$ and $v_A \neq v_C$.

A *series* connection of resistors is shown in Fig. 3.3. Note that the current through each resistor must be the same:

$$i = i_1 = i_2 = \cdots = i_n \tag{3.1}$$

but that the overall voltage is the sum of resistor voltages:

$$v = v_1 + v_2 + \cdots + v_n \tag{3.2}$$

Substituting Ohm's law for each resistor into Eq. (3.2), we obtain

$$
\begin{aligned}
v &= R_1 i + R_2 i + \cdots + R_n i \\
&= (R_1 + R_2 + \cdots + R_n)i
\end{aligned}
\tag{3.3}
$$

Thus, *from the standpoint of terminals a-b, the series combination is equivalent to a single equivalent resistor whose value is the sum of the values of the resistors*:

$$R_{eq} = R_1 + R_2 + \cdots + R_n \tag{3.4}$$

Now let us consider the *parallel* combination of resistors, as shown in Fig. 3.4. Note that for this case, as opposed to the series connection, the overall current entering the terminals of the combination is the sum of the individual currents:

$$i = i_1 + i_2 + \cdots + i_n \tag{3.5}$$

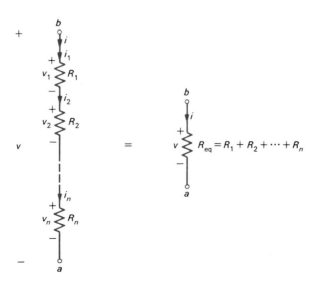

FIGURE 3.3
The equivalent resistance of resistors in series

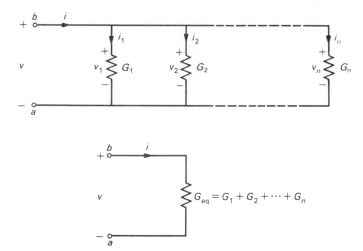

FIGURE 3.4
The equivalent
conductance of
resistors in parallel.

but the individual resistor voltages must all be equal:

$$v = v_1 = v_2 = \cdots = v_n \tag{3.6}$$

Substituting Ohm's law in terms of conductance into Eq. (3.5) yields

$$
\begin{aligned}
i &= G_1 v + G_2 v + \cdots + G_n v \\
&= (G_1 + G_2 + \cdots + G_n)v
\end{aligned}
\tag{3.7}
$$

Thus, *a parallel combination of resistors is equivalent (at the terminals) to a single resistor whose conductance is the sum of the conductances of each of the resistors in the parallel connection.* Note in both the series and parallel reductions that once we replace the combination with an equivalent resistor, we no longer have retained the individual voltages and currents of the resistors in the combination.

One very common case which we frequently use is the parallel combination of two resistors shown in Fig. 3.5. The equivalent conductance is the sum of the

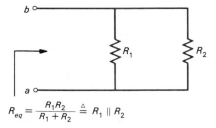

FIGURE 3.5
A very common
case of equivalent
resistance—two
resistors in parallel.

FIGURE 3.6
Illustration of
frequently
encountered
cases: (*a*) two
resistors of equal
value in parallel;
(*b*) two resistors
of widely
dissimilar values in
parallel (the
equivalent
resistance is
smaller than and
approximately
equal to the
smaller resistance).

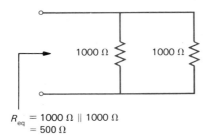

R_{eq} = 1000 Ω || 1000 Ω
= 500 Ω

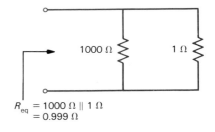

R_{eq} = 1000 Ω || 1 Ω
= 0.999 Ω

conductances of the two resistors, and the equivalent resistance is the reciprocal of this:

$$R_{eq} = \frac{1}{G_{eq}}$$

$$= \frac{1}{G_1 + G_2}$$

$$= \frac{1}{1/R_1 + 1/R_2}$$

$$= \frac{R_1 R_2}{R_1 + R_2}$$

$$\triangleq R_1 || R_2 \tag{3.8}$$

This is denoted as $R_1 || R_2$.

In a parallel combination of resistors, the equivalent resistance is smaller than the smallest resistance in the parallel combination, and the parallel combination of two identical resistors yields an equivalent resistance equal to one-half the value of one of the resistors:

$$R || R = \tfrac{1}{2} R \tag{3.9}$$

These points are illustrated in Fig. 3.6.

Example 3.1   Consider the parallel-series combination of resistors shown in Fig. 3.7. Determine an equivalent resistance at terminals *a-b*.

Solution   First replace the parallel combination of R_2 and R_3 with $R_2 || R_3 = 1.5$ Ω. Then R_{eq} becomes the series combination of R_1, $R_2 || R_3$, and R_4, so that

$$R_{eq} = R_1 + R_2 || R_3 + R_4$$
$$= 7.5 \ \Omega$$

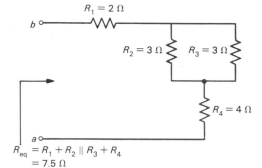

FIGURE 3.7
Example 3.1:
reduction of
series-parallel
combinations.

$R_{eq} = R_1 + R_2 \| R_3 + R_4$
$= 7.5\ \Omega$

These series-parallel reduction rules can be quite useful in determining the voltages and currents of a circuit, as the following examples show.

Example 3.2

Determine the current i delivered by the 10-V voltage source in Fig. 3.8a.

Solution First reduce the parallel combination of 2-Ω resistors to an equivalent 2-$\Omega\|$2-$\Omega = $ 1-Ω combination at terminal c. Then add this to the 1-Ω resistor in series with it. Then reduce the parallel combination of 1 Ω + 2 $\Omega\|$2 $\Omega = 2\ \Omega$ and 3 Ω, which is equivalent to 2 $\Omega\|$3 $\Omega = \frac{6}{5}\ \Omega$. Add this to the 2-$\Omega$ resistor which is in series with the voltage source so that the voltage source "sees" a resistance of 2 $\Omega + \frac{6}{5}\ \Omega = 2\frac{6}{5}\ \Omega$. Replacing the circuit attached to the 10-V voltage source with this equivalent resistance at terminals a-b, we then compute

$$i = \frac{10\ \text{V}}{2\frac{6}{5}\ \Omega} = 3.125\ \text{A}$$

This example illustrates an important point. When we use these series-parallel reduction rules, it is important to label points in the circuit at which a replacement is being made. Note in the last step in Fig. 3.8c that node c has disappeared. However, as far as computing i is concerned, this is unimportant since the effect of the resistors attached to node c has been included in the 2$\frac{6}{5}$-Ω equivalent resistance. Note also that the voltage v across the 3-Ω resistor has disappeared in the last step (Fig. 3.8c). To compute v, we must return to Fig. 3.8b. Note in Fig. 3.8b that current i passes through the 2-Ω resistor and produces a voltage across it of $i \times 2\ \Omega = 6.25$ V. The unknown voltage v can be determined by writing KVL around the outside loop:

$$10 - 2i - v = 0$$

or

$$v = 10 - 2i$$
$$= 3.75\ \text{V}$$

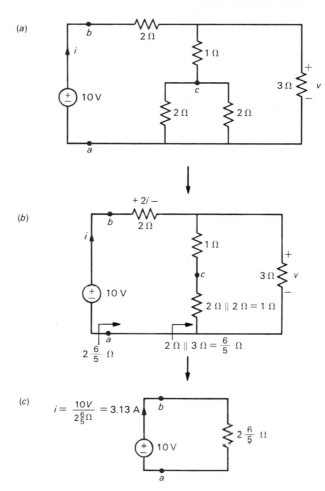

FIGURE 3.8
Example 3.2:
solution by
equivalent resistor
reductions.

Alternatively, we may compute v by multiplying i by the equivalent resistance across which v is located, which is $\frac{6}{5}\,\Omega$:

$$v = \tfrac{6}{5}\,\Omega \times i$$
$$= 3.75 \text{ V}$$

Example 3.3   Determine voltages v, v_x, and v_y and current i_y in Fig. 3.9a.

Solution   First reduce the parallel combination of 2- and 1-Ω resistors to an equivalent $2\text{-}\Omega \| 1\text{-}\Omega = \frac{2}{3}\text{-}\Omega$ resistor, as shown in Fig. 3.9b. Now it is clear than the 1-A current passes through each resistor in Fig. 3.9b so that

$$v_x = -2\,\Omega \times 1\text{ A} \qquad v_y = -\tfrac{2}{3}\,\Omega \times 1\text{ A}$$
$$= -2\text{ V} \qquad\qquad = -\tfrac{2}{3}\text{ V}$$

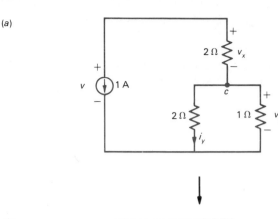

FIGURE 3.9
Example 3.3:
solution by
equivalent resistor
reduction.

(Note the signs.) Next v can be found in two ways. One way is to note that the 1-A current source sees an equivalent resistance of $2\,\Omega + \frac{2}{3}\Omega = \frac{8}{3}\Omega$ and that v is across this equivalent resistance. Thus

$$v = -\tfrac{8}{3}\Omega \times 1\,\text{A}$$
$$= -\tfrac{8}{3}\,\text{V}$$

The other way is to write KVL around the outside loop to obtain

$$v - v_y - v_x = 0$$

or

$$v = v_y + v_x$$
$$= -2 - \tfrac{2}{3}$$
$$= -\tfrac{8}{3}\,\text{V}$$

Knowing v_y, we may now determine i_y from Fig. 3.9a as

$$i_y = \frac{v_y}{2\,\Omega}$$

$$= \frac{-\tfrac{2}{3}}{2}$$

$$= -\tfrac{1}{3}\,\text{A}$$

Note in Fig. 3.9b that i_y has disappeared at this reduction stage.

3.2   VOLTAGE AND CURRENT DIVISION

There are two additional solution techniques which, like series and parallel combinations of resistors, are useful in solving for the voltages and currents in a circuit. These are the rules of voltage and current division.

First let us consider *voltage division*. Suppose that a known voltage is applied across a series combination of resistors, as shown in Fig. 3.10. Current i through the series combination is

$$i = \frac{v_s}{R_1 + R_2 + \cdots + R_n} \tag{3.10}$$

since the voltage source sees an equivalent resistance $R_{eq} = R_1 + R_2 + \cdots + R_n$. The individual voltages across the individual resistors are

$$v_1 = R_1 i$$
$$v_2 = R_2 i$$
$$\vdots$$
$$v_n = R_n i \tag{3.11}$$

Substituting Eq. (3.10), we find that

$$v_i = \frac{R_i}{R_1 + R_2 + \cdots + R_n} v_s \tag{3.12}$$

Thus, *the voltage divides according to the ratio of the resistor in question to the sum of the resistors in the series combination*. This is called the *voltage-division rule*.

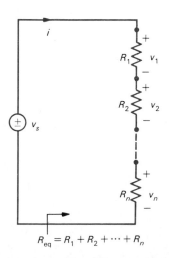

FIGURE 3.10
Illustration of the voltage-division rule.

Example 3.4

Consider the example shown in Fig. 3.11. Determine the voltage v.

Solution   First combine the two 2-Ω resistors into an equivalent 1-Ω resistor. Then voltage v is, by voltage division,

$$v = \frac{1\,\Omega}{1\,\Omega + 3\,\Omega}\,10\text{ V}$$

$$= 2.5\text{ V}$$

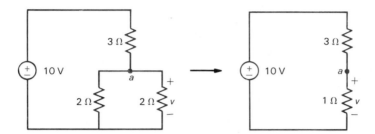

FIGURE 3.11
Example 3.4:
illustration of
voltage division.

The voltage across the series combination need not be provided by an ideal source. It may be a result of some other portion of the circuit.

Example 3.5

Consider the circuit in Fig. 3.12. Determine voltages v_x and v_y.

Solution   First reduce the circuit connected to node a to an equivalent resistance of $\frac{4}{3}\,\Omega$, and redraw the circuit slightly. Note that node b and v_x have disappeared in the reduction.

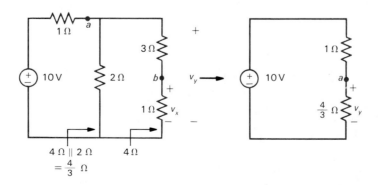

FIGURE 3.12
Example 3.5:
illustration of
voltage division.

Nevertheless, by voltage division we may obtain

$$v_y = \frac{\frac{4}{3}\,\Omega}{\frac{4}{3}\,\Omega + 1\,\Omega}\, 10\ \text{V}$$

$$= \tfrac{40}{7}\ \text{V}$$

Now from the original circuit we have, by voltage division,

$$v_x = \frac{1\,\Omega}{1\,\Omega + 3\,\Omega}\, v_y$$

$$= \tfrac{10}{7}\ \text{V}$$

The next rule, *current division*, is similar to the voltage-division rule. Consider the parallel combination of resistors shown in Fig. 3.13. The equivalent conductance seen by the current source is

$$G_{\text{eq}} = G_1 + G_2 + \cdots + G_n \tag{3.13}$$

and the resulting voltage across the current source and parallel combination is

$$v = \frac{1}{G_{\text{eq}}}\, i_s \tag{3.14}$$

Now each current is given by

$$i_i = G_i v$$

$$= \frac{G_i}{G_1 + G_2 + \cdots + G_n}\, i_s \tag{3.15}$$

This is the current-division rule. *The current entering a parallel combination of resistors divides according to the ratio of the conductance in question to the sum of the conductances in the parallel combination.*

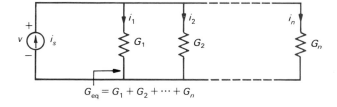

FIGURE 3.13
Illustration of the
current-division
rule.

Example 3.6

Consider the circuit shown in Fig. 3.14. Determine i_x.

FIGURE 3.14
Example 3.6:
illustration of
current division.

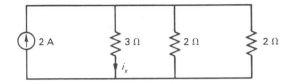

Solution

$$i_x = \frac{\frac{1}{3}S}{\frac{1}{3}S + \frac{1}{2}S + \frac{1}{2}S} \, 2 \, A$$

$$= \frac{1}{2} A$$

Quite often we tend to deal in resistance rather than conductance. A common case where this can be easily done when using current division is the case of two parallel resistances, shown in Fig. 3.15. The currents are, by current division,

$$i_1 = \frac{G_1}{G_1 + G_2} \, i_s \qquad i_2 = \frac{G_2}{G_1 + G_2} \, i_s$$

$$= \frac{R_2}{R_1 + R_2} \, i_s \qquad = \frac{R_1}{R_1 + R_2} \, i_s$$

(3.16)

Note that when we are dealing with resistance instead of conductance, the current divides according to the ratio of the resistance *opposite* the one in question to the sum of the resistances in the combination. This technique can be adapted to cases involving more than two parallel resistors, as the following example shows.

FIGURE 3.15
Illustration of
current division for
an important
case—two resistors
in parallel.

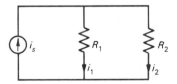

Example 3.7

Reconsider the circuit in Fig. 3.14. Determine i_x.

Solution   The two 2-Ω resistors are in parallel with each other, and we may combine them into an equivalent 1-Ω resistor. Then we have 1 Ω in parallel with 3 Ω and wish to find the current through the 3-Ω resistor:

$$i_x = \frac{1 \, \Omega}{3 \, \Omega + 1 \, \Omega} \, 2 \, A$$

$$= \frac{1}{2} A$$

3.3 SUPERPOSITION

Perhaps the most powerful principle in analyzing *linear* circuits is the principle of *superposition*. This principle does *not* apply to *nonlinear* circuits. It is therefore important to determine what constitutes a linear circuit: *A linear circuit is one which contains only linear circuit elements.* The circuit elements which we have considered—the resistor and the ideal voltage and current source—are linear elements. All the ideal sources are linear, so we need to focus on the resistors to determine whether a circuit is nonlinear. An example of a nonlinear resistor is one in which the terminal voltage-current characteristic is not a straight line or does not pass through the origin, such as is shown in Fig. 3.16. For nonlinear resistors, the terminal voltage and current are not linearly related. For example, $v = Ri^2$ is a nonlinear relationship since the voltage does not depend directly on the current but depends on the square of the current.

The principle of superposition is quite simple. *If a linear circuit has N ideal sources present, any branch voltage or current is composed of the sum of N contributions, each of which is due to each source acting individually when all others are set equal to zero (deactivated).* When we set the value of an ideal *voltage source* to zero (deactivate it), the voltage of that branch becomes zero. Thus, we replace the voltage source with a *short circuit*, as shown in Fig. 3.17a. Conversely, when we set an ideal *current source* to zero (deactivate it), the current of that branch becomes zero. Thus, we replace it with an *open circuit*, as shown in Fig. 3.17b.

To illustrate the principle of superposition, let us consider the (linear) circuit in Fig. 3.18a. The voltage across R_1 consists of two components, v' due to the ideal voltage source and v'' due to the ideal current source; similarly, current i through resistor R_2 consists of two components. Each of these contributions is computed

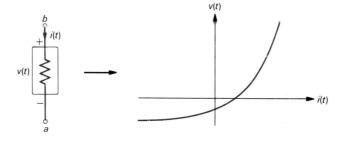

FIGURE 3.16 A general nonlinear resistor.

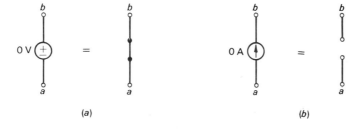

FIGURE 3.17 Replacement of "deactivated" sources with (*a*) a short circuits and (*b*) an open circuit.

(*a*)   (*b*)

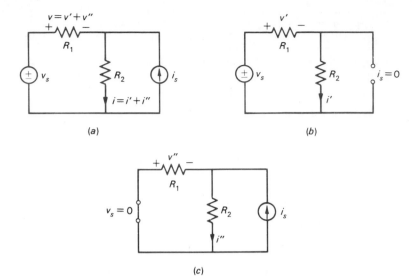

FIGURE 3.18
Illustration of the
superposition
principle.

from the circuits in Fig. 3.18b and c. The contributions due to the voltage source, with the current source deactivated, are computed from Fig. 3.18b as

$$v' = \frac{R_1}{R_1 + R_2} v_s$$

$$i' = \frac{v_s}{R_1 + R_2}$$

The contributions due to the current source, with the voltage source deactivated, are computed from Fig. 3.18c as

$$v'' = -R_1 || R_2 i_s = -\frac{R_1 R_2}{R_1 + R_2} i_s$$

$$i'' = \frac{R_1}{R_1 + R_2} i_s$$

The reader should verify these results, using voltage and current division and equivalent resistances. The total voltage v and current i due to both sources activated becomes

$$v = v' + v'' = \frac{R_1}{R_1 + R_2} v_s - R_1 || R_2 i_s$$

$$i = i' + i'' = \frac{v_s}{R_1 + R_2} + \frac{R_1}{R_1 + R_2} i_s$$

Example 3.8   Using superposition, compute the current i and voltage v in the circuit of Fig. 3.19a. Show that these results satisfy KVL and KCL.

Solution   We determine from Fig. 3.19b and c that

$$v' = 4 \text{ V} \qquad v'' = -\tfrac{36}{5} \text{ V}$$
$$i' = 2 \text{ A} \qquad i'' = \tfrac{12}{5} \text{ A}$$

so that

$$v = v' + v''$$
$$= -\tfrac{16}{5} \text{ V}$$
$$i = i' + i''$$
$$= \tfrac{22}{5} \text{ A}$$

To verify that these solutions satisfy KVL and KCL, we must compute all element voltages and currents. This is done in Fig. 3.19d using only Ohm's law. Applying KVL around the loop containing the two resistors and the voltage source, we have

$$10 - (-\tfrac{16}{5}) - \tfrac{66}{5} \overset{?}{=} 0$$

which is true. Applying KCL at the upper node to which the current source is attached, we have

$$6 - \tfrac{22}{5} + (-\tfrac{16}{10}) \overset{?}{=} 0$$

which is true. Since the branch voltages and currents satisfy KVL and KCL for the circuit, we have found the solution.

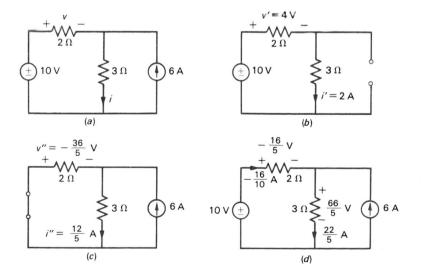

FIGURE 3.19
Example 3.8:
illustration of
superposition.

In applying superposition, it is not necessary to consider each source individually. We may group several sources and determine the contribution due to this group of sources.

Example 3.9   Determine voltage v in the circuit of Fig. 3.20a.

Solution   To illustrate this concept, we consider the two voltage sources as jointly contributing v' and the current sources as individually contributing v'' and v'''. These solutions are shown in Fig. 3.20b, c, and d:

$$v' = \frac{2}{2+3}(5-10)$$

$$= -2\text{ V}$$

$$v'' = 0$$

$$v''' = -\tfrac{18}{5}\text{ V}$$

(Note that, by current division, all of the 2-A current source in Fig. 3.20c goes through the short circuit caused by deactivating the 10-V voltage source.) Thus

$$v = v' + v'' + v'''$$

$$= -2 + 0 - \tfrac{18}{5}$$

$$= -\tfrac{28}{5}\text{ V}$$

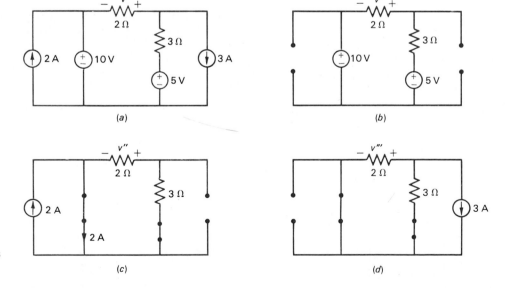

FIGURE 3.20
Example 3.9:
illustration of
superposition by
combining groups
of sources.

It is worthwhile at this stage to point out that superposition does *not* apply to the computation of power.

Example 3.10   Determine the power delivered to the 3-Ω resistor in Fig. 3.19.

Solution   Since $i = \frac{22}{5}$ A (from Example 3.8), we find that

$$p_{3\Omega} = 3i^2$$
$$= 58.08 \text{ W}$$

If we try to compute this by superposition, we find that

$$p_{3\Omega} \overset{?}{=} 3(i')^2 + 3(i'')^2$$
$$= 3(2)^2 + 3(\tfrac{12}{5})^2$$
$$= 12 + 17.28$$
$$= 29.28$$
$$\neq 58.08 \text{ W}$$

It is important to note that an ideal voltage or current source may actually be absorbing power (having power delivered to it). Whether the element absorbs or delivers power depends solely on the circuit to which the source is connected.

Example 3.11   Determine the power delivered by the 2- and 3-A current sources in Fig. 3.20a.

Solution   To do this, we need only determine the terminal voltage of each current source and its polarity. The voltage across the 2-A current source is simply the value of the 10-V voltage source. Because of the polarity of this voltage and the current of the current source, the 2-A current source is delivering 2 A × 10 V = 20 W. The terminal voltage across the 3-A current source is, by KVL, the sum 10 V + $v = 10 - \frac{28}{5} = \frac{22}{5}$ V. But this voltage has the polarity + at the top of the current source and − at the bottom. Because of this terminal voltage polarity and the direction of the current source, it is delivering $-(\frac{22}{5})(3) = -\frac{66}{5}$ W. Thus, the 3-A current source is actually absorbing $\frac{66}{5}$ W of power!

Now we will show, very simply, why superposition works. The branch voltages and currents are governed by (must simultaneously satisfy) KVL, KCL, and the element relations. For example, consider the circuit in Fig. 3.21a. KVL only needs to be written around one loop, L, and KCL needs to be written at only one node, a. These two equations, along with the resistor relations, become

KVL loop L:

$$v_1 + v_2 = v_s$$

KCL node a:

$$i_2 - i_1 = i_s \tag{3.17}$$

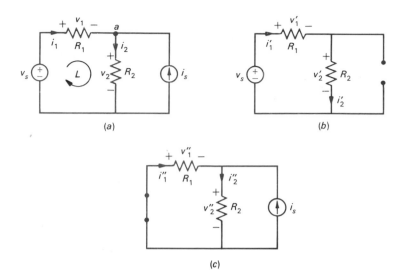

FIGURE 3.21
Illustration of the
proof of the
superposition
principle.

Ohm's law for R_1:

$$v_1 - R_1 i_1 = 0$$

Ohm's law for R_2:

$$v_2 - R_2 i_2 = 0$$

Note that we have placed the values of the ideal sources (the known quantities in these equations) on the right-hand side and all other quantities on the left. We now have four equations which can be solved for the four unknowns: v_1, i_1, v_2, i_2. We would not, of course, resort to solving these simultaneously, since simpler methods of analysis will do (as illustrated in previous examples); nevertheless, they illustrate why superposition works. The solutions, by superposition, for the contributions of the two sources to these four branch voltages and currents are shown in Fig. 3.21b and c. The principle of superposition states that if these contributions satisfy the circuit equations individually, then their sum will satisfy the circuit equations. From Fig. 3.21b, the contributions due to the voltage source satisfy [set $i_s = 0$ in the circuit equations of Eq. (3.17)]

$$
\begin{aligned}
v_1' + v_2' &= v_s \\
i_2' - i_1' &= 0 \\
v_1' - R_1 i_1' &= 0 \\
v_2' - R_2 i_2' &= 0
\end{aligned}
\tag{3.18}
$$

Similarly, from Fig. 3.21c, the contributions due to the current source satisfy [set $v_s = 0$ in the circuit equations of Eq. (3.17)]

$$
\begin{aligned}
v_1'' + v_2'' &= 0 \\
i_2'' - i_1'' &= i_s \\
v_1'' - R_1 i_1'' &= 0 \\
v_2'' - R_2 i_2'' &= 0
\end{aligned}
\tag{3.19}
$$

The question now is, given that v_1', v_2', i_1', i_2' satisfy Eq. (3.18) and that v_1'', v_2'', i_1'', i_2'' satisfy Eq. (3.19), is it true that $v_1' + v_1''$, $v_2' + v_2''$, $i_1' + i_1''$, $i_2' + i_2''$ *automatically* satisfy Eq. (3.17)? Let us see. Substitute the sum variables into Eq. (3.17) to yield

$$
\begin{aligned}
(v_1' + v_1'') + (v_2' + v_2'') &= v_s + 0 \\
(i_2' + i_2'') - (i_1' + i_1'') &= 0 + i_s \\
(v_1' + v_1'') - R_1(i_1' + i_1'') &= 0 + 0 \\
(v_2' + v_2'') - R_2(i_2' + i_2'') &= 0 + 0
\end{aligned}
\tag{3.20}
$$

Note that we have added a zero to the right-hand side of all (3.17) equations. Now it is easy to see that Eq. (3.20) can be factored as the *sum* of Eqs. (3.18) and (3.19), which were assumed to be true, so Eq. (3.20) is true.

Now it is easy to see why superposition does not work for nonlinear circuits. Suppose that resistor R_1 in the above circuit were nonlinear and characterized by

$$
v_1 = R_1 i_1^2
\tag{3.21}
$$

The prime variables due to the voltage source satisfy

$$
v_1' - R_1 i_1'^2 = 0
\tag{3.22}
$$

and the double-prime variables due to the current source satisfy

$$
v_1'' - R_1 i_1''^2 = 0
\tag{3.23}
$$

Now can

$$
v_1' + v_1'' - R_1(i_1' + i_1'')^2 = 0
\tag{3.24}
$$

be factored into the sum of Eqs. (3.22) and (3.23)? Let us try:

$$
v_1' + v_1'' - R_1 i_1'^2 - 2R_1 i_1' i_1'' - R_1 i_1''^2 = 0
\tag{3.25}
$$

In Eq. (3.25) we identify Eqs. (3.22) and (3.23), but there is an additional term, $-2R_1 i_1' i_1''$, which is not accounted for. Thus, we cannot say that Eq. (3.25) is satisfied solely because Eqs. (3.22) and (3.23) were presumed to be satisfied. Now we clearly see why superposition works for linear circuits and does not work for nonlinear circuits. All it takes to make an entire circuit nonlinear is for one resistor to be nonlinear! In our future use of superposition we will begin to see its extraordinary usefulness in solving linear circuits.

3.4   THÉVENIN AND NORTON EQUIVALENT CIRCUITS

The Thévenin and Norton equivalent circuits allow us to replace a portion of a circuit which is connected to two terminals with another circuit which has the same effect on the rest of the circuit. The principle is quite simple and, again, is valid only for a linear portion of a circuit. In other words, a circuit may be nonlinear due to the presence of a nonlinear resistor—but we may *replace* that linear portion of the circuit which is attached to the terminals of the nonlinear resistor with an equivalent, and often simpler, circuit.

To illustrate this principle and show how sensible it is, let us consider an electric circuit which has been partitioned into parts connected by wires, as shown in Fig. 3.22a. Now let us ask the question, How are the voltage v and current i at the terminals of the linear portion related? The obvious answer is that they are related as a linear equation such as

$$v = Ai + B \tag{3.26}$$

where A and B are two constants.

How else would they be related? A relation such as $v = Ai^2 + B$ would not be logical, since the portion of the circuit was stipulated to be linear. Now all we need to do to obtain an equivalent circuit with which to replace this linear portion at terminals a-b is to find one which has the same terminal voltage and current relation as in Eq. (3.26).

The Thévenin equivalent circuit is shown in Figure 3.22b. Writing KVL around the loop containing V_{OC} and R_{TH}, we obtain

$$v = R_{TH}i + V_{OC} \tag{3.27}$$

Thus, if we know A and B in Eq. (3.26), perhaps from some measurements of the circuit, we identify $V_{OC} = B$ and $R_{TH} = A$. The quantity V_{OC} is called the *open-circuit voltage*, because when we open-circuit the terminals, $i = 0$, we obtain from Eq. (3.27)

$$V_{OC} = v|_{i=0} \tag{3.28}$$

Obviously, V_{OC} is due to the ideal sources present in this portion of the circuit; if there were no ideal sources present (some may be present in the remainder circuit),

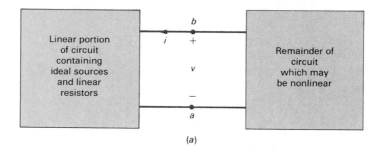

(a)

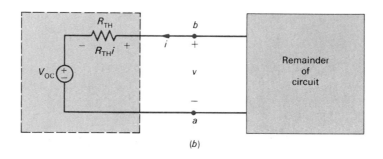

(b)

FIGURE 3.22
Equivalent
representation of
two-terminal
circuits:
(a) two-part
circuit connected
by wires;
(b) Thévenin
equivalent;
(c) Norton
equivalent.

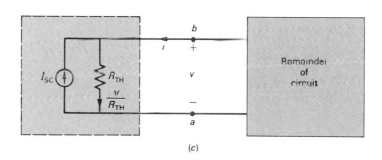

(c)

then V_{OC} would be zero. Similarly, R_{TH} is called the *Thévenin equivalent resistance*, because when we set $V_{OC} = 0$ (deactivate all ideal sources in this portion of the circuit), from Eq. (3.27)

$$R_{TH} = \left. \frac{v}{i} \right|_{V_{OC} = 0} \qquad (3.29)$$

These concepts are illustrated in Fig. 3.23a and b.

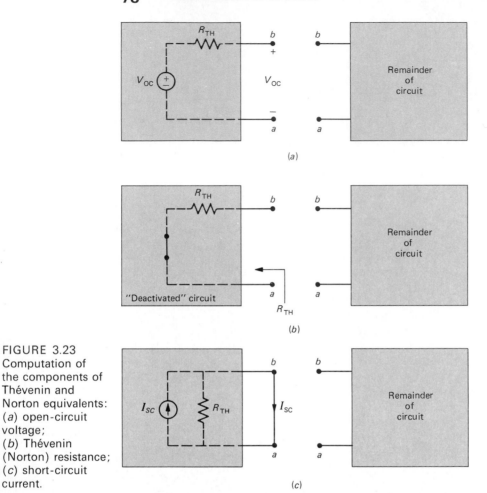

FIGURE 3.23
Computation of
the components of
Thévenin and
Norton equivalents:
(a) open-circuit
voltage;
(b) Thévenin
(Norton) resistance;
(c) short-circuit
current.

Example 3.12   Consider the circuit shown in Fig. 3.24a. Determine current i.

Solution   Reduce the portion of the circuit to the left of terminals a-b to a Thévenin equivalent. First, remove the 5-Ω resistor (the remainder circuit) and compute the open-circuit voltage, as shown in Fig. 3.24b; by superposition this is

$$V_{OC} = 2\,\text{V} - 10\,\text{V}$$
$$= -8\,\text{V}$$

Note that the 3-Ω resistor has no effect on this, since no current flows through it when terminals a-b are open-circuited. Second, to obtain the Thévenin resistance, deactivate all independent sources in this portion of the circuit—replace ideal voltage sources with short circuits and ideal current sources with open circuits. This is shown in Fig. 3.24c. Then R_{TH} is the resistance, seen looking back into the terminals of this deactivated portion of the circuit:

$$R_{TH} = 2\,\Omega \| 2\,\Omega + 3\,\Omega$$
$$= 4\,\Omega$$

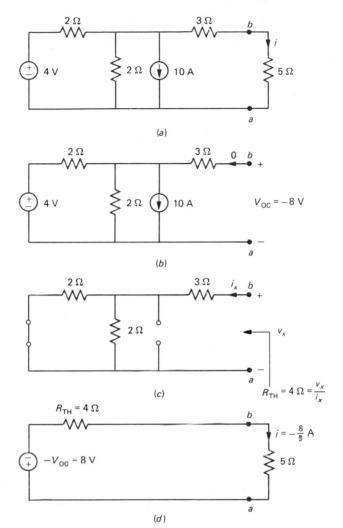

FIGURE 3.24
Example 3.12:
illustrating the use
of a Thévenin
equivalent.

It is also the ratio

$$R_{TH} = \frac{v_x}{i_x}$$

in this circuit. Now attach the Thévenin equivalent circuit to the remainder of the circuit (the 5-Ω resistor), taking care to observe the proper polarity of V_{OC} and the terminal labels. This is shown in Fig. 3.24d. From this we obtain

$$i = -\frac{1}{5\,\Omega + 4\,\Omega}\, 8\,V = -\tfrac{8}{9}\,A$$

Note in the previous example that it is vital to label the terminals at which we wish to make a replacement, to define V_{OC} with respect to these, and to attach the Thévenin equivalent circuit in such a way that the polarity of V_{OC} conforms to this chosen designation.

An alternative form is the Norton equivalent circuit shown in Fig. 3.22c. Writing KCL at node b in Fig. 3.22c, we obtain

$$i = \frac{1}{R_{TH}} v - I_{SC} \tag{3.30}$$

If we rewrite Eq. (3.27) in this form, we have

$$i = \frac{1}{R_{TH}} v - \frac{V_{OC}}{R_{TH}} \tag{3.31}$$

Comparing Eqs. (3.30) and (3.31), we find that

$$I_{SC} = \frac{V_{OC}}{R_{TH}}$$

$$= -i|_{v=0} \tag{3.32}$$

From Eq. (3.32) it is clear that I_{SC} is the current through a short circuit placed across terminals a-b after the portion of the circuit is disconnected from the remainder circuit. As in the case of V_{OC}, I_{SC} is caused by ideal sources; if no ideal sources are present in this portion of the circuit, I_{SC} will be zero. This is illustrated in Fig. 3.23c.

Example 3.13

Compute i in Example 3.12 by using a Norton equivalent.

Solution Resistor R_{TH} was computed in Example 3.12 to be $R_{TH} = 4\,\Omega$. Thus, we need to find only I_{SC}. Removing the remainder circuit (the 5-Ω resistor) and placing a short circuit across terminals a-b as shown in Fig. 3.25a, we obtain, by superposition,

$$I_{SC} = \tfrac{1}{2}\,A - \tfrac{5}{2}\,A$$
$$= -2\,A$$

Note that

$$I_{SC} = \frac{V_{OC}}{R_{TH}}$$

Now reattaching the Norton equivalent circuit to the remainder circuit as shown in Fig. 3.25b, we find (by current division) that, just as in Example 3.12,

$$i = -\frac{4\,\Omega}{4\,\Omega + 5\,\Omega}\,2\,A$$
$$= -\tfrac{8}{9}\,A$$

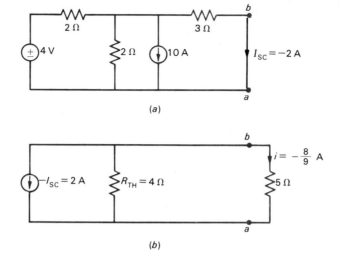

FIGURE 3.25
Example 3.13:
illustrating the use
of a Norton
equivalent.

Note that in the examples of this section we have had numerous occasions to use the analytical techniques of previous sections. It is very important that the reader understand those techniques since they are used on numerous occasions throughout this text.

3.5 NODE VOLTAGE AND MESH CURRENT ANALYSIS

The previous analysis techniques allow us to solve rather simple circuits. They also provide considerable insight into the solution. In this section we will investigate two additional methods which may be used to solve more complicated circuits.

3.5.1 Node Voltage Analysis

Node voltage analysis is a method for obtaining a set of equations to be solved for a set of circuit voltages—the node voltages. Once these node voltages are obtained, we can easily obtain all branch voltages and currents from them.

For example, consider the circuit shown in Fig. 3.26a, which contains three nodes. Select one of the three nodes as a reference node. (The particular choice is arbitrary.) Next define the node voltages as the voltages of each of the remaining nodes *with respect to the reference node*, as shown in Fig. 3.26b. Next write KCL at each node in terms of these node voltages. For example, at node *a* in Fig. 3.26a we have

$$\frac{1}{R_1} V_a + \frac{1}{R_2} (V_a - V_b) = i_{s1} - i_{s2} \tag{3.33}$$

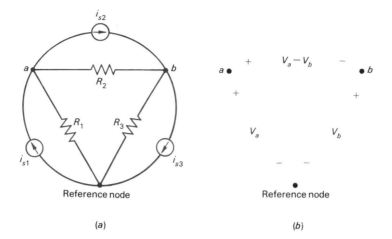

FIGURE 3.26
Definition of node
voltages.

(a)   (b)

and at node b

$$\frac{1}{R_3} V_b + \frac{1}{R_2} (V_b - V_a) = i_{s2} - i_{s3} \qquad (3.34)$$

Rewriting Eqs. (3.33) and (3.34) by grouping terms multiplying each node voltage, we have

node a:

$$\left(\frac{1}{R_1} + \frac{1}{R_2}\right) V_a - \frac{1}{R_2} V_b = i_{s1} - i_{s2} \qquad (3.35)$$

node b:

$$-\frac{1}{R_2} V_a + \left(\frac{1}{R_3} + \frac{1}{R_2}\right) V_b = i_{s2} - i_{s3}$$

These equations can be solved by Gauss elimination or Cramer's rule (see Appendix E) for node voltages V_a and V_b. Once these are obtained, all other branch voltages may be obtained in terms of the node voltages. For example, the branch voltage across R_1 is V_a and across R_3 is V_b. The branch voltage across R_2 is the difference $V_a - V_b$. Once all branch voltages are determined, the branch currents are obtained by Ohm's law.

We may set up the node voltage equations for a circuit by inspection. Note the pattern in Eq. (3.35). When we write the equation at a particular node, the coefficient of that node voltage is the sum of the conductances connected to that node. The coefficient of one of the other node voltages in that equation is negative and is equal to the conductance connected between the two nodes. The right-hand side of that equation is the sum of the ideal current sources attached to that node. They appear

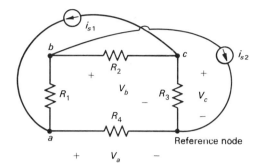

FIGURE 3.27
Illustration of
writing node
voltage equations.

as positive in this sum if they are pointing to the node and as negative if they are
pointing away from this node.

This result can be generalized to larger circuits. For example, consider the circuit
in Fig. 3.27. In terms of the node voltages shown, we write, by inspection,

node a:

$$\left(\frac{1}{R_1} + \frac{1}{R_4}\right)V_a - \frac{1}{R_1}V_b = i_{s1}$$

node b:

$$-\frac{1}{R_1}V_a + \left(\frac{1}{R_1} + \frac{1}{R_2}\right)V_b - \frac{1}{R_2}V_c = -i_{s2} \tag{3.36}$$

node c:

$$-\frac{1}{R_2}V_b + \left(\frac{1}{R_2} + \frac{1}{R_3}\right)V_c = -i_{s1}$$

Apply the general rule developed above to obtain Eq. (3.36).

Example 3.14   As an example, write and solve the node voltage equations for the circuit of Fig. 3.28a and
determine the current i_x.

Solution   The nodes and node voltages are designated as shown and the node voltage
equations become

node a:

$$(\tfrac{1}{1} + \tfrac{1}{2})V_a - (\tfrac{1}{2})V_b = 2 + 3 = 5$$

node b:

$$-(\tfrac{1}{2})V_a + (\tfrac{1}{2} + \tfrac{1}{3})V_b = 4 - 3 - 1 = 0$$

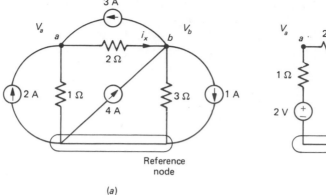

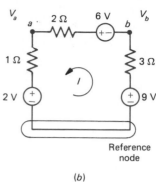

FIGURE 3.28
Example 3.14.

(a)

(b)

Reference
node

Reference
node

Solving these by Cramer's rule gives

$$V_a = \frac{\begin{vmatrix} 5 & -\frac{1}{2} \\ 0 & \frac{5}{6} \end{vmatrix}}{\begin{vmatrix} \frac{3}{2} & -\frac{1}{2} \\ -\frac{1}{2} & \frac{5}{6} \end{vmatrix}} = \frac{25}{6} \text{ V}$$

$$V_b = \frac{\begin{vmatrix} \frac{3}{2} & 5 \\ -\frac{1}{2} & 0 \end{vmatrix}}{\begin{vmatrix} \frac{3}{2} & -\frac{1}{2} \\ -\frac{1}{2} & \frac{5}{6} \end{vmatrix}} = \frac{5}{2} \text{ V}$$

Therefore

$$i_x = \frac{V_a - V_b}{2} = \frac{5}{6} \text{ A}$$

This problem could also be solved by converting the current source–parallel resistor combinations to voltage source–series resistor combinations using Norton to Thévenin transformations as shown in Fig. 3.28b. From that circuit we may obtain, in terms of I,

$$I = \frac{9 + 6 - 2}{1 + 2 + 3} = \frac{13}{6} \text{ A}$$

and

$$V_a = 2 \text{ V} + I1 \text{ } \Omega = \tfrac{25}{6} \text{ V}$$

$$V_b = 9 \text{ V} - I3 \text{ } \Omega = \tfrac{5}{2} \text{ V}$$

Note that i_x has disappeared from the circuit of Fig. 3.28b. However, the node voltages remain the same as in Fig. 3.28a so that i_x can be found from the circuit of Fig. 3.28a. Also we should realize that

$$i_x = 3 - I = \tfrac{5}{6} \text{ A}$$

The preceding has assumed that only current sources were present in the circuit for which we desired to write the node voltage equations. When the circuit contains voltage sources, node voltage equations can still be written, and the number of resulting simultaneous equations to be solved is reduced by the number of voltage sources. In order to illustrate this technique, consider the circuits of Fig. 3.29 which have voltage sources in addition to current sources. The circuit of Fig. 3.29a has a voltage source v_s connecting node b to the chosen reference node. In this case, the node voltage of node b is constrained by this voltage source, $V_b = v_s$, and KCL needs to be written only at node a:

node a:

$$\frac{V_a}{R_1} + \frac{V_a - V_b}{R_2} = i_s \tag{3.37}$$

But

$$V_b = v_s \tag{3.38}$$

Substituting (3.38) into (3.37) gives

$$\left(\frac{1}{R_1} + \frac{1}{R_2}\right)V_a = i_s + \frac{v_s}{R_2} \tag{3.39}$$

which may be solved for V_a.

Figure 3.29b shows a case where the voltage source is not connected to the chosen reference node. This case is not significantly different from that of Fig. 3.29a. The voltage source relates the two node voltages so that only one of them is unknown:

$$V_b = V_a + v_s \tag{3.40}$$

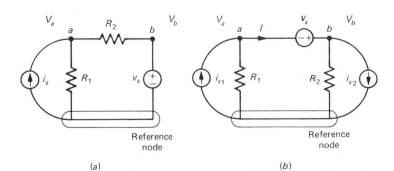

FIGURE 3.29   (a)   (b)

In order to write the node voltage equations, we define an unknown current, I, through the voltage source and write KCL at either node:

node a:

$$\frac{V_a}{R_1} + I = i_{s1} \tag{3.41}$$

Writing KCL at the other node gives

node b:

$$\frac{V_b}{R_2} - I = -i_{s2} \tag{3.42}$$

Adding (3.41) and (3.42) to eliminate I and substituting (3.40) gives

$$\frac{V_a}{R_1} + \frac{V_a + v_s}{R_2} = i_{s1} - i_{s2} \tag{3.43}$$

which gives one equation in V_a:

$$\left(\frac{1}{R_1} + \frac{1}{R_2} \right) V_a = i_{s1} - i_{s2} - \frac{v_s}{R_2} \tag{3.44}$$

which can be solved for V_a.

Example 3.15   Write and solve the node voltage equations for the circuit of Fig. 3.30a and determine the current i_x.

Solution   The nodes are labeled as shown. Writing KCL at nodes b and c gives

node b:

$$\frac{V_b - V_a}{1} + i_x = 2 + 3$$

node c:

$$\frac{V_c}{2} - i_x = -1$$

Adding these gives

$$V_b - V_a + \frac{V_c}{2} = 4$$

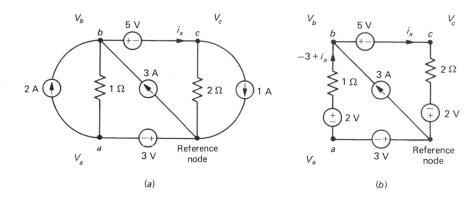

FIGURE 3.30
Example 3.15.

(a)   (b)

But the voltage sources constrain the node voltages as

$$V_a = -3 \text{ V}$$
$$V_b = 5 \text{ V} + V_c$$

Substituting these constraints gives

$$5 + V_c + 3 + \frac{V_c}{2} = 4$$

or

$$V_c = \quad \tfrac{8}{3} \text{ V}$$

From the circuit

$$i_x = \frac{V_c}{2} + 1$$

$$= -\tfrac{1}{3} \text{ A}$$

The circuit can also be solved by using a Norton to Thévenin transformation as shown in Fig. 3.30b. Note that the 3-A current source cannot be converted since there is no resistance in parallel with it. Nevertheless, from this circuit we may write KCL at node b giving the current through the left branch as $-3 + i_x$. Writing KVL around the outer loop gives

$$1(-3 + i_x) + 2i_x = -3 + 2 - 5 + 2$$

or

$$3i_x = -1$$

giving $i_x = -\tfrac{1}{3}$ A again.

3.5.2 Mesh Current Analysis

The mesh current analysis method, like the node voltage analysis method, generates a set of simultaneous equations. The solution to this set of equations can be used to find the solutions for all branch voltages and currents of the circuit. In the mesh current method we use a set of mesh currents, whereas in the node voltage method we used a set of node voltages.

A mesh current is a fictitious current which is defined to circulate around a mesh of the circuit. A circuit mesh can be thought of as panes in a window. If we draw the circuit on a piece of paper with no branches crossing each other, the meshes are circuit loops which do not encircle other circuit elements; examples are given in Fig. 3.31a. Not all circuits can be laid out to contain only meshes. Those which cannot are called nonplanar circuits; an example is shown in Fig. 3.31b. For nonplanar circuits, the mesh current analysis method which we discuss cannot be used, but the node voltage method discussed previously can be used.

We will choose to define the mesh currents to circulate clockwise around only that mesh. Counterclockwise circulation could have been used, but we will always assume clockwise circulation for the mesh currents. An example is shown in Fig. 3.32. Note that the current through R_2 is the difference of the two mesh currents, since both mesh currents flow through this branch but in opposite directions. Now write KVL around each mesh:

mesh 1:

$$R_1 I_1 + R_2(I_1 - I_2) = v_{s1}$$

mesh 2: (3.45)

$$R_3 I_2 + R_2(I_2 - I_1) = -v_{s2}$$

Group coefficients of I_1 and I_2 and write Eq. (3.45) as

mesh 1:

$$(R_1 + R_2)I_1 - R_2 I_2 = v_{s1}$$

mesh 2: (3.46)

$$-R_2 I_1 + (R_2 + R_3)I_2 = -v_{s2}$$

Note that each equation for a mesh has, on the left-hand side of the equation, the sum of the resistances common to that mesh multiplying that mesh current minus the product of the adjacent mesh current and the resistance common to both meshes—and on the right-hand side the sum of any voltage sources encountered in traversing the mesh. If the voltage source tends to "push" in the direction of the mesh current, it appears in the sum on the right-hand side as a positive quantity,

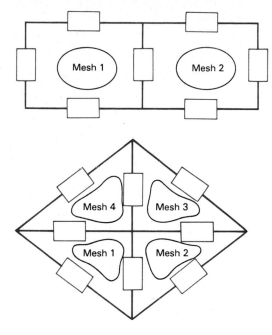

(a) Meshes of a circuit

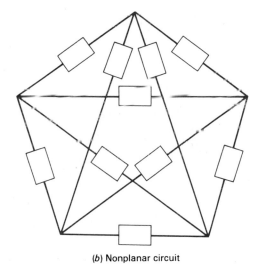

FIGURE 3.31
Defining meshes
of a circuit:
(a) examples of
mesh definition;
(b) a nonplanar
circuit for which
meshes cannot be
defined.

(b) Nonplanar circuit

FIGURE 3.32
Illustration of
writing mesh
current equations.

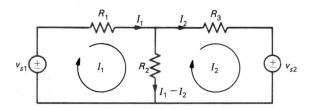

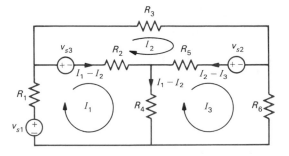

FIGURE 3.33
Illustration of
writing mesh
current equations
for more
complicated
circuits.

and as a negative quantity if it tends to oppose the mesh current. We will find this to be a general rule.

A slightly more complicated example is shown in Fig. 3.33. For this circuit, with the mesh currents as indicated, we obtain

mesh 1:

$$(R_1 + R_2 + R_4)I_1 - R_2 I_2 - R_4 I_3 = v_{s1} - v_{s3}$$

mesh 2:

$$-R_2 I_1 + (R_2 + R_3 + R_5)I_2 - R_5 I_3 = v_{s2} + v_{s3} \qquad (3.47)$$

mesh 3:

$$-R_4 I_1 - R_5 I_2 + (R_4 + R_5 + R_6)I_3 = -v_{s2}$$

Note that the rule for writing the mesh equations by inspection, as developed previously for the two-mesh circuit in Fig. 3.32, works here. For example, in writing the mesh equation for mesh 3, $R_4 + R_5 + R_6$ is the common resistance around this mesh, while R_4 is shared with mesh 1 and R_5 is shared with mesh 2. Also note that v_{s2} is encountered in traversing the mesh but that it tends to oppose I_3; it is therefore entered as a minus quantity on the right-hand side of that mesh equation.

Example 3.16   Using mesh currents, determine the current i in the circuit of Fig. 3.34.

Solution   The mesh currents are defined as shown. The mesh current equations become

mesh 1:

$$(2 + 1 + 1)I_1 - 1I_2 = -2$$

mesh 2:

$$-1I_1 + (2 + 1 + 1)I_2 = -4$$

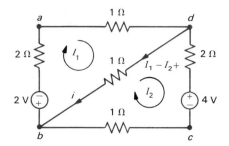

FIGURE 3.34
Example 3.16.

Here we must solve for both I_1 and I_2 since

$$i = I_1 - I_2$$

By Cramer's rule we obtain

$$I_1 = \frac{\begin{vmatrix} -2 & -1 \\ -4 & 4 \end{vmatrix}}{\begin{vmatrix} 4 & -1 \\ -1 & 4 \end{vmatrix}}$$

$$= \frac{-12}{15}$$

$$= -\tfrac{4}{5}$$

$$I_2 = \frac{\begin{vmatrix} 4 & -2 \\ -1 & 4 \end{vmatrix}}{\begin{vmatrix} 4 & -1 \\ -1 & 4 \end{vmatrix}}$$

$$= \frac{-18}{15}$$

$$= -\tfrac{6}{5}$$

Thus

$$i = I_1 - I_2$$

$$= \tfrac{2}{5}\, A$$

The preceding use of mesh current analysis has presumed that the circuit did not contain any current sources. However, the mesh current analysis method can also be applied to circuits that contain current sources, and the resulting number of simultaneous equations is reduced by the number of current sources. In order to illustrate the adaptation of mesh current analysis to circuits that contain current sources, consider the circuit shown in Fig. 3.35a. The mesh currents are defined

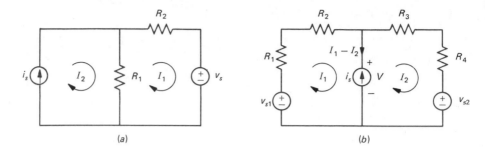

FIGURE 3.35

(a)   (b)

as shown. Observe that the presence of the current source i_s in an outside mesh constrains that mesh current as

$$I_2 = i_s \tag{3.48}$$

KVL needs to be written only around mesh 1, giving

mesh 1:

$$(R_1 + R_2)I_1 - R_1 I_2 = -v_s \tag{3.49}$$

Substituting (3.48) gives

mesh 1:

$$(R_1 + R_2)I_1 = -v_s + R_1 i_s \tag{3.50}$$

which can be solved for I_1.

The previous example illustrated the method for the case where the current source is in an outside mesh. The case of a current source in an inside mesh is similar. Consider the circuit shown in Fig. 3.35b. A current source i_s is common to mesh 1 and mesh 2 and as such constrains the relation between the two mesh currents as

$$I_1 - I_2 = -i_s \tag{3.51}$$

To handle this case we define an unknown voltage V across the current source and write the mesh current equations in the usual manner by writing KVL around the meshes:

mesh 1:

$$(R_1 + R_2)I_1 + V = v_{s1} \tag{3.52}$$

mesh 2:

$$(R_3 + R_4)I_2 - V = -v_{s2} \tag{3.53}$$

Adding these to eliminate V gives

$$(R_1 + R_2)I_1 + (R_3 + R_4)I_2 = v_{s1} - v_{s2} \qquad (3.54)$$

Substituting the constraint between the mesh currents given in (3.51) gives

$$(R_1 + R_2 + R_3 + R_4)I_2 = v_{s1} - v_{s2} + (R_1 + R_2)i_s \qquad (3.55)$$

which can be solved for I_2.

Example 3.17   Write and solve the mesh current equations for the circuit shown in Fig. 3.36a.

Solution   The 2-A current source in mesh 1 constrains that mesh current as

$$I_1 = -2\,\text{A}$$

Defining voltage V across the 3-A current source and writing KVL around mesh 2 and mesh 3 gives

mesh 2:

$$(2 + 3)I_2 - 2I_1 + V = -4$$

mesh 3:

$$2I_3 - V = 2$$

Adding these to eliminate V gives

$$-2I_1 + (2 + 3)I_2 + 2I_3 = -2$$

Substituting the constraints imposed by the current sources,

$$I_1 = -2\,\text{A}$$

and

$$I_2 - I_3 = 3\,\text{A}$$

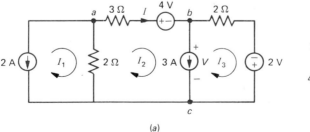

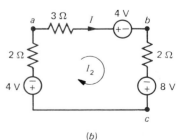

FIGURE 3.36
Example 3.17.   *(a)*   *(b)*

gives

$$(2 + 2 + 3)I_3 = -2 - (2 + 3)3 - (2)2$$
$$= -21$$

giving

$$I_3 = -3 \text{ A}$$

Thus

$$I = I_2 = I_3 + 3 = 0$$

As with all other problems, there are many ways to solve this problem. Figure 3.36*b* illustrates an alternative solution obtained with Norton and Thévenin transformations. The 2-A and 2-Ω combination is converted to 4 V in series with 2 Ω. The 2 V in series with 2 Ω is first converted to 1 A in parallel with 2 Ω. This current source is added to the 3-A source and the result converted back to 8 V in series with 2 Ω. Note that I and I_2 in this circuit are the same as in the original circuit. From this reduction it becomes clear that

$$I = I_2 = \frac{8 \text{ V} - 4 \text{ V} - 4 \text{ V}}{(2 + 3 + 2)\,\Omega} = 0 \text{ A}$$

as before.

3.6   MAXIMUM POWER TRANSFER

At any instant of time the power delivered *by* the ideal sources must equal the power delivered *to* the resistors of a circuit. This is a rather obvious but important result. Thus, KVL and KCL must be such that this is satisfied, and they are therefore not independent laws but are somehow related.

Example 3.18

Show that conservation of power is satisfied for the circuit in Fig. 3.37*a*.

Solution   The resistor voltages and currents can be obtained by using superposition, Ohm's law, and current division, as shown in Fig. 3.37*b*. The 10-V voltage source delivers

FIGURE 3.37
Example 3.18:
illustration of
conservation of
power in resistive
circuits.

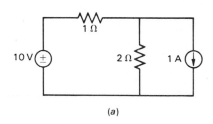

(a)

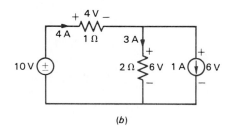

(b)

10 V × 4 A = 40 W to the circuit, whereas the 1-A current source delivers −6 V × 1 A = 6 W to the circuit and is actually absorbing 6 W. The two resistors absorb (4 A)² × 1 Ω + (3 A)² × 2 Ω = 34 W of power and conservation of power is achieved.

With regard to power, one final point should be mentioned. All practical sources have internal resistance, represented as R_S in Fig. 3.38. Therefore, when this source is connected to a load, represented by R_L, not all the open-circuit voltage of the source v_S will appear across R_L. Some of v_S will be dropped across R_S, $i_L R_S$. The question now arises as to what would be the optimum choice for R_L (if we have a choice) such that maximum power will be delivered from the source to the load. Ordinarily, we do not have any choice over R_S since it is the equivalent internal resistance of the source. Let us determine the optimum value of R_L to achieve maximum power transfer from the source to the load. The load current is

$$i_L = \frac{v_S}{R_S + R_L} \tag{3.56}$$

and the power delivered to the load is

$$p_L = i_L^2 R_L$$

$$= \frac{R_L}{(R_S + R_L)^2} v_S^2 \tag{3.57}$$

The important question here is, What value of R_L will maximize this expression? This is found by differentiating p_L with respect to R_L and setting the result equal to zero:

$$\frac{dp_L}{dR_L} = 0$$

$$= \frac{(R_S + R_L)^2 - 2R_L(R_S + R_L)}{(R_S + R_L)^4} v_S^2 \tag{3.58}$$

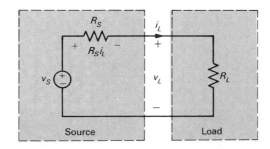

FIGURE 3.38
Maximum power transfer from a source to a load.

To find the value of R_L which satisfies this, we simply set the numerator equal to zero and obtain

$$R_L = R_S \tag{3.59}$$

Thus, for maximum power transfer to a load, we should choose the load resistor to be equal to the internal resistance of the source.

PROBLEMS

3.1   Determine R_{eq} for the circuit of Fig. P.3.1.

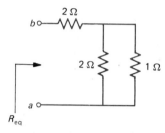

FIGURE P3.1

3.2   Determine R_{eq} for the circuit of Fig. P.3.2.

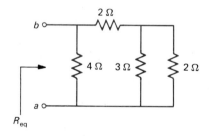

FIGURE P3.2

3.3   Determine R_{eq} for the circuit of Fig. P.3.3.

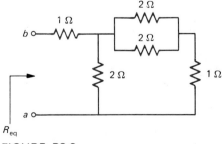

FIGURE P3.3

3.4   Determine R_{eq} for the circuit of Fig. P.3.4.

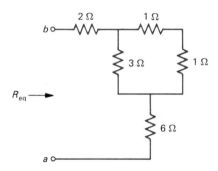

FIGURE P3.4

3.5   Determine R_{eq} for the circuit of Fig. P.3.5.

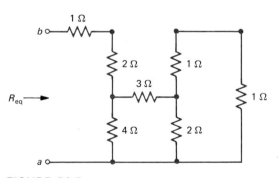

FIGURE P3.5

3.6   Determine R_{eq} for the circuit of Fig. P.3.6.

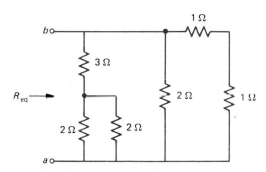

FIGURE P3.6

3.7   Determine voltage v_x using voltage division and equivalent resistor reductions for the circuit in Fig. P3.7.

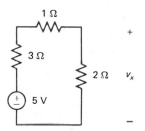

FIGURE P3.7

3.8   Determine voltage v_x using voltage division and equivalent resistor reductions for the circuit in Fig. P3.8.

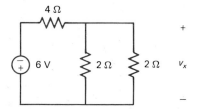

FIGURE P3.8

3.9   Determine voltage v_x using voltage division and equivalent resistor reductions for the circuit in Fig. P3.9.

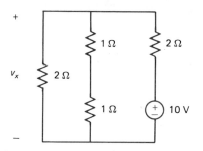

FIGURE P3.9

3.10   Determine voltage v_x using voltage division and equivalent resistor reductions for the circuit in Fig. P3.10.

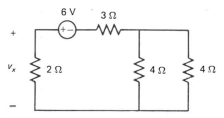

FIGURE P3.10

3.11   Determine voltage v_x using voltage division and equivalent resistor reductions for the circuit in Fig. P3.11.

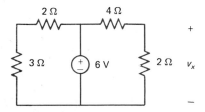

FIGURE P3.11

3.12 Determine voltage v_x using voltage division and equivalent resistor reductions for the circuit in Fig. P3.12.

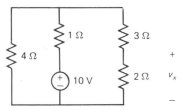

FIGURE P3.12

3.13 Determine the current i_x using voltage division and equivalent resistor reductions for the circuit of Fig. P3.13.

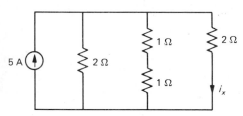

FIGURE P3.13

3.14 Determine the current i_x using current division and equivalent resistor reductions for the circuit of Fig. P3.14.

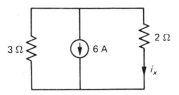

FIGURE P3.14

3.15 Determine the current i_x using current division and equivalent resistor reductions for the circuit of Fig. P3.15.

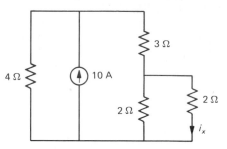

FIGURE P3.15

3.16 Determine the current i_x using current division and equivalent resistor reductions for the circuit of Fig. P3.16.

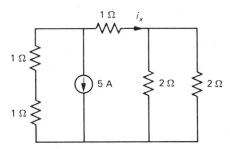

FIGURE P3.16

3.17 Determine the current i_x using current division and equivalent resistor reductions for the circuit of Fig. P3.17.

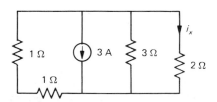

FIGURE P3.17

3.18   Determine the current i_x using current division and equivalent resistor reductions for the circuit of Fig. P3.18.

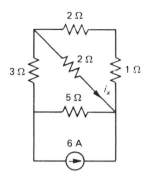

FIGURE P3.18

3.19   Determine voltage v_x and current i_x using superposition for the circuit in Fig. P3.19.

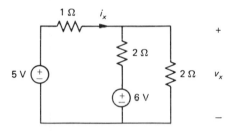

FIGURE P3.19

3.20   Determine voltage v_x and current i_x using superposition for the circuit in Fig. P3.20.

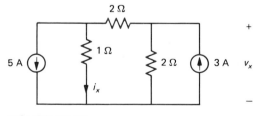

FIGURE P3.20

3.21   Determine voltage v_x and current i_x using superposition for the circuit in Fig. P3.21.

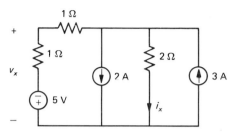

FIGURE P3.21

3.22   Determine voltage v_x and current i_x using superposition for the circuit in Fig. P3.22.

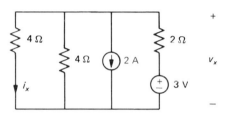

FIGURE P3.22

3.23   Determine voltage v_x and current i_x using superposition for the circuit in Fig. P3.23.

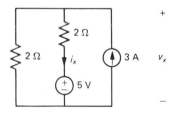

FIGURE P3.23

3.24 Determine voltage v_x and current i_x using superposition for the circuit in Fig. P3.24.

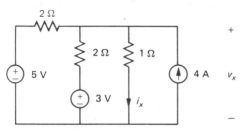

FIGURE P3.24

3.25 Determine voltage v_x and current i_x using superposition for the circuit in Fig. P3.25.

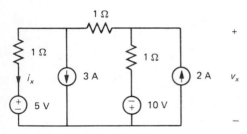

FIGURE P3.25

3.26 Reduce the circuit shown in Fig. P3.26 to a Thévenin equivalent and a Norton equivalent circuit.

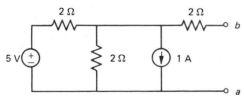

FIGURE P3.26

3.27 Reduce the circuit shown in Fig. P3.27 to a Thévenin equivalent and a Norton equivalent circuit.

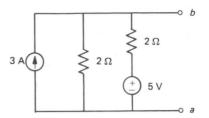

FIGURE P3.27

3.28 Reduce the circuit shown in Fig. P3.28 to a Thévenin equivalent and a Norton equivalent circuit.

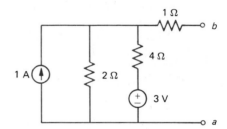

FIGURE P3.28

3.29 Reduce the circuit shown in Fig. P3.29 to a Thévenin equivalent and a Norton equivalent circuit.

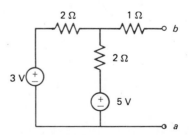

FIGURE P3.29

3.30 Reduce the circuit shown in Fig. P3.30 to a Thévenin equivalent and a Norton equivalent circuit.

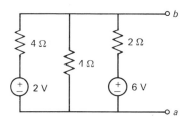

FIGURE P3.30

3.31 Reduce the circuit shown in Fig. P3.31 to a Thévenin equivalent and a Norton equivalent circuit.

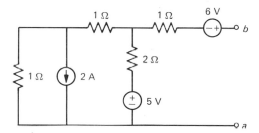

FIGURE P3.31

3.32 Attach a 1-Ω resistor to the terminals of the circuit of Fig. P3.27 and solve for the current through the resistor, using (a) superposition, (b) the Thévenin equivalent circuit, and (c) the Norton equivalent circuit.

3.33 Attach a 1-Ω resistor to the terminals of the circuit of Fig. P3.28 and solve for the current through the resistor, using (a) superposition, (b) the Thévenin equivalent circuit, and (c) the Norton equivalent circuit.

3.34 Attach a 1-Ω resistor to the terminals of the circuit of Fig. P3.29 and solve for the current through the resistor, using (a) superposition, (b) the Thévenin equivalent circuit, and (c) the Norton equivalent circuit.

3.35 Attach a 1-Ω resistor to the terminals of the circuit of Fig. P3.30 and solve for the current through the resistor, using (a) superposition, (b) the Thévenin equivalent circuit, and (c) the Norton equivalent circuit.

3.36 Attach a 1-Ω resistor to the terminals of the circuit of Fig. P3.31 and solve for the current through the resistor, using (a) superposition, (b) the Thévenin equivalent circuit, and (c) the Norton equivalent circuit.

3.37 Determine voltage v_x and current i_x in Fig. P3.37 by using node voltage equations.

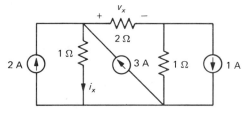

FIGURE P3.37

3.38 Determine voltage v_x and current i_x in Fig. P3.38 by using node voltage equations.

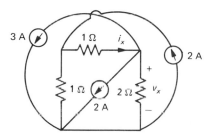

FIGURE P3.38

3.39 Determine voltage v_x and current i_x in Fig. P3.39 by using node voltage equations.

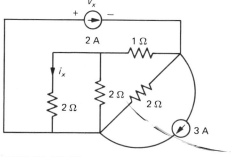

FIGURE P3.39

3.40 Determine voltage v_x and current i_x in Fig. P3.40 by using node voltage equations.

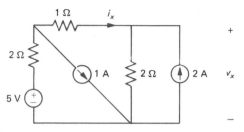

FIGURE P3.40

3.41 Determine voltage v_x and current i_x in Fig. P3.41 by using node voltage equations.

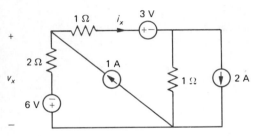

FIGURE P3.41

3.42 Determine voltage v_x and current i_x in Fig. P3.42 by using mesh current equations.

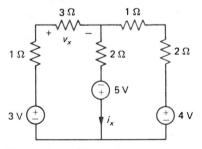

FIGURE P3.42

3.43 Determine voltage v_x and current i_x in Fig. P3.43 by using mesh current equations.

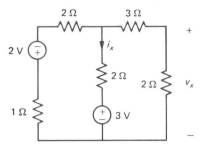

FIGURE P3.43

3.44 Determine voltage v_x and current i_x in Fig. P3.44 by using mesh current equations.

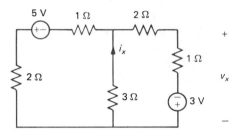

FIGURE P3.44

3.45 Determine voltage v_x and current i_x in Fig. P3.45 by using mesh current equations.

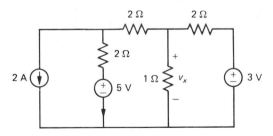

FIGURE P3.45

3.46 Determine voltage v_x and current i_x in Fig. P3.46 by using mesh current equations.

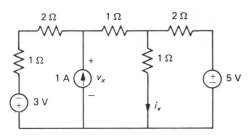

FIGURE P3.46

3.47 Determine R_L in Fig. P3.47 such that maximum power will be delivered to it. Determine that maximum power.

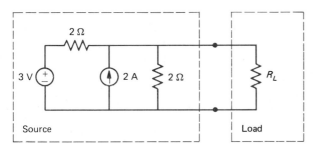

FIGURE P3.47

3.48 Determine R_L in Fig. P3.48 such that maximum power will be delivered to it. Determine that maximum power.

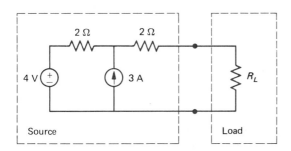

FIGURE P3.48

3.49 Determine R_L in Fig. P3.49 such that maximum power will be delivered to it. Determine that maximum power.

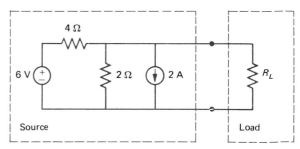

FIGURE P3.49

Energy Storage Elements

In the previous two chapters we studied circuits composed only of ideal voltage and current sources and resistors. The circuit equations governing the branch voltages and currents—KVL, KCL, and the element relations—are algebraic equations because the resistor voltage and current are related by an algebraic equation: $v = Ri$. Recall also that the resistor constantly absorbs power: $p = i^2R = v^2/R$.

In this chapter we study the two remaining basic circuit elements—the capacitor and the inductor. Unlike the resistor, these elements store energy, and their voltages and currents are related by differential equations. Therefore, whenever we include one of these elements in a circuit, the circuit equations governing the branch voltages and currents become differential equations, which are more difficult to solve than the algebraic circuit equations encountered with resistive circuits.

Although circuits containing these energy storage elements are more difficult to analyze than resistive circuits, they allow us to build circuits which provide a greater variety of signal processing than is attainable with resistive circuits. An example of this benefit is in the construction of electric filters, which are considered in Chap. 5.

4.1 THE CAPACITOR

The first energy storage element that we consider is the capacitor. Suppose we place two rectangular metal plates in close proximity and, via wires, attach a battery to the plates, as shown in Fig. 4.1. When the battery is connected to the plates, charge Q is transferred from the battery to the plates; a positive amount is present on the plate attached to the positive terminal of the battery, and an equal but negative amount is placed on the other plate. Thus, the plates separate charge. This charge and voltage between the plates (provided here by the battery) are related by the capacitance of this capacitor:

$$Q = CV \tag{4.1}$$

104

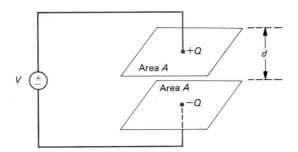

FIGURE 4.1
A capacitor.

The quantity C is the capacitance of the capacitor whose units are, according to Eq. (4.1), coulombs per volt. This combination of units is given the name of farads (F), where $1\ F = 1\ C/V$.

For the parallel-plate capacitor shown in Fig. 4.1 with plates of area A and separation d in air, the capacitance may be computed from the approximate relation

$$C \doteq 8.84 \times 10^{-12}\, \frac{A}{d} \tag{4.2}$$

The units of capacitance for practical capacitors are usually given in microfarads $(1\ \mu F = 10^{-6}\ F)$ or picofarads $(1\ pF = 10^{-12}\ F)$ since a 1-F capacitor would be equivalent to a separation of $1\ m$ between two plates, each having an area of approximately 44,000 square miles (mi^2). It is possible, however, to construct 1-F electrolytic capacitors that can be held in one's hand.

Capacitors are constructed in several ways. One common method is to roll two sheets of metal foil which are separated by a dielectric such as paper into a tubular shape to conserve space, as shown in Fig. 4.2. Leads (wires) are attached to each sheet and form the terminals of the capacitor. Very large values of capacitance, such as $100\ \mu F$, are constructed as electrolytic capacitors. Modern integrated-circuit technology allows the construction of a wide variety of capacitance values in an extremely small space.

The element symbol for a capacitor is shown in Fig. 4.3. The current entering one terminal of the capacitor is equal to the rate of buildup of charge on the plate attached to this terminal:

$$i(t) = \frac{dQ}{dt} \tag{4.3}$$

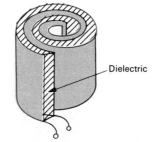

FIGURE 4.2
Physical
construction of a
capacitor to
occupy a small
space.

Dielectric

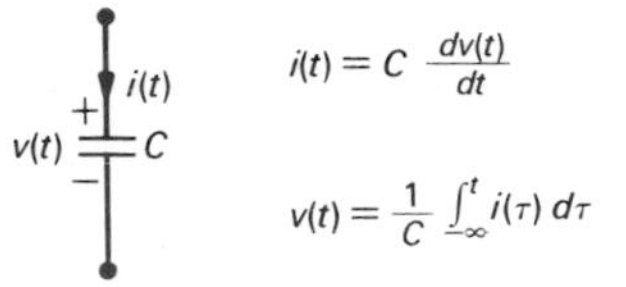

$$i(t) = C \frac{dv(t)}{dt}$$

$$v(t) = \frac{1}{C} \int_{-\infty}^{t} i(\tau)\, d\tau$$

FIGURE 4.3   The terminal v-i relationship for a capacitor.

If we differentiate Eq. (4.1) with respect to t and substitute the result into Eq. (4.3), we obtain the terminal voltage-current relationship for the capacitor:

$$i(t) = C \frac{dv(t)}{dt} \tag{4.4}$$

Here we have assumed that the capacitance is fixed. If the separation distance between the plates changed with time, then we would write $C(t)$, since the capacitance would vary with time and we would have to take this into account when differentiating Eq. (4.1) with respect to time. Note also that in order for Eq. (4.4) to be true, $v(t)$ and $i(t)$ must be labeled on the element with the passive sign convention. If they were not labeled with the passive sign convention, a negative sign would be needed in Eq. (4.4). If, on the other hand, $v(t)$ and $i(t)$ were labeled on the element with the passive sign convention and we found that $i(t) = -C[dv(t)/dt]$, then this would be a true negative capacitor. (We consider only positive capacitors.) Examples are shown in Fig. 4.4. This proper labeling of the element voltage and current to conform to the passive sign convention so that Eq. (4.4) is the correct terminal relation should be carefully studied. If the terminal voltage and current of a positive capacitor are not labeled with the passive sign convention but Eq. (4.4) is used to relate them nevertheless, the result is incorrect.

The terminal relationship in Eq. (4.1) is said to describe a linear capacitor, since the terminal voltage and charge on the plates are linearly related. If, on the other hand, the charge and terminal voltage of a capacitor were related by a nonlinear equation such as $Q = CV^2$, then the capacitor would be classified as nonlinear. We consider only linear capacitors.

The terminal voltage-current relationship in Eq. (4.4) can be written in an alternative form by integrating both sides of the equation to yield

$$v(t) = \frac{1}{C} \int_{-\infty}^{t} i(\tau)\, d\tau \tag{4.5}$$

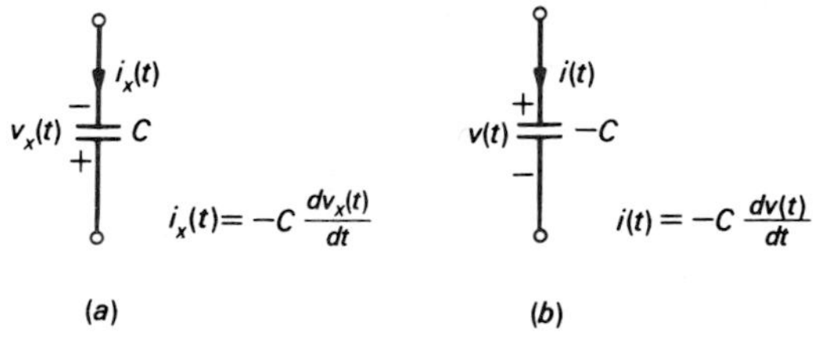

$$i_x(t) = -C \frac{dv_x(t)}{dt}$$

(a)

$$i(t) = -C \frac{dv(t)}{dt}$$

(b)

FIGURE 4.4   Terminal relations for capacitors, and the passive sign convention: (a) element not labeled with the passive sign convention; (b) a negative capacitor (note that the terminal voltage and current are labeled with the passive sign convention).

where τ is the dummy variable of integration. This result shows that the terminal voltage depends on the past values of the terminal current. This makes sense because the integral of the current over all previous time yields the present value of charge on the capacitor plates.

Usually, we designate some origin in time for our investigations as $t = 0$. If we separate the integral in Eq. (4.5) into a portion for $t < 0$ and a portion for $t > 0$, we have

$$v(t) = \frac{1}{C} \int_0^t i(\tau)\, d\tau + \frac{1}{C} \int_{-\infty}^0 i(\tau)\, d\tau \qquad (4.6)$$

Note that the portion to $t < 0$ is the value of the capacitor voltage at $t = 0$:

$$v(0) = \frac{1}{C} \int_{\infty}^0 i(\tau)\, d\tau \qquad (4.7)$$

Thus, Eq. (4.6) can be written in terms of this "initial" capacitor voltage as

$$v(t) = \frac{1}{C} \int_0^t i(\tau)\, d\tau + v(0) \qquad (4.8)$$

The energy stored in a capacitor at a particular time can be found by integrating the expression for the instantaneous power delivered to the capacitor

$$p(t) = v(t)i(t)$$
$$= Cv(t)\frac{dv(t)}{dt} \qquad (4.9)$$

to give

$$w(t) = \int_{-\infty}^t p(\tau)\, d\tau$$
$$= C \int_{-\infty}^t v(\tau)\frac{dv(\tau)}{d\tau}\, d\tau$$
$$= \tfrac{1}{2}Cv^2(t) - \tfrac{1}{2}Cv^2(-\infty) \qquad (4.10)$$

Now if we assume that the capacitor voltage was zero at $t = -\infty$, the energy stored in the capacitor at some time t depends on only the voltage of the capacitor at that time

$$w(t) = \tfrac{1}{2}Cv^2(t) \qquad (4.11)$$

an interesting result.

This energy is stored in the electric field between the plates by the separation of charge, as discussed in Chap. 2. We will find that the next energy storage element which we consider, the inductor, stores energy in the dual field—the magnetic field. These duality concepts will occur quite frequently, and the reader should be alert to find them.

Example 4.1

Consider the voltage waveform $v_s(t)$ applied to the 5-F capacitor in Fig. 4.5a. Sketch the capacitor current and stored energy as a function of time.

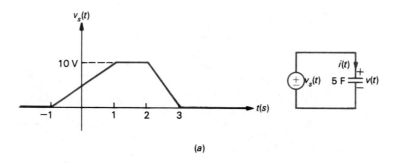

(a)

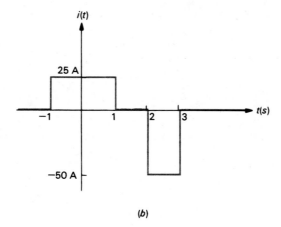

(b)

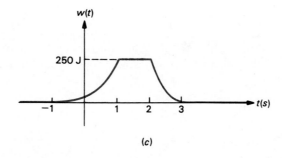

(c)

FIGURE 4.5
Example 4.1:
illustration of a
capacitor driven
by a voltage
source.

Solution   The capacitor voltage equals $v_s(t)$, and the capacitor current is given by Eq. (4.4):

$$i(t) = 5 \frac{dv_s(t)}{dt}$$

Now $v_s(t)$ is characterized by

$$
\begin{aligned}
v_s(t) &= 0 & t &\leq -1 \\
&= 5(t + 1) \quad \text{V} & -1 &\leq t \leq 1 \\
&= 10 \text{ V} & 1 &\leq t \leq 2 \\
&= -10(t - 3) \quad \text{V} & 2 &\leq t \leq 3 \\
&= 0 & 3 &\leq t
\end{aligned}
$$

Thus, the capacitor current is found from Eq. (4.4) by differentiating $v_s(t)$ and multiplying the result by $C = 5$ F:

$$
\begin{aligned}
i(t) &= 0 & t &\leq -1 \\
&= 25 \text{ A} & -1 &\leq t \leq 1 \\
&= 0 & 1 &\leq t \leq 2 \\
&= -50 \text{ A} & 2 &\leq t \leq 3 \\
&= 0 & 3 &\leq t
\end{aligned}
$$

which is sketched in Fig. 4.5b. The energy stored at any instant is given by Eq. (4.11):

$$
\begin{aligned}
w(t) &= \tfrac{1}{2}(5)v_s^2(t) \\
&= 0 & t &\leq -1 \\
&= 62.5(t^2 + 2t + 1) \quad \text{J} & -1 &\leq t \leq 1 \\
&= 250 \text{ J} & 1 &\leq t \leq 2 \\
&= 250(t^2 - 6t + 9) \quad \text{J} & 2 &\leq t \leq 3 \\
&= 0 & 3 &\leq t
\end{aligned}
$$

which is sketched in Fig. 4.5c.

Example 4.2   Suppose a current source is attached to a 5-F capacitor as shown in Fig. 4.6a. Sketch the waveform of the capacitor voltage.

Solution   The capacitor voltage is given by Eq. (4.5):

$$v(t) = \tfrac{1}{5} \int_{-\infty}^{t} i_s(\tau)\, d\tau$$

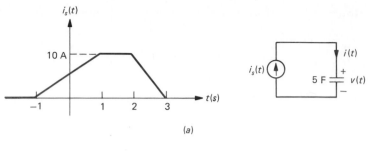

(a)

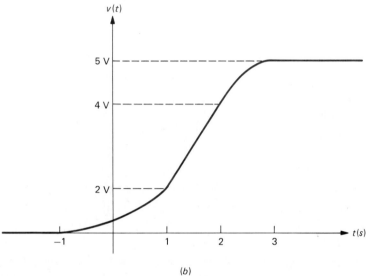

FIGURE 4.6
Example 4.2:
illustration of a
capacitor driven
by a current
source.

(b)

The waveform of the current source is the same shape as the waveform of the voltage source in the previous example:

$$
\begin{aligned}
i_s(t) &= 0 & t \leq -1 \\
&= 5(t + 1) \quad \text{A} & -1 \leq t \leq 1 \\
&= 10 \text{ A} & 1 \leq t \leq 2 \\
&= -10(t - 3) \quad \text{A} & 2 \leq t \leq 3 \\
&= 0 & 3 \leq t
\end{aligned}
$$

Thus, the capacitor voltage is found by integrating the waveform of the current source and multiplying the result by $\frac{1}{5}$ F. The result is

$$
\begin{aligned}
v(t) &= 0 & t \leq -1 \\
&= \frac{t^2}{2} + t + \frac{1}{2} \quad \text{V} & -1 \leq t \leq 1 \\
&= 2t \quad \text{V} & 1 \leq t \leq 2 \\
&= -t^2 + 6t - 4 \quad \text{V} & 2 \leq t \leq 3 \\
&= 5 \text{ V} & 3 \leq t
\end{aligned}
$$

which is sketched in Fig. 4.6b. For example, for the interval $2 \le t \le 3$ we integrate Eq. (4.5) to obtain

$$v(t) = \tfrac{1}{5} \int_{-\infty}^{t} i_s(\tau)\, d\tau$$

$$= \tfrac{1}{5} \int_{-\infty}^{t} [-10(\tau - 3)]\, d\tau$$

$$= \tfrac{1}{5} \int_{2}^{t} [-10(\tau - 3)]\, d\tau + v(2)$$

$$= -2\left(\frac{\tau^2}{2} - 3\tau\right)\Big|_{2}^{t} + v(2)$$

$$= -t^2 + 6t - 8 + 4$$

$$= -t^2 + 6t - 4$$

The energy stored in the capacitor can be obtained from Eq. (4.11) by squaring the waveform for $v(t)$ determined above and multiplying the result by $C/2 = 2.5$.

Several capacitors in series or parallel may be combined into an equivalent capacitance. First consider the parallel connection of capacitors shown in Fig. 4.7. The terminal voltage of the combination is equal to each of the capacitor terminal voltages:

$$v(t) = v_1(t) = v_2(t) = \cdots = v_n(t) \tag{4.12}$$

and the terminal current of the combination is the sum of the terminal currents of each one:

$$i(t) = i_1(t) + i_2(t) + \cdots + i_n(t) \tag{4.13}$$

Substituting the terminal relation for each capacitor given in Eq. (4.4) into Eq. (4.13) gives

$$i(t) = C_1 \frac{dv(t)}{dt} + C_2 \frac{dv(t)}{dt} + \cdots + C_n \frac{dv(t)}{dt}$$

$$= (C_1 + C_2 + \cdots + C_n) \frac{dv(t)}{dt} \tag{4.14}$$

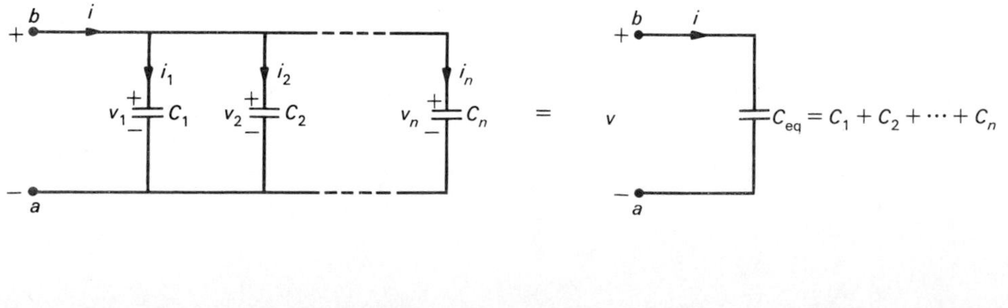

FIGURE 4.7
The equivalent
capacitance of
capacitors in
parallel.

Thus, an equivalent capacitor at terminals *a-b* is

$$C_{eq} = C_1 + C_2 + \cdots + C_n \tag{4.15}$$

and capacitors in parallel combine as resistors in series.

Next consider the series connection of capacitors shown in Fig. 4.8. Here the terminal current of the combination is equal to each of the capacitor currents:

$$i(t) = i_1(t) = i_2(t) = \cdots = i_n(t) \tag{4.16}$$

and the terminal voltage of the combination is the sum of the capacitor voltages:

$$v(t) = v_1(t) + v_2(t) + \cdots + v_n(t) \tag{4.17}$$

Substituting the terminal relation for each capacitor in the form given in Eq. (4.8) into Eq. (4.17) yields

$$
\begin{aligned}
v(t) &= \frac{1}{C_1} \int_0^t i(\tau)\, d\tau + v_1(0) + \frac{1}{C_2} \int_0^t i(\tau)\, d\tau + v_2(0) \\
&\quad + \cdots + \frac{1}{C_n} \int_0^t i(\tau)\, d\tau + v_n(0) \\
&= \left(\frac{1}{C_1} + \frac{1}{C_2} + \cdots + \frac{1}{C_n} \right) \int_0^t i(\tau)\, d\tau \\
&\quad + [v_1(0) + v_2(0) + \cdots + v_n(0)]
\end{aligned}
\tag{4.18}
$$

Thus, the equivalent capacitor at terminals *a-b* has the value

$$\frac{1}{C_{eq}} = \frac{1}{C_1} + \frac{1}{C_2} + \cdots + \frac{1}{C_n} \tag{4.19}$$

which we symbolically denote as

$$C_{eq} = C_1 \| C_2 \| \cdots \| C_n \tag{4.20}$$

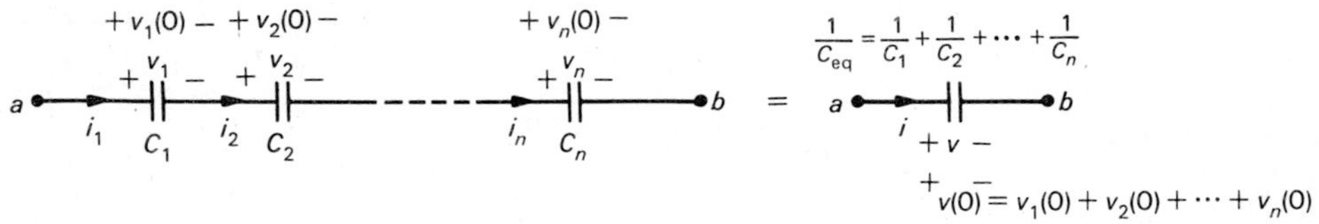

FIGURE 4.8
The equivalent
capacitance of
capacitors in
series.

with an initial voltage which is the sum of the initial voltages of each capacitor:

$$v(0) = v_1(0) + v_2(0) + \cdots + v_n(0) \tag{4.21}$$

and capacitors in series combine as resistors in parallel.

4.2 THE INDUCTOR

The second energy storage element we consider is the inductor. Inductors store energy in a magnetic field just as capacitors store energy in an electric field.

Suppose we form a length of wire into a loop and pass a current through this wire, as shown in Fig. 4.9a. A magnetic field is generated around the wire in the form of magnetic flux ψ whose units are webers (Wb). This magnetic field has the capability to attract ferrous metals, such as iron. The magnetic flux ψ and the current i producing it are linearly related by a quantity called the *inductance* of the loop, L:

$$\psi = Li \tag{4.22}$$

The units of inductance L are webers per ampere (Wb/A). This combination of units is called the *henry* (H), where $1\ \mathrm{H} = 1\ \mathrm{Wb/A}$. If N of these loops are wound about some central core, as shown in Fig. 4.9b, the total magnetic flux encircling, or "linking," current i is N times Eq. (4.22). Thus, a current produces magnetic flux

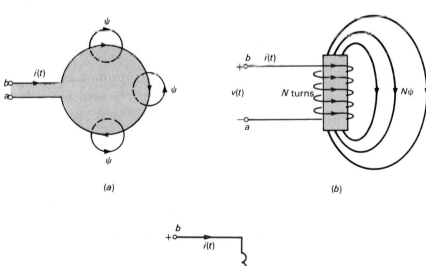

(a)   (b)

FIGURE 4.9
Inductance of current loops: (a) a single loop; (b) multiple loops around a common core; (c) the inductor symbol.

(c)

$$v(t) = L\frac{di(t)}{dt}$$

$$i(t) = \frac{1}{L}\int_{-\infty}^{t} v(\tau)\,d\tau$$

which links that current, and the current and resulting flux are related by the inductance L.

Faraday's law states that the time-varying magnetic flux in the above examples will induce a voltage in the loops which tends to produce a current in the loop. This induced current and associated magnetic flux tend to oppose the change in the original current and its associated magnetic flux. This induced voltage is related to the time rate of change of the magnetic flux:

$$v(t) = \frac{d\psi}{dt} \tag{4.23}$$

Substituting Eq. (4.22), we obtain, for an inductance which does not change with time (i.e., the physical shape of the loop is constant),

$$v(t) = L\frac{di(t)}{dt} \tag{4.24}$$

This is illustrated in Fig. 4.9c, where the circuit symbol for the inductor is shown.

Mathematically, the inductor is the dual of the capacitor. For example, compare the terminal relationship for the inductor in Eq. (4.24) to the terminal relationship for the capacitor given in Eq. (4.4). One can be obtained from the other by interchanging v and i and interchanging L and C. We will find many other relationships for the inductor which are duals of the corresponding capacitor relationships.

The alternative characterization of the inductor is obtained by integrating both sides of Eq. (4.24) to yield

$$i(t) = \frac{1}{L}\int_{-\infty}^{t} v(\tau)\,d\tau$$

$$= \frac{1}{L}\int_{0}^{t} v(\tau)\,d\tau + i(0) \tag{4.25}$$

where $i(0)$ is the inductor current at $t = 0$:

$$i(0) = \frac{1}{L}\int_{-\infty}^{0} v(\tau)\,d\tau \tag{4.26}$$

Note that the current through an inductor at a particular time depends on the values of the inductor voltage at all prior instants of time.

It is once again important that the terminal voltage and current for the inductor are labeled in Fig. 4.9c with the passive sign convention. If they are not, a minus sign is required in the terminal relationships in Eqs. (4.24) and (4.25).

If the inductor, or coil, in Fig. 4.9b is wound about a ferrous core such as iron, the relationship between the magnetic flux ψ and current i will be a nonlinear one,

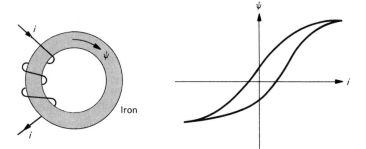

FIGURE 4.10
A nonlinear
inductor—a
toroid with a
ferrous core.

as shown in Fig. 4.10, and the inductor is said to be nonlinear. We consider only linear inductors in this text.

Typical values for inductors are in the millihenry (1 mH = 10^{-3} H) or microhenry (1 μH = 10^{-6} H) range. An approximate value for an air-core inductor consisting of N turns each having an area of A which is wound on a toroid with mean circumference l is

$$L \doteq 4\pi \times 10^{-7} \frac{N^2 A}{l} \tag{4.27}$$

The energy stored in an inductor (in its magnetic field) is obtained by integrating the instantaneous power

$$p(t) = v(t)i(t)$$
$$= Li(t)\frac{di(t)}{dt} \tag{4.28}$$

to yield

$$w(t) = \int_{-\infty}^{t} p(\tau)\, d\tau$$
$$= \tfrac{1}{2}Li^2(t) - \tfrac{1}{2}Li^2(-\infty)$$
$$= \tfrac{1}{2}Li^2(t) \tag{4.29}$$

where we have assumed that the inductor current at $t = -\infty$ is zero. Thus, the energy stored in an inductor at an instant of time depends on only the inductor current at that time. This is the dual of the result for the energy stored in a capacitor, which depended only on the capacitor voltage.

The other results which are duals of corresponding results for the capacitor are the series and parallel reductions of inductors shown in Fig. 4.11. For these we can easily show, by substituting the terminal relations for the individual inductors given in Eq. (4.24) or (4.25) into

$$v = v_1 + v_2 + \cdots + v_n$$
$$i = i_1 = i_2 = \cdots = i_n \tag{4.30}$$

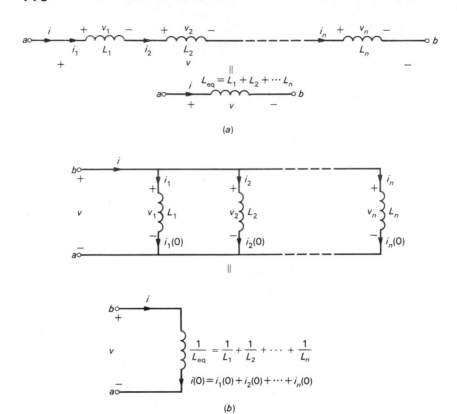

FIGURE 4.11
The equivalent
inductance of
inductors in
series and
parallel:
(*a*) series
combination;
(*b*) parallel
combination.

for the series combination or

$$v = v_1 = v_2 = \cdots = v_n$$
$$i = i_1 + i_2 + \cdots + i_n \tag{4.31}$$

for the parallel combination, that inductors in series combine like resistors in series:

$$L_{eq} = L_1 + L_2 + \cdots + L_n \tag{4.32}$$

and that inductors in parallel combine like resistors in parallel:

$$\frac{1}{L_{eq}} = \frac{1}{L_1} + \frac{1}{L_2} + \cdots + \frac{1}{L_n} \tag{4.33}$$

denoted as $L_{eq} = L_1 || L_2 || \cdots || L_n$ with an initial current equal to the sum of the initial currents of each inductor:

$$i(0) = i_1(0) + i_2(0) + \cdots + i_n(0) \tag{4.34}$$

Example 4.3   Consider the current source applied to the 5-H inductor shown in Fig. 4.12a. Sketch the inductor voltage and energy stored in the inductor.

Solution   The inductor voltage is found from Eq. (4.24) as the product of the inductance and the slope of the current waveform of the inductor:

$$v(t) = 5 \frac{di_s(t)}{dt}$$

which is sketched in Fig. 4.12b. The stored energy at any instant is related to the square of the inductor current at that time by Eq. (4.29):

$$w(t) = \tfrac{1}{2} 5 i_s^2(t)$$

which is sketched in Fig. 4.12c.

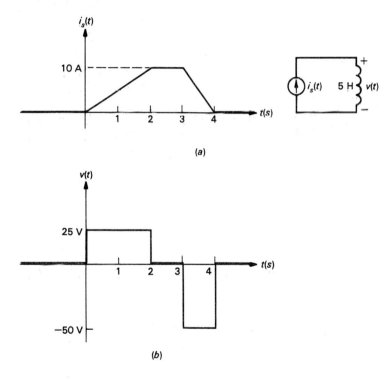

(a)

(b)

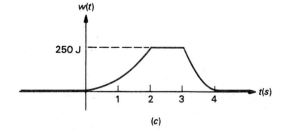

(c)

FIGURE 4.12
Example 4.3:
illustration of an
inductor driven
by a current
source.

Example 4.4   Sketch the inductor current for the configuration in Fig. 4.13*a*.

Solution   The inductor voltage waveform is $v_s(t)$, and the inductor current is related to the integral of this current as the ratio of the area under the curve up to this t and the inductance, as in Eq. (4.25):

$$i(t) = \tfrac{1}{5} \int_{-\infty}^{t} v_s(\tau)\, d\tau$$

This is sketched in Fig. 4.13*b*.

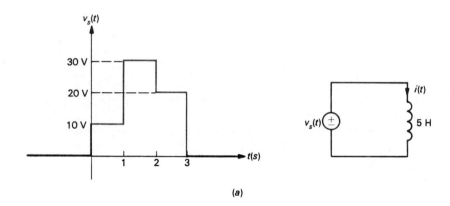

(a)

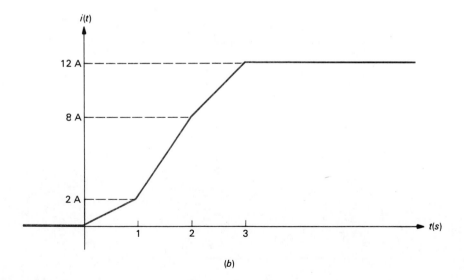

(b)

FIGURE 4.13
Example 4.4:
illustration of an
inductor driven
by a voltage
source.

4.3 CONTINUITY OF CAPACITOR VOLTAGES AND INDUCTOR CURRENTS

The voltage across a capacitor cannot change value instantaneously. This can be seen in several ways. From the relation between the terminal voltage and current given in Eq. (4.4), we see that if the terminal voltage were to change value instantaneously, the capacitor current would be infinite—an unrealistic situation. Alternatively, from the integral relation in Eq. (4.5) we see that the terminal voltage is related to the integral of the terminal current. Realistic currents $i(t)$, when integrated, will yield a continuous function. Thus, the voltage cannot change value instantaneously.

Another way of seeing this result is to examine the expression for stored energy in Eq. (4.11). If the capacitor voltage changed value instantaneously, the stored energy would also change instantaneously—an unrealistic situation.

However, there is no reason to rule out an instantaneous change in the capacitor current. This was illustrated in Example 4.1, where the capacitor current shown in Fig. 4.5b changes value instantaneously at $t = -1, 1, 2, 3$.

Similarly, *the current through an inductor cannot change value instantaneously.* This is easily seen from the terminal relations for the inductor in Eqs. (4.24) and (4.25) as well as from the relation for the stored energy in Eq. (4.29). However, there is no reason to rule out an instantaneous change in the value of the inductor voltage (see Fig. 4.12b).

This latter property of inductors is used in generating large voltages in an automobile ignition system to fire the spark plugs. A current is passed through an inductor, the "ignition coil." A cam opens and closes a switch connected in series with this ignition coil, thus reducing the current through the coil to zero in a very short time. From the terminal equation for the ignition coil in Eq. (4.24), we see that changing the current in the coil in a short period can generate large voltages dependent on di/dt and the inductance of the coil.

4.4 TRANSFORMERS AND COUPLED COILS

We have seen how a time-varying current passing through the windings of an inductor can produce a time-varying magnetic flux which links this current and produces an induced voltage at the terminals of the inductor; this is given by Eq. (4.24). If some of this time-varying magnetic flux also links the turns of some other inductor, it will also induce a voltage at the terminals of that inductor. We then say that there is *mutual inductance* between the two inductors and refer to them as being *coupled.*

For example, consider the two inductors which are wound on a common core (such as iron) in Fig. 4.14a. A current $i_1(t)$ entering terminals a-b of one inductor will produce a magnetic flux ψ which will link not only its own turns but also those of the second inductor. Thus, this time-varying current $i_1(t)$ produces an induced voltage at terminals a-b of

$$v'_1(t) = L_1 \frac{di_1(t)}{dt} \tag{4.35}$$

FIGURE 4.14
Transformers and
mutual
inductance:
(*a*) physical
configuration;
(*b*) circuit
symbol.

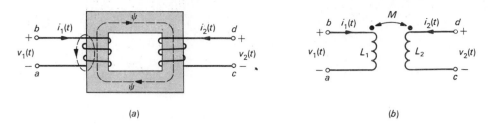

(*a*)   (*b*)

The resulting magnetic flux also produces an induced voltage at terminals *c-d* of the second inductor of

$$v_2'(t) = M \frac{di_1(t)}{dt} \tag{4.36}$$

where M is said to be the mutual inductance between the two inductors. Not all the magnetic flux generated in the first coil will pass through, or "link," the turns of the second coil. Some of this flux will leak out through the air, as shown in Fig. 4.14*a*. Therefore the mutual inductance is no greater than the self-inductance: $M \leq \sqrt{L_1 L_2}$. For ferrous cores most of the flux is confined to the core and the value of mutual inductance approaches that of the self-inductance. Because of the way in which the inductors have been wound on the common core, the flux ψ induces at terminals *c-d* a voltage [Eq. (4.36)] which has the same polarity as shown for $v_2(t)$. This is the meaning of the dots on the circuit symbol in Fig. 4.14*b*. A current entering a dotted terminal produces an induced voltage which is positive at the dotted terminals. Similarly, a time-varying current $i_2(t)$ entering terminal *d* will induce a voltage at the terminals of the first inductor of

$$v_1''(t) = M \frac{di_2(t)}{dt} \tag{4.37}$$

(positive at terminal *b*) and at the terminals of the second inductor of

$$v_2''(t) = L_2 \frac{di_2(t)}{dt} \tag{4.38}$$

(positive at terminal *d*). Combining Eqs. (4.35) and (4.37) and also combining Eqs. (4.36) and (4.38), we have

$$\begin{aligned} v_1(t) &= v_1'(t) + v_1''(t) \\[4pt] &= L_1 \frac{di_1(t)}{dt} + M \frac{di_2(t)}{dt} \\[6pt] v_2(t) &= v_2'(t) + v_2''(t) \\[4pt] &= M \frac{di_1(t)}{dt} + L_2 \frac{di_2(t)}{dt} \end{aligned} \tag{4.39}$$

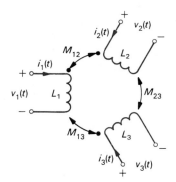

FIGURE 4.15
Coupled coils.

Equations (4.39) are the circuit equations for two coupled inductors. It is very important to observe the dot notation when writing the terminal equations. This result can be extended to any number of coupled coils. An example of three coupled coils is shown in Fig. 4.15. The circuit equations become

$$v_1 = L_1 \frac{di_1}{dt} + M_{12} \frac{di_2}{dt} + M_{13} \frac{di_3}{dt}$$

$$v_2 = M_{12} \frac{di_2}{dt} + L_2 \frac{di_2}{dt} + M_{23} \frac{di_3}{dt} \qquad (4.40)$$

$$v_3 = M_{13} \frac{di_1}{dt} + M_{23} \frac{di_2}{dt} + L_3 \frac{di_3}{dt}$$

Note the dot notation and resulting polarity of each component of induced voltage in Eq. (4.40).

The terminal voltage-current relationships of coupled inductors can be quickly written by observing the dots and the defined voltage and current polarities relative to those dots. *A current that enters a dotted terminal induces a voltage L di/dt, via self-inductance, that is positive at that dotted terminal and also induces a voltage M di/dt, via mutual inductance, that is positive at the dotted terminal of the other inductance.* This statement is equally correct if we replace the word "enters" with "leaves" and replace the word "positive" with "negative."

Example 4.5

Consider the set of two coupled inductors shown in Fig. 4.16a. Write the terminal equations in terms of the voltages and currents defined on the circuit diagram. Repeat this for the coupled set of three inductors shown in Fig. 4.16b.

Solution The self-contributions are dependent only on whether the voltage and the associated current are labeled with the passive sign convention. (The dots have nothing to do with these self-contributions.) So i_a produces a self-contribution to v_a that is

$$v_a' = -L_a \frac{di_a}{dt}$$

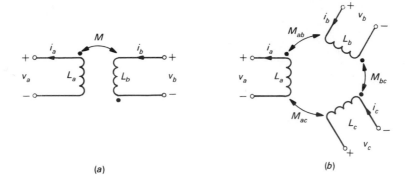

FIGURE 4.16
Example 4.5:
writing terminal
equations for
coupled coils.

(a)   *(b)*

and i_b produces a self-contribution to v_b that is

$$v_b' = L_b \frac{di_b}{dt}$$

Note that because v_a and i_a are not labeled with the passive sign convention, the self-contribution due to i_a is opposite to the desired voltage v_a. The self-contribution due to i_b is in the direction of the desired voltage v_b.

The current i_a produces a mutual inductance contribution to v_b

$$v_b'' = M \frac{di_a}{dt}$$

which is negative at the dotted terminal of coil b since i_a leaves the dotted terminal of coil a. But this polarity corresponds to the desired polarity of v_b, so it is entered as a positive contribution. Current i_b also leaves the dotted terminal of coil b, so it produces a voltage

$$v_a'' = -M \frac{di_b}{dt}$$

that is negative at the dotted terminal of coil a. But this is opposite to the desired polarity of v_a so that it is entered as a negative contribution. Therefore the terminal equations become

$$v_a = v_a' + v_a''$$

$$= -L_a \frac{di_a}{dt} - M \frac{di_b}{dt}$$

$$v_b = v_b' + v_b''$$

$$= M \frac{di_a}{dt} + L_b \frac{di_b}{dt}$$

The technique can be expanded to any number of coupled inductors if each voltage is viewed as the sum of a self-term and mutual terms due to other currents that are coupled to this coil via mutual inductance. For example, consider the three coupled inductors shown in Fig. 4.16b. The self-term for coil *a* is

$$v'_a = -L_a \frac{di_a}{dt}$$

and for coils *b* and *c* the self-terms are

$$v'_b = L_b \frac{di_b}{dt}$$

$$v'_c = -L_c \frac{di_c}{dt}$$

since only coil *b* has its voltage and current labeled with the passive sign convention. Current i_b leaves the dotted terminal of coil *b* and therefore produces contributions

$$v''_a = -M_{ab} \frac{di_b}{dt}$$

$$v''_c = M_{bc} \frac{di_b}{dt}$$

Since these contributions are opposite to v_a but in the direction of v_c, we enter $M_{ab}\, di_b/dt$ as a negative quantity in the equation for v_a and we enter $M_{bc}\, di_b/dt$ as a positive quantity in the equation for v_c. Similarly, current i_a produces contributions

$$v'''_b = M_{ab} \frac{di_a}{dt}$$

$$v'''_c = M_{ac} \frac{di_a}{dt}$$

that are in the direction of v_b and v_c, respectively. Thus they are entered in those equations as positive quantities. Finally, current i_c enters the dotted terminal of coil *c* and therefore produces contributions

$$v''''_a = M_{ac} \frac{di_c}{dt}$$

$$v''''_b = -M_{bc} \frac{di_c}{dt}$$

The first contribution is in the direction of v_a, while the second contribution is opposite to the direction of v_b. Therefore the first contribution is entered as a positive contribution in the equation for v_a, and the second contribution is entered as a negative contribution in the equation for v_b. Thus the terminal equations become

$$v_a = v_a' + v_a'' + v_a'''$$

$$= -L_a \frac{di_a}{dt} - M_{ab} \frac{di_b}{dt} + M_{ac} \frac{di_c}{dt}$$

$$v_b = v_b''' + v_b' + v_b''''$$

$$= M_{ab} \frac{di_a}{dt} + L_b \frac{di_b}{dt} - M_{bc} \frac{di_c}{dt}$$

$$v_c = v_c''' + v_c'' + v_c'$$

$$= M_{ac} \frac{di_a}{dt} + M_{bc} \frac{di_b}{dt} - L_c \frac{di_c}{dt}$$

Coupled coils are often referred to as transformers for the following reason. Consider the pair of coupled coils shown in Fig. 4.17. Let us assume an ideal case in which all the flux generated in the first coil links both the N_1 turns of the first coil and the N_2 turns of the second coil. In other words, there is no "leakage flux" in the coils. Ferrous cores such as iron tend to produce a close approximation to this ideal situation. The first coil has N_1 turns and the second has N_2 turns. A current i_1 in the first coil produces magnetic flux ψ in each turn of the coil which links the N_2 turns of the second coil. The voltage induced in the first coil by i_1 is equal to

$$v_1' = \frac{d}{dt}(N_1 \psi)$$

$$= N_1 \frac{d\psi}{dt} \tag{4.41}$$

while the voltage induced in the second coil by i_1 is equal to

$$v_2' = \frac{d}{dt}(N_2 \psi)$$

$$= N_2 \frac{d\psi}{dt} \tag{4.42}$$

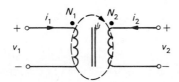

FIGURE 4.17
The ideal
transformer.

[Each turn in coil 2 has a voltage $(d/dt)(\psi)$ induced in it since flux ψ links each turn.] Now the ratio of these two voltages is

$$\frac{v_2'}{v_1'} = \frac{N_2}{N_1} \tag{4.43}$$

Thus, suppose a voltage v_1 is applied to coil 1; then a voltage

$$v_2 = \frac{N_2}{N_1} v_1 \tag{4.44}$$

will be induced at the terminals of coil 2. If $N_2 > N_1$, then v_2 will be larger than v_1, and we say that the voltage v_1 has been "stepped up" by the "turns ratio" N_2/N_1. Of course, if the voltage had been applied to the terminals of coil 2, the induced voltage at the terminals of coil 1 would have been "stepped down" by the ratio N_1/N_2 if $N_1 < N_2$. Step-up transformers are used to obtain from a much lower voltage the very high voltage (as high as 365,000 V) used in power transmission lines.

Note that the instantaneous power delivered to the first coil is

$$p_1 = v_1 i_1 \tag{4.45}$$

and that the power delivered to the second coil at that time is

$$p_2 = v_2 i_2 \tag{4.46}$$

If we assume that the transformer is lossless, then the total power delivered to the transformer must be zero:

$$p_1 + p_2 = 0 \tag{4.47}$$

or

$$v_1 i_1 + v_2 i_2 = 0 \tag{4.48}$$

Substituting Eq. (4.44) into Eq. (4.48) yields

$$v_1 i_1 + \frac{N_2}{N_1} v_1 i_2 = 0 \tag{4.49}$$

or

$$i_2 = -\frac{N_1}{N_2} i_1 \tag{4.50}$$

Thus, for current i_1 entering the dotted terminal of coil 1, current i_2 must leave the dotted terminal of coil 2 and is $(N_1/N_2)i_1$. The current ratio in Eq. (4.50) is the

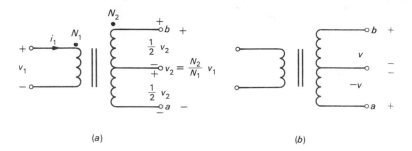

FIGURE 4.18
Center-tapped
transformers.

(a)   (b)

reciprocal of the voltage ratio in Eq. (4.44). Thus, if a transformer steps up a voltage applied to coil 1, it steps down the current entering coil 1 by the same ratio, so that the transformer remains lossless. Such a transformer is said to be *ideal*.

Quite often the center-tapped ideal transformers illustrated in Fig. 4.18a are used. The second coil has a tap placed at its center so that the voltage induced across the second coil $(N_2/N_1)v_1$ is divided equally between both halves of that coil. Note that this transformer can be used to provide two voltages which are opposite to each other in polarity, as shown in Fig. 4.18b. The voltage of terminal a with respect to the center tap is opposite in polarity to the voltage of terminal b with respect to the center tap.

One final point should be mentioned—reflection of resistance through a transformer. Consider the two-coil transformer in Fig. 4.19 which has the second coil attached to a resistor. The resistance seen, looking into the terminals of the first coil, is

$$R_{eq} = \frac{v_1}{i_1} \tag{4.51}$$

Substituting

$$v_1 = \frac{N_1}{N_2} v_2$$

$$i_1 = -\frac{N_2}{N_1} i_2 \tag{4.52}$$

and

$$v_2 = -Ri_2 \tag{4.53}$$

FIGURE 4.19
Illustration of
resistance
reflection
through a
transformer.

we obtain

$$R_{eq} = \left(\frac{N_1}{N_2}\right)^2 R \tag{4.54}$$

Thus, the resistor R appears at the terminals of the first coil larger or smaller by the ratio $(N_1/N_2)^2$, depending on whether $N_1 > N_2$ or $N_1 < N_2$. This can be useful in matching for maximum power transfer, as the following example shows.

Example 4.6

Consider the source and load shown in Fig. 4.20a. Design a circuit to provide maximum power transfer from the source to the load. Determine the average power delivered to the load.

Solution   If the load resistance has been 100 Ω, maximum power transfer to the load would take place. If we place an ideal transformer with a turns ratio of 5:1 between the source and load, the source will see a resistance of

$$(\tfrac{5}{1})^2 \times 4\,\Omega = 100\,\Omega$$

Thus, the source will deliver maximum power to the transformer and the 4-Ω load. But the transformer is assumed to be lossless (ideal), so all this power is delivered to the 4-Ω resistor. From Chap. 2, the average power delivered to the equivalent 100-Ω resistor seen by the source is

$$P_{AV} = \frac{V_{RMS}^2}{100\,\Omega}$$

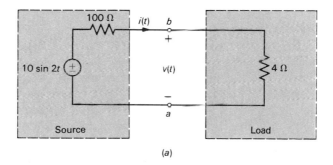

(a)

FIGURE 4.20
Example 4.6:
maximum power
transfer, utilizing
a matching
transformer.

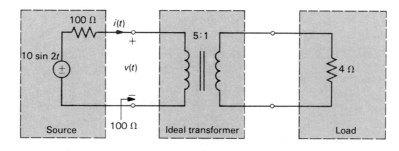

(b)

where V_{RMS} is the RMS value of the sinusoidal voltage $v(t)$. But, by voltage division,

$$v(t) = \frac{100\ \Omega}{100\ \Omega + 100\ \Omega}\, 10 \sin 2t$$

$$= 5 \sin 2t \quad \text{V}$$

which has an RMS value of

$$V_{\text{RMS}} = \frac{5}{\sqrt{2}}\ \text{V}$$

Thus, the average power delivered to the 4-Ω load in Fig. 4.20b is

$$P_{\text{AV}} = \frac{(5/\sqrt{2}\ \text{V})^2}{100\ \Omega}$$

$$= 0.125\ \text{W}$$

4.5 THE CIRCUIT DIFFERENTIAL EQUATIONS FOR CIRCUITS CONTAINING ENERGY STORAGE ELEMENTS

The equations governing the branch voltages and currents consist of KVL, KCL, and the element relations; including energy storage elements does not change this fact. However, if the circuit contains at least one energy storage element, then these circuit equations will become differential equations, which are more difficult to solve than the algebraic equations encountered with the purely, resistive circuits in Chap. 3. The solution of these differential equations is considered in the following two chapters. Nevertheless, as long as the elements are linear, the resulting equations are still linear and the currents and voltages can be found by using the principle of superposition.

Example 4.7   Write the circuit equations for the circuit shown in Fig. 4.21. Show that the currents and voltages can be obtained by using the principle of superposition.

FIGURE 4.21
Example 4.7:
writing circuit
equations for
circuits
containing
energy storage
elements.

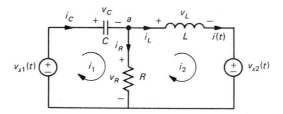

Solution Using the labeled element voltages and currents, we need six simultaneous equations in the six unknown branch voltages and circuits: v_L, i_L, v_C, i_C, v_R, i_R. Writing KCL at node a, KVL around loops L_1 and L_2, and the three element equations yields those six equations:

KCL node a:

$$i_L - i_C + i_R = 0$$

KVL loop L_1:

$$v_R + v_C = v_{s1}(t)$$

KVL loop L_2:

$$v_R - v_L = v_{s2}(t)$$

resistor:

$$v_R - Ri_R = 0$$

inductor:

$$v_L - L\frac{di_L}{dt} = 0$$

capacitor:

$$i_C - C\frac{dv_C}{dt} = 0$$

Note that the two energy storage elements have introduced derivatives of two of the element unknowns, i_L and v_C. Thus, the six equations have become differential equations.

Nevertheless, superposition still holds for these "dynamic" circuits containing energy storage elements as long as the energy storage elements (and all resistors in the circuit) are linear. This can easily be demonstrated for the circuit in Fig. 4.21. Suppose that v'_L, i'_L, v'_C, i'_C, v'_R, i'_R are due to $v_{s1}(t)$ with $v_{s2}(t) = 0$ and that v''_L, i''_L, v''_C, i''_C, v''_R, i''_R are due to $v_{s2}(t)$ with $v_{s1}(t) = 0$. KVL and KCL are linear and are satisfied by the sums of the prime and double-prime solutions by superposition. Now if the prime variables satisfy the element relations:

$$\left.\begin{array}{c} v'_R - Ri'_R = 0 \\[2mm] v'_L - L\dfrac{di'_L}{dt} = 0 \\[2mm] i'_C - C\dfrac{dv'_C}{dt} = 0 \end{array}\right\} \quad \text{due to } v_{s1}(t) \tag{4.55}$$

and the double-prime variables satisfy the element relations, too:

$$\left.\begin{array}{c} v_R'' - Ri_R'' = 0 \\[2mm] v_L'' - L\dfrac{di_L''}{dt} = 0 \\[2mm] i_C'' - C\dfrac{dv_C''}{dt} = 0 \end{array}\right\} \quad \text{due to the } v_{s2}(t) \tag{4.56}$$

then the sums of the prime and double-prime variables satisfy these element relations, since

$$(v_R' + v_R'') - R(i_R' + i_R'') = 0$$

$$(v_L' + v_L'') - L\frac{d}{dt}(i_L' + i_L'') = 0 \tag{4.57}$$

$$(i_C' + i_C'') - C\frac{d}{dt}(v_C' + v_C'') = 0$$

can be factored into the sum of Eqs. (4.55) and (4.56), which were presumed to be true.

The circuit differential equations obtained above using KVL, KCL, and the element relations can be reduced to a single differential equation relating any element current or voltage to any source.

Example 4.8   Determine the differential equation relating $i(t)$ to $v_{s1}(t)$ and $v_{s2}(t)$ for the circuit of Fig. 4.21.

Solution   We could write the two mesh current equations using the indicated mesh current directions. Then since $i(t) = i_2(t)$, we only need to eliminate $i_1(t)$ from these equations. The mesh current equations are

Mesh 1: $\dfrac{1}{C}\displaystyle\int_{-\infty}^{t} i_1(\tau)\,d\tau + R[i_1(t) - i_2(t)] = v_{s1}(t)$

Mesh 2: $L\dfrac{di_2(t)}{dt} + R[i_2(t) - i_1(t)] = -v_{s2}(t)$

Solving the second mesh current equation for $i_1(t)$ gives

Mesh 2: $i_1(t) = i_2(t) + \dfrac{L}{R}\dfrac{di_2(t)}{dt} + \dfrac{1}{R}v_{s2}(t)$

Differentiating the first mesh current equation gives

Mesh 1: $\dfrac{1}{C}i_1(t) + R\dfrac{di_1(t)}{dt} - R\dfrac{di_2(t)}{dt} = \dfrac{d}{dt}v_{s1}(t)$

Substituting and simplifying give the differential equation relating $i(t) = i_1(t)$ to the two voltage sources as

$$\frac{d^2}{dt^2}i(t) + \frac{1}{RC}\frac{d}{dt}i(t) + \frac{1}{LC}i(t) = \frac{1}{L}\frac{d}{dt}v_{s1}(t) - \frac{1}{RLC}v_{s2}(t) - \frac{1}{L}\frac{d}{dt}v_{s2}(t)$$

This is a linear second-order constant-coefficient ordinary differential equation relating the desired "output," $i(t)$, to the two "inputs," $v_{s1}(t)$ and $v_{s2}(t)$. The solution to these types of differential equations will be studied in Chap. 6.

4.6 RESPONSE OF THE ENERGY STORAGE ELEMENTS TO DC SOURCES

We have seen in the previous section that the principle of superposition applies to linear circuits which contain linear resistors, inductors, and capacitors. In this section we investigate how to determine the portions of the branch voltages and currents of a circuit which are due to the dc, or constant-value, sources in the circuit. We will find a simple technique for doing so—*replace all inductors with short circuits and all capacitors with open circuits* and solve the resulting resistive circuit.

For example, consider the circuit shown in Fig. 4.22*a* which contains a dc voltage source V_s and a dc current source I_s, a resistor R, an inductor L, and a capacitor C.

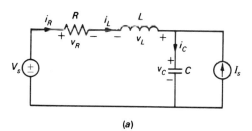

(a)

FIGURE 4.22
Illustration of the principle of superposition for circuits containing energy storage elements.

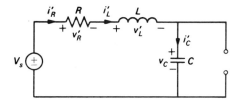 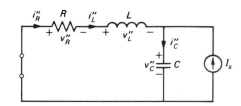

(b)

The solution for each branch voltage and current in the circuit—for example, v_R, i_R, v_L, i_L, v_C, and i_C—consists of two components, and each component is due to each source acting individually, as shown in Fig. 4.22b. In each circuit in Fig. 4.22b there is only one source and it is dc. The important question is, What would we reasonably expect the form of the branch currents and voltages due to the dc sources to be? The obvious answer is that they will be dc (constant), too.

Consider the voltage-current relationship for the capacitor shown in Fig. 4.23a:

$$i_C(t) = C \frac{dv_C}{dt} \tag{4.58}$$

Now suppose that both the current and the voltage are dc (constant): $i_C(t) = I_C$ and $v_C(t) = V_C$. Then Eq. (4.58) becomes

$$I_C = C \frac{dV_C}{dt}$$
$$= 0 \tag{4.59}$$

since the derivative of a constant is zero. Thus, the dc value of the capacitor current is zero, which is equivalent to replacing the capacitor with an open circuit!

Similarly, consider the inductor shown in Fig. 4.23b. If the current and voltage are dc, $i_L(t) = I_L$ and $v_L(t) = V_L$. Then the terminal relation of the inductor,

$$v_L(t) = L \frac{di_L(t)}{dt} \tag{4.60}$$

shows that

$$V_L = L \frac{dI_L}{dt}$$
$$= 0 \tag{4.61}$$

(a)   (b)

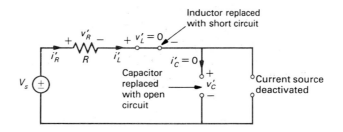

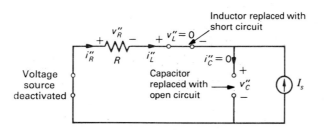

FIGURE 4.24
Illustration of the application of the principle of superposition for circuits containing energy storage elements and dc sources.

Thus, the dc value of the inductor voltage is zero, which is equivalent to replacing the inductor with a short circuit. Therefore, to determine the response due to a dc source, replace inductors with short circuits and capacitors with open circuits and solve the resulting resistive circuit. This is illustrated for the circuit of Fig. 4.22 in Fig. 4.24.

Example 4.9   Determine voltage v_x and current i_x in the circuit of Fig. 4.25a.

Solution   By superposition, the solution is shown in Fig. 4.25b where, because both sources are dc, we have replaced the inductors with short circuits and the capacitors with open circuits. We can easily solve these, using our resistive-circuit-analysis techniques developed in Chap. 3, to yield

$$v'_x = i'_x = 0$$
$$v''_x = -2 \text{ V}$$
$$i''_x = -1 \text{ A}$$

as the reader should verify. Thus

$$v_x = -2 \text{ V}$$
$$i_x = -1 \text{ A}$$

Note that the 2-F capacitor in series with the 10-V voltage source effectively blocks any effect of this source on the circuit.

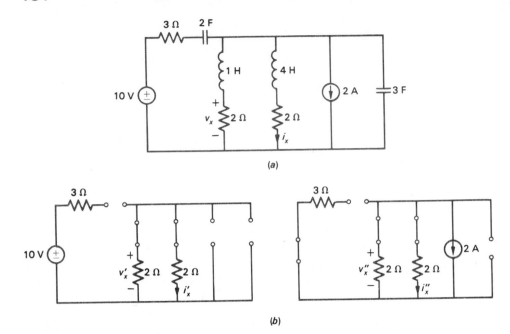

FIGURE 4.25
Example 4.9:
illustration of
superposition.

We have seen in this section that computing the contribution of a dc source to a branch voltage or current in a circuit is very simple—replace inductors with short circuits and capacitors with open circuits, and solve the resulting resistive circuit. The two forms of ideal sources which we study exclusively in this text are the dc source and the ac, or sinusoidal, source. In Chap. 5 we determine a method of solving for that contribution to a branch voltage or current of a circuit which is due to a sinusoidal source in the circuit. We will find that a simple method evolves there which also makes use of the resistive-circuit-analysis techniques we developed in Chap. 3. Now we are beginning to see why a thorough understanding of those resistive-circuit-analysis methods is so important—we will utilize them over and over.

PROBLEMS

4.1   Determine current $i(t)$ at $t = \frac{1}{2}$ s, $t = 1.5$ s, $t = 3$ s, and $t = 5$ s in the circuit of Fig. P4.1.

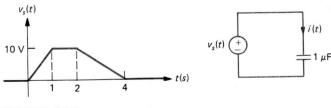

FIGURE P4.1

4.2 Determine voltage $v(t)$ at $t = 1$ s, $t = 2$ s, $t = 2.5$ s, $t = 3$ s, $t = 4$ s, and $t = 5$ s in the circuit of Fig. P4.2.

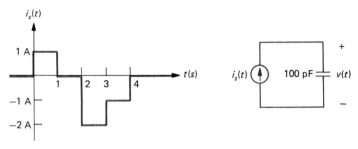

FIGURE P4.2

4.3 Sketch the voltage $v(t)$ and energy stored in the capacitor as a function of time for the circuit of Fig. P4.3.

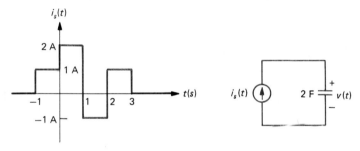

FIGURE P4.3

4.4 Sketch the current $i(t)$ and energy stored in the capacitor as a function of time for the circuit of Fig. P4.4.

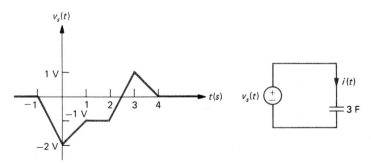

FIGURE P4.4

4.5   Sketch the current $i(t)$ and the energy stored in the capacitor for the circuit of Fig. P4.5.

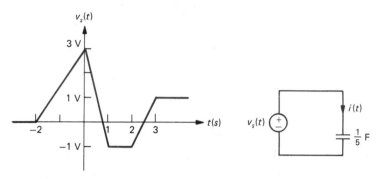

FIGURE P4.5

4.6   Sketch the current $i(t)$ for the circuit of Fig. P4.6.

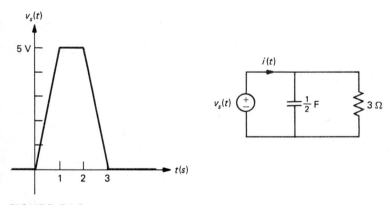

FIGURE P4.6

4.7   Sketch the voltage $v(t)$ for the circuit of Fig. P4.7.

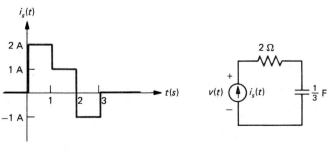

FIGURE P4.7

4.8 For the circuit of Fig. P4.8, determine the equivalent capacitance at terminals *a-b*.

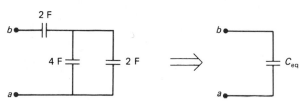

FIGURE P4.8

4.9 For the circuit of Fig. P4.9, determine the equivalent capacitance at terminals *a-b*.

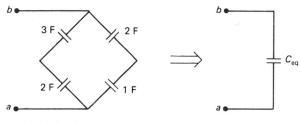

FIGURE P4.9

4.10 For the circuit of Fig. P4.10, determine the equivalent capacitance at terminals *a-b*.

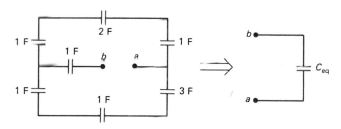

FIGURE P4.10

4.11 The energy stored in a 6-F capacitor is given by $w_c(t) = 5e^{-t/3}$ J for $t \geq 0$. Determine the capacitor voltage and current at $t = 2$ s.

4.12   Determine the voltage $v(t)$ at $t = \frac{1}{2}$ s, $t = 1.5$ s, $t = 2.5$ s, $t = 3.5$ s, $t = 5$ s in the circuit of Fig. P4.12.

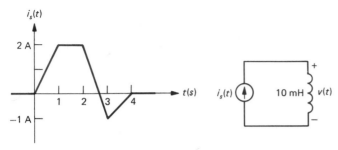

FIGURE P4.12

4.13   Determine the current $i(t)$ at $t = 0$ s, $t = 1$ s, $t = 2$ s, $t = 3$ s, $t = 4$ s in the circuit of Fig. P4.13.

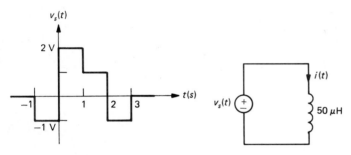

FIGURE P4.13

4.14   Sketch the current $i(t)$ and energy stored in the inductor for the circuit of Fig. P4.14.

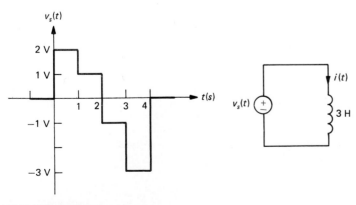

FIGURE P4.14

4.15 Sketch the voltage $v(t)$ and energy stored in the inductor for the circuit of Fig. P4.15.

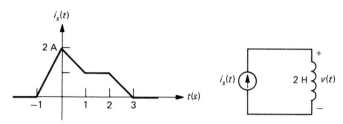

FIGURE P4.15

4.16 Sketch the voltage $v(t)$ and energy stored in the inductor for the circuit of Fig. P4.16.

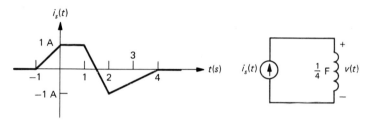

FIGURE P4.16

4.17 Sketch the voltage $v(t)$ in the circuit of Fig. P4.17.

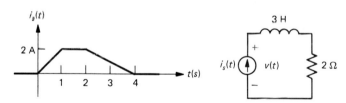

FIGURE P4.17

4.18   Sketch the current $i(t)$ in the circuit of Fig. P4.18.

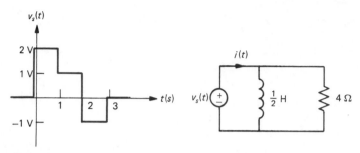

FIGURE P4.18

4.19   For the circuit of Fig. P4.19, determine the equivalent inductance at terminals *a-b*.

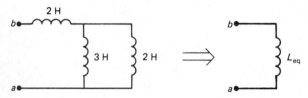

FIGURE P4.19

4.20   For the circuit of Fig. P4.20, determine the equivalent inductance at terminals *a-b*.

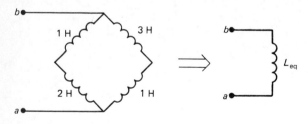

FIGURE P4.20

4.21 For the circuit of Fig. P4.21, determine the equivalent inductance at terminals *a-b*.

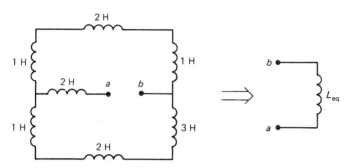

FIGURE P4.21

4.22 Write the voltage-current relationship for the coupled inductors of Fig. P4.22.

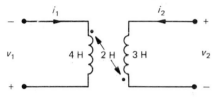

FIGURE P4.22

4.23 Write the voltage-current relationship for the coupled inductors of Fig. P4.23.

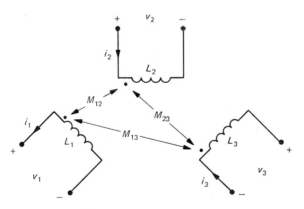

FIGURE P4.23

4.24 Write the voltage-current relationship for the coupled inductors of Fig. P4.24.

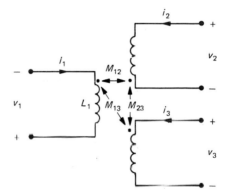

FIGURE P4.24

4.25   For the circuit of Fig. P4.25 determine $v_{out}(t)$.

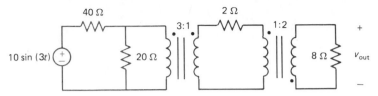

FIGURE P4.25

4.26   Determine the turns ratio N_1/N_2 of the ideal transformer of Fig. P4.26 such that maximum power is delivered to the speaker.

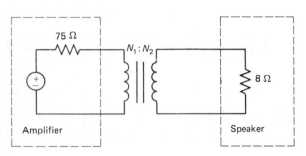

FIGURE P4.26

4.27   Write the differential equation relating i_{out} to v_s in the circuit of Fig. P4.27.

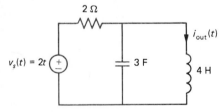

FIGURE P4.27

4.28   Write the differential equation relating v_{out} to v_s in the circuit of Fig. P4.28.

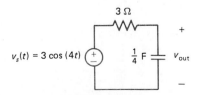

FIGURE P4.28

4.29   Write the differential equation relating v_{out} to v_s in the circuit of Fig. P4.29.

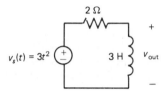

FIGURE P4.29

4.30   Write the differential equation relating i_{out} to i_s in the circuit of Fig. P4.30.

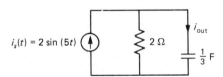

FIGURE P4.30

4.31   Write the differential equation relating i_{out} to i_s in the circuit of Fig. P4.31.

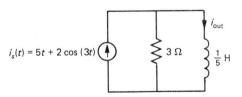

FIGURE P4.31

4.32   Write the differential equation relating v_{out} to v_s in the circuit of Fig. P4.32.

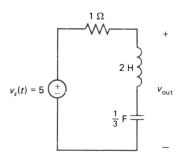

FIGURE P4.32

4.33   Write the differential equation relating i_{out} to i_s in the circuit of Fig. P4.33.

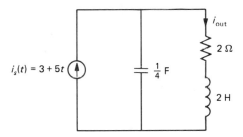

FIGURE P4.33

4.34   Write the differential equation relating v_{out} to v_s in the circuit of Fig. P4.34.

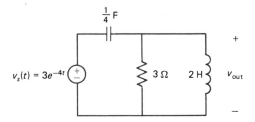

FIGURE P4.34

4.35   Determine i_{out} in the circuit of Fig. P.4.35.

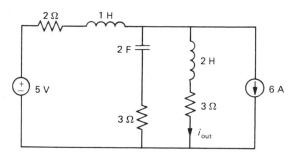

FIGURE P4.35

4.36 Determine i_{out} in the circuit of Fig. P.4.36.

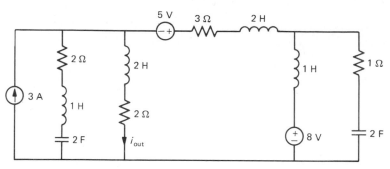

FIGURE P4.36

4.37 Determine v_{out} in the circuit of Fig. P.4.37.

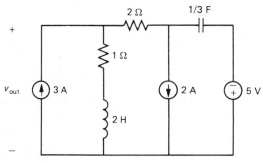

FIGURE P4.37

4.38 Determine v_{out} in the circuit of Fig. P.4.38.

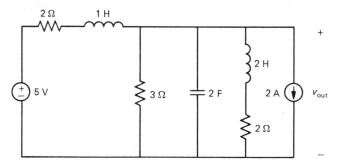

FIGURE P4.38

AC Circuits

In a circuit containing several ideal current and voltage sources, we may determine a particular branch voltage or current by superposition, i.e., by computing the contributions to this branch voltage or current of each source acting individually, with the other sources set to zero (deactivated). A deactivated voltage source is replaced with a short circuit, whereas a deactivated current source is replaced with an open circuit. In Sec. 4.6 we determined a simple method for computing the contribution to a branch voltage or current of a dc (constant-value) source—replace all inductors with short circuits and all capacitors with open circuits, and solve the resulting resistive circuit.

In this chapter we examine this problem for ac, or sinusoidal, sources. We will find that the contribution to a particular branch voltage or branch current due to a sinusoidal source of value either $A \sin(\omega t + \phi)$ or $A \cos(\omega t + \phi)$ is of the same form and frequency. We will also determine a simple technique for computing this contribution; the method makes use of our resistive-circuit-analysis techniques developed in Chap. 3 with only one small, additional complication—the method requires the manipulation of complex numbers rather than real numbers. However, this requirement to deal with complex numbers is a small price to pay for the tremendous solution advantages of the method.

Most of our later work will require the analysis of circuits containing sinusoidal sources. The results of this chapter are therefore very important, since we will be using them frequently throughout the remainder of this text. Hence, they should be studied very carefully.

Sinusoidal sources are very important for a number of reasons. One reason is that electric power is transmitted as a sinusoidal voltage and current at a frequency of 60 Hz (50 Hz in Europe). But perhaps the most important reason for their study is that any periodic function of time can be represented as an infinite sum of sinusoids with a Fourier series representation, as is shown in App. B; therefore, any ideal voltage or current source whose value is a periodic function of time may be represented as a sum of sinusoidal sources, and superposition can be employed to find the contribution of each of these sinusoidal components to a particular branch voltage or current of the circuit. Furthermore, nonperiodic functions of time can also be represented as an infinite sum of sinusoids with a Fourier integral; any arbitrary function of time can thus be viewed as being composed of sinusoidal components. Therefore, we do not need to investigate methods of finding the solutions for branch voltages or currents of a circuit due to sources which are arbitrary functions of time; we only need to investigate methods of finding the solution for sinusoidal sources. This chapter is devoted exclusively to that problem.

5.1 SINUSOIDAL (AC) SOURCES

Since the sinusoidal function is basic to our study, we devote this introductory section to a close examination of its properties. The sine function $A \sin \omega t$ is plotted in Fig. 5.1a. Here ω denotes the radian frequency, and $\omega = 2\pi f$. The units of ω are radians per second (rad/s), whereas the units of f are cycles per second or hertz. Thus the product of ωt has the unit of radians (rad). Although the argument of the sine function ωt is radians, we often refer to it in degrees. Since 2π rad is equivalent to $360°$, 1 rad is equivalent to $57.3°$. The sine function $A \sin \omega t$ achieves maximum values of A at $\omega t = \ldots, \pi/2, 5\pi/2, \ldots$ and minimum values of $-A$ at $\omega t = \ldots, -\pi/2, 3\pi/2, \ldots$. Note that the maxima as well as the minima are separated by 2π, as are all other corresponding points. Since $\omega t = 2\pi f t = 2\pi t/T$, the function is periodic with a period of $T = 1/f$. If the function were plotted versus t instead of ωt, we would divide the values shown by $\omega = 2\pi f$ or, equivalently, $2\pi/T$. For example, $\omega t = \pi$ corresponds to $t = T/2$, or one half period. Similarly, $\omega t = 2\pi$ corresponds to $t = T$, or one full period.

The cosine function $A \cos \omega t$ is plotted in Fig. 5.1b. The cosine is essentially the sine function advanced in time or shifted backward by $\omega t = \pi/2$, or $90°$, or $t = T/4$.

Adding a phase angle ϕ to the argument of the sine function shifts it *backward* by that phase angle. This is illustrated in Fig. 5.1c. This is easy to see and remember because of the following important observation. Note that points on two sinusoidal waveforms are identical where the arguments of the sinusoidal functions are identical, that is, $\sin \phi_1 = \sin \phi_2$ when $\phi_1 = \phi_2$. (Of course, we could also add multiples of 2π to ϕ_1 or ϕ_2 and not change this observation.) So adding ϕ to ωt to change $\sin \omega t$ to $\sin (\omega t + \phi)$ means that we must shift $\sin \omega t$ *backward* by ϕ in order to track corresponding points on the two waveforms (whose functions have the same values of the arguments at those two points). For example, $\omega t + \phi = 0$ at $\omega t = -\phi$. So $\sin \omega t = 0$ at $\omega t = 0$ and $\sin (\omega t + \phi) = 0$ at $\omega t = -\phi$. We use this observation on numerous occasions, so it is important that it be firmly understood.

Similar observations apply to the cosine function. In fact, since the cosine function is the sine function shifted backward by $\pi/2$, we can convert one to the other

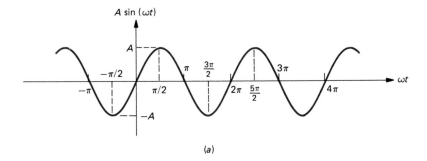

(a)

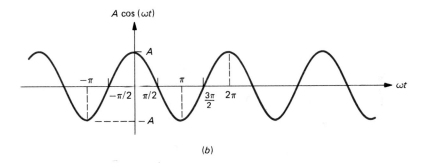

(b)

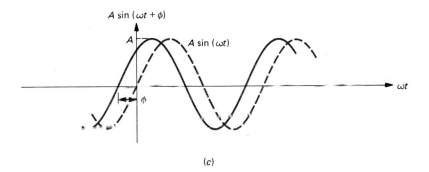

(c)

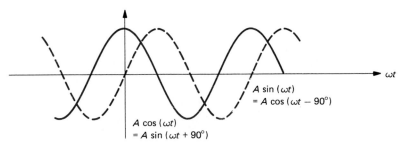

FIGURE 5.1

(d)

by adding or subtracting 90°:

$$A \cos \theta = A \sin \left(\theta + \frac{\pi}{2} \right)$$

$$= A \sin (\theta + 90°) \tag{5.1}$$

$$A \sin \theta = A \cos \left(\theta - \frac{\pi}{2} \right)$$

$$= A \cos (\theta - 90°) \tag{5.2}$$

These very important relations are illustrated in Fig. 5.1*d*. So an important result is that *adding* a phase angle to the argument of the sine or cosine function shifts the function *backward* (advances it in time) by that angle. Similarly, *subtracting* an angle from the argument of a sine or cosine function shifts the function *forward* (delays it in time) by that angle. We need to convert a cosine function to its equivalent sine function and vice versa on numerous occasions, so it is very important that we be able to use this result without error.

Also

$$- A \cos \theta = A \cos (\theta \pm \pi)$$

$$= A \cos (\theta \pm 180°) \tag{5.3}$$

and

$$- A \sin \theta = A \sin (\theta \pm \pi)$$

$$= A \sin (\theta \pm 180°) \tag{5.4}$$

Relations (5.3) and (5.4) are easily seen if we first invert the function (to produce $-A \sin \omega t$ or $+A \cos \omega t$) and then determine the required angle needed to shift the other function to match it.

Trigonometric identities can also be used to show equivalent forms. These are

$$\sin (\omega t \pm \phi) = \sin \omega t \cos \phi \pm \cos \omega t \sin \phi \tag{5.5}$$

$$\cos (\omega t \pm \phi) = \cos \omega t \cos \phi \mp \sin \omega t \sin \phi \tag{5.6}$$

Remember also that $\sin (-\phi) = -\sin \phi$ but $\cos (-\phi) = \cos \phi$. Therefore we can write $A \sin (\omega t + \phi)$ in an alternative form as

$$A \sin (\omega t + \phi) = A (\sin \omega t \cos \phi + \cos \omega t \sin \phi)$$

$$= A_1 \sin \omega t + A_2 \cos \omega t \tag{5.7}$$

where $A_1 = A \cos \phi$ and $A_2 = A \sin \phi$. Similarly,

$$B \cos (\omega t + \phi) = B(\cos \omega t \cos \phi - \sin \omega t \sin \phi)$$
$$= B_1 \cos \omega t + B_2 \sin \omega t \tag{5.8}$$

where $B_1 = B \cos \phi$ and $B_2 = -B \sin \phi$.

5.2 RESPONSE OF CIRCUITS TO AC SOURCES

Consider the circuit shown in Fig. 5.2. The ideal voltage source is sinusoidal at a radian frequency of $\omega = 3$ rad/s. Let us examine the determination of the solution for the inductor current $i(t)$. First we derive an equation relating $i(t)$ and the source. To do so, we note that $i(t)$ also passes through the 3-Ω resistor, developing a voltage across it of $Ri(t) = 3i(t)$. The voltage across the inductor is $L\ di(t)/dt = 2\ di(t)/dt$. Applying KVL around the loop, we have

$$2\frac{di(t)}{dt} + 3i(t) = 10 \sin 3t \tag{5.9}$$

Now let us find the solution to this equation. Let us guess the form of the solution. If we can obtain a guess which will satisfy Eq. (5.9), then we need look no further. A reasonable choice would be to assume $i(t)$ in the form of another sinusoid at a radian frequency of 3 rad/s also:

$$i(t) = I \sin 3t \tag{5.10}$$

Substituting into Eq. (5.9), we have

$$6I \cos 3t + 3I \sin 3t \overset{?}{=} 10 \sin 3t \tag{5.11}$$

But this equation cannot be satisfied by any choice of I. So we try

$$i(t) = I_s \sin 3t + I_c \cos 3t \tag{5.12}$$

Substituting into Eq. (5.9) gives

$$6I_s \cos 3t - 6I_c \sin 3t + 3I_s \sin 3t + 3I_c \cos 3t \overset{?}{=} 10 \sin 3t \tag{5.13}$$

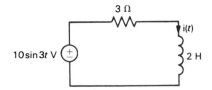

FIGURE 5.2
Illustration of the response of circuits to sinusoidal sources.

Matching coefficients of $\sin 3t$ and $\cos 3t$, we have

$$3I_s - 6I_c = 10$$
$$6I_s + 3I_c = 0 \tag{5.14}$$

By Cramer's rule we obtain

$$I_s = \frac{\begin{vmatrix} 10 & -6 \\ 0 & 3 \end{vmatrix}}{\begin{vmatrix} 3 & -6 \\ 6 & 3 \end{vmatrix}}$$

$$= \frac{30}{45}$$

$$= 0.67 \tag{5.15}$$

$$I_c = \frac{\begin{vmatrix} 3 & 10 \\ 6 & 0 \end{vmatrix}}{45}$$

$$= -\frac{60}{45}$$

$$= -1.33$$

and the solution has been found. Substituting Eq. (5.15) into (5.12) yields the solution

$$i(t) = 0.67 \sin 3t - 1.33 \cos 3t \tag{5.16}$$

This may be placed in an alternative form with the trigonometric identity

$$A \sin \theta + B \cos \theta = M \sin (\theta + \phi)$$

where $M = \sqrt{A^2 + B^2}$ and $\phi = \tan^{-1}(B/A)$ so that

$$i(t) = 1.49 \sin (3t - 63.43°) \tag{5.17}$$

For this rather simple circuit we needed to (1) derive the differential equation relating the unknown branch current to the source and (2) solve that differential equation. Step 2 required the solution of two simultaneous equations. In the next sections we develop a much simpler method for finding this solution which avoids both these steps. The tradeoff is that we must deal with complex numbers in that method. However, this is a small price to pay for the method's considerable computational advantage.

5.3 COMPLEX NUMBERS AND COMPLEX ALGEBRA

As indicated in the previous section, the method which we will develop involves the manipulation of complex numbers. We therefore need to become very familiar with complex numbers and the rules for their manipulation (complex algebra).

A complex number consists of two parts: a real part and an imaginary part. Complex numbers are indicated by boldface throughout this text. For example, the complex number **C** may be represented as

$$\mathbf{C} = a + jb \tag{5.18}$$

where a and b are real numbers and j symbolizes the square root of -1:

$$j = \sqrt{-1} \tag{5.19}$$

The real part of **C** is a and is denoted by

$$a = \text{Re } \mathbf{C} \tag{5.20}$$

Similarly, the imaginary part of **C** is b and is denoted by

$$b = \text{Im } \mathbf{C} \tag{5.21}$$

The complex number **C** can be represented as a vector in the complex plane by plotting the real part on the horizontal axis and the imaginary part on the vertical axis, as shown in Fig. 5.3. The length of this vector is referred to as the *magnitude* of the complex number and is denoted as $|\mathbf{C}|$ or C, where

$$C = \sqrt{a^2 + b^2} \tag{5.22}$$

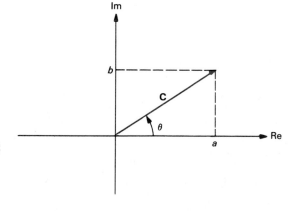

FIGURE 5.3
Illustration of the complex number **C** as a two-dimensional vector.

The angle θ, measured counterclockwise from the positive, horizontal axis is

$$\theta = \tan^{-1}\frac{b}{a} \tag{5.23}$$

The representation of the complex number in terms of its real and imaginary parts as in Eq. (5.18) is referred to as the *rectangular form*. Alternatively, the complex number may be represented in terms of its magnitude and angle in *polar form*:

$$\mathbf{C} = C\underline{/\theta} \tag{5.24}$$

From Fig. 5.3

$$a = C\cos\theta$$
$$b = C\sin\theta \tag{5.25}$$

Thus

$$\mathbf{C} = C\cos\theta + jC\sin\theta$$
$$= C(\cos\theta + j\sin\theta) \tag{5.26}$$

The last portion of Eq. (5.26) is written as

$$e^{j\theta} \triangleq \cos\theta + j\sin\theta \tag{5.27}$$

This is referred to as *Euler's* (pronounced "oiler's") *identity*. Thus

$$\mathbf{C} = Ce^{j\theta} \tag{5.28}$$

Note that the magnitude of $e^{j\theta}$ is unity, since

$$|e^{j\theta}| = \sqrt{\cos^2\theta + \sin^2\theta} \tag{5.29}$$
$$= 1$$

The angle of $e^{j\theta}$ is simply

$$\underline{/e^{j\theta}} = \tan^{-1}\frac{\sin\theta}{\cos\theta}$$
$$= \theta \tag{5.30}$$

Thus, the polar form of $e^{j\theta}$ is

$$e^{j\theta} = 1\underline{/\theta} \tag{5.31}$$

Now consider the addition, subtraction, multiplication, and division of any two complex numbers. If

$$\mathbf{A} = a + jb$$
$$\mathbf{B} = c + jd \tag{5.32}$$

then the sum of these two is simply the vector addition of the two complex vectors, as shown in Fig. 5.4a. The result is

$$
\begin{aligned}
A + B &= (a + jb) + (c + jd) \\
&= (a + c) + j(b + d)
\end{aligned} \tag{5.33}
$$

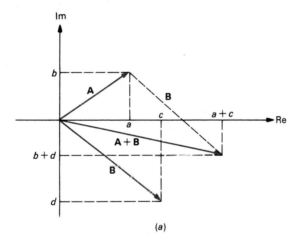

(a)

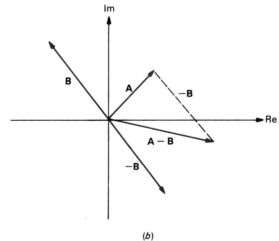

(b)

FIGURE 5.4
(a) Addition and
(b) subtraction of
complex numbers.

Thus

$$\text{Re} (\mathbf{A} + \mathbf{B}) = \text{Re} \, \mathbf{A} + \text{Re} \, \mathbf{B}$$
$$\text{Im} (\mathbf{A} + \mathbf{B}) = \text{Im} \, \mathbf{A} + \text{Im} \, \mathbf{B} \tag{5.34}$$

The subtraction of two complex numbers is defined as

$$\mathbf{A} - \mathbf{B} = \mathbf{A} + (-\mathbf{B})$$
$$= (a - c) + j(b - d) \tag{5.35}$$

as shown in Fig. 5.4b.

The multiplication of two complex numbers is defined by the following. First we represent the two numbers in their polar form:

$$\mathbf{A} = A\underline{/\theta_A}$$
$$\mathbf{B} = B\underline{/\theta_B} \tag{5.36}$$

The product is defined by

$$\mathbf{AB} = AB\underline{/\theta_A + \theta_B} \tag{5.37}$$

Thus, the magnitude of the product is the product of the magnitudes, and the angle of the product is the sum of the angles. Note that

$$j^2 = \sqrt{-1}\sqrt{-1}$$
$$= -1$$
$$j^3 = -j$$
$$j^4 = 1 \tag{5.38}$$
$$j^5 = j$$
$$\vdots$$

Thus, the product can be obtained in rectangular form as

$$\mathbf{AB} = (a + jb)(c + jd)$$
$$= ac + jad + jbc + j^2bd$$
$$= (ac - bd) + j(ad + bc) \tag{5.39}$$

Division is defined as

$$\frac{\mathbf{A}}{\mathbf{B}} = \frac{A}{B}\underline{/\theta_A - \theta_B} \tag{5.40}$$

so that the magnitude is the magnitude of the numerator A divided by the magnitude of the denominator B, and the angle is obtained by subtracting the angle of the denominator from the angle of the numerator.

The conjugate of a complex number $\mathbf{C}$ is denoted with a star (*) as $\mathbf{C}^*$ and is obtained by changing the sign of the imaginary part. For example, if

$$\mathbf{C} = a + jb \tag{5.41}$$

then

$$\mathbf{C}^* = a - jb \tag{5.42}$$

In terms of the polar form,

$$\begin{aligned}\mathbf{C} &= C\underline{/\theta} \\ \mathbf{C}^* &= C\underline{/-\theta}\end{aligned} \tag{5.43}$$

as is easily seen from Fig. 5.5. In terms of the conjugate we may obtain some other interesting results. For example, for the complex numbers $\mathbf{A}$ and $\mathbf{B}$ defined in Eq. (5.32), we may write

$$\begin{aligned}\frac{\mathbf{A}}{\mathbf{B}} &= \frac{a+jb}{c+jd} \\[2mm] &= \frac{a+jb}{c+jd}\frac{c-jd}{c-jd} \\[2mm] &= \frac{ac+bd}{c^2+d^2} + j\,\frac{bc-ad}{c^2+d^2} \\[2mm] &= \frac{\mathbf{A}}{\mathbf{B}}\cdot\frac{\mathbf{B}^*}{\mathbf{B}^*}\end{aligned} \tag{5.44}$$

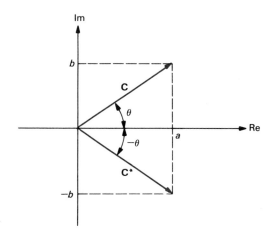

FIGURE 5.5
Illustration of the conjugate $\mathbf{C}^*$ of a complex number $\mathbf{C}$.

Note that

$$\begin{aligned}
\mathbf{BB^*} &= (c + jd)(c - jd) \\
&= c^2 + d^2 \\
&= |\mathbf{B}|^2
\end{aligned} \tag{5.45}$$

Thus, the square of the magnitude of a complex number is the product of itself and its conjugate. This is a very important result and is used on numerous occasions.

Several quite useful identities may now be obtained. For example,

$$(\mathbf{A^*})^* = \mathbf{A}$$

$$(\mathbf{AB})^* = \mathbf{A^*B^*}$$

$$\left(\frac{\mathbf{A}}{\mathbf{B}}\right)^* = \frac{\mathbf{A^*}}{\mathbf{B^*}} \tag{5.46}$$

$$(\mathbf{A} + \mathbf{B})^* = \mathbf{A^*} + \mathbf{B^*}$$

The reader should study these identities and prove each one.

Example 5.1   If

$$\mathbf{A} = 2 + j3 = 3.61\underline{/56.31°}$$
$$\mathbf{B} = 3 + j5 = 5.83\underline{/59.04°}$$

find $\mathbf{C}$ in rectangular and polar form when

(a) $\mathbf{C} = \mathbf{A} + \mathbf{B}$

(b) $\mathbf{C} = \mathbf{A} - \mathbf{B}$

(c) $\mathbf{C} = \mathbf{AB}$

(d) $\mathbf{C} = \dfrac{\mathbf{A}}{\mathbf{B}}$

(e) $\mathbf{C} = \mathbf{A} + \mathbf{B^*} = (\mathbf{A^*} + \mathbf{B})^*$

(f) $\mathbf{C} = \mathbf{A^*B} = (\mathbf{AB^*})^*$

(g) $\mathbf{C} = \dfrac{\mathbf{A}}{\mathbf{B^*}} = \left(\dfrac{\mathbf{A^*}}{\mathbf{B}}\right)^*$

Solution

(a) $\mathbf{C} = \mathbf{A} + \mathbf{B}$
$$\begin{aligned}
&= (2 + j3) + (3 + j5) \\
&= 5 + j8 \\
&= 9.43\underline{/58°}
\end{aligned}$$

(*b*) $\mathbf{C} = \mathbf{A} - \mathbf{B}$

$$= (2 + j3) - (3 + j5)$$

$$= -1 - j2$$

$$= 2.24\underline{/-116.57°}$$

(*c*) $\mathbf{C} = \mathbf{AB}$

$$= (2 + j3)(3 + j5)$$

$$= -9 + j19$$

$$\mathbf{C} = 3.61\underline{/56.31°}\ 5.83\underline{/59.04°}$$

$$= 21.02\underline{/115.35°}$$

Note: $-9 + j19 = 21.02\underline{/115.35°}$

(*d*) $\mathbf{C} = \dfrac{\mathbf{A}}{\mathbf{B}}$

$$= \frac{2 + j3}{3 + j5}$$

$$= \frac{2 + j3}{3 + j5}\frac{3 - j5}{3 - j5}$$

$$= 0.62 - j0.03$$

$$\mathbf{C} = \frac{3.61\underline{/56.31°}}{5.83\underline{/59.04°}}$$

$$= 0.62\underline{/-2.73°}$$

Note: $0.62 - j0.03 = 0.62\underline{/-2.73°}$

(*e*) $\mathbf{C} = \mathbf{A} + \mathbf{B}^{*}$

$$= (2 + j3) + (3 - j5)$$

$$= 5 - j2$$

Note: $\mathbf{C} = \mathbf{A} + \mathbf{B}^{*}$

$$= (\mathbf{A}^{*} + \mathbf{B})^{*}$$

$$= (2 - j3 + 3 + j5)^{*}$$

$$= (5 + j2)^{*}$$

$$= 5 - j2$$

(*f*) $\mathbf{C} = \mathbf{A}^{*}\mathbf{B}$

$$= (2 - j3)(3 + j5)$$

$$= 21 + j1$$

$$= 21.02\underline{/2.73°}$$

$$= 3.61\underline{/-56.31°}\ 5.83\underline{/59.04°}$$

$$= 21.02\underline{/2.73°}$$

$$= (\mathbf{AB}^{*})^{*}$$

$$= (3.61\underline{/56.31°}\ 5.83\underline{/-59.04°})^{*}$$

$$= (21.02\underline{/-2.73°})^{*}$$

$$= 21.02\underline{/2.73°}$$

(g) $\mathbf{C} = \dfrac{\mathbf{A}}{\mathbf{B}^*}$

$\qquad = \dfrac{2 + j3}{3 - j5}$

$\qquad = \dfrac{2 + j3}{3 - j5}\dfrac{3 + j5}{3 + j5}$

$\qquad = \dfrac{-9 + j19}{34}$

$\qquad = -0.26 + j0.56$

$\qquad = 0.62\underline{/115.35^\circ}$

$\qquad = \dfrac{3.61\underline{/56.31^\circ}}{5.83\underline{/-59.04^\circ}}$

$\qquad = 0.62\underline{/115.34}$

$\qquad = \left(\dfrac{\mathbf{A}^*}{\mathbf{B}}\right)^*$

$\qquad = \left(\dfrac{3.61\underline{/-56.31^\circ}}{5.83\underline{/59.04^\circ}}\right)^*$

$\qquad = (0.62\underline{/-115.35^\circ})^*$

$\qquad = 0.62\underline{/115.35^\circ}$

The above complex operations can be used repeatedly to reduce a complex expression to a real and an imaginary part or, as the following example shows, to a magnitude and angle.

Example 5.2

Suppose

$$\mathbf{A} = 1 + j2$$
$$\mathbf{B} = 3 - j5$$
$$\mathbf{C} = 2 + j4$$
$$\mathbf{D} = 1 - j3$$
$$\mathbf{E} = 4 + j1$$

Evaluate

$$\mathbf{F} = \dfrac{\mathbf{A} + \mathbf{B}/\mathbf{C}}{\mathbf{DE}}$$

Solution First we evaluate **B/C**. We write in polar form

$$\frac{\mathbf{B}}{\mathbf{C}} = \frac{5.83\underline{/-59.04°}}{4.47\underline{/63.43°}}$$

$$= 1.30\underline{/-122.47°}$$

Converting to rectangular form gives

$$\frac{\mathbf{B}}{\mathbf{C}} = -0.70 - j1.10$$

Now

$$\mathbf{A} + \frac{\mathbf{B}}{\mathbf{C}} = (1 + j2) + (-0.7 - j1.1)$$

$$= 0.3 + j0.9$$

$$= 0.95\underline{/71.57°}$$

$$\mathbf{F} = \frac{\mathbf{A} + \mathbf{B/C}}{\mathbf{DE}}$$

$$= \frac{0.95\underline{/71.57°}}{(1 - j3)(4 + j1)}$$

$$= \frac{0.95\underline{/71.57°}}{3.16\underline{/-71.57°}\ 4.12\underline{/14.04°}}$$

$$= \frac{0.95\underline{/71.57°}}{13.02\underline{/-57.53°}}$$

$$= 0.07\underline{/129.10°}$$

The manipulation of complex number expressions, as in the preceding example, generally requires several conversions of complex numbers between rectangular and polar forms. An electronic calculator having the ability to convert between these forms is particularly advantageous in reducing these expressions, and we will have numerous occasions in the remainder of this chapter to reduce such expressions.

5.4 USE OF THE COMPLEX EXPONENTIAL SOURCE

With this knowledge of complex numbers and algebra, let us reexamine the solution of the circuit in Fig. 5.2 which we considered in Sec. 5.2. The differential equation relating $i(t)$ to the source was

$$2\frac{di(t)}{dt} + 3i(t) = 10 \sin 3t \tag{5.47}$$

Instead of solving Eq. (5.47), let us replace the sinusoidal source with

$$10 \sin 3t \rightarrow 10e^{j3t} \tag{5.48}$$

as shown in Fig. 5.6a. Thus, Eq. (5.47) becomes

$$2\frac{dI}{dt} + 3I = 10e^{j3t} \tag{5.49}$$

A reasonable guess for the form of the solution to this equation would be to assume I to be of the same form as the right-hand side:

$$I = \mathbf{I}e^{j3t} \tag{5.50}$$

Substitute Eq. (5.50) into Eq. (5.49):

$$2(j3)\mathbf{I}e^{j3t} + 3\mathbf{I}e^{j3t} = 10e^{j3t} \tag{5.51}$$

Canceling e^{j3t}, which is common to both sides, we obtain the solution for $\mathbf{I}$ as

$$\mathbf{I} = \frac{10}{3 + j6}$$
$$= 1.49\underline{/-63.43^\circ} \tag{5.52}$$

Thus, the solution to Eq. (5.49) is

$$I = \mathbf{I}e^{j3t}$$
$$= 1.49\underline{/-63.43^\circ}\, e^{j3t}$$
$$= 1.49e^{j(3t-63.43^\circ)} \tag{5.53}$$

Now let us examine how the solution I to Eq. (5.49) is related to the solution $i(t)$ to the original equation in Eq. (5.47). The solution for I in Eq. (5.53) can be written, with Euler's identity, as

$$I = 1.49e^{j(3t-63.43^\circ)}$$
$$= 1.49 \cos (3t - 63.43^\circ) + j1.49 \sin (3t - 63.43^\circ) \tag{5.54}$$

FIGURE 5.6
An important
example—
illustration of
superposition and
the use of complex
exponentials.

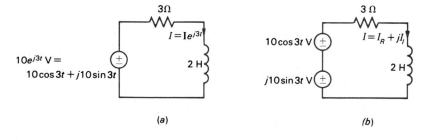

(a)   (b)

Also the right-hand side of Eq. (5.49) can be written, using Euler's identity, as

$$10e^{j3t} = 10 \cos 3t + j10 \sin 3t \tag{5.55}$$

Thus, Eq. (5.49) can be written as

$$2 \frac{d}{dt} [1.49 \cos (3t - 63.43°) + j1.49 \sin (3t - 63.43°)]$$

$$+ 3[1.49 \cos (3t - 63.43°) + j1.49 \sin (3t - 63.43°)]$$

$$= 10 \cos 3t + j10 \sin 3t \quad (5.56)$$

Equating the real and imaginary parts of this equation yields two equations:

$$2 \frac{d}{dt} [1.49 \cos (3t - 63.43°)] + 3[1.49 \cos (3t - 63.43°)] = 10 \cos 3t \tag{5.57a}$$

$$2 \frac{d}{dt} [1.49 \sin (3t - 63.43°)] + 3[1.49 \sin (3t - 63.43°)] = 10 \sin 3t \tag{5.57b}$$

Now compare the original equation, Eq. (5.47), with Eq. (5.57b), and we see that the solution to this original equation in Eq. (5.47) is

$$i(t) = 1.49 \sin (3t - 63.43°) \tag{5.58}$$

as was obtained in Sec. 5.2 by guessing a form of solution and substituting that guess into the equation to find the values of the unknowns in the assumed form of the solution.

Right now, to perceive a method in this, it is important to review what we have done. Note that the right-hand side of the original equation, 10 sin 3t (the value of the source), is the imaginary part of the right-hand side of Eq. (5.49), $10e^{j3t}$ (the new value of the source):

$$10 \sin 3t = \text{Im} \, 10e^{j3t} \tag{5.59}$$

Note, too, that the solution to the original equation in Eq. (5.47) is also the imaginary part of the solution to the new equation in Eq. (5.49):

$$i(t) = \text{Im} \, \mathbf{I}e^{j3t} \tag{5.60}$$

At this point it is easy to see why this happens. Since

$$10e^{j3t} = 10 \cos 3t + j10 \sin 3t \tag{5.61}$$

we see that we may represent the $10e^{j3t}$ source as the series combination of two sources, 10 cos 3t and j10 sin 3t, as shown in Fig. 5.6b, when we replace the original sinusoidal source in Fig. 5.2 with this new source, as in Fig. 5.6a. Thus, by superposition, the solution for I will also consist of two components, one due to the real part of $10e^{j3t}$,

$$10 \cos 3t \rightarrow I_R \tag{5.62}$$

and one due to the imaginary part of $10e^{j3t}$,

$$j10 \sin 3t \rightarrow jI_I \tag{5.63}$$

Therefore, if we replace the original source with $10e^{j3t}$, the solution for the current with the original source present is the imaginary part of the solution due to the source $10e^{j3t}$. Superposition has come into play once again. It is also clear that if the circuit had been a nonlinear one (at least one circuit element had been nonlinear), this method (which relied on the validity of superposition) *would not work*.

Suppose that the ideal voltage source had been cosinusoidal with a value of $10 \cos 3t$. How would we find the solution for $i(t)$? The answer is now quite clear; the solution would be the real part of the solution due to the $10e^{j3t}$ source, I_R, since $10 \cos 3t$ is the real part of $10e^{j3t}$.

Now it is easy to see that this method will work for any linear circuit. To compute the contribution to a branch current or voltage due to a sinusoidal source $A \sin(\omega t + \phi)$ or $A \cos(\omega t + \phi)$, replace the source with $Ae^{j\phi}e^{j\omega t}$ and solve the resulting circuit for the complex branch current or voltage of interest:

$$\begin{aligned} I &= \mathbf{I}e^{j\omega t} \\ V &= \mathbf{V}e^{j\omega t} \end{aligned} \tag{5.64}$$

where

$$\begin{aligned} \mathbf{I} &= I\underline{/\theta_I} \\ \mathbf{V} &= V\underline{/\theta_V} \end{aligned} \tag{5.65}$$

Now if the source was a sine source, the solution is the imaginary part of Eq. (5.64),

$$A \sin(\omega t + \phi) \rightarrow \begin{cases} I \sin(\omega t + \theta_I) \\ V \sin(\omega t + \theta_V) \end{cases} \tag{5.66}$$

and if the source was a cosine source, the solution is the real part of Eq. (5.64),

$$A \cos(\omega t + \phi) \rightarrow \begin{cases} I \cos(\omega t + \theta_I) \\ V \cos(\omega t + \theta_V) \end{cases} \tag{5.67}$$

Example 5.3   Consider the circuit in Fig. 5.7a. Determine $v(t)$.

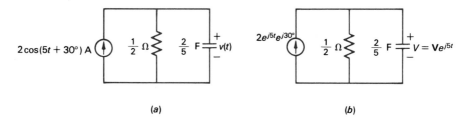

FIGURE 5.7
Example 5.3:
example of the
solution technique.

(a)   (b)

Solution   The differential equation relating $v(t)$ to the source is obtained by applying KCL at the upper node. The current passing through the $\frac{1}{2}$-Ω resistor is $2v(t)$, and the current passing through the $\frac{2}{5}$-F capacitor is $\frac{2}{5}\,dv(t)/dt$, so that

$$\frac{2}{5}\frac{dv(t)}{dt} + 2v(t) = 2\cos(5t + 30°)$$

$$= \text{Re } 2e^{j5t}e^{j30°}$$

Thus, replacing the source with $2e^{j5t}e^{j30°}$, as in Fig. 5.7b, and substituting

$$V = \mathbf{V}e^{j5t}$$

into the above equation yield

$$\tfrac{2}{5}j5\mathbf{V}e^{j5t} + 2\mathbf{V}e^{j5t} = 2e^{j5t}e^{j30°}$$

Solving for $\mathbf{V}$ gives

$$\mathbf{V} = \frac{2\underline{/30°}}{\tfrac{2}{5}j5 + 2}$$

$$= \frac{2\underline{/30°}}{2\sqrt{2}\underline{/45°}}$$

$$= 0.707\underline{/-15°}$$

Thus

$$V = \mathbf{V}e^{j5t}$$
$$= 0.707e^{-j15°}e^{j5t}$$
$$= 0.707e^{j(5t-15°)}$$

Since the original source was the real part of the replacement source,

$$2\cos(5t + 30°) = \text{Re } 2\underline{/30°}\,e^{j5t}$$

then

$$v(t) = \text{Re } (V = \mathbf{V}e^{j5t})$$
$$= 0.707 \cos (5t - 15°)$$

The reader should check, by direct substitution into the original differential equation, that this is indeed the solution.

5.5   THE PHASOR CIRCUIT

From the previous example, it is clear that a considerable simplification in the solution of an ac circuit has been obtained. In finding the component of a branch voltage or current due to an ideal voltage or current source which has the form

$$A \cos (\omega t + \phi) \tag{5.68a}$$

or

$$A \sin (\omega t + \phi) \tag{5.68b}$$

we replace that source with

$$Ae^{j(\omega t + \phi)} = A\underline{/\phi}e^{j\omega t} \tag{5.69}$$

Next we assume that the unknown branch voltages and currents are of the form

$$Ie^{j(\omega t + \theta_I)} = I\underline{/\theta_I}e^{j\omega t} \tag{5.70a}$$

or

$$Ve^{j(\omega t + \theta_V)} = V\underline{/\theta_V}e^{j\omega t} \tag{5.70b}$$

where the ω in the assumed forms of the solutions for the branch currents and voltages in Eqs. (5.70) is the same as the ω of the source in Eqs. (5.68). Since $e^{j\omega t}$ is common to all quantities in the circuit, we may discard it and solve the resulting phasor circuit for $I\underline{/\theta_I}$ or $V\underline{/\theta_V}$. If the source was cosinusoidal, as shown in Eq. (5.68a), then the solution is the real part of the solution in Eqs. (5.70):

$$I \cos (\omega t + \theta_I)$$
$$V \cos (\omega t + \theta_V) \tag{5.71a}$$

and if the source was sinusoidal, as in Eq. (5.68b), then the solution is the imaginary part of the solution in Eqs. (5.70):

$$I \sin (\omega t + \theta_I)$$
$$V \sin (\omega t + \theta_V) \tag{5.71b}$$

When we replace the value of the actual source in Eqs. (5.68) with the value of Eq. (5.69), replace the actual branch voltages and currents with their assumed forms in this circuit, as given in Eqs. (5.70), and remove the common $e^{j\omega t}$ from all quantities, then the resulting circuit is referred to as the *phasor circuit*. The assumed complex solutions in Eqs. (5.70) without the common $e^{j\omega t}$ terms, $\mathbf{I} = I\underline{/\theta_I}$ and $\mathbf{V} = V\underline{/\theta_V}$, are referred to as the *phasor* currents and voltages of the circuit. The original circuit is referred to as the *time-domain circuit*.

Now let us examine the solution of this phasor circuit *without* deriving the differential equation relating the branch variable of interest to the source. First let us examine how the voltage and current for each type of element are related. For the resistor,

$$v_R(t) = Ri_R(t) \tag{5.72}$$

Replacing $v_R(t)$ and $i_R(t)$ with their phasor equivalents, we have

$$\mathbf{V}_R e^{j\omega t} = R\mathbf{I}_R e^{j\omega t} \tag{5.73}$$

or

$$\mathbf{V}_R = R\mathbf{I}_R \tag{5.74}$$

Thus, the phasor voltage and current for a resistor are related again by Ohm's law. Next consider the inductor:

$$v_L(t) = L\frac{di_L(t)}{dt} \tag{5.75}$$

Replacing with phasor equivalents, we have

$$\mathbf{V}_L e^{j\omega t} = L\frac{d}{dt}\mathbf{I}_L e^{j\omega t} \tag{5.76}$$

or

$$\mathbf{V}_L = j\omega L\mathbf{I}_L \tag{5.77}$$

Note that the phasor voltage and phasor current for an inductor are also related in a form similar to Ohm's law as

$$\mathbf{V}_L = \mathbf{Z}_L\mathbf{I}_L \tag{5.78}$$

where

$$\mathbf{Z}_L = j\omega L \tag{5.79}$$

The quantity $\mathbf{Z}_L$ is called the *impedance* of the inductor. For the resistor,

$$\mathbf{V}_R = \mathbf{Z}_R \mathbf{I}_R \tag{5.80}$$

where the impedance of the resistor is the value of the resistance. Similarly, for the capacitor,

$$i_C(t) = C\,\frac{dv_C(t)}{dt} \tag{5.81}$$

Time-domain circuit

Phasor (sinusoidal, steady-state) circuit

$\mathbf{Z}_R = R$ $\mathbf{V}_R = R\mathbf{I}_R$

$\mathbf{Z}_L = j\omega L$ $\mathbf{V}_L = j\omega L\mathbf{I}_L$

$\mathbf{Z}_C = \dfrac{1}{j\omega C}$ $\mathbf{V}_C = \dfrac{1}{j\omega C}\,\mathbf{I}_C$

$V_S \sin(\omega t + \phi)$
$V_S \cos(\omega t + \phi)$

$\mathbf{V}_S = V_S e^{j\phi}$
$= V_S \angle\phi$

$I_S \sin(\omega t + \phi)$
$I_S \cos(\omega t + \phi)$

$\mathbf{I}_S = I_S e^{j\phi}$
$= I_S \angle\phi$

FIGURE 5.8
Circuit
components in the
phasor circuit.

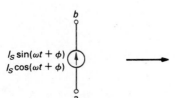

Assuming phasor quantities, we have

$$\mathbf{I}_c e^{j\omega t} = j\omega C \mathbf{V}_c e^{j\omega t} \tag{5.82}$$

or

$$\mathbf{V}_c = \frac{1}{j\omega C} \mathbf{I}_c \tag{5.83}$$

Thus

$$\mathbf{V}_c = \mathbf{Z}_c \mathbf{I}_c \tag{5.84}$$

and the impedance of the capacitor is

$$\mathbf{Z}_c = \frac{1}{j\omega C} \tag{5.85}$$

The reciprocals of these impedances are referred to as the *admittances* of the elements and are denoted by $\mathbf{Y}$. These concepts are summarized in Fig. 5.8.

Example 5.4   Determine the current $i(t)$ in Fig. 5.9a

Solution   The phasor circuit is shown in Fig. 5.9b ($\omega = 4$). Solving, we obtain

$$\begin{aligned}
\mathbf{I} &= \frac{2\underline{/30^\circ}}{6 + j12 + 3/j} \\
&= \frac{2\underline{/30^\circ}}{6 + j12 - j3} \\
&= \frac{2\underline{/30^\circ}}{6 + j9} \\
&= 0.18\underline{/-26.31^\circ}
\end{aligned}$$

FIGURE 5.9
Example 5.4: an
example of the use
of the phasor
circuit.

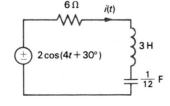

(a) Time-domain circuit

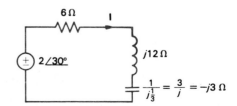

(b) Phasor circuit

Thus, the current is

$$i(t) = \text{Re } \mathbf{I}e^{j4t}$$
$$= 0.18 \cos (4t - 26.31°)$$

because the original source was the real part of $2e^{j(4t + 30°)}$:

$$2 \cos (4t + 30°) = \text{Re } 2e^{j(4t + 30°)}$$

If the circuit contains several sinusoidal sources, not all of which are of the same frequency, the result can be obtained by superposition in the time domain.

Example 5.5   Determine the current $i(t)$ in the circuit of Fig. 5.10*a*.

Solution   The phasor circuit diagram for the current source ($\omega = 3$) is shown in Fig. 5.10*b*. We obtain the phasor current due to this source by current division as

$$\mathbf{I}' = \frac{2 + j12}{(2 + j12) + 2/j3} \, 2\underline{/10°}$$
$$= 2.11\underline{/10.55°}$$

and

$$i'(t) = 2.11 \sin (3t + 10.55°)$$

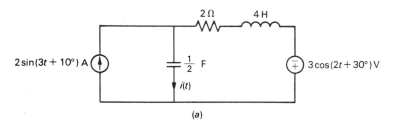

(a)

FIGURE 5.10
Example 5.5: use
of phasor circuits
with the
superposition
principle.

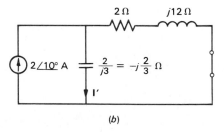

(b)

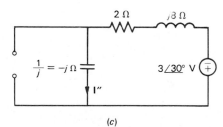

(c)

The phasor circuit for the voltage source ($\omega = 2$) is shown in Fig. 5.10c. The phasor current due to this source is

$$\mathbf{I}'' = -\frac{3\underline{/30°}}{2 + j8 + 1/j}$$

$$= -0.41\underline{/-44.05°}$$

and

$$i''(t) = -0.41 \cos(2t - 44.05°)$$

Thus, the current due to both sources is

$$i(t) = i'(t) + i''(t)$$
$$= 2.11 \sin(3t + 10.55°) - 0.41 \cos(2t - 44.05°)$$

It should be clear that the phasor circuit can be treated as a resistive circuit having "complex-valued resistors" when we are solving for the phasor branch variable of interest. All the various techniques which we discovered in Chap. 3 for solving resistive circuits—voltage division, current division, Thévenin and Norton equivalents (using a complex Thévenin "impedance" $\mathbf{Z}_{TH}$ with a phasor open-circuit voltage $\mathbf{V}_{OC}$ or short-circuit current $\mathbf{I}_{SC}$), node voltage analysis and mesh current analysis—can be used in solving the phasor circuit for the phasor branch variable of interest. Thus, the analysis of phasor circuits is no more difficult, theoretically, than that of resistive circuits; the manipulation of complex numbers, however, adds to the computational difficulty.

Example 5.6   Solve for the voltage $v(t)$ in the circuit of Fig. 5.11a by reducing the circuit at terminals a-b to a Thévenin equivalent.

Solution   The phasor circuit is shown in Fig. 5.11b. The phasor open-circuit voltage is found from Fig. 5.11c as

$$\mathbf{V}_{OC} = \frac{2}{2 + j9}\, 2\underline{/10°}$$

$$= 0.43\underline{/-67.47°}$$

The Thévenin impedance is found from Fig. 5.11d as

$$\mathbf{Z}_{TH} = \frac{4}{j3} + 2||j9$$

$$= 2.11\underline{/-25.52°}$$

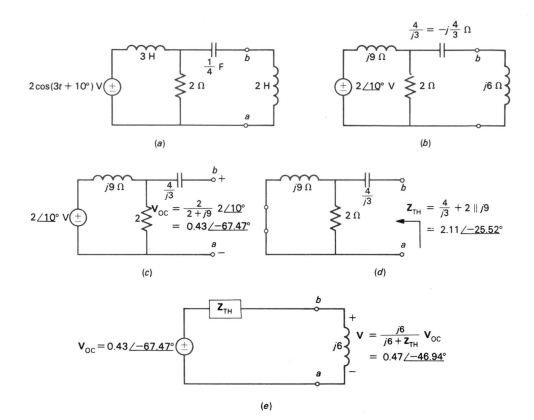

FIGURE 5.11
Example 5.6:
Thévenin
equivalent
reduction in
phasor circuits.

Reattaching, the complete circuit is shown in Fig. 5.11e, from which we obtain, by voltage division,

$$\mathbf{V} = \frac{j6}{j6 + \mathbf{Z}_{TH}} \mathbf{V}_{OC}$$

$$= 1.10\underline{/20.53°} \; 0.43\underline{/-67.47°}$$

$$= 0.47\underline{/-46.94°}$$

so that

$$v(t) = \text{Re } \mathbf{V}e^{j3t}$$

$$= 0.47 \cos(3t - 46.94°)$$

Sinusoidal sources (cosine or sine) may be included in one phasor circuit as long as (1) they have the same radian frequency ω and (2) they are all either cosine or sine. The first restriction on including several sources in one phasor circuit has to do with the need to compute the impedances of the inductors $j\omega L$ and capacitors $1/(j\omega C)$. If two sources having different radian frequencies are to be included in the same phasor circuit, which ω is used in computing the impedances of the inductors and capacitors? For the case of two or more sources which have different radian frequencies, we have no recourse but to use superposition in the time domain and compute the time-domain responses, using separate phasor circuits. This was illustrated in Example 5.5.

Those sources having the same radian frequency can be included in the same phasor circuit if they are all either sine or cosine form. This requirement to include several sources in the same phasor circuit only if they are all sine or all cosine results from the requirement that once this phasor circuit is solved, the desired time-domain current or voltage is to be obtained by multiplying it by $e^{j\omega t}$ and taking either the real part or the imaginary part of this result. Suppose, for example, that two sources have the same radian frequency but one is cosine and the other is sine. If we include both sources in one phasor circuit, solve for the desired phasor current or voltage, and multiply it by $e^{j\omega t}$, do we take the real or the imaginary part of the result? There is no satisfactory answer to this, so we must convert all sources to be included in the same phasor circuit to either cosine or sine form. This can be easily done by recalling the trigonometric identities

$$\sin (\omega t + \phi) = \cos (\omega t + \phi - 90°)$$

$$\cos (\omega t + \phi) = \sin (\omega t + \phi + 90°)$$

Once all sources of the same radian frequency are converted to sine or cosine form, they may be included in one phasor circuit. Then all our previous resistive-circuit-analysis techniques, superposition, Thévenin and Norton equivalents, mesh current and node voltage analysis, etc., can be used to solve this phasor circuit. The following example illustrates this procedure.

Example 5.7

Solve the circuit of Fig. 5.12a for $I(t)$ by including both sources in one phasor circuit and using mesh current analysis to solve for the phasor current.

Solution We choose (arbitrarily) to convert the cosine source to a sine source, as shown in Fig. 5.12b. Since both sources are sine and have the same radian frequency ($\omega = 3$), they can be included in one phasor circuit, as shown in Fig. 5.12c. Phasor mesh currents are written for that circuit:

$$(4 + j3 + 3 - j3)I_1 - (3 - j3)I_2 = 10\underline{/90°}$$

$$-(3 - j3)I_1 + (2 + j3 + 3 - j3)I_2 = 5\underline{/30°}$$

Solving gives

$$I_2 = 1.95\underline{/11°}$$

Since both sources were converted to sine form, the time-domain current becomes

$$i(t) = i_2(t) = \text{Im } 1.95\underline{/11°}e^{j3t} = 1.95 \sin (3t + 11°)$$

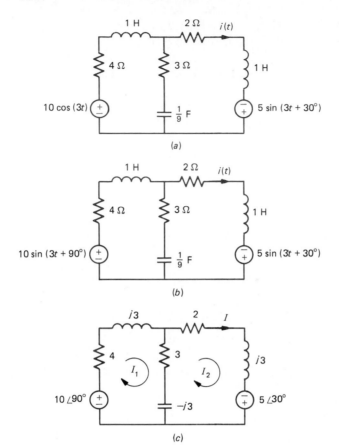

FIGURE 5.12
Example 5.7:
mesh current
analysis in phasor
circuits.

5.6 AVERAGE POWER, REACTIVE POWER, AND POWER FACTOR

Now that we have considered how to obtain the solution for any branch voltage or current of a circuit due to a sinusoidal source, let us consider how to calculate the power resulting from sinusoidal sources. The instantaneous power delivered to the element shown in Fig. 5.13a is

$$p(t) = v(t)i(t) \tag{5.86}$$

Since each variable will be sinusoidal, we may write

$$\begin{aligned} p(t) &= V \sin(\omega t + \theta_V)\, I \sin(\omega t + \theta_I) \\ &= VI \sin(\omega t + \theta_V) \sin(\omega t + \theta_I) \end{aligned} \tag{5.87}$$

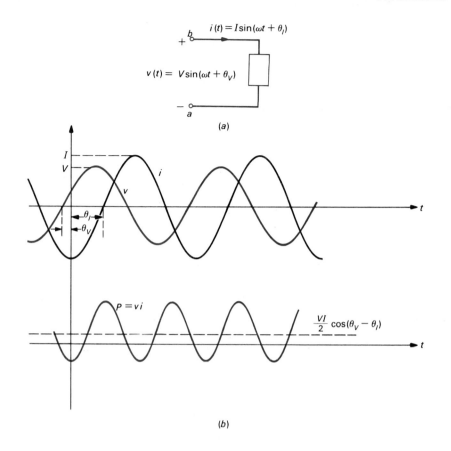

(a)

(b)

FIGURE 5.13
Average power
dissipation in
(absorbed by) an
element because
of sinusoidal
excitation.

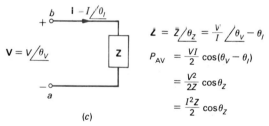

(c)

Denoting the difference in the phase angles of the voltage and current as

$$\theta = \theta_V - \theta_I \qquad (5.88)$$

we may write, using trigonometric identities,

$$p(t) = \frac{VI}{2} \cos \theta - \frac{VI}{2} \cos (2\omega t + \theta_V + \theta_I) \qquad (5.89)$$

which is plotted in Fig. 5.13b. Thus, the instantaneous power consists of a dc component, $(VI/2) \cos \theta$, and of a sinusoidal component, $(VI/2) \cos (2\omega t + \theta_V + \theta_I)$, which varies at twice the frequency f of the voltage or current. The average power is the integral of this instantaneous power over a period of $p(t)$, $T = 1/f$, averaged over that period:

$$P_{AV} = \frac{1}{T} \int_0^T p(t)\, dt$$

$$= \frac{VI}{2} \cos(\theta_V - \theta_I) \tag{5.90}$$

which becomes obvious from the plot in Fig. 5.13b. The phasor representation is shown in Fig. 5.13c. Note that

$$\mathbf{Z} = Z\underline{/\theta_Z}$$

$$= \frac{\mathbf{V}}{\mathbf{I}}$$

$$= \frac{V\underline{/\theta_V}}{I\underline{/\theta_I}}$$

$$= \frac{V}{I}\,\underline{/\theta_V - \theta_I} \tag{5.91}$$

where $\mathbf{Z}$ is the phasor impedance of the element. Therefore,

$$Z = \frac{V}{I} \tag{5.92}$$

and

$$\theta_Z = \theta_V - \theta_I \tag{5.93}$$

Thus, we may write the average power in Eq. (5.90) as

$$P_{AV} = \frac{V^2}{2Z} \cos \theta_Z$$

$$= \tfrac{1}{2} I^2 Z \cos \theta_Z \tag{5.94}$$

In Chap. 2 we define the effective, or RMS, value of a sinusoidal voltage or current as

$$V_{RMS} = \frac{V}{\sqrt{2}}$$

$$\tag{5.95}$$

$$I_{RMS} = \frac{I}{\sqrt{2}}$$

Therefore Eq. (5.94) can be written as

$$P_{AV} = \frac{V_{RMS}^2}{Z} \cos \theta_Z$$

$$= I_{RMS}^2 Z \cos \theta_Z \tag{5.96}$$

which provides a convenient method for computing the average power dissipated in a circuit element.

The average power dissipated in an element can be written as a more general result:

$$P_{AV} = \tfrac{1}{2} \operatorname{Re} \mathbf{VI}^*$$

$$= \tfrac{1}{2} \operatorname{Re} \mathbf{V}^*\mathbf{I} \tag{5.97}$$

where $\mathbf{I}^*$ is the conjugate of phasor $\mathbf{I}$ and Re($\cdot$) denotes "real part." To show that Eq. (5.97) reduces to Eq. (5.90), we write

$$\mathbf{V} = V\underline{/\theta_V}$$

$$\mathbf{I} = I\underline{/\theta_I} \tag{5.98}$$

Substituting Eqs. (5.98) into Eq. (5.97), we obtain

$$P_{AV} = \tfrac{1}{2} \operatorname{Re} V\underline{/\theta_V}I\underline{/-\theta_I}$$

$$= \tfrac{1}{2} \operatorname{Re} VI\underline{/\theta_V - \theta_I}$$

$$= \frac{VI}{2} \operatorname{Re} [\cos (\theta_V - \theta_I) + j \sin (\theta_V - \theta_I)]$$

$$= \frac{VI}{2} \cos (\theta_V - \theta_I) \tag{5.99}$$

Equation (5.97) is a very important result used in numerous instances. The phasor power is

$$\mathbf{P} = \tfrac{1}{2}\mathbf{VI}^*$$

$$= \mathbf{V}_{RMS}\mathbf{I}_{RMS}^*$$

$$= \frac{VI}{2} \cos (\theta_V - \theta_I) + j \frac{VI}{2} \sin (\theta_V - \theta_I) \tag{5.100}$$

The real part of Eq. (5.100) is, of course, the average power dissipated in the element. The imaginary part is referred to as the *reactive power* and is denoted as

$$Q = \tfrac{1}{2} \operatorname{Im} \mathbf{V}\mathbf{I}^*$$

$$= \operatorname{Im} \mathbf{V}_{\mathrm{RMS}} \mathbf{I}^*_{\mathrm{RMS}}$$

$$= \frac{VI}{2} \sin(\theta_V - \theta_I) \tag{5.101}$$

so that

$$\mathbf{P} = P_{\mathrm{AV}} + jQ \tag{5.102}$$

The reactive power does not represent real power loss in the element and only indicates temporary energy storage, as we shall see.

Now let us investigate the application of these concepts to the R, L, and C elements. For the resistor shown in Fig. 5.14, the phasor voltage and current are in phase:

$$\mathbf{V} = R\mathbf{I} \tag{5.103}$$

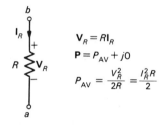

$$\mathbf{V}_R = R\mathbf{I}_R$$
$$\mathbf{P} = P_{\mathrm{AV}} + j0$$
$$P_{\mathrm{AV}} = \frac{V_R^2}{2R} = \frac{I_R^2 R}{2}$$

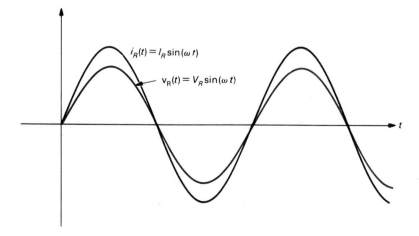

$$i_R(t) = I_R \sin(\omega t)$$
$$v_R(t) = V_R \sin(\omega t)$$

FIGURE 5.14
Complex power of a resistor due to sinusoidal excitation.

so that

$$P_{\text{AV}} = \frac{V^2}{2R}$$

$$= \frac{I^2 R}{2} \tag{5.104}$$

The reactive power Q for this element is zero since $\theta_V = \theta_I$ and $\sin(\theta_V - \theta_I) = 0$. Therefore, no reactive power is associated with the resistor.

For the inductor shown in Fig. 5.15, the phasor voltage and current are related by

$$\mathbf{V}_L = j\omega L \mathbf{I}_L$$

$$= \omega L \underline{/90°}\ \mathbf{I}_L \tag{5.105}$$

and the phasor voltage leads the current by 90°. Thus

$$P_{\text{AV}} = \frac{V_L I_L}{2}\cos 90°$$

$$= 0 \tag{5.106a}$$

$$Q = \frac{V_L I_L}{2}\sin 90°$$

$$= \frac{V_L I_L}{2} \tag{5.106b}$$

The inductor current absorbs no (real) power but has a reactive power associated with it. These results are shown in Fig. 5.15.

For the capacitor shown in Fig. 5.16, the phasor voltage and current are related by

$$\mathbf{V}_C = \frac{1}{j\omega C}\ \mathbf{I}_C \tag{5.107}$$

or

$$\mathbf{I}_C = j\omega C \mathbf{V}_C$$

$$= \omega C \underline{/90°}\ \mathbf{V}_C \tag{5.108}$$

Thus, the phasor current leads the voltage by 90°—as opposed to the inductor, in which the voltage leads the current by 90°. This difference can be remembered by the mnemonic device "*ELI* and *ICE* man," where E corresponds to voltage and I

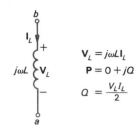

$$V_L = j\omega L I_L$$
$$P = 0 + jQ$$
$$Q = \frac{V_L I_L}{2}$$

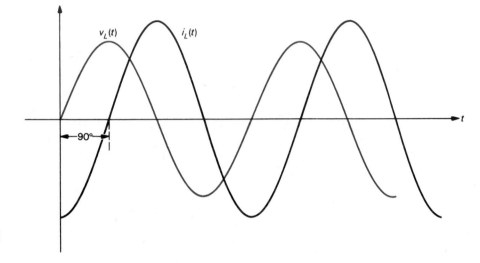

FIGURE 5.15
Complex power
for an inductor
due to sinusoidal
excitation.

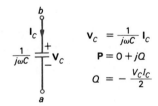

$$V_C = \frac{1}{j\omega C} I_C$$
$$P = 0 + jQ$$
$$Q = -\frac{V_C I_C}{2}$$

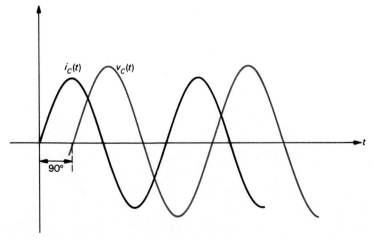

FIGURE 5.16
Complex power
for a capacitor due
to sinusoidal
excitation.

corresponds to current. As in the case of an inductor, the capacitor does not dissipate power but instead stores energy. This becomes clear since

$$P_{\text{AV}} = \frac{V_C I_C}{2} \cos(-90°)$$

$$= 0 \qquad\qquad (5.109a)$$

$$Q = \frac{V_C I_C}{2} \sin(-90°)$$

$$= -\frac{V_C I_C}{2} \qquad\qquad (5.109b)$$

Note that Q is negative for a capacitor—as opposed to an inductor, for which Q is positive.

5.6.1   Power Factor Correction

We have seen that the energy storage elements dissipate no power ($P_{\text{AV}} = 0$) but store energy ($Q \neq 0$). In power distribution, the net energy transfer from the power plant to the consumer is the average power. The reactive power is not power which is converted to a useful function and therefore represents no net energy transfer. It is desirable to minimize this reactive power in a load, as the following shows. Consider the series RL load shown in Fig. 5.17a, representing a consumer load. Electric motors and other typical consumer loads are typically represented in this fashion. The source is represented as an ideal voltage source v_S in series with a source resistance R_S. The transmission line is represented as a simple resistance R_{line} corresponding to the resistance of the line conductors. The phasor circuit is shown in Fig. 5.17b, and the phasor load current $\mathbf{I}_L$ is

$$\mathbf{I}_L = \frac{\mathbf{V}_S}{R_S + R_{\text{line}} + R + j\omega L} \qquad\qquad (5.110)$$

The total average power dissipated in the load is the power dissipated in the load resistance R:

$$P_{\text{AV}} = \frac{I_L^2 R}{2} \qquad\qquad (5.111)$$

The magnitude of the phasor current is, from Eq. (5.110),

$$I_L = \frac{V_S}{\sqrt{(R_S + R_{\text{line}} + R)^2 + (\omega L)^2}} \qquad\qquad (5.112)$$

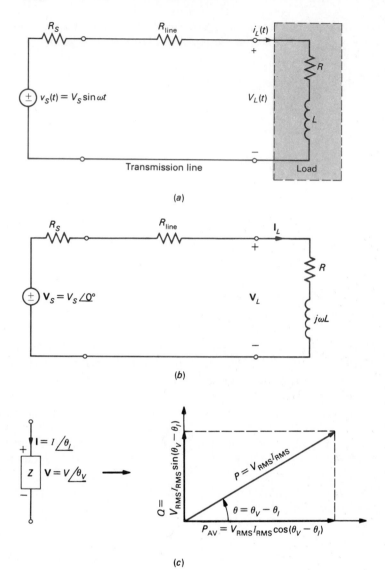

FIGURE 5.17
Power distribution
to an inductive
load.

For a fixed value of source voltage V_S, the magnitude of the load current and consequently the average power delivered to the load are obviously smaller than if there were no inductance in the load ($L = 0$). In power distribution, the load voltage $V_L(t)$ is usually fixed (e.g., 120 V). Therefore, if the load has an inductive component, the required current is increased for a given average power consumption. But a higher current increases the line losses ($\frac{1}{2}I_L^2 R_{\text{line}}$) and generator losses ($\frac{1}{2}I_L^2 R_S$).

This suitability of a particular impedance for absorption of average power is given by its power factor:

$$\cos(\theta_V - \theta_I) = \frac{P_{\text{AV}}}{V_{\text{RMS}} I_{\text{RMS}}} \tag{5.113}$$

where V_{RMS} and I_{RMS} are the RMS voltage and current associated with the terminals of the impedance. The phasor power associated with the impedance has a magnitude

$$|\mathbf{P}| = P$$
$$= V_{RMS} I_{RMS} \qquad (5.114)$$

and

$$P_{AV} = P \cos (\theta_V - \theta_I) \qquad (5.115a)$$
$$Q = P \sin (\theta_V - \theta_I) \qquad (5.115b)$$

as shown in Fig. 5.17c. Thus, the power factor (pf)

$$pf = \cos (\theta_V - \theta_I)$$
$$= \frac{P_{AV}}{P} \qquad (5.116)$$

gives a measure of the element's ability to absorb power. For the resistor, pf = 1, and for the inductor and capacitor, pf = 0. For any other element which is a combination of R, L or C elements,

$$0 \le pf \le 1 \qquad (5.117)$$

Example 5.8

For the circuit shown in Fig. 5.18a, find the average and reactive power associated with the load and compute the power factor of the load.

Solution The phasor circuit is shown in Fig. 5.18b ($\omega = 3$). The phasor voltage and current of the load are

$$\mathbf{V}_L = \frac{5 + j9 - j2}{4 + j6 + 5 + j9 - j2} \, 100\underline{/0^\circ}$$

$$= \frac{5 + j7}{9 + j13} \, 100\underline{/0^\circ}$$

$$= 54.41\underline{/-0.84^\circ}$$

$$\mathbf{I}_L = \frac{100\underline{/0^\circ}}{4 + j6 + 5 + j9 - j2}$$

$$= 6.32\underline{/-55.3^\circ}$$

The average power delivered to the load can be found by simply computing the average power dissipated in the 5-Ω resistor of the load:

$$P_{AV} = \tfrac{1}{2} I_L^2 5$$
$$= \tfrac{1}{2} (6.32)^2 5$$
$$= 100 \text{ W}$$

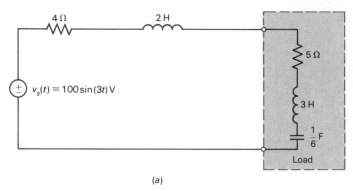

(a)

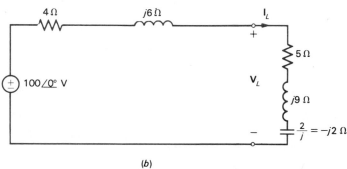

(b)

FIGURE 5.18
Example 5.7:
illustration of
average and
reactive power
delivered to a
complex load.

Both the average and reactive power can be found from the phasor power of the load:

$$\mathbf{P} = \tfrac{1}{2}\mathbf{V}_L\mathbf{I}_L^*$$
$$= \tfrac{1}{2}54.41\underline{/-0.84^\circ}\ 6.32\underline{/55.3^\circ}$$
$$= 172.05\underline{/54.46^\circ}$$
$$= 100 + j140$$

as

$$P_{\mathrm{AV}} = \mathrm{Re}\ \mathbf{P}$$
$$= 100\ \mathrm{W}$$
$$Q = \mathrm{Im}\ \mathbf{P}$$
$$= 140$$

Thus, the load appears to have a net inductive component (since Q is nonzero and positive), as is clear from the phasor circuit. The power factor of the load is

$$\mathrm{pf} = \cos\theta_Z$$
$$= \cos(\theta_V - \theta_I)$$
$$= \cos 54.46^\circ$$
$$= 0.58$$

5.6.2 Maximum Power Transfer

Consider the circuit shown in Fig. 5.19a. A source having phasor voltage $\mathbf{V}_S$ and source impedance $\mathbf{Z}_S$ is connected to a load impedance $\mathbf{Z}_L$. Suppose that the source voltage and impedance, $\mathbf{V}_S$ and $\mathbf{Z}_S$, are fixed, but that we have a choice as to the value of the load impedance $\mathbf{Z}_L$. We wish to find the value of $\mathbf{Z}_L$ which will result in the maximum average power being transferred from the source to the load. If we represent each impedance with a real and imaginary part

$$\mathbf{Z}_S = R_S + jX_S$$
$$\mathbf{Z}_L = R_L + jX_L$$

(5.118)

then the phasor load current $\mathbf{I}_L$ is

$$\mathbf{I}_L = \frac{\mathbf{V}_S}{\mathbf{Z}_S + \mathbf{Z}_L}$$

(5.119)

and its magnitude becomes

$$I_L = \frac{V_S}{\sqrt{(R_S + R_L)^2 + (X_S + X_L)^2}}$$

(5.120)

The average power delivered to the load is

$$P_{\text{AV}} = \tfrac{1}{2} I_L^2 R_L$$

(5.121)

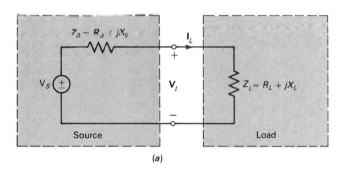

(a)

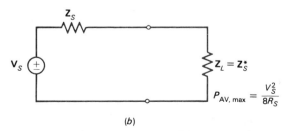

(b)

FIGURE 5.19
Illustration of maximum power transfer for complex source and load impedances.

which could also be determined from

$$P_{AV} = \tfrac{1}{2} \operatorname{Re} \mathbf{V}_L \mathbf{I}_L^*$$
$$= \tfrac{1}{2} \operatorname{Re} \mathbf{I}_L \mathbf{I}_L^* \mathbf{Z}_L$$
$$= \tfrac{1}{2} \operatorname{Re} I_L^2 \mathbf{Z}_L$$
$$= \frac{I_L^2}{2} \operatorname{Re} \mathbf{Z}_L$$
$$= \frac{I_L^2}{2} R_L \tag{5.122}$$

Substituting Eq. (5.120) into Eq, (5.121), we obtain

$$P_{AV} = \frac{V_S^2}{2} \frac{R_L}{(R_S + R_L)^2 + (X_S + X_L)^2} \tag{5.123}$$

If we can vary only X_L, then maximum power is delivered to the load if

$$X_L = -X_S \tag{5.124}$$

If only R_L can be varied, the result is a bit more difficult to see. If we form dP_{AV}/dR_L and set this equal to zero, we find that maximum power transfer occurs for

$$R_L = \sqrt{R_S^2 + (X_S + X_L)^2} \tag{5.125}$$

Now if both R_L and X_L can be varied, we obtain maximum power transfer to the load by substituting Eq. (5.124) into Eq. (5.125) to yield

$$\begin{aligned} X_L &= -X_S \\ R_L &= R_S \end{aligned} \tag{5.126}$$

or

$$\begin{aligned} \mathbf{Z}_L &= R_S - jX_S \\ &= \mathbf{Z}_S^* \end{aligned} \tag{5.127}$$

Therefore, maximum power is delivered from a source to a load when the load impedance is the conjugate of the source, as shown in Fig. 5.19b. If the source and load are purely resistive, $\mathbf{Z}_S = R_S$ and $\mathbf{Z}_L = R_L$, the maximum power is delivered to the load when $R_L = R_S$ (as we found in Chap. 3) and, once again, $\mathbf{Z}_L = \mathbf{Z}_S^*$.

When the load impedance is the conjugate of the source impedance, the load and source are said to be *matched*. Under matched conditions, the total average power delivered to $\mathbf{Z}_S$ and $\mathbf{Z}_L$ is equally divided. Denoting $\mathbf{Z}_S = R + jX$ and $\mathbf{Z}_L = \mathbf{Z}_S^* = R - jX$, the average power delivered to the load is, from Eq. (5.123),

$$P_{\mathrm{AV,load}} = \frac{V_S^2}{8R} \qquad \text{matched} \tag{5.128}$$

5.7 FREQUENCY RESPONSE OF CIRCUITS

In previous sections we have examined the response of circuits to sinusoids that have a fixed frequency. We now examine the response of a circuit to a sinusoidal source whose frequency takes on a range of values. The response of a particular current or voltage of the circuit to this variable-frequency source is referred to as the *frequency response* of the circuit. Usually when this frequency response is computed, the amplitude of the sinusoidal source is held constant at a value of unity while the frequency is varied.

There are a number of practical reasons why we may be interested in the response of a circuit to a sinusoidal source whose frequency varies over a range of frequencies. Time-domain signals can be thought of as being composed of a large number of sinusoidal frequency components. If we desire the response of the circuit to this time-domain signal and the circuit is linear, we may use superposition to pass each frequency component of this general signal through the circuit and add the responses to these components to give the response to the general waveform. This gives insight into how the circuit affects the various frequency components of the signal. We use this concept in this chapter when we examine the design of simple filters which extract certain desired frequency components of the signal and reject the other components.

To illustrate this important concept, consider the periodic "square wave" pulse train shown in Fig. 5.20a. This signal is typical of clock and data waveforms in digital computers. This is a periodic waveform having period T. It is said to be periodic since points on the waveform are identical multiples of T earlier or later: $x(t \pm nT) = x(t)$ for $n = 1, 2, 3, \dots$. Any periodic waveform can be written as the sum of sinusoidal components whose frequencies are multiples of the reciprocal of the period, $f_0 = 1/T$, using the Fourier series (see App. B). For example, the Fourier series of the square wave in Fig. 5.20a is

$$x(t) = \frac{A}{2} + \frac{2A}{\pi} \sin \omega_0 t + \frac{2A}{3\pi} \sin 3\omega_0 t + \frac{2A}{5\pi} \sin 5\omega_0 t + \cdots \tag{5.129}$$

The leading term $A/2$ is the average or dc value of the waveform. The next term is a sinusoid at the fundamental frequency of the waveform, $\omega_0 = 2\pi f_0 = 2\pi/T$. The second sinusoid has a frequency of $3\omega_0$ which is 3 times the fundamental and is referred to as the *third harmonic*. The second harmonic, $2\omega_0$, is not present for this waveform. In order that (5.129) exactly represents the waveform, we must use an infinite number of harmonics. In practice, use of a finite number gives a reasonable approximation. This is illustrated in Fig. 5.20b for a square wave having $A = 1$ and $T = 1$ where we have shown the first seven harmonics and their sum. The result

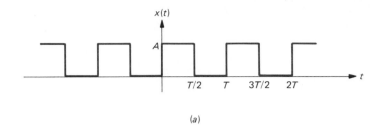

(a)

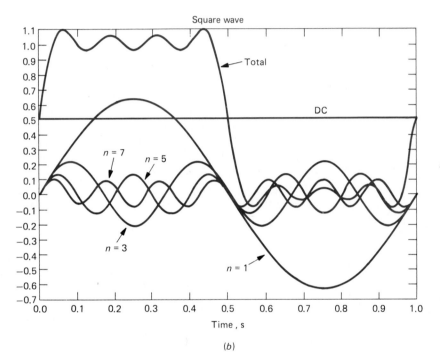

FIGURE 5.20
Contributions of
Fourier series
components to
total waveform for
square wave.

(b)

using only the dc term and the first seven harmonics gives a reasonable approximation to the true waveform.

In general, we may represent any periodic waveform as the infinite sum of sinusoids with the Fourier series as

$$x(t) = X_0 + X_1 \sin(\omega_0 t + \phi_1) + X_2 \sin(2\omega_0 t + \phi_2) + X_3 \sin(3\omega_0 t + \phi_3) + \cdots$$

$$(5.130)$$

If the waveform is not periodic, a similar result applies with the Fourier transform, but the individual frequency components are not separated by multiples of the fundamental frequency but form a continuous spectrum. These important observations suggest that to find the response of a circuit to some general waveform, we may alternatively determine that response as the superposition of the responses to

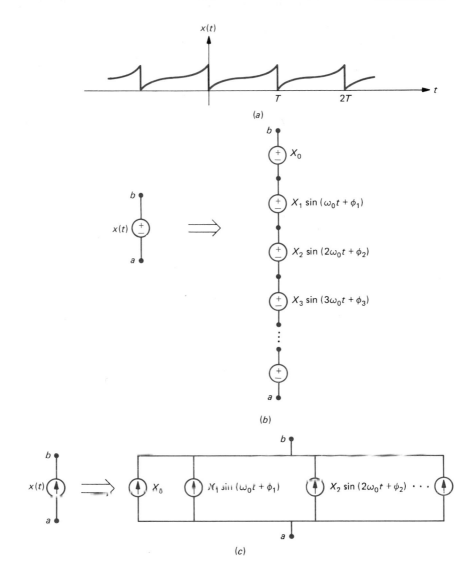

FIGURE 5.21

the individual sinusoidal components of that waveform. For example, consider the general periodic waveform shown in Fig. 5.21a. This waveform may be represented alternatively as in (5.130). If the waveform is that of an ideal voltage source, (5.130) suggests that we may replace it with the series combination of sinusoidal voltage sources, as illustrated in Fig. 5.21b. If the waveform is that of an ideal current source, (5.130) suggests that we may replace it with the parallel combination of sinusoidal current sources, as illustrated in Fig. 5.21c. Superposition of these components in order to determine the response to $x(t)$ is then clearly evident. This is a very powerful principle and is called upon at various places throughout this text. The utility of the result relies on the linearity of the circuit; otherwise superposition does not apply.

5.7.1 Transfer Functions

These ideas show that we need a method of determining the response of a circuit current or voltage to a source whose frequency is variable. A simple way of determining this frequency response is with the use and idea of a transfer function. A *transfer function* is the ratio of the magnitude and phase of a circuit variable to the magnitude and phase of the sinusoidal source that caused it. This is generally denoted with the symbol $H(j\omega)$. For example, consider the simple circuit shown in Fig. 5.22a where the voltage source frequency is variable. The phasor circuit is shown in Fig. 5.22b. To simplify the notation, we have replaced $j\omega$ with the symbol p. Thus the impedance of an inductor is pL, and that of a capacitor is $1/(pC)$. Once the transfer function $H(p)$ is derived, we simply replace the symbol p with $j\omega$ to give $H(j\omega)$. This use of the alternative symbol p to denote $j\omega$ is simply a convenient notational tool and serves no other functional purpose. From the phasor circuit of Fig. 5.22b we obtain the transfer function as the ratio of the desired response variable $\mathbf{V}_{\text{out}}$ and the source $\mathbf{V}_S$ as

$$H(p) = \frac{\mathbf{V}_{\text{out}}}{\mathbf{V}_S}$$

$$= \frac{1/(pC)}{R + 1/(pC)}$$

$$= \frac{1}{1 + pRC} \tag{5.131}$$

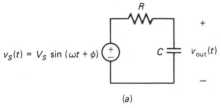

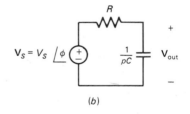

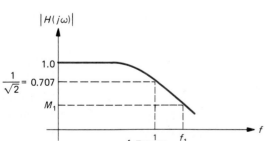

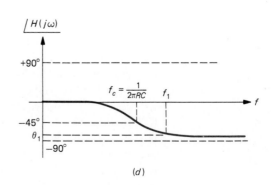

FIGURE 5.22

The transfer function for sinusoidal excitation is obtained by substituting $p = j\omega$ to give

$$H(j\omega) = \frac{1}{1 + j\omega RC} \tag{5.132}$$

The magnitude and phase become

$$|H(j\omega)| = \frac{1}{\sqrt{1 + (\omega RC)^2}} \tag{5.133a}$$

$$\underline{/H(j\omega)} = -\tan^{-1}\omega RC \tag{5.133b}$$

The magnitude of $H(j\omega)$ as ω is varied from $\omega = 0$ (dc) to $\omega \to \infty$ is plotted in Fig. 5.22c, and the angle is plotted in Fig. 5.22d. Note that we have plotted these versus f rather than radian frequency ω, where $\omega = 2\pi f$. This is more conventional although use of ω would have been acceptable. To produce these plots, we should first plot the value at dc and the asymptote as $\omega \to \infty$. Then we should determine other values of ω which yield significant values of $H(j\omega)$. For example, for $\omega RC = 1$ or $f = 1/(2\pi RC)$, the denominator of $H(j\omega)$ is $1 + j1$:

$$H\left(j\omega = j\frac{1}{RC}\right) = \frac{1}{1 + j1} \tag{5.134}$$

$$= \frac{1}{\sqrt{2}\underline{/45°}}$$

$$= 0.707\underline{/-45°}$$

Once the $\omega = 0$ and $\omega \to \infty$ points are plotted (as are any critical frequencies between those frequencies), it is a simple matter to sketch the complete frequency response for this transfer function.

Given the frequency-response plots in Fig. 5.22c and d, we can determine the response to some general sinusoidal input, say $v_S(t) = V_S \sin(\omega_1 t + \phi)$. The result is

$$\begin{aligned}
\mathbf{V}_{\text{out}} &= H(j\omega) \times \mathbf{V}_S \\
&= |H(j\omega_1)|\underline{/H(j\omega_1)} \times V_S\underline{/\phi} \\
&= M_1\underline{/\theta_1} \times V_S\underline{/\phi} \\
&= M_1 V_S\underline{/\theta_1 + \phi}
\end{aligned} \tag{5.135}$$

where M_1 and θ_1 are determined from the frequency-response plots in Fig. 5.22c and d at $\omega_1 = 2\pi f_1$. Thus the time-domain output is

$$v_{\text{out}}(t) = M_1 V_S \sin(\omega_1 t + \theta_1 + \phi) \tag{5.136}$$

An additional and more important use of the frequency-response plots was discussed earlier. They provide an easy and rapid visual depiction of how this circuit would process a band of frequencies. For example, the magnitude plot in Fig. 5.22c shows that for a signal containing a range of frequency components (such as voice, music, digital data, noise, etc.), most components of that signal at frequencies above $f_c = 1/(2\pi RC)$ would not be passed to the output (we say that these would be filtered out) and most of those below f_c would be passed virtually unchanged. Electric filters, discussed next, utilize this important concept.

5.7.2 Filters

Electric filters are circuits that selectively pass certain frequency components of a signal (ideal source) and reject other frequency components of that signal. These are useful in a number of instances. For example, radio stations broadcast on a carrier frequency. The information (voice or music) is superimposed on a narrow frequency band about that carrier frequency. When we tune the radio, in effect we are varying the tuned frequency of a bandpass filter such that the information around the station's carrier frequency is processed and passed to the speaker and all the other radio station transmissions that are present at the antenna but are at different carrier frequencies are "filtered out" by the filter. Filters have many other useful purposes. Suppose a narrow band of frequencies such as the 60-Hz power frequency is causing interference with a desired signal whose frequency components are at a higher frequency range. A high-pass filter can be used to reject or filter out the 60-Hz signal but pass the desired higher-frequency signal. The remainder of this section is devoted to examining some simple circuits that implement the various types of filters.

Suppose that an ideal voltage source $x(t)$ with source resistance R_S is connected to a resistive load R_L as shown in Fig. 5.23a. Suppose that we wish to eliminate certain frequency components of $x(t)$ by inserting a filter between the source and load, as shown in Fig. 5.23b. Let us now investigate some rather simple filters for doing this.

The first filter is known as a low-pass filter which is to pass only certain low frequencies and reject higher ones. What we need, then, is a device which provides a direct connection between the source and load at low frequencies and provides isolation between the source and load at higher frequencies. A possibility consisting simply of a series inductor is shown in Fig. 5.24a. We have replaced $x(t)$ with a general sinusoid $V\sin(\omega t + \phi)$ representing one of its frequency components. Note that at direct current ($\omega = 0$) the inductor is a short circuit ($Z_L = j\omega L$), but that as the frequency ω is increased, its impedance increases, thus isolating the source and load at higher frequencies. The phasor circuit is shown in Fig. 5.24b, and the phasor load current is

$$\mathbf{I}_L = \frac{V\underline{/\phi}}{R_S + R_L + j\omega L} \tag{5.137}$$

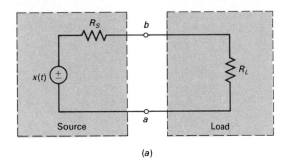

(a)

FIGURE 5.23
Illustration of the
use of a filter to
eliminate certain
frequency
components of a
signal.

(b)

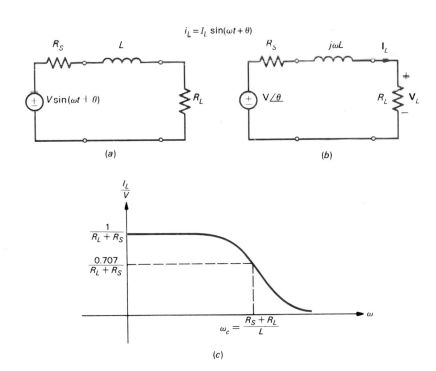

(a)

(b)

(c)

FIGURE 5.24
A low-pass filter.

The magnitude is

$$I_L = \frac{V}{\sqrt{(R_S + R_L)^2 + (\omega L)^2}} \tag{5.138}$$

which is plotted in Fig. 5.24c in terms of the ratio

$$\begin{aligned}
\frac{I_L}{V} &= \frac{1}{\sqrt{(R_S + R_L)^2 + (\omega L)^2}} \\
&= \frac{1}{R_S + R_L} \frac{1}{\sqrt{1 + [\omega L/(R_S + R_L)]^2}}
\end{aligned} \tag{5.139}$$

At $\omega = 0$ (dc) this ratio is $1/(R_S + R_L)$, as is clear from the circuit diagram. As the frequency increases, this ratio eventually decreases asymptotically to zero. At a frequency ω_c such that

$$\omega_c = \frac{R_S + R_L}{L} \tag{5.140}$$

Eq. (5.139) has been reduced by a factor of $1/\sqrt{2}$ of its value at $\omega = 0$ (dc). The average power delivered to R_L is

$$P_{AV} = \tfrac{1}{2} I_L^2 R_L \tag{5.141}$$

and at ω_c the average power delivered to R_L is one-half of its value at dc. Thus, ω_c is referred to as the *half-power point* of the low-pass filter.

Now let us consider the design of a high-pass filter, one which passes high-frequency components of the signal and rejects low-frequency ones. The obvious choice is to substitute a capacitor for the inductor of the low-pass filter, since the impedance of a capacitor $Z_C = 1/(j\omega C)$ is infinite at direct current but is reduced as the frequency is increased. The circuit and frequency response are shown in Fig. 5.25. The ratio of load current to source voltage is

$$\frac{I_L}{V} = \frac{1}{R_S + R_L} \frac{1}{\sqrt{1 + \left[\dfrac{1}{\omega C(R_S + R_L)}\right]^2}} \tag{5.142}$$

At a frequency

$$\omega_c = \frac{1}{C(R_S + R_L)} \tag{5.143}$$

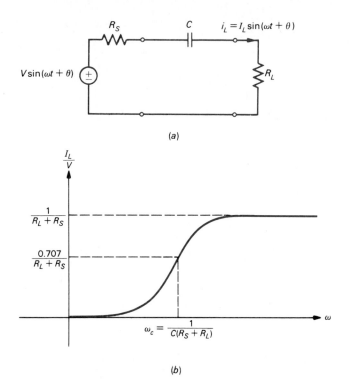

FIGURE 5.25
A high-pass filter.

the ratio in Eq. (5.142) has been reduced from its high-frequency value by $1/\sqrt{2}$, and the average power delivered to R_L has been reduced by $\frac{1}{2}$.

The low-pass (high-pass) filter passes low frequencies (high frequencies) and rejects high frequencies (low frequencies). Filters can also be designed to pass or reject certain bands of frequencies. These filters make use of more than one energy storage element and utilize the phenomenon of resonance. A bandpass filter is shown in Fig. 5.26a. At dc the capacitor presents an infinite impedance, and at $\omega = \infty$ the inductor presents an infinite impedance. At intermediate frequencies, the impedance of the series LC combination

$$\mathbf{Z} = j\omega L + \frac{1}{j\omega C}$$

$$= \frac{1 - \omega^2 LC}{j\omega C} \tag{5.144}$$

is less than infinite. At a frequency of

$$\omega_0 = \frac{1}{\sqrt{LC}} \tag{5.145}$$

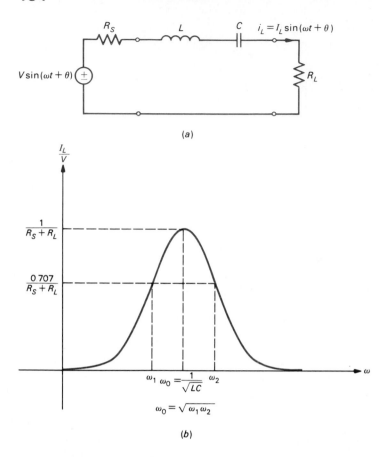

FIGURE 5.26
A bandpass filter.

the impedance is zero and the source and load are directly connected. The ratio of the load current and source voltage can easily be obtained from the phasor circuit and becomes

$$
\begin{aligned}
\frac{I_L}{V} &= \frac{1}{\sqrt{(R_S + R_L)^2 + [\omega L - 1/(\omega C)]^2}} \\
&= \frac{1}{R_S + R_L} \frac{1}{\sqrt{1 + \left[\dfrac{1 - \omega^2 LC}{\omega C(R_S + R_L)} \right]^2}}
\end{aligned}
\tag{5.146}
$$

At frequency

$$
\omega_0 = \frac{1}{\sqrt{LC}}
\tag{5.147}
$$

Eq. (5.146) becomes

$$\frac{I_L}{V} = \frac{1}{R_S + R_L} \qquad \omega = \omega_0 \tag{5.148}$$

which is the largest value this ratio can attain at any frequency. As the frequency is lowered from ω_0, the ratio eventually decreases to zero at dc, as is clear from the circuit, since at dc the capacitor presents an open circuit. As the frequency is raised above ω_0, the ratio also eventually decreases to zero at $\omega = \infty$, as is clear from the circuit, since at $\omega = \infty$ the inductor presents an open circuit. At $\omega = \omega_0$, the inductor and capacitor are said to resonate such that their series impedance $\omega_0 L - 1/(\omega_0 C)$ is zero.

The half-power points are frequencies such that the ratio I_L/V is reduced to $1/\sqrt{2}$ of its value at ω_0. From Eq. (5.146) this occurs when

$$\frac{1 - \omega^2 LC}{\omega C(R_S + R_L)} = \pm 1 \tag{5.149}$$

or

$$\omega^2 \pm \frac{R_S + R_L}{L}\omega - \frac{1}{LC} = 1 \tag{5.150}$$

There are two values of positive frequency which satisfy this equation:

$$\omega_u = \frac{R_S + R_L}{2L} + \sqrt{\left(\frac{R_S + R_L}{2L}\right)^2 + \frac{1}{LC}} \tag{5.151a}$$

$$\omega_l = -\frac{R_S + R_L}{2L} + \sqrt{\left(\frac{R_S + R_L}{2L}\right)^2 + \frac{1}{LC}} \tag{5.151b}$$

There are two other solutions of (5.150), but these yield negative frequencies and are therefore of no utility. The frequencies ω_u and ω_l are the points where the response is down from its maximum at ω_0 by $1/\sqrt{2}$. The power delivered to the load is therefore down by 3 decibels (dB) at these frequencies from that delivered at ω_0. Thus ω_u and ω_l are referred to as the *upper* and *lower half-power* or 3-dB *frequencies*.

The bandwidth B is the frequency separation between the half-power points

$$B = \omega_u - \omega_l$$

$$= \frac{R_S + R_L}{L} \tag{5.152}$$

and represents the effective passband of the filter. It can readily be shown that ω_0, ω_l, and ω_u are related such that

$$\omega_0 = \sqrt{\omega_l \omega_u} \tag{5.153}$$

[Substitute Eqs. (5.151) into Eq. (5.153).]

The ratio of the center frequency ω_0 to the bandwidth B of the filter is a measure of the frequency discrimination of the filter; the narrower the bandwidth, the greater the ability of the filter to isolate or separate out a single frequency component of the signal. This ratio is defined as the Q of the filter:

$$Q = \frac{\omega_0}{B}$$

$$= \frac{\omega_0 L}{R_S + R_L} \tag{5.154}$$

The smaller the source and load resistances, the larger the Q and the resulting sharpness of the filter.

A band-reject filter can be similarly constructed to reject a narrow band of frequencies and pass the remaining ones. This type of filter is useful in preventing unwanted frequency components from interfering with transmissions in radio circuits. Figure 5.27 shows a simple example and its frequency response. At dc the inductor acts as a short circuit, and at $\omega = \infty$ the capacitor acts as a short circuit. However, at intermediate frequencies the impedance of the parallel LC combination

$$\mathbf{Z} = \frac{1}{j\omega C + 1/(j\omega L)}$$

$$= \frac{j\omega L}{1 - \omega^2 LC} \tag{5.155}$$

is larger than zero. In particular, at

$$\omega_0 = \frac{1}{\sqrt{LC}} \tag{5.156}$$

the impedance is infinite, as seen from Eq. (5.155), and the LC combination isolates the source from the load. From the phasor circuit we can obtain

$$\frac{I_L}{V} = \frac{1}{\sqrt{(R_S + R_L)^2 + \left(\dfrac{\omega L}{1 - \omega^2 LC}\right)^2}}$$

$$= \frac{1}{R_S + R_L} \frac{1}{\sqrt{1 + \left[\dfrac{\omega L/(R_S + R_L)}{1 - \omega^2 LC}\right]^2}} \tag{5.157}$$

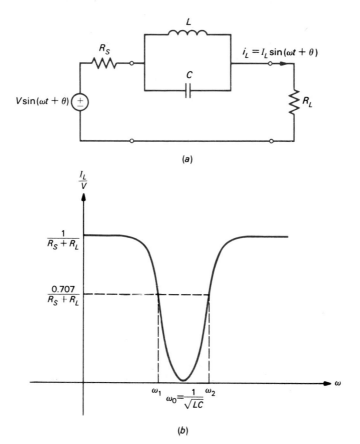

FIGURE 5.27
A band-reject filter.

At a frequency such that

$$\frac{\omega L/(R_S + R_L)}{1 - \omega^2 LC} = \pm 1 \tag{5.158}$$

the frequency response is reduced by $1/\sqrt{2}$ from its maximum value, and the average power delivered to the load is reduced by $\frac{1}{2}$ from its maximum value. Solving Eq. (5.158), we obtain

$$\omega^2 \pm \frac{1}{C(R_S + R_L)} \omega - \frac{1}{LC} = 0 \tag{5.159}$$

Solving Eq. (5.159), we again obtain two positive frequencies

$$\omega_l = -\alpha + \sqrt{\alpha^2 + \omega_0^2} \tag{5.160a}$$

$$\omega_u = \alpha + \sqrt{\alpha^2 + \omega_0^2} \tag{5.160b}$$

where

$$\alpha = \frac{1}{2C(R_S + R_L)} \qquad (5.161)$$

The bandwidth becomes

$$B = \omega_u - \omega_l$$

$$= 2\alpha$$

$$= \frac{1}{C(R_S + R_L)} \qquad (5.162)$$

and the Q of the filter becomes

$$Q = \frac{\omega_0}{B}$$

$$= \omega_0 C(R_S + R_L) \qquad (5.163)$$

Obviously, the response characteristics and the other properties of these filters depend on the source and load. Therefore, we cannot speak of the properties of a filter without considering the particular source and load with which it will be used. The simple filters considered in this section show via their frequency-response characteristics how a particular frequency component of the source waveform will be affected. Other, more elaborate filters have corresponding properties but are not as easily obtained.

Example 5.9   A square-wave voltage source having an amplitude of 5 V, a frequency of 1 kilohertz (kHz), a pulse width of 0.5 ms, and internal source resistance of 50 Ω is applied to a load which is represented at its input terminals as a 100-Ω resistor; this is shown in Fig. 5.28a. The problem is that the abrupt rise and fall of the leading and trailing edges is troublesome for the receiver. We would like to smooth these pulses somewhat to provide a less abrupt transition. The abruptness of the transition is primarily due to the high-frequency components of the wave; to slow these transitions, we insert a filter between the source and load to remove some of these high-frequency components. Suppose that we arbitrarily attempt to reduce all components above 5 kHz.

Design a low-pass filter to accomplish this. Determine the resulting voltage $v_L(t)$ across the load, and show its amplitude spectrum.

Solution   The Fourier series of the square wave is (see App. B)

$$v_S(t) = 2.5 + \frac{10}{\pi} \sin \omega_0 t + \frac{10}{3\pi} \sin 3\omega_0 t + \frac{10}{5\pi} \sin 5\omega_0 t + \cdots$$

where $\omega_0 = 2\pi f_0$ and $f_0 = 1/T = 1$ kHz. The lowest frequency component, $f_0 = 1$ kHz, is called the *fundamental frequency* of the wave; the next frequency, $3f_0 = 3$ kHz, is called the

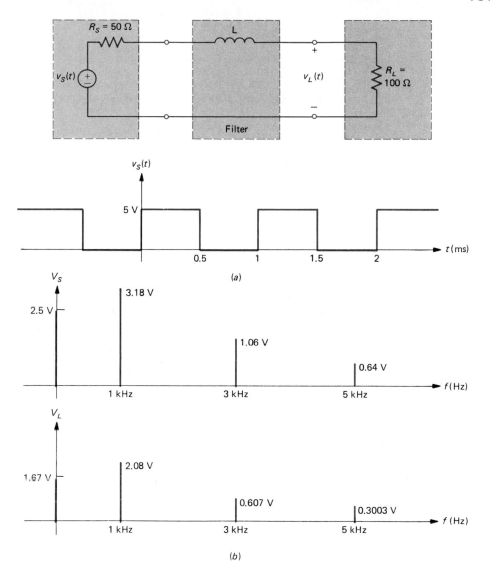

FIGURE 5.28
Example 5.9:
illustration of the
use of a filter to
reduce the
amplitudes of a
signal's frequency
components.

third harmonic of the wave; etc. The term 2.5 is the average value (dc component) of the waveform, which is obvious from the sketch of $v_S(t)$. Suppose we select the half-power point of the filter to be 5 kHz. Thus from Eq. (5.140)

$$f_C = 5 \text{ kHz}$$

$$= \frac{1}{2\pi} \frac{R_S + R_L}{L}$$

$$= \frac{1}{2\pi} \frac{150}{L}$$

Solving for L gives

$$L = \frac{1}{2\pi} \frac{150}{5 \times 10^3}$$

$$= 4.77 \text{ mH}$$

Replacing the circuit with its phasor equivalent gives the load voltage as

$$\mathbf{V}_L = \frac{R_L}{R_S + R_L + j\omega L} \mathbf{V}_S$$

The dc component ($f = 0$) is reduced to

$$\frac{R_L}{R_S + R_L} \times 2.5 = 1.67 \text{ V}$$

The fundamental frequency component ($f = 1$ kHz) of $10/\pi = 3.18$ V becomes

$$\mathbf{V}_{L_1} = \frac{100}{150 + j2\pi \times 10^3 \times 4.77 \times 10^{-13}} \times 3.18\underline{/0^\circ}$$

$$= 2.08\underline{/-11.3^\circ}$$

Similarly, the third harmonic ($f = 3$ kHz) becomes

$$\mathbf{V}_{L_3} = 0.607\underline{/-31^\circ}$$

while the fifth harmonic becomes

$$\mathbf{V}_{L_5} = 0.3003\underline{/-45^\circ}$$

and the seventh harmonic becomes

$$\mathbf{V}_{L_7} = 0.176\underline{/-54.5^\circ}$$

The amplitude spectra of $v_S(t)$ and $v_L(t)$ are sketched in Fig. 5.28b. A sketch of $v_L(t)$ in the time domain can be obtained from the Fourier series of $v_L(t)$ by combining the above phasor voltages:

$$\begin{aligned}
v_L(t) = \ & 1.67 + 2.08 \sin (2\pi \times 10^3 t - 11.3^\circ) \\
& + 0.607 \sin (6\pi \times 10^3 t - 31^\circ) \\
& + 0.3003 \sin (10\pi \times 10^3 t - 45^\circ) \\
& + 0.176 \sin (14\pi \times 10^3 t - 54.5^\circ) \\
& + \cdots
\end{aligned}$$

PROBLEMS

5.1 Convert the following functions to equivalent functions by determining θ.
(a) $3 \sin (2t - 30°) = 3 \cos (2t + \theta)$
(b) $2 \sin (3t + 45°) = 2 \cos (3t + \theta)$
(c) $\cos (t - 135°) = \sin (3t + \theta)$
(d) $2 \cos (2t + 215°) = 2 \sin (2t + \theta)$
(e) $3 \cos (3t - 65°) = 3 \sin (3t + \theta)$
(f) $2 \sin (t - 215°) = 2 \cos (t + \theta)$
(g) $2 \sin (3t - 30°) = 2 \cos (3t + \theta)$
(h) $\cos (2t - 250°) = \sin (2t + \theta)$
Sketch your results to verify that they are correct.

5.2 Convert the following trigonometric expressions to equivalent forms by giving M and θ or A and B as required.
(a) $2 \cos 3t - 3 \sin 3t = M \cos (3t + \theta)$
(b) $\cos 2t + 2 \sin 2t = M \sin (2t + \theta)$
(c) $-\cos 2t + 4 \sin 2t = M \cos (2t + \theta)$
(d) $2 \cos 4t - 6 \sin 4t = M \sin (2t + \theta)$
(e) $3 \cos (t + 75°) = A \cos t + B \sin t$
(f) $-2 \sin (t - 30°) = A \cos t + B \sin t$
(g) $-4 \cos (t + 135°) = A \cos t + B \sin t$
(h) $2 \cos (2t + 30°) - 3 \sin (2t - 75°) = M \cos (2t + \theta)$
(i) $-\sin (t + 135°) + 2 \cos (t - 30°) = M \sin (t + \theta)$
Verify your answer in (i) by choosing $t = 1.3$ rad.

5.3 For the series RLC circuit shown in Fig. P5.3 derive a differential equation relating $i(t)$ to $v_S(t)$. Determine $i(t)$ in the indicated form from that differential equation when
(a) $v_S(t) = 10 \sin 2t$ V, $R = 2 \Omega$, $L = 2$ H, $C = \frac{1}{6}$ F
$i(t) = M \sin (2t + \theta)$
(b) $v_S(t) = 10 \cos (3t + 30°)$, $R = 2 \Omega$, $L = 1$ H, $C = \frac{1}{4}$ F
$i(t) = M \cos (3t + \theta)$

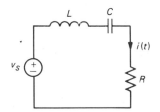

FIGURE P5.3

5.4 For the circuit of Fig. P5.4 derive a differential equation relating $v(t)$ to $i_S(t)$. Determine $v(t)$ in the indicated form from that differential equation when
(a) $i_S(t) = 2 \cos (2t - 60°)$, $R = 2 \Omega$, $L = 1$ H. $C = \frac{1}{2}$ F
$v(t) = M \cos (2t + \theta)$
(b) $i_S(t) = \sin (3t + 150°)$, $R = 2 \Omega$, $L = 1$ H, $C = \frac{1}{3}$ F
$v(t) = M \sin (3t + \theta)$

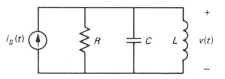

FIGURE P5.4

5.5 For the circuit of Fig. P5.5 derive a differential equation relating $i(t)$ to $i_S(t)$. Determine $i(t)$ as $i(t) = M \sin (3t + \theta)$ from that differential equation when $R = 3\Omega$, $C = \frac{1}{6}$ F, and $i_S(t) = 2 \sin (3t - 150°)$.

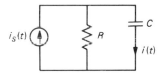

FIGURE P5.5

5.6 For the following complex numbers **A** and **B** calculate $\mathbf{A} + \mathbf{B}$, $\mathbf{A}/\mathbf{B}$, and $\mathbf{AB}$. Express your answer both in rectangular form and in polar form.

	A	B		A	B
(a)	$1 + j2$	$-1 - j3$	(h)	$-3 + j10$	$8 - j2$
(b)	$-1 + j1$	$2 - j1$	(i)	$6 + j8$	$-3 + j9$
(c)	$5 - j4$	$10 + j8$	(j)	$-2 - j10$	$-12 + j20$
(d)	$-3 + j7$	$8 - j14$	(k)	$6 - j7$	$30 - j20$
(e)	$2 - j6$	$-2 - j10$	(l)	$-8 - j2$	$10 + j15$
(f)	$-10 - j1$	$-4 + j2$	(m)	$-8 + j10$	$-5 - j10$
(g)	$5 - j8$	$-2 - j6$			

5.7 For complex numbers $\mathbf{A} = 1 - j3$, $\mathbf{B} = 2\underline{/-30°}$, $\mathbf{C} = 2 + j1$, and $\mathbf{D} = 3\underline{/150°}$ evaluate the following expressions. Place your answer in both rectangular and polar form
(*a*) $\mathbf{A} + \mathbf{B}$; (*b*) $\mathbf{B} + \mathbf{D}$; (*c*) $\mathbf{AB}$; (*d*) $\mathbf{AC}$;
(*e*) $\mathbf{A}/\mathbf{B}$; (*f*) $\mathbf{A}/\mathbf{C}$; (*g*) $1/\mathbf{A}$; (*h*) $\mathbf{AC}^*$;
(*i*) $\mathbf{B} + \dfrac{\mathbf{AD}}{\mathbf{C}}$; (*j*) $\mathbf{BC} - \dfrac{\mathbf{AD}}{\mathbf{C}}$.

5.8 For the complex numbers $\mathbf{A} = 20\underline{/-135°}$, $\mathbf{B} = 10 - j8$, $\mathbf{C} = 2 + j6$, and $\mathbf{D} = 15\underline{/60°}$ evaluate

(*a*) $\mathbf{F} = \dfrac{\mathbf{A} + \mathbf{B}}{\mathbf{CD}}$; (*b*) $\mathbf{F} = \dfrac{\mathbf{D}/\mathbf{B} + \mathbf{A}}{\mathbf{C}}$; (*c*) $\mathbf{G} = \dfrac{\mathbf{B}}{\mathbf{C} + \mathbf{D}} + \mathbf{A}$.

5.9 Solve Prob. 5.3 by replacing $v_S(t)$ with the exponential source.
5.10 Solve Prob. 5.4 by replacing $i_S(t)$ with the exponential source.
5.11 Solve Prob. 5.5 by replacing $i_S(t)$ with the exponential source.
5.12 Solve Prob. 5.3 by using the phasor circuit.
5.13 Solve Prob. 5.4 by using the phasor circuit.
5.14 Solve Prob. 5.5 by using the phasor circuit.
5.15 Determine $i(t)$ in the circuit of Fig. P5.15.

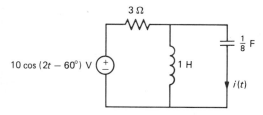

FIGURE P5.15

5.16 Determine $i(t)$ in the circuit of Fig.P5.16, using three methods: (*a*) voltage and current division; (*b*) reduce the circuit at terminals *a* and *b* to a Thévenin equivalent; (*c*) reduce the circuit at terminals *a* and *b* to a Norton equivalent.

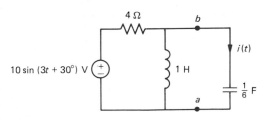

FIGURE P5.16

5.17 Determine $v(t)$ in the circuit of Fig. P5.17.

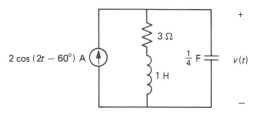

FIGURE P5.17

5.18 Determine $i(t)$ in the circuit of Fig. P5.18 using three methods: (*a*) voltage and current division; (*b*) reduce the circuit attached to terminals *a* and *b* to a Thévenin equivalent; (*c*) reduce the circuit attached to terminals *a* and *b* to a Norton equivalent.

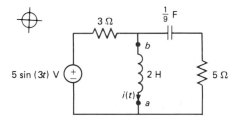

FIGURE P5.18

5.19 For the circuit of Fig. P5.19, determine $v(t)$.

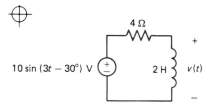

FIGURE P5.19

5.20 For the circuit of Fig. P5.20, determine $i(t)$.

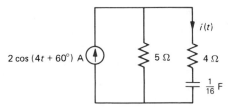

FIGURE P5.20

5.21   For the circuit of Fig. P5.21, determine $v(t)$.

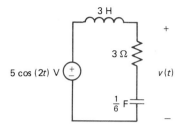

FIGURE P5.21

5.22   For the circuit of Fig. P5.22, determine $v(t)$.

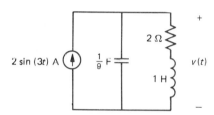

FIGURE P5.22

5.23   For the circuit of Fig. P5.23, determine $v(t)$

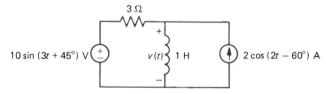

FIGURE P5.23

5.24   For the circuit of Fig. P5.24, determine $i(t)$.   **5.25**   For the circuit of Fig. P5.25, determine $v(t)$.

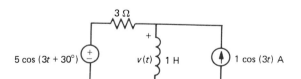

FIGURE P5.24   **FIGURE P5.25**

5.26   For the circuit of Fig. P5.26, determine $i(t)$.

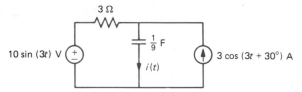

FIGURE P5.26

5.27   For the circuit of Fig. P5.27, determine $i(t)$.

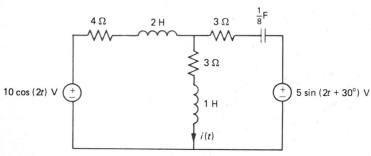

FIGURE P5.27

5.28   For the circuit of Fig. P5.28, determine $i(t)$.

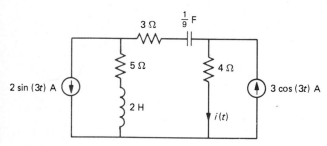

FIGURE P5.28

5.29   For the circuit of Fig. P5.29, determine $i(t)$.

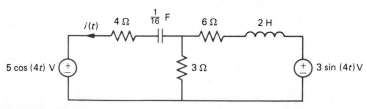

FIGURE P5.29

5.30 For the circuit of Fig. P5.30, determine the average power delivered to each element, and show that conservation of power is achieved.

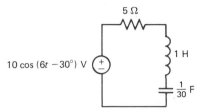

FIGURE P5.30

5.33 For the circuit of Fig. P5.33, determine the average power delivered to each element, and show that conservation of power is achieved.

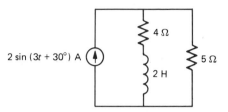

FIGURE P5.33

5.31 For the circuit of Fig. P5.31, determine the average power delivered to each element, and show that conservation of power is achieved.

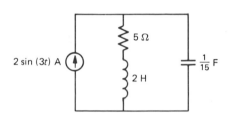

FIGURE P5.31

5.32 For the circuit of Fig. P5.32, determine the average power delivered to each element, and show that conservation of power is achieved.

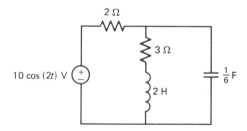

FIGURE P5.32

5.34   For the circuit of Fig. P5.34, determine
R and C such that the maximum average power is
delivered to the load.

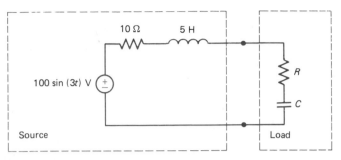

FIGURE P5.34

5.35   For the circuit of Fig. P5.35, determine
C such that the power factor of the impedance seen by
the source is unity.

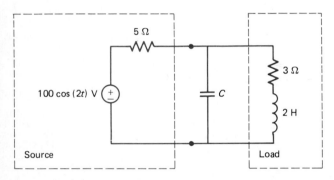

FIGURE P5.35

5.36   For the circuit of Fig. P5.36, determine
L such that the power factor of the impedance seen by
the source is unity.

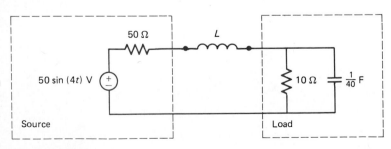

FIGURE P5.36

5.37 For the circuit of Fig. P5.37, sketch the frequency response (magnitude and phase) of $v(t)$.

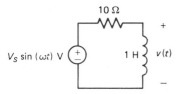

FIGURE P5.37

5.38 For the circuit of Fig. P5.38, sketch the frequency response (magnitude and phase) of $i(t)$.

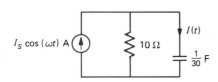

FIGURE P5.38

5.39 For the circuit of Fig. P5.39, sketch the frequency response (magnitude and phase) of $v(t)$.

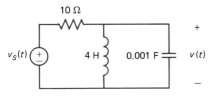

FIGURE P5.39

5.40 For the circuit of Fig. P5.40, sketch the frequency response (magnitude and phase) of $v(t)$.

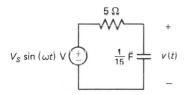

FIGURE P5.40

5.41 Design the low-pass filter in Fig. P5.41 to have a half-power frequency of 10 kHz.

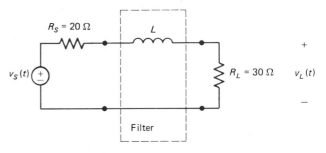

FIGURE P5.41

5.42 Design the high-pass filter of Fig. P5.42 to have a half-power frequency of 1 megahertz (MHz).

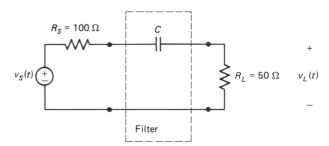

FIGURE P5.42

5.43   Design the bandpass filter of Fig. P5.43 to have a center frequency of 1 MHz and a bandwidth of 10 kHz. What is the Q of this filter?

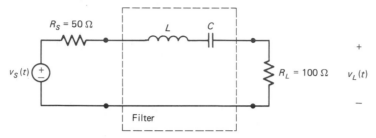

FIGURE P5.43

5.44   Design the band-reject filter of Fig. P5.44 to have a center frequency of 100 kHz and a bandwidth of 5 kHz. What is the Q of this filter?

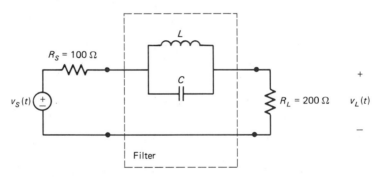

FIGURE P5.44

5.45   The circuit of Fig. P5.45 is another version of a low-pass filter. Confirm this by investigating the behavior at direct current and at infinite frequency. Derive the transfer function, and design a filter having a half-power frequency of 1 MHz.

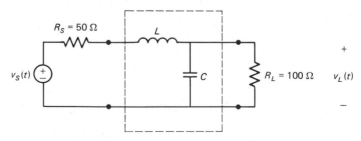

FIGURE P5.45

5.46   The circuit of Fig. P5.46 is another
version of a high-pass filter. Confirm this by
investigating the behavior at direct current and at
infinite frequency. Derive the transfer function, and
design a filter having a half-power frequency of 500 kHz.

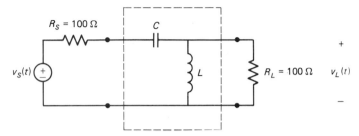

FIGURE P5.46

5.47   In order to investigate the effects of a filter
on the waveform of a signal, consider the problem
shown in Fig. 5.28 which was solved in Example 5.9.
The time-domain load voltage was obtained as

$$v_L(t) = 1.67 + 2.08 \sin (2\pi \times 10^3 t - 11.3°)$$
$$+ 0.607 \sin (6\pi \times 10^3 t - 31°)$$
$$+ 0.3003 \sin (10\pi \times 10^3 t - 45°)$$
$$+ 0.176 \sin (14\pi \times 10^3 t - 54.5°)$$
$$+ \cdots$$

Plot this waveform using only the first seven
harmonics and compare to $v_L(t)$ with the filter (the
inductor) removed (replaced with a short circuit).

Transients

In previous chapters we investigated the primary circuit elements—the resistor R, the capacitor C, the inductor L, and the ideal voltage and current sources. We determined the circuit element voltages and currents when the circuits contained dc (constant-value) or ac (sinusoidal waveform) ideal voltage and circuit sources. These solutions are said to be the steady-state values of the element voltages and currents.

In this chapter we investigate the transient values of the element voltages and currents. In circuits containing energy storage elements, the inductor current $i_L(t)$ and capacitor voltage $v_C(t)$ cannot change value instantaneously; otherwise, their stored energies $w_L(t) = \frac{1}{2}Li_L^2(t)$ and $w_C = \frac{1}{2}Cv_C^2(t)$ would change instantaneously. Thus, there is a certain amount of "inertia" associated with these elements. Suppose, for example, that we find the solutions for the element voltages and currents for a particular circuit and then, at some later time, open or close a switch located somewhere in the circuit; activating the switch *changes the circuit*, and now we must solve this new circuit. The element voltages and currents will have changed because we have a new circuit caused by the switching operation. The object of this chapter is to determine the behavior of the element voltages and currents in the intermediate, or transient, time interval while they are adjusting to their new values.

This idea is illustrated in Fig. 6.1. The original circuit consists of A and B connected together. Suppose we solve for the element voltages and currents in the B part of the circuit. At $t = 0$ a switch opens, disconnecting A and B. Now the element currents and voltages in the B part which we solved for previously must have new values since the A part has been removed by the switch. There must be a time interval when the element currents and voltages are adjusting to their new values. We call this the *transient period*, and the solution is called the *transient* (exists only momentarily) *solution*.

The final solution after this transient period is called the *steady-state solution* for the new circuit. The portion of this steady-state solution due to dc sources in B is found by replacing all inductors with short circuits and all capacitors with open

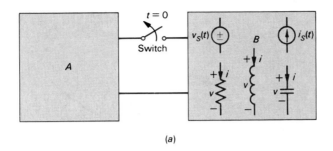

(a)

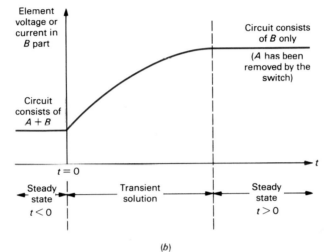

(b)

FIGURE 6.1
Illustration of the
transient period of
the circuit solution
following a
switching
operation.

circuits. This is due to the fact that all currents and voltages in the circuit due to a constant-value source must also be constant in value. Since

$$v_L = L \frac{di_L}{dt} \tag{6.1}$$

if the inductor currents i_L are constant, then $v_L = 0$, which is a short circuit. Similarly, if the capacitor voltages v_C are constant, then

$$i_C = C \frac{dv_C}{dt} \tag{6.2}$$

and $i_C = 0$, which is equivalent to an open circuit. If the source is sinusoidal (ac), we may use the phasor principles developed in the previous chapter to find the portion of the steady-state solution due to this type of source.

These steady-state computational methods also apply to the circuit $A + B$ before the switch is closed at $t = 0$. In other words, we assume that for $t < 0$, A and B are connected together and no switching operation or other disturbance has occurred in either part *A* or part *B* for a very long time prior to the switching operation at $t = 0$. Thus, all currents and voltages in $A + B$ have reached steady state at $t = 0^-$; that is, they have reached steady state immediately prior to the switching operation.

6.1 THE *RL* CIRCUIT

Consider the circuit shown in Fig. 6.2*a* consisting of a dc voltage source, a resistor, and an inductor. Before $t = 0$, the switch is open. The value of the inductor current at $t = 0^-$ (immediately prior to the closing of the switch) is obviously zero since no sources are attached to it for $t < 0$, as shown in Fig. 6.2*b*. At $t = 0$ the switch closes, thus attaching the voltage source and resistor to the inductor, as shown in Fig. 6.2*c*. The steady-state (ss) value of the inductor current for $t > 0$ is found in Fig. 6.2*d* by replacing the inductor with a short circuit since the source is dc. Therefore,

$$i_{L,ss} = \frac{V_S}{R}$$

$$= 5\,\text{A} \tag{6.3}$$

We found in Chap. 4 that this inductor current cannot change instantaneously from its value at $t = 0^-$, 0 A, to its steady-state value for $t > 0$ of 5 A, or else the inductor voltage given by Eq. (6.1) would become infinite. To examine this transitional behavior of the inductor current, let us obtain an equation relating the inductor current i_L and the voltage source V_S. Applying KVL around the loop in the $t > 0$ circuit of Fig. 6.2*c*, we obtain

$$v_L(t) + v_R(t) = V_S \tag{6.4a}$$

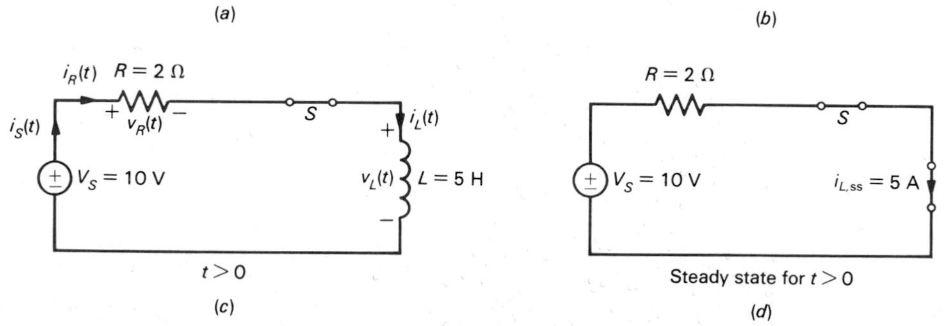

FIGURE 6.2 The *RL* circuit and its complete solution.

or

$$L \frac{di_L}{dt} + Ri_L = V_S \tag{6.4b}$$

Rewriting Eqs. (6.4) by dividing both sides by L, we have

$$\frac{di_L}{dt} + \frac{R}{L} i_L = \frac{V_S}{L} \tag{6.5}$$

This type of equation is a differential equation (DE) since it involves a derivative of the unknown variable i_L. It is the simplest possible type of differential equation and has a very simple solution; other types of differential equations are more difficult to solve. It is said to be an ordinary DE since only ordinary derivatives are involved; a partial differential equation would involve partial derivatives. Note that since the unknown $i_L(t)$ is a function of only one variable, time t, no partial derivatives are necessary. It is also said to be a first-order differential equation since the highest derivative is first order. The equation is a constant-coefficient one since the coefficients of the unknown 1 and R/L, are independent of t. If the resistor had been a time-varying one such as $R(t) = 2t$, then the equation would have been a nonconstant-coefficient one, which would be very difficult to solve. The final and most important property of the DE is that it is linear. A DE is linear if the unknown i_L and any of its derivatives are not raised to a power other than unity or do not appear as products of each other. For example, terms such as $i_L^2(t)$ and $i_L \, di_L/dt$ would make the equation nonlinear. Having a term such as $d^2i_L(t)/dt^2$ makes the equation second-order (if it is the highest derivative), but a term such as $\lfloor di_L(t)/dt \rfloor^2$ makes the equation nonlinear.

We will encounter this type of differential equation in numerous places, so it is important to outline a simple method of solution. To illustrate this method, let us consider a general form of a linear first-order ordinary constant-coefficient DE:

$$\frac{dx(t)}{dt} + ax(t) = f(t) \tag{6.6}$$

Here $x(t)$ denotes the unknown, $f(t)$ denotes the known right-hand side "forcing function," and a, which is a constant, is the coefficient of $x(t)$. Note in our example in Eq. (6.5) that $x(t) = i_L(t)$, $f(t) = V_S/L$, and $a = R/L$. If the source is not a dc one, $f(t)$ will be a function of time. The solution to Eq. (6.6) is easily obtained if we do something to Eq. (6.6) which may appear to be inconsequential—add a zero to the right-hand side:

$$\frac{dx(t)}{dt} + ax(t) = 0 + f(t) \tag{6.7}$$

The addition of this zero to the right-hand side, along with the important property of linearity, allows us to see that the solution for $x(t)$ will be the sum of two parts.

The transient or natural solution, $x_T(t)$, is the portion of the solution to Eq. (6.7) due to the zero part of the right-hand side:

$$\frac{dx_T(t)}{dt} + ax_T(t) = 0 \tag{6.8}$$

The steady-state or forced solution, $x_{ss}(t)$, is the portion of the solution to Eq. (6.7) due to $f(t)$:

$$\frac{dx_{ss}(t)}{dt} + ax_{ss}(t) = f(t) \tag{6.9}$$

In terms of these two solutions, the solution to Eq. (6.7) is

$$x(t) = x_T(t) + x_{ss}(t) \tag{6.10}$$

This is easy to see by substituting Eq. (6.10) into Eq. (6.7) and factoring as

$$\frac{d}{dt}(x_T + x_{ss}) + a(x_T + x_{ss}) = 0 + f(t) \tag{6.11}$$

or

$$\underbrace{\frac{dx_T}{dt} + ax_T} + \underbrace{\frac{dx_{ss}}{dt} + ax_{ss}} = \underbrace{0} + \underbrace{f(t)} \tag{6.12}$$

The factorization in Eq. (6.12) shows that the result is the sum of two equalities, Eqs. (6.8) and (6.9), which is an equality. Superposition and linearity come into play once again.

First consider the transient solution to Eq. (6.8). To satisfy Eq. (6.8), $x_T(t)$ must be such that, when differentiated with respect to t, the result must be of the same form as $x_T(t)$ if the sum of dx_T/dt and ax_T is to equal zero. A possible form for $x_T(t)$ which achieves this objective is the exponential e^{pt}. If we differentiate e^{pt}, we obtain pe^{pt}, which is of the same form as e^{pt}. Thus, let us try a solution of the form

$$x_T(t) = Ae^{pt} \tag{6.13}$$

where A and p are, as yet, undetermined constants. Substituting Eq. (6.13) into Eq. (6.8), we obtain

$$pAe^{pt} + aAe^{pt} = 0 \tag{6.14}$$

or

$$(p + a)Ae^{pt} = 0 \tag{6.15}$$

Now A cannot be zero, or x_T would be zero for all t: a trivial solution. Also e^{pt} cannot be zero for all t regardless of p. Thus, we conclude that

$$p + a = 0 \tag{6.16}$$

or

$$p = -a \tag{6.17}$$

Thus, the transient solution is

$$x_T(t) = Ae^{-at} \tag{6.18}$$

where A is, as yet, some undetermined constant. We will evaluate A later.

We have already determined how to find the steady-state or forced solution to Eq. (6.9) without, perhaps, knowing that we have done so. The steady-state solution is any function which satisfies Eq. (6.9). Since Eq. (6.9) is simply a mathematical statement of the circuit constraints, we may as well find $x_{ss}(t)$ directly from the circuit. Each ideal source will contribute a distinct portion of $f(t)$. For example, suppose a circuit has N ideal sources. Then $f(t)$ will consist of N pieces with each piece due to only one of the sources:

$$f(t) = f_1(t) + f_2(t) + \cdots + f_N(t) \tag{6.19}$$

Here $f_i(t)$ is due to the ith ideal source with all other sources deactivated. Since the DE in Eq. (6.9) is linear, there will also be N contributions to $x_{ss}(t)$:

$$x_{ss}(t) = x_{ss1}(t) + x_{ss2}(t) + \cdots + x_{ssN}(t) \tag{6.20}$$

where $x_{ssi}(t)$ is due to $f_i(t)$ with the other sources contributing to the other $f(t)$'s deactivated. Thus, we may find the steady-state solution directly from the circuit by superposition. To find the portion of the steady-state solution due to a dc source, we first deactivate all the other sources and replace inductors with short circuits and capacitors with open circuits, and then we solve the resulting circuit for the element voltage or current of interest, $x_{ssi}(t)$. To find the portion of the steady-state solution due to an ac source, we deactivate all the other sources and use the phasor method of the previous chapter to obtain $x_{ssi}(t)$ directly from the circuit. Therefore, we already know how to find the steady-state solution for dc or ac sources.

The solution to Eq. (6.7) is the sum of the transient and steady-state solutions:

$$x(t) = x_T(t) + x_{ss}(t)$$
$$= Ae^{-at} + x_{ss}(t) \tag{6.21}$$

Two points about the total solution in Eq. (6.21) should be made. The constant A is, as yet, unknown. To evaluate it we need some additional information. Suppose we wish to obtain the solution for $x(t)$ for all times greater than some initial time t_0; that is, we wish to find $x(t)$ such that Eq. (6.21) is correct for all $t \geq t_0$. Without any loss of generality, we may select this intial time as $t_0 = 0$. Evaluating Eq. (6.21) immediately after $t = 0$, or $t = 0^+$, we have

$$x(0^+) = A + x_{ss}(0^+) \tag{6.22}$$

so that

$$A = x(0^+) - x_{ss}(0^+) \tag{6.23}$$

The value of $x(t)$ at the initial time, $x(0^+)$, is called the *initial condition* on $x(t)$. We will have found $x_{ss}(t)$ for a particular $f(t)$, so it is simple to obtain the constant $x_{ss}(0^+)$ by substituting $t = 0$ into $x_{ss}(t)$. Substituting Eq. (6.23) into Eq. (6.21), we have the solution for $x(t)$ for all $t \geq 0$ in terms of the initial value of $x(t)$:

$$x(t) = \underbrace{[x(0^+) - x_{ss}(0^+)]e^{-at}}_{\text{transient}} + \underbrace{x_{ss}(t)}_{\substack{\text{steady} \\ \text{state}}} \tag{6.24}$$

As time progresses, e^{-at} decreases (assuming a is positive) so that the transient solution eventually goes to zero. Thus, the name transient solution is quite appropriate: It is only present momentarily. At a time $t = 1/a$, the transient solution has decayed to $1/e$, or 37 percent of its initial value at $t = 0^+$, which is $x(0^+) - x_{ss}(0^+)$. Clearly, the constant a must have the dimension of reciprocal time; i.e., the units of $1/a$ are seconds. Thus, the reciprocal of a is known as the time constant and is denoted by

$$T = \frac{1}{a} \quad \text{s} \tag{6.25}$$

so that Eq. (6.24) may be written as

$$x(t) = [x(0^+) - x_{ss}(0^+)]e^{-t/T} + x_{ss}(t) \tag{6.26}$$

This final result is very important and should be committed to memory. We will no longer need to repeat the solution process for a first-order DE; we simply put it in the form of Eq. (6.6) where the coefficient of dx/dt is unity, identify a, determine $x_{ss}(t)$ and $x(0^+)$ from the circuit, and write the solution in the form of Eq. (6.26).

With this solution to a general linear first-order constant-coefficient DE, let us return to the example shown in Fig. 6.2. The DE relating the inductor current $i_L(t)$

to the dc source V_S is

$$\frac{di_L(t)}{dt} + \frac{R}{L} i_L(t) = \frac{V_S}{L} \tag{6.27}$$

Comparing Eqs. (6.27) and (6.6), we identify $a = R/L$ so that the time constant is

$$T = \frac{1}{a}$$

$$= \frac{L}{R}$$

$$= \tfrac{5}{2} \text{ s} \tag{6.28}$$

The steady-state solution can be found directly from Eq. (6.27) by substituting an assumed solution which is a constant [since the right-hand side of Eq. (6.27) is constant] and obtaining

$$i_{L,ss} = \frac{V_S}{R}$$

$$= 5 \text{ A} \tag{6.29}$$

Note that this can also be obtained from the $t > 0$ circuit by replacing the inductor with a short circuit (since the source is a dc one), as shown in Fig. 6.2d. Therefore, the solution for the inductor current is of the form of Eq. (6.26):

$$i_r(t) = [i_L(0^+) - i_{L,ss}(0^+)]e^{-t/T} + i_{L,ss}(t) \tag{6.30}$$

where $T = L/R = \tfrac{5}{2}$. Therefore,

$$i_L(t) = [i_L(0^+) - 5]e^{-2t/5} + 5 \tag{6.31}$$

To finish the solution, we need $i_L(0^+)$; that is, we need the value of the inductor current immediately after the switch closes. We know that the inductor current cannot change value instantaneously; otherwise, the inductor voltage

$$v_L(t) = L\frac{di_L}{dt}$$

would become infinite. Therefore, the inductor current immediately before the switch closes $i_L(0^-)$ and the inductor current immediately after the switch closes $i_L(0^+)$ must be the same:

$$i_L(0^+) = i_L(0^-)$$

$$= 0 \tag{6.32}$$

Substituting into Eq. (6.31), we have the total solution for the inductor current for $t > 0$:

$$i_L(t) = (0 - 5)e^{-2t/5} + 5$$
$$= 5(1 - e^{-2t/5}) \qquad A \qquad t > 0 \tag{6.33}$$

The total solution in Eq. (6.33) is plotted in Fig. 6.3a. Note that the solution never actually reaches the steady-state value of 5 A but that it comes very close to 5 A after a short time. At $t = T = L/R = \frac{5}{2}$ s, the inductor current is $5(1 - e^{-1}) = 5(0.63) = 3.16$ A; thus, after one time constant has elapsed, the current has risen

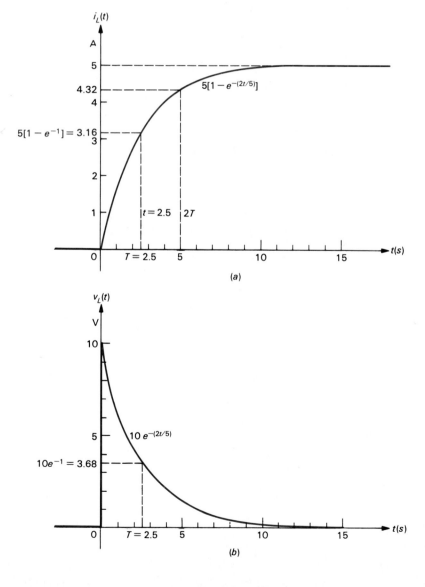

FIGURE 6.3
The solution for the circuit of Fig. 6.2 for $t > 0$:
(*a*) inductor current;
(*b*) inductor voltage.

to 63 percent of its final, or steady-state, value. After two time constants, it has risen to 86 percent of its steady-state value, and after five time constants have elapsed ($t = 5T = 12.5$ s), the inductor current is 99 percent of its final value; thus, after five time constants have elapsed, the inductor current has essentially reached its steady-state value. This "long" time (12.5 s) is a result of the values chosen for our circuit elements. Suppose we had chosen more typical values such as $R = 200\,\Omega$ and $L = 5$ mH; then the time constant would have been $L/R = 2.5 \times 10^{-5}$, or 25 μs. If the source voltage had been chosen as $V_S = 100$ V, then the inductor current would have risen to 99 percent of its steady-state value of 0.5 A in about 125 μs, or 0.125 ms.

The inductor voltage can be obtained by simply differentiating the solution for the inductor current in Eq. (6.33):

$$v_L = L \frac{di_L}{dt}$$

$$= 10e^{-2t/5} \qquad t > 0 \tag{6.34}$$

This is shown in Fig. 6.3b. Note that at $t = 0^+$, $v_L(0^+) = 10$ V and the inductor voltage has *changed instantaneously* from its value at $t = 0^-$ of $v_L(0^-) = 0$ V. Therefore, *an inductor voltage may change value instantaneously, but the inductor current immediately prior to and immediately after a switching operation must be the same:*

$$i_L(0^+) = i_L(0^-) \tag{6.35}$$

Also note that the inductor voltage eventually decays to zero. This is obviously correct, since in the steady state the inductor behaves as a short circuit for dc sources and $v_{L,ss} = 0$. After one time constant, the inductor voltage has decayed to $10(e^{-1}) = 10(0.37) = 3.7$, or 37 percent of its initial value.

Once we have found the solution for the inductor voltage and current, we may work back into the $t > 0$ circuit to find any other element voltages and currents of interest for $t > 0$. For example, from Fig. 6.2c the current associated with the voltage source is

$$i_S(t) = i_L(t)$$

$$= 5(1 - e^{-2t/5}) \qquad t > 0 \tag{6.36}$$

The resistor voltage and resistor current are

$$i_R(t) = i_L(t)$$

$$= 5(1 - e^{-2t/5}) \qquad t > 0 \tag{6.37}$$

$$v_R(t) = Ri_R(t)$$

$$= 10(1 - e^{-2t/5}) \qquad t > 0 \tag{6.38}$$

Thus, finding the solution for the inductor current is the crucial part of the process.

Once that is done, all other element voltages and currents in the circuit can be obtained in terms of the inductor current.

It is worthwhile to explain what is happening in qualitative terms. The inductor stores $w_L(t) = \frac{1}{2}Li_L^2(t)$ energy in its magnetic field. At $t = 0^+$, $i_L(0^+) = 0$ and no magnetic field is established by the inductor current. For $t > 0$, the inductor current and resulting magnetic field are building up to their final (steady-state) values.

Example 6.1

For the circuit in Fig. 6.4a, determine the solution for the current in resistor R_1, $i_{R1}(t)$, for $t > 0$. The switch opens at $t = 0$.

Solution   Our basic method will be to solve for the inductor current for $t > 0$ and then find all other element voltages and currents from that. We again choose to solve for the inductor current, rather than its voltage, for the important reason that we will need to know the initial condition at $t = 0^+$ on the variable we are solving for in terms of its value at $t = 0^-$. Since the inductor current cannot change value instantaneously

$$i_L(0^+) = i_L(0^-)$$

we can easily obtain the initial condition $i_L(0^+)$. If we had tried to solve instead for the inductor voltage, we would need its initial value at $t = 0^+$, $v_L(0^+)$. However,

$$v_L(0^+) \neq v_L(0^-)$$

necessarily, so we could not assume that the inductor voltage was continuous in order to establish that initial condition.

Since we will eventually need $i_L(0^+)$, let us solve for it now. The $t < 0$ circuit is shown in Fig. 6.4b. Since the source is a dc one, we replace the inductor with a short circuit and obtain

$$i_L(0^-) = \frac{V_S}{R_1 \| R_2}$$

$$= 10 \text{ A}$$

$$= i_L(0^+)$$

The next step is to find the time constant for the inductor current for $t > 0$. The $t > 0$ circuit is shown in Fig. 6.4c. From this we identify the time constant as

$$T = \frac{L}{R_1}$$

$$= 2.5 \text{ s}$$

The steady-state solution is found from the $t > 0$ circuit by replacing the inductor with a short circuit (since the source is a dc one), as shown in Fig. 6.4d:

$$i_{L,\text{ss}} = \frac{V_S}{R_1}$$

$$= 5 \text{ A} \qquad t > 0$$

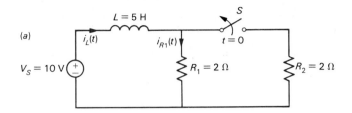

(a)

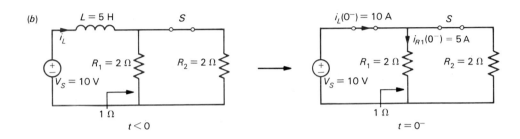

(b)

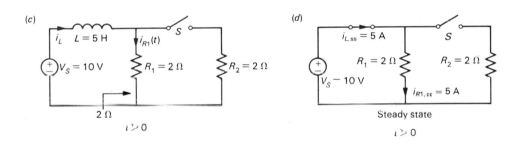

(c)   (d)

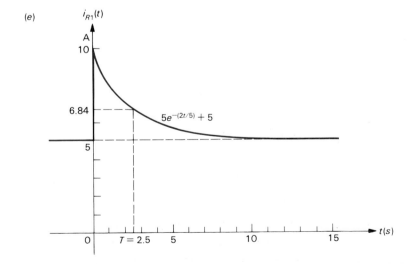

(e)

FIGURE 6.4
Example 6.1:
illustration of the
complete solution
for an *RL* circuit
with nonzero
initial condition.

Therefore, the total solution is

$$
\begin{aligned}
i_L(t) &= [i_L(0^+) - i_{L,\text{ss}}(0^+)]e^{-R_1 t/L} + i_{L,\text{ss}}(t) \\
&= (10 - 5)e^{-2t/5} + 5 \\
&= 5e^{-t/2.5} + 5 \qquad \text{A} \qquad t > 0
\end{aligned}
$$

Now to find the resistor current $i_{R1}(t)$ for $t > 0$, we merely need to note that, in the $t > 0$ circuit in Fig. 6.4c,

$$
\begin{aligned}
i_{R1}(t) &= i_L(t) \\
&= 5e^{-t/2.5} + 5 \qquad \text{A}
\end{aligned}
$$

The result is plotted in Fig. 6.4e. Note that the value of the resistor current at $t = 0^-$ is found, by current division, from Fig. 6.4b to be

$$
\begin{aligned}
i_{R1}(0^-) &= \frac{R_2}{R_1 + R_2}\, i_L(0^-) \\
&= 5 \text{ A}
\end{aligned}
$$

Thus, the resistor current changes value instantaneously at $t = 0$.

6.2   THE RC CIRCUIT

As a second example, consider a similar circuit shown in Fig. 6.5a containing a dc source, a resistor, and a capacitor. At $t = 0$ the switch closes, connecting the capacitor to the current source and resistor. For $t < 0$, the value of the capacitor voltage is zero since no sources are attached to it, as shown in Fig. 6.5b. The steady-state value for the capacitor voltage for $t > 0$ is found by replacing the capacitor with an open circuit (since the source is a dc one), as shown in Fig. 6.4d:

$$
v_{C,\text{ss}} = 20 \text{ V} \qquad t > 0 \tag{6.39}
$$

(Since the source is a dc one, the capacitor voltage and all other steady-state element voltages and currents are constant. Thus, the capacitor current is zero: $i_C = C\, dv_C/dt = 0$, an open circuit.) The complete solution for the capacitor voltage can be found by deriving the differential equation relating v_C and the source I_S. Applying KCL at the upper node in the $t > 0$ circuit in Fig. 6.5c, we obtain

$$
i_C + i_R = I_S \tag{6.40a}
$$

or

$$
C\frac{dv_C}{dt} + \frac{v_C}{R} = I_S \tag{6.40b}
$$

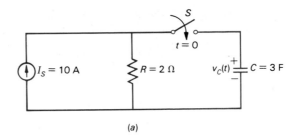

(a)

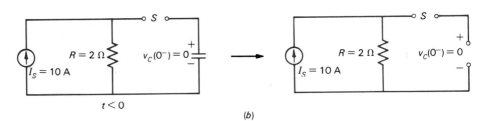

(b)

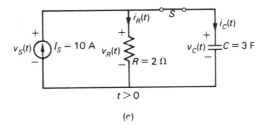

(c)

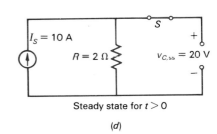

FIGURE 6.5
The *RC* circuit and
its complete
solution.

Dividing both sides by C gives

$$\frac{dv_C(t)}{dt} + \frac{1}{RC} v_C(t) = \frac{I_S}{C} \tag{6.41}$$

This equation is identical in form to the one derived for the inductor current in the previous example. Comparing Eq. (6.41) to Eq. (6.6), we identify $a = 1/(RC)$, $f(t) = I_S/C$. Thus, the total solution is of the form of Eq. (6.26):

$$v_C(t) = [v_C(0^+) - v_{C,\text{ss}}(0^+)]e^{-t/(RC)} + v_{C,\text{ss}}(t) \tag{6.42}$$

where we identify the time constant as

$$T = \frac{1}{a}$$

$$= RC$$

$$= 6 \text{ s} \tag{6.43}$$

The steady-state solution was found previously as

$$v_{C,ss}(t) = RI_S$$
$$= 20 \text{ V} \tag{6.44}$$

Note that Eq. (6.44) satisfies Eq. (6.41) since I_S is constant (direct current). Thus, the total solution is

$$v_C(t) = [v_C(0^+) - 20]e^{-t/6} + 20 \quad \text{V} \quad t > 0 \tag{6.45}$$

The value of the capacitor voltage immediately prior to the closure of the switch, $t = 0^-$, is zero. This must also be the value of the capacitor voltage immediately after the switch closure

$$v_C(0^+) = v_C(0^-)$$
$$= 0 \text{ V} \tag{6.46}$$

since the capacitor voltage cannot change value instantaneously; otherwise, the capacitor current

$$i_C = C\frac{dv_C}{dt}$$

would become infinite. Also the energy stored in the capacitor $w_C = \frac{1}{2}Cv_C^2$ would change instantaneously if the capacitor voltage did so. Applying Eq. (6.46) to Eq. (6.45) gives

$$v_C(t) = (0 - 20)e^{-t/6} + 20$$
$$= 20(1 - e^{-t/6}) \quad t > 0 \tag{6.47}$$

This solution is plotted in Fig. 6.6a. The capacitor voltage has reached 63 percent of its final value at $t = T = 6$ s and 86 percent after two time constants have elapsed.

The capacitor current can be obtained by differentiating the solution for the capacitor voltage in Eq. (6.47):

$$i_C = C\frac{dv_C}{dt}$$
$$= 10e^{-t/6} \quad t > 0 \tag{6.48}$$

This is plotted in Fig. 6.6b. Note that the capacitor current eventually decays to zero, which is a sensible result since the capacitor appears as an open circuit to a dc source in steady state. Note also that the capacitor current has changed value instantaneously at $t = 0$ from its value at $t = 0^-$ of 0 A to its value at $t = 0^+$ of 10 A. Therefore, *a capacitor current may change value instantaneously, but the*

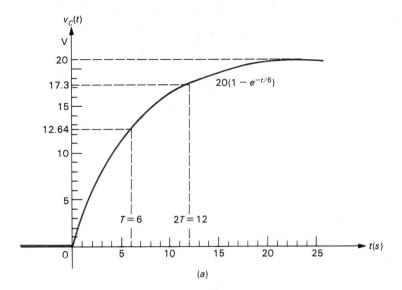

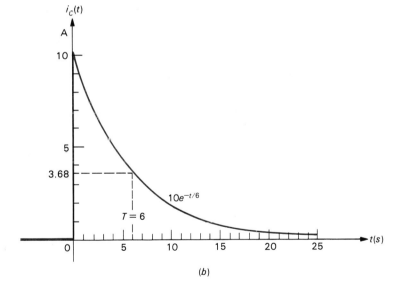

FIGURE 6.6
The solution for
the circuit of Fig.
6.5: (*a*) capacitor
voltage;
(*b*) capacitor
current.

capacitor voltage immediately prior to and immediately after a switching operation must be the same:

$$v_C(0^+) = v_C(0^-) \tag{6.49}$$

Again, as was the case for the inductor, once we find the solution for the voltage and current of the energy storage element, we may work back into the remainder of the $t > 0$ circuit shown in Fig. 6.5c to find all other element voltages and currents for $t > 0$:

$$v_S(t) = v_C(t)$$
$$= 20(1 - e^{-t/6}) \qquad t > 0 \tag{6.50}$$

$$v_R(t) = v_C(t)$$
$$= 20(1 - e^{-t/6}) \qquad t > 0 \tag{6.51}$$

$$i_R(t) = \frac{v_R(t)}{R}$$
$$= 10(1 - e^{-t/6}) \qquad t > 0 \tag{6.52}$$

Note that KVL and KCL are satisfied by these solutions. For example, KCL at the upper node is

$$i_C(t) + i_R(t) = I_S \tag{6.53}$$

or

$$10e^{-t/6} + 10(1 - e^{-t/6}) = 10 \tag{6.54}$$

Again it is worthwhile to explain in qualitative terms what is happening. The capacitor stores energy $w_C(t) = \frac{1}{2}Cv_C^2(t)$ in its electric field established by the electric charge on the capacitor plates. At $t = 0^+$, $v_C(0^+) = 0$ and no electric field exists. As t increases, the charge is being transferred to the capacitor plates by the capacitor current "charging up" the capacitor to its steady-state value. At steady state, the charge transferred to the capacitor is

$$Q_{ss} = CV_{C,ss}$$
$$= (3 \text{ F})(20 \text{ V})$$
$$= 60 \text{ C} \tag{6.55}$$

Example 6.2   For the circuit shown in Fig. 6.7a, determine $i(t)$ for $t > 0$. The switch opens at $t = 0$.

Solution   Our method is again to solve for the capacitor voltage $v_C(t)$ for $t > 0$ and then to find $i(t)$ from that. The $t < 0$ circuit is shown in Fig. 6.7b. Since the source is dc, we replace the capacitor with an open circuit and obtain

$$v_C(0^-) = (R_1 \| R_2)I_S$$
$$= 5 \text{ V}$$
$$= v_C(0^+)$$

The $t > 0$ circuit is shown in Fig. 6.7c. We see that the time constant is

$$T = R_1 C$$
$$= 6 \text{ s}$$

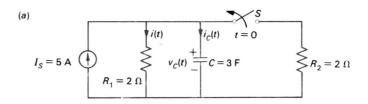

(a)

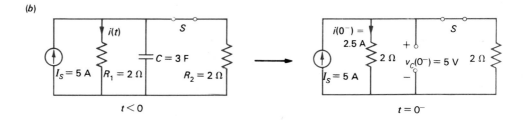

(b)

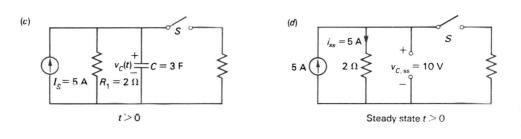

(c)   (d)

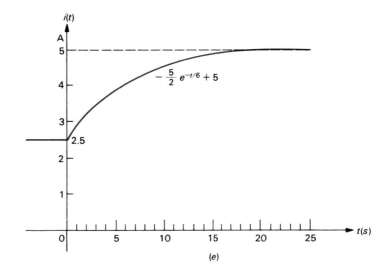

(e)

FIGURE 6.7
Example 6.2:
illustration of the
complete solution
for an *RC* circuit
with nonzero
initial conditions.

The steady-state solution is found from the $t > 0$ circuit by replacing the capacitor with an open circuit, as shown in Fig. 6.7d:

$$v_{C,\text{ss}} = I_S R_1$$
$$= 10 \text{ V}$$

Thus, for $t > 0$,

$$v_{C(t)} = [v_C(0^+) - v_{C,\text{ss}}(0^+)]e^{-t/(RC)} + v_{C,\text{ss}}(t)$$
$$= (5 - 10)e^{-t/6} + 10$$
$$= -5e^{-t/6} + 10 \quad \text{V} \quad t > 0$$

In the $t > 0$ circuit of Fig. 6.7c we see that

$$i(t) = \frac{v_C(t)}{R_1}$$
$$= -\tfrac{5}{2}e^{-t/6} + 5 \quad t > 0$$

This is plotted in Fig. 6.7e. [Note that $i(0^-) = 2.5$ A, which can also be obtained from Fig. 6.7b.] The resistor current rises to its steady-state value of $i(\infty) = 5$ A after approximately five time constants have elapsed. This steady-state value for the resistor current can also be obtained from Fig. 6.7d. This type of final check on the answer using the steady-state circuits for $t < 0$ and for $t > 0$ should always be made.

6.3   WRITING THE SOLUTION BY INSPECTION FOR CIRCUITS CONTAINING ONE ENERGY STORAGE ELEMENT

We now have the necessary tools to solve any circuit which contains only one energy storage element—an inductor or a capacitor—in which a switching operation takes place. The key is to put the $t > 0$ circuit in the form of the basic circuits of the previous sections—RL or RC.

For example, suppose the energy storage element is an inductor. Draw the $t > 0$ circuit (after the switching operation has taken place). Then reduce the remainder of the $t > 0$ circuit (which contains only ideal sources and resistors) attached to the inductor terminals to a Thévenin equivalent, as shown in Fig. 6.8. From this reduction it is clear that the time constant of the inductor current is

$$T = \frac{L}{R_{\text{TH}}} \tag{6.56}$$

Then from the $t > 0$ circuit find the steady-state value of the inductor current $i_{L,\text{ss}}(t)$. Next draw the $t < 0$ circuit as it appears prior to the switching operation.

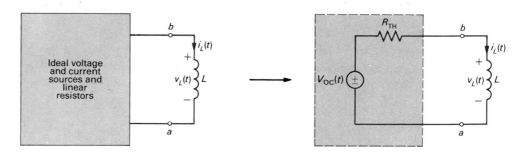

FIGURE 6.8
Solution of a
circuit containing
one inductor by
use of a Thévenin
equivalent.

Then find in this $t < 0$ circuit the steady-state value of the inductor current $i_L(0^-)$. By continuity of inductor currents,

$$i_L(0^+) = i_L(0^-) \tag{6.57}$$

Then write the total solution for the inductor current for $t > 0$:

$$i_L(t) = [i_L(0^+) - i_{L,\text{ss}}(0^+)]e^{-R_{\text{TH}}t/L} + i_{L,\text{ss}}(t) \tag{6.58}$$

The process is very straightforward and simple since the only circuit analysis required is in the determination of the steady-state value of the inductor current in the $t > 0$ circuit, $i_{L,\text{ss}}(t)$, and the steady-state value of the inductor current in the $t < 0$ circuit, $i_L(0^-)$. In previous chapters we have examined methods for obtaining the steady-state solutions due to dc sources and due to ac sources.

Note that we really don't need to find $V_{\text{OC}}(t)$ in the Thévenin equivalent. The time constant $T = L/R_{\text{TH}}$ and the form of the transient solution depend only on the equivalent Thévenin resistance seen by the inductor. The steady-state solution may be found directly from the circuit without the need to compute $V_{\text{OC}}(t)$.

To reiterate, the solution steps are as follows:

1 Draw the $t > 0$ circuit.
2 Find the Thévenin resistance seen by the inductor R_{TH} in this $t > 0$ circuit.
3 Find the steady-state value of the inductor current in the $t > 0$ circuit $i_{L,\text{ss}}(t)$.
4 Draw the $t < 0$ circuit, and find the steady-state value of the inductor current in this circuit $i_L(0^-)$.
5 By continuity of inductor currents, obtain

$$i_L(0^+) = i_L(0^-) \tag{6.59}$$

6 Write the solution for the inductor current for $t > 0$:

$$i_L(t) = [i_L(0^+) - i_{L,\text{ss}}(0^+)]e^{-R_{\text{TH}}t/L} + i_{L,\text{ss}}(t) \qquad t > 0 \tag{6.60}$$

7 Work back into the circuit attached to the inductor terminals to find any other variable of interest.

Example 6.3   Find the complete solution for the voltage $v_x(t)$ for $t > 0$ for the circuit in Fig. 6.9.

Solution   The Thévenin resistance seen by the inductor for $t > 0$ can be found in the usual fashion by drawing the $t > 0$ circuit and setting all ideal sources to zero, as shown in Fig. 6.9b, with the result

$$R_{\text{TH}} = 3\ \Omega$$

The time constant of the inductor current is

$$T = \frac{L}{R_{\text{TH}}}$$
$$= \tfrac{4}{3}\ \text{s}$$

The steady-state value of the inductor current for $t > 0$ is found by replacing the inductor with a short circuit in the $t > 0$ circuit, as shown in Fig. 6.9c, since the sources are both dc. The result, as determined by any of the previous analysis techniques for resistive circuits, is

$$i_{L,\text{ss}} = -1\ \text{A}$$

The initial current at $t = 0^-$ is found from the $t < 0$ circuit as the steady-state value of the inductor current for $t < 0$. Since both sources are dc, we may replace the inductor with a short circuit in the $t < 0$ circuit, as shown in Fig. 6.9d, and obtain (by superposition)

$$i_L(0^-) = \tfrac{1}{4}\ \text{A}$$
$$= i_L(0^+)$$

Thus, the solution for $i_L(t)$ for $t > 0$ is

$$i_L(t) = (\tfrac{1}{4} + 1)e^{-3t/4} - 1$$
$$= \tfrac{5}{4}e^{-3t/4} - 1$$

which is sketched in Fig. 6.10a. The unknown voltage $v_x(t)$ can be found as usual by working back in the $t > 0$ circuit shown in Fig. 6.9e. Note that the current through the 2-Ω resistor across which $v_x(t)$ is defined is equal to $i_L(t)$ for $t > 0$. Thus

$$v_x(t) = -2i_L(t)$$
$$= -\tfrac{5}{2}e^{-3t/4} + 2$$

This is sketched in Fig. 6.10b. Note that $v_x(0^-)$ can be found from the $t = 0^-$ circuit in Fig. 6.9d to be

$$v_x(0^-) = \tfrac{13}{4}\ \text{V}$$

and v_x changes instantaneously at $t = 0$, as shown in Fig. 6.10b. Note that both solutions asymptotically approach their steady-state values: $i_{L,\text{ss}} = -1\ \text{A}$ and $v_{x,\text{ss}} = 2\ \text{V}$.

(a)

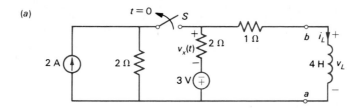

(b)

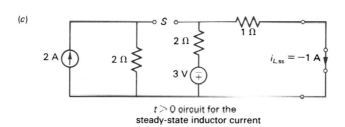

$t > 0$ circuit for R_{TH}

(c)

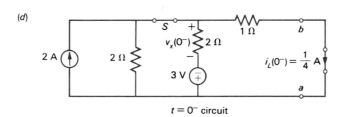

$t > 0$ circuit for the
steady-state inductor current

(d)

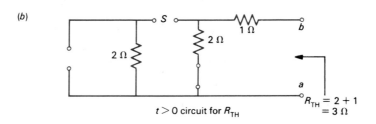

$t = 0^-$ circuit

(e)

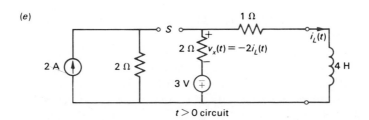

$t > 0$ circuit

FIGURE 6.9
Example 6.3.

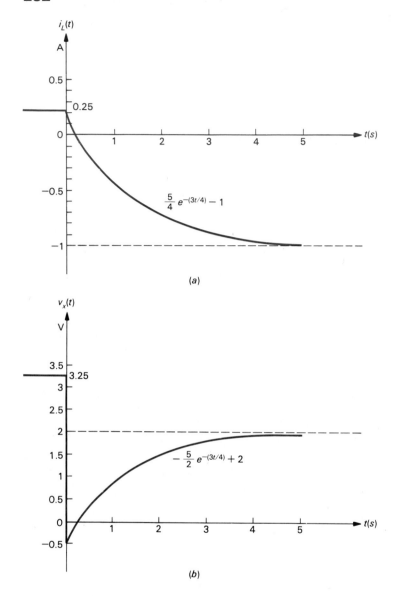

FIGURE 6.10
Example 6.3
solution.

If the energy storage element is a capacitor, we reduce the remainder of the circuit attached to the capacitor terminals to a Norton equivalent, as shown in Fig. 6.11, and essentially we repeat the process used above for the inductor problem. Here we solve for the capacitor voltage $v_C(t)$. We do so because we will need to use the continuity of capacitor voltages to find the initial condition $v_C(0^+) = v_C(0^-)$. The only difference between this problem and the inductor problem is (1) in finding the steady-state values $v_C(0^-)$ and $v_{C,ss}(t)$ and (2) the time constant is

$$T = R_{TH}C \qquad (6.61)$$

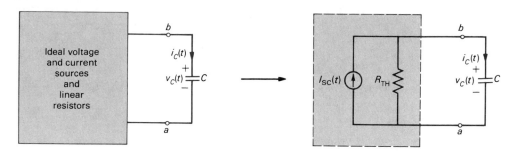

Once again, we do not actually need to find $I_{SC}(t)$ since the steady-state value $v_{C,ss}(t)$ may be found directly from the $t > 0$ circuit. The total solution for the capacitor voltage for $t > 0$ is written

$$v_C(t) = [v_C(0^+) - v_{C,ss}(0^+)]e^{-t/(R_{TH}C)} + v_{C,ss}(t) \qquad t > 0 \tag{6.62}$$

Again we solve for the capacitor voltage rather than the capacitor current since we can use the continuity of capacitor voltages:

$$v_C(0^+) = v_C(0^-) \tag{6.63}$$

Example 6.4   Determine the solution for the voltage $v_x(t)$ across the 2-Ω resistor in Fig. 6.12a.

Solution   The $t > 0$ circuit is shown in Fig. 6.12b. Deactivating the source in this circuit, we find that $R_{TH} = 3\,\Omega$, as shown in Fig. 6.12c. Thus, the time constant is

$$T = R_{TH}C$$
$$= 15\ \text{s}$$

The steady-state circuit for $t > 0$ is shown in Fig. 6.12d, where we have replaced the capacitor with an open circuit since the source in the $t > 0$ circuit is dc. From this we obtain

$$v_{C,ss} = -4\ \text{V}$$

Drawing the steady-state $t < 0$ circuit at $t = 0^-$, as in Fig. 6.4e, we obtain

$$v_C(0^-) = -2\ \text{V}$$

Thus, by the continuity of the capacitor voltage, we obtain

$$v_C(0^+) = v_C(0^-)$$
$$= -2\ \text{V}$$

Thus, for $t > 0$

$$v_C(t) = [v_C(0^+) - v_{C,ss}(0^+)]e^{-t/(R_{TH}C)} + v_{C,ss}(t)$$
$$= (-2 + 4)e^{-t/15} - 4$$
$$= 2e^{-t/15} - 4 \qquad t > 0$$

(a)

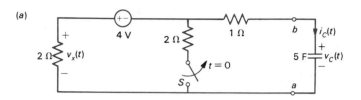

(b)

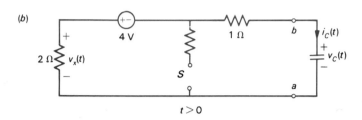

(c)

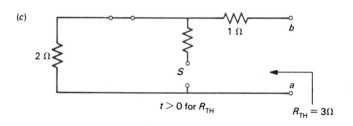

(d)

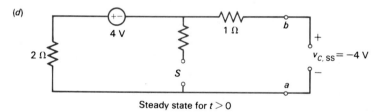

FIGURE 6.12
Example 6.4.

Now in the $t > 0$ circuit in Fig. 6.12b we observe that

$$v_x(t) = -2i_C(t)$$

So we need to find the capacitor current:

$$i_C(t) = C\frac{dv_C(t)}{dt}$$

$$= -\tfrac{2}{3}e^{-t/15}$$

and

$$v_x(t) = \tfrac{4}{3}e^{-t/15} \qquad t > 0$$

This is plotted in Fig. 6.12f. Note in the $t = 0^-$ circuit in Fig. 6.12e that

$$v_x(0^-) = 2 \text{ V}$$

From the above solution

$$v_x(0^+) = \tfrac{4}{3}$$

Thus, the resistor voltage has changed value instantaneously at $t = 0$. The solution for $v_x(t)$ asymptotically approaches zero. This can be seen to be the correct result from the steady-state $t > 0$ circuit in Fig. 6.12d.

6.4   CIRCUITS INVOLVING TWO ENERGY STORAGE ELEMENTS

Our previous results on circuits containing one energy storage element are of more benefit than we perhaps realize. Most of these results and techniques may be easily extended to handle a large number of circuits which contain two energy storage elements. We will only consider the cases in which one energy storage element is an inductor and the other is a capacitor and these are either in series with each other or in parallel. Other cases can be handled by similar methods.

6.4.1   Series *LC* and Parallel *LC* Circuits

First consider the series *LC* case shown in Fig. 6.13a. The remainder of the circuit containing only resistors and ideal voltage and current sources can be represented as a Thévenin equivalent in Fig. 6.13b. In Fig. 6.13b, we may write a differential equation in terms of the inductor current by applying KVL around the loop to yield

$$v_L(t) + v_C(t) = V_{OC}(t) - R_{TH}i_L(t) \tag{6.64}$$

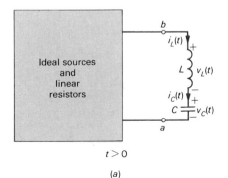

$t > 0$

(a)

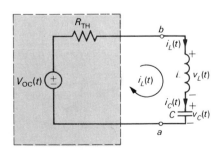

(b)

FIGURE 6.13
A series *LC* circuit.

Note that the current around the loop is the same as $i_L(t)$ so that

$$v_L(t) = L\frac{di_L(t)}{dt} \tag{6.65}$$

$$v_C(t) = \frac{1}{C}\int_{-\infty}^{t} i_C(\tau)\,d\tau$$

$$= \frac{1}{C}\int_{-\infty}^{t} i_L(\tau)\,d\tau \tag{6.66}$$

Substituting Eqs. (6.66) and (6.65) into Eq. (6.64), we obtain

$$L\frac{di_L(t)}{dt} + \frac{1}{C}\int_{-\infty}^{t} i_L(\tau)\,d\tau = V_{OC}(t) - R_{TH}i_L(t) \tag{6.67}$$

Differentiating Eq. (6.67) and dividing both sides by L, we obtain

$$\frac{d^2 i_L(t)}{dt^2} + \frac{R_{TH}}{L}\frac{di_L(t)}{dt} + \frac{1}{LC}i_L(t) = \frac{1}{L}\frac{dV_{OC}(t)}{dt} \tag{6.68}$$

This equation is a second-order constant-coefficient linear ordinary DE. Once we solve Eq. (6.68) for $i_L(t)$, we may work back into the remainder of the circuit to find any other variables of interest. The inductor voltage $v_L(t)$ is found from $i_L(t)$ via Eq. (6.65). The total voltage across terminals a-b, $v_L + v_C$, can be found from

$$v_L(t) + v_C(t) = V_{OC}(t) - R_{TH}i_L(t) \tag{6.69}$$

and therefore the capacitor voltage is, from Eq. (6.69),

$$v_C(t) = V_{OC}(t) - R_{TH}i_L(t) - v_L(t) \tag{6.70}$$

The capacitor current equals the inductor current:

$$i_C(t) = C\frac{dv_C(t)}{dt}$$

$$= i_L(t) \tag{6.71}$$

Therefore, it is sufficient to find the solution for the inductor current from Eq. (6.68) and then use this solution to find any other variable of interest.

For the parallel LC circuit shown in Fig. 6.14a, we may obtain a Norton equivalent representation, as shown in Fig. 6.14b. From this circuit, we may write a differential equation in terms of the capacitor voltage $v_C(t)$ by writing KCL at node b:

$$i_L(t) + i_C(t) = I_{sc}(t) - \frac{v_C(t)}{R_{TH}} \tag{6.72}$$

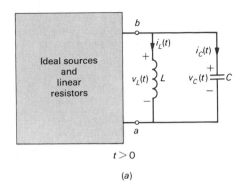

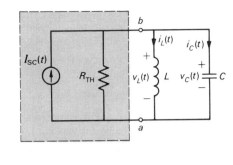

FIGURE 6.14
A parallel LC
circuit.

$t > 0$

(a)

(b)

Substituting

$$i_C(t) = C \frac{dv_C(t)}{dt} \tag{6.73}$$

and

$$i_L(t) = \frac{1}{L} \int_{-\infty}^{t} v_C(\tau) \, d\tau \tag{6.74}$$

[since $v_L(t) = v_C(t)$] into Eq. (6.72) and rearranging, we obtain

$$\frac{d^2 v_C(t)}{dt} + \frac{1}{CR_{TH}} \frac{dv_C(t)}{dt} + \frac{1}{LC} v_C(t) = \frac{1}{C} \frac{dI_{SC}(t)}{dt} \tag{6.75}$$

This is also a second-order constant-coefficient linear ordinary DE of the same form as that obtained for the series LC problem in Eq. (6.68). Once we solve Eq. (6.75) for $v_C(t)$, we may obtain any other branch variable in the remainder of the network by working back into this remainder network since the voltage of terminals a-b is $v_C(t)$. The capacitor current is found from Eq. (6.73). The inductor current is then found from Eq. (6.72). The inductor voltage equals the capacitor voltage:

$$v_L(t) = L \frac{di_L(t)}{dt}$$

$$= v_C(t) \tag{6.76}$$

Thus, the solution for $v_C(t)$ is the key to finding the solution for the other variables of interest.

For either case, we must solve a second-order constant-coefficient linear ordinary DE of the form

$$\frac{d^2 x(t)}{dt^2} + a \frac{dx(t)}{dt} + bx(t) = f(t) \tag{6.77}$$

[Compare Eq. (6.77) to Eqs. (6.68) and (6.75).] For the series LC problem

$$a = \frac{R_{\text{TH}}}{L}$$

$$b = \frac{1}{LC}$$

(6.78)

For the parallel LC problem

$$a = \frac{1}{R_{\text{TH}}C}$$

$$b = \frac{1}{LC}$$

(6.79)

Note that for both problems b is the same. For the series LC circuit, $1/a$ is the time constant of the RL problem of the previous sections. For the parallel LC circuit, $1/a$ is the time constant of the RC problem. Again, because of the linearity of the equation, the solution will consist of the sum of a transient solution $x_T(t)$ and a steady-state solution $x_{ss}(t)$

$$x(t) = x_T(t) + x_{ss}(t)$$

(6.80)

where the transient solution satisfies Eq. (6.77) with $f(t) = 0$:

$$\frac{d^2 x_T(t)}{dt^2} + a \frac{dx_T(t)}{dt} + bx_T(t) = 0$$

(6.81)

and the steady-state solution satisfies Eq. (6.77) with $f(t) \neq 0$:

$$\frac{d^2 x_{ss}(t)}{dt^2} + a \frac{dx_{ss}(t)}{dt} + bx_{ss}(t) = f(t)$$

(6.82)

First let us discuss the steady-state solution. Once again, $x_{ss}(t)$ is any function which, when substituted into Eq. (6.82), identically satisfies it. In Chap. 4 we considered a simple method for obtaining the steady-state solution due to a dc source (replace inductors with short circuits and capacitors with open circuits), and in Chap. 5 we considered simple ways of obtaining the steady-state solution due to a sinusoidal source. In this chapter we consider only circuits containing dc ideal sources, so that $V_{\text{OC}}(t)$ and $I_{\text{SC}}(t)$ will be constant. Therefore, again, *to find the steady-state solution for dc sources for any branch voltage of any energy storage element, we replace the inductor with a short circuit and the capacitor with an open circuit*, as shown in Fig. 6.15. Note that in the series LC problem only the steady-state capacitor voltage is nonzero and is determined by the remainder circuit. Similarly, for the parallel LC problem, only the steady-state inductor current will be nonzero. Note that in the

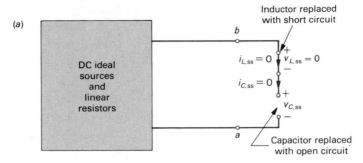

Series *LC* circuit

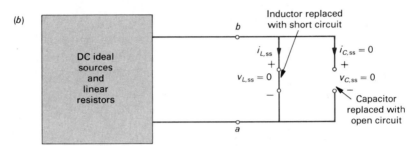

Parallel *LC* circuit

FIGURE 6.15
Determining
steady-state
solutions for dc
sources: (*a*) series
LC circuit;
(*b*) parallel *LC*
circuit.

series *LC* circuit Eq. (6.68) shows that since $V_{OC}(t)$ will be dc if the circuit contains only dc sources, $i_{L,ss} = 0$. Similarly, for the parallel *LC* circuit containing only dc sources, Eq. (6.75) shows that $v_{C,ss} = 0$. These facts need not be memorized since they will become evident when one draws the steady-state circuit for a particular problem, as shown in Fig. 6.15.

Now for the transient solution. The solution to Eq. (6.81) must again be of the form

$$x_T(t) = Ae^{pt} \qquad (6.83)$$

so that it can have the possibility of satisfying Eq. (6.81). Substituting Eq. (6.83) into Eq. (6.81), we obtain

$$p^2 Ae^{pt} + apAe^{pt} + bAe^{pt} = 0 \qquad (6.84)$$

or

$$(p^2 + ap + b)Ae^{pt} = 0 \qquad (6.85)$$

This can be true only if

$$p^2 + ap + b = 0 \qquad (6.86)$$

which is called the *characteristic equation* for the differential equation.

Depending on the values of the circuit elements R_{TH}, L, and C, we have three possibilities for the two roots p_1 and p_2 of the characteristic equation—Eq. (6.86). The two roots are given by

$$p_1, p_2 = -\frac{a}{2} \pm \frac{1}{2}\sqrt{a^2 - 4b} \tag{6.87}$$

which can be written, by factoring out $-a/2$, as

$$p_1, p_2 = -\frac{a}{2}\left(1 \pm \sqrt{1 - \frac{4b}{a^2}}\right) \tag{6.88}$$

The two roots will be real and unequal if

$$\frac{4b}{a^2} < 1$$

and will be equal if

$$\frac{4b}{a^2} = 1$$

The two roots will be complex if

$$\frac{4b}{a^2} > 1$$

Now let us investigate each case in more detail.

Case 1: Roots Real and Distinct   For $4b/a^2 < 1$, the roots of the characteristic equation will be real and unequal (distinct). From Eq. (6.88) it is clear that if a is positive, both these roots will be negative. For circuits containing positive resistors, inductors, and capacitors, R_{TH}, L, and C will be positive and so will a. Therefore, for realistic circuits, both roots will be negative, and we designate them as

$$
\begin{aligned}
p_1 &= -\alpha_1 \\
p_2 &= -\alpha_2
\end{aligned}
\tag{6.89}
$$

where α_1 and α_2 are positive numbers. Therefore, the transient solution will be of the form

$$x_T(t) = A_1 e^{-\alpha_1 t} + A_2 e^{-\alpha_2 t} \tag{6.90}$$

where A_1 and A_2 are, as yet, undetermined constants. Thus, the transient solution consists of the sum of two decaying exponentials, as shown in Fig. 6.16a. The solution

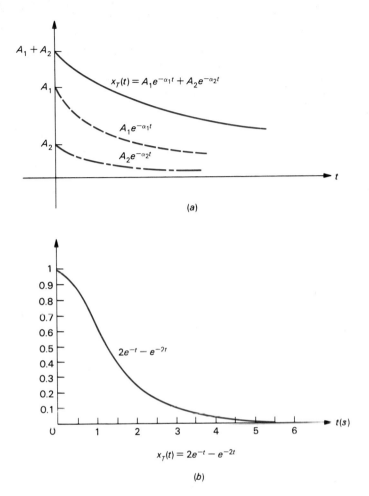

FIGURE 6.16
The overdamped
solution.

for the specific case $A_1 = 2$, $\alpha_1 = 1$, $A_2 = -1$, $\alpha_2 = 2$ is shown in Fig. 6.16*b*. For this case, the transient solution is said to be *overdamped*.

Case 2: Roots Real and Equal If $4b = a^2$, the radical in Eq. (6.88) will be zero and the roots will be equal:

$$p_1, p_2 = -\frac{a}{2}$$

$$= -\alpha \tag{6.91}$$

If we attempt to write the form of the transient solution as previously done for distinct roots, we find that

$$\begin{aligned} x_T(t) &= A_1 e^{-\alpha t} + A_2 e^{-\alpha t} \\ &= (A_1 + A_2)e^{-\alpha t} \\ &= Ae^{-\alpha t} \end{aligned} \tag{6.92}$$

In other words, this form of the transient solution reduces to the same as that for a first-order differential equation. Thus, for the case of equal roots, we suspect that Eq. (6.92) is not the proper form of the transient solution.

Let us now reexamine our procedure for determining the form of the transient solution. Recall that to find a solution of

$$\frac{d^2x_T(t)}{dt^2} + a\frac{dx_T(t)}{dt^2} + bx_T(t) = 0 \tag{6.93}$$

we assumed a form for $x_T(t)$ as

$$x_T(t) = Ae^{pt} \tag{6.94}$$

substituted into Eq. (6.93), and determined p such that this form would satisfy Eq. (6.93). For equal roots, another possible form which will satisfy Eq. (6.93) is

$$x_T(t) = At^{pt} \tag{6.95}$$

To see this, we substitute Eq. (6.95) into Eq. (6.93) to obtain

$$(pAe^{pt} + pAe^{pt} + p^2Ate^{pt}) + a(Ae^{pt} + pAte^{pt}) + b(Ate^{pt}) = 0 \tag{6.96}$$

Grouping terms, we obtain

$$(2p + a)Ae^{pt} + \underbrace{(p^2 + ap + b)}_{0}Ate^{pt} = 0 \tag{6.97}$$

The coefficient of Ate^{pt} is the characteristic equation and thus is automatically equal to zero. The question remains as to whether

$$2p + a \stackrel{?}{=} 0 \tag{6.98}$$

for these real, repeated roots. But this real, repeated root given in Eq. (6.91) exactly satisfies Eq. (6.98) so that Eq. (6.95) is a valid form of the transient solution. By superposition, we may combine Eqs. (6.94) and (6.95) to obtain the form of the transient solution which satisfies Eq. (6.93):

$$\begin{aligned} x_T(t) &= A_1e^{-\alpha t} + A_2te^{-\alpha t} \\ &= (A_1 + A_2t)e^{-\alpha t} \end{aligned} \tag{6.99}$$

Note that this solution cannot be reduced any further and that it contains exactly two undetermined constants. For this case, the transient solution is said to be critically damped, and Eq. (6.99) is plotted in Fig. 6.17a. The result for $A_1 = 2$, $A_2 = -1$, and $\alpha = 2$ is shown in Fig. 6.17b.

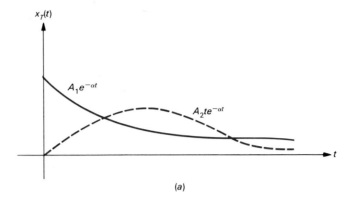

(a)

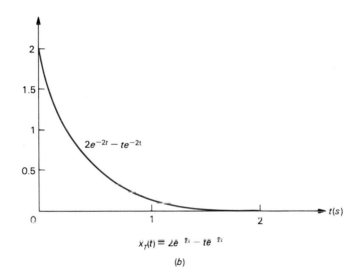

$$x_T(t) = 2e^{-2t} - te^{-2t}$$

(b)

FIGURE 6.17
The critically
damped solution.

Case 3: Complex Roots   If $4b/a^2 > 1$ in Eq. (6.88), the roots will be complex. Denoting

$$j = \sqrt{-1}$$

we may write Eq. (6.88) as

$$p_1, p_2 = -\frac{a}{2}\left(1 \pm j\sqrt{\frac{4b^2}{a} - 1}\right) \tag{6.100}$$

so that we will denote the roots as

$$p_1, p_2 = -\alpha \pm j\beta \tag{6.101}$$

The two roots are conjugates. As we found in the previous chapter, the conjugate of a complex number $\mathbf{c}$ is denoted by $\mathbf{c}^*$ and is obtained by replacing all imaginary terms in $\mathbf{c}$ with their negatives. For example, if $\mathbf{c} = 2 + j3$, then $\mathbf{c}^* = 2 - j3$. The transient solution becomes

$$
\begin{aligned}
x_T(t) &= A_1 e^{(-\alpha + j\beta)t} + A_2 e^{(-\alpha - j\beta)t} \\
&= e^{-\alpha t}(A_1 e^{j\beta t} + A_2 e^{-j\beta t})
\end{aligned}
\tag{6.102}
$$

Although this is a valid transient solution, we will find it convenient to write Eq. (6.102) in an alternative form:

$$
x_T(t) = e^{-\alpha t}(C_1 \cos \beta t + C_2 \sin \beta t)
\tag{6.103}
$$

This can be easily shown by substituting Euler's identity, which we studied in Chap. 5,

$$
e^{j\beta t} = \cos \beta t + j \sin \beta t
$$

into Eq. (6.102) to yield

$$
\begin{aligned}
x_T(t) &= e^{-\alpha t}[A_1(\cos \beta t + j \sin \beta t) + A_2(\cos \beta t - j \sin \beta t)] \\
&= e^{-\alpha t}[(A_1 + A_2) \cos \beta t + j(A_1 - A_2) \sin \beta t]
\end{aligned}
\tag{6.104}
$$

Comparing Eqs. (6.103) and (6.104), we identify

$$
C_1 = A_1 + A_2
\tag{6.105a}
$$

$$
C_2 = j(A_1 - A_2)
\tag{6.105b}
$$

Since A_1 and A_2 are (as yet) undetermined constants, we may as well use the form given in Eq. (6.103). Since the transient solution given in Eq. (6.102) or (6.103) must be real, the undetermined constants in Eq. (6.103) must be real. However, the constants in Eq. (6.103), C_1 and C_2, are related to the constants in Eq. (6.102), A_1 and A_2, by Eqs. (6.105). But C_2 which is real is $j(A_1 - A_2)$. In order that C_2 be real, A_1 and A_2 must be complex and conjugates of each other. For example, if we write

$$
\begin{aligned}
A_1 &= A + jB \\
A_2 &= A - jB
\end{aligned}
\tag{6.106}
$$

substitution in Eqs. (6.105) yields

$$
\begin{aligned}
C_1 &= A_1 + A_2 \\
&= 2A \\
C_2 &= j(A_1 - A_2) \\
&= j(j2B) \\
&= -2B
\end{aligned}
\tag{6.107}
$$

Therefore, the undetermined constants A_1 and A_2 in Eq. (6.102) are complex and conjugates.

An alternative form of the solution in Eq. (6.103) can be written as

$$x_T(t) = Ce^{-\alpha t} \sin (\beta t + \phi) \tag{6.108}$$

To show this, we expand $C \sin (\beta t + \phi)$ in Eq. (6.108) as

$$C \sin (\beta t + \phi) = C \sin \beta t \cos \phi + C \cos \beta t \sin \phi \tag{6.109}$$

Thus, we identify

$$\begin{aligned} C_1 &= C \cos \phi \\ C_2 &= C \sin \phi \end{aligned} \tag{6.110}$$

and

$$\begin{aligned} \frac{\sin \phi}{\cos \phi} &= \tan \phi \\ &= \frac{C_2}{C_1} \end{aligned} \tag{6.111}$$

or

$$\phi = \tan^{-1} \frac{C_2}{C_1} \tag{6.112}$$

The transient solution for the case of complex roots is said to be *underdamped*, and Eq. (6.108) is sketched in Fig. 6.18a. The specific case for $C = 1$, $\alpha = 2$, $\beta = 20$, $\phi = 30°$ is sketched in Fig. 6.18b.

Note that in all three cases the transient solution will eventually decay to zero, leaving only the steady-state solution. For the overdamped case,

$$x_T(t) = A_1 e^{-\alpha_1 t} + A_2 e^{-\alpha_2 t} \tag{6.113}$$

will go to zero as time progresses. For the critically damped case,

$$x_T(t) = A_1 e^{-\alpha t} + A_2 t e^{-\alpha t} \tag{6.114}$$

will also go to zero since $e^{-\alpha t}$ goes to zero faster than t goes to infinity as time progresses. For the underdamped case,

$$x_T(t) = Ce^{-\alpha t} \sin (\beta t + \phi) \tag{6.115}$$

also goes to zero as time progresses. We have assumed, however, that the real parts of the roots α_1 and α_2 for the underdamped case, or α for the critically damped

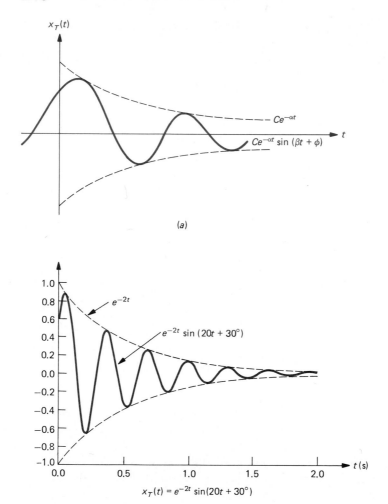

FIGURE 6.18
The underdamped
solution.

$$x_T(t) = e^{-2t}\sin(20t + 30°)$$

(b)

and underdamped cases, are nonzero. From the form of the roots in Eq. (6.88) it is easy to see that the real parts of these roots will not be zero if $R_{TH} \neq 0$. Therefore, if the circuit contains some resistance, the transient solution will eventually decay to zero. This is a realistic case, since all physical circuits contain some resistance even though it may be only the resistance of the connecting wires. Thus, we consider only circuits for which the transient solution decays to zero. Such circuits are said to be *stable circuits*.

6.4.2   Switching Operations and Initial Conditions

Again we wish to obtain complete solutions for circuit voltages or circuits after some switching operation has taken place at $t = 0$. The procedure is almost identical to the case of one energy storage element, except that now we must obtain two initial conditions (at $t = 0^+$) on the variable of interest (i_L, v_L, i_C, or v_C) to evaluate the two undetermined constants in the transient solution.

Again we consider circuits containing only dc ideal voltage and current sources. We first draw the $t < 0$ circuit—the circuit as it exists prior to the switching operator. Next we assume that it has been in this state for a very long time so that it has reached steady state and all transients have essentially vanished. *Only the inductor current and capacitor voltage are continuous at $t = 0$: the time at $t = 0^-$ and $t = 0^+$.* Thus, since the $t < 0$ circuit contains only dc sources and is in steady state, we replace the inductor with a short circuit and the capacitor with an open circuit to obtain $i_L(0^-)$ and $v_C(0^-)$. We thus have

$$i_L(0^+) = i_L(0^-) \tag{6.116}$$

$$v_C(0^+) = v_C(0^-) \tag{6.117}$$

Next we draw the $t > 0$ circuit and write the form of the solution for either the inductor current or the capacitor voltage by inspection. If L and C are in series,

$$\begin{aligned} a &= \frac{R_{TH}}{L} \\ b &= \frac{1}{LC} \end{aligned} \tag{6.118}$$

If L and C are in parallel,

$$\begin{aligned} a &= \frac{1}{CR_{TH}} \\ b &= \frac{1}{LC} \end{aligned} \tag{6.119}$$

Next we find the roots of

$$p^2 + ap + b = 0 \tag{6.120}$$

and write the form of the transient solution by inspection, depending on whether the roots are real and distinct, real and repeated, or complex conjugates.

Next we find the steady-state solution by drawing the $t > 0$ circuit (which contains only dc sources) and replace the inductor with a short circuit and the capacitor with an open circuit. Then we form the total solution for $t > 0$ by adding this steady-state solution to the transient solution:

$$i_L(t) = i_{L,T}(t) + i_{L,ss}(t) \qquad t > 0 \tag{6.121}$$

$$v_C(t) = v_{C,T}(t) + v_{C,ss}(t) \qquad t > 0 \tag{6.122}$$

There are two unknowns in the transient solution regardless of the form of the roots. If we wish to find the solution for the inductor current, we have one initial condition $i_L(0^+)$ to be used, but we need one more condition on $i_L(0^+)$ to evaluate the two constants. If we knew the derivative of i_L at $t = 0^+$, $di_L/dt|_{t=0^+}$, we could evaluate the two constants since we could evaluate Eq. (6.121) at $t = 0^+$ as

$$i_L(0^+) = i_{L,T}(0^+) + i_{L,ss}(0^+) \tag{6.123}$$

and then we could differentiate Eq. (6.121) and evaluate the result at $t = 0^+$:

$$\left.\frac{di_L}{dt}\right|_{t=0^+} = \left.\frac{di_{L,T}}{dt}\right|_{t=0^+} + \left.\frac{di_{L,ss}}{dt}\right|_{t=0^+} \tag{6.124}$$

Thus, Eqs. (6.123) and (6.124) would allow us to evaluate the two undetermined constants in the transient solution. Since we know only $i_L(0^+)$ and $v_C(0^+)$, how do we obtain $di_L/dt|_{t=0^+}$? Note that the inductor voltage is

$$v_L(t) = L\frac{di_L(t)}{dt} \tag{6.125}$$

Therefore,

$$\left.\frac{di_L}{dt}\right|_{t=0^+} = \frac{1}{L}v_L(0^+) \tag{6.126}$$

If we can find the inductor voltage at $t = 0^+$, we can find the derivative of the inductor current at $t = 0^+$ with Eq. (6.126). To find the inductor voltage (which is *not* necessarily continuous at $t = 0$), we draw the $t = 0^+$ circuit. Knowing $i_L(0^+)$ and $v_C(0^+)$ in this circuit, we may use KVL, KCL, and Ohm's law to find $v_L(0^+)$. A similar process would apply if we had chosen to obtain the capacitor voltage equation in Eq. (6.122); that is, find $v_C(0^+)$ and $dv_C/dt|_{t=0^+} = (1/C)i_C(0^+)$.

Example 6.5   Consider the circuit in Fig. 6.19a. Find the form of the inductor current.

Solution   The $t = 0^-$ circuit is shown in Fig. 6.19b and

$$v_C(0^+) = v_C(0^-) = 2\text{ V}$$
$$i_L(0^+) = i_L(0^-) = 0\text{ A}$$

The $t > 0$ circuit is shown in Fig. 6.19c. From this we obtain

$$R_{\text{TH}} = 2\ \Omega$$

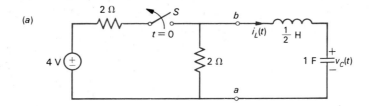

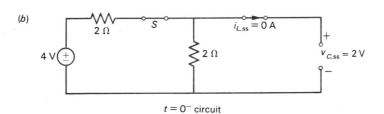

$t = 0^-$ circuit

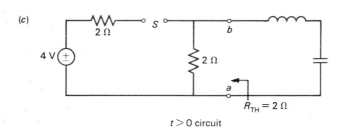

$t > 0$ circuit

FIGURE 6.19
Example 6.5:
illustration of
determining initial
conditions and the
complete solution
for $t > 0$.

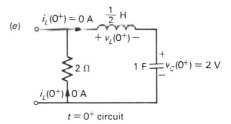

Steady-state $t > 0$ circuit $t = 0^+$ circuit

and since it is a series LC circuit,

$$a = \frac{R_{TH}}{L}$$

$$= 4$$

$$b = \frac{1}{LC}$$

$$= 2$$

so that the characteristic equation is

$$p^2 + 4p + 2 = 0$$

The roots are

$$p_1, p_2 = -2 \pm \tfrac{1}{2}\sqrt{16 - 8}$$
$$= -2 \pm \sqrt{2}$$
$$= -3.414, \ -0.586$$

Thus, the transient solution is

$$i_{L,T}(t) = A_1 e^{-3.414t} + A_2 e^{-0.586t}$$

The steady-state circuit for $t > 0$ is shown in Fig. 6.19d. From it we obtain

$$i_{L,\text{ss}} = 0 \qquad t > 0$$

Thus, the total solution is

$$i_L(t) = A_1 e^{-3.414t} + A_2 e^{-0.586t}$$

The $t = 0^+$ circuit is shown in Fig. 6.19e. From it we obtain, by writing a KVL equation around the loop,

$$v_L(0^+) = -2i_L(0^+) - v_C(0^+)$$
$$= -2 \ \text{V}$$

But

$$v_L(0^+) = L \left.\frac{di_L}{dt}\right|_{t=0^+}$$

so

$$\left.\frac{di_L}{dt}\right|_{t=0^+} = \frac{1}{L} v_L(0^+)$$
$$= -4 \ \text{A/s}$$

The total solution at $t = 0^+$ is

$$i_L(0^+) = A_1 + A_2$$
$$= 0 \ \text{A}$$

Differentiating the total solution and evaluating at $t = 0^+$, we obtain

$$\left.\frac{di_L}{dt}\right|_{t=0^+} = -3.414 A_1 - 0.586 A_2$$
$$= -4$$

or

$$A_1 + A_2 = 0$$
$$-3.414A_1 - 0.586A_2 = -4$$

Thus, we solve for A_1 and A_2 by Cramer's rule as

$$A_1 = \frac{\begin{vmatrix} 0 & 1 \\ -4 & -0.586 \end{vmatrix}}{\begin{vmatrix} 1 & 1 \\ -3.414 & -0.586 \end{vmatrix}}$$

$$= 1.414$$

$$A_2 = \frac{\begin{vmatrix} 1 & 0 \\ -3.414 & -4 \end{vmatrix}}{\begin{vmatrix} 1 & 1 \\ -3.414 & -0.586 \end{vmatrix}}$$

$$= -1.414$$

Thus, the total solution is

$$i_L(t) = 1.414e^{-3.414t} - 1.414e^{-0.586t} \qquad t > 0$$

This is plotted in Fig. 6.20.

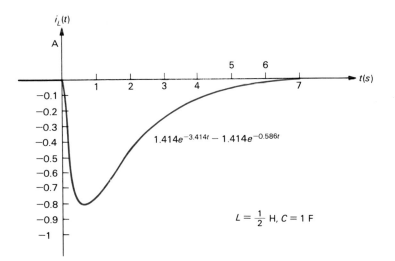

FIGURE 6.20
Plot of the inductor current for $t \geq 0$ for the circuit of Fig. 6.19.

Example 6.6   Repeat Example 6.5 but with $L = 1$ H, $C = 1$ F.

Solution   Several of the previous example results remain the same. Changing the values of L and C only affects (1) the roots and (2)

$$\frac{di_L}{dt}\bigg|_{t=0^+} = \frac{1}{L} v_L(0^+)$$

Note that the initial values of $i_L(0^+)$ and $v_C(0^+)$ remain unchanged. Also $v_L(0^+)$ remains unchanged since this depended on $i_L(0^+)$ and the circuit structure at $t = 0^+$. For this case

$$a = \frac{R_{\text{TH}}}{L}$$

$$= 2$$

$$b = \frac{1}{LC}$$

$$= 1$$

and

$$p^2 + 2p + 1 = 0$$

The roots are repeated

$$p_1, p_2 = -1$$

and the solution is

$$i_L(t) = A_1 e^{-t} + A_2 t e^{-t}$$

Now $i_L(0^+)$ again is 0 A. Evaluating i_L at $t = 0^+$, we obtain

$$i_L(0^+) = A_1$$

$$= 0 \text{ A}$$

Thus, $A_1 = 0$. Now differentiating the solution, we obtain

$$\frac{di_L}{dt} = -A_1 e^{-t} - A_2 t e^{-t} + A_2 e^{-t}$$

and

$$\frac{di_L}{dt}\bigg|_{t=0^+} = -A_1 + A_2$$

$$= A_2$$

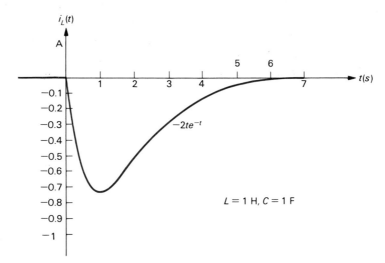

FIGURE 6.21
Example 6.6: Fig.
6.19 with $L = 1$ H,
$C = 1$ F.

But

$$\left.\frac{di_L}{dt}\right|_{t=0^+} = \frac{1}{L} v_L(0^+)$$

$$= -2 \text{ V}$$

So

$$A_2 = -2$$

and

$$i_L(t) = -2te^{-t}$$

This is plotted in Fig. 6.21.

Example 6.7   Repeat Example 6.5 with $L = 5$ H, $C = 1$ F.

Solution   Again, only certain parts of the solution remain unchanged. For this case

$$a = \frac{R_{TH}}{L}$$

$$= \frac{2}{5}$$

$$b = \frac{1}{LC}$$

$$= \frac{1}{5}$$

and

$$p^2 + \tfrac{2}{5}p + \tfrac{1}{5} = 0$$

The roots are

$$
\begin{aligned}
p_1, p_2 &= -\tfrac{1}{5} \pm \tfrac{1}{2}\sqrt{\tfrac{4}{25} - \tfrac{4}{5}} \\
&= -\tfrac{1}{5} \pm j\tfrac{2}{5}
\end{aligned}
$$

Thus, the solution is

$$i_L(t) = e^{-t/5}(C_1 \cos \tfrac{2}{5}t + C_2 \sin \tfrac{2}{5}t)$$

Again

$$i_L(0^+) = 0 \text{ A}$$

but

$$
\left. \frac{di_L}{dt} \right|_{t=0^+} = \frac{1}{L} v_L(0^+)
$$

$$
= -\frac{2}{5}
$$

since $v_L(0^+) = -2$ V again.

Evaluating the solution, we obtain

$$
\begin{aligned}
i_L(0^+) &= C_1 \\
&= 0
\end{aligned}
$$

or $C_1 = 0$. The derivative of i_L at $t = 0^+$ is a bit more complicated:

$$
\begin{aligned}
\frac{di_L}{dt} = &-\frac{1}{5} C_1 e^{-t/5} \cos \frac{2}{5}t - \frac{2}{5} C_1 e^{-t/5} \sin \frac{2}{5}t \\
&- \frac{1}{5} C_2 e^{-t/5} \sin \frac{2}{5}t + \frac{2}{5} C_2 e^{-t/5} \cos \frac{2}{5}t
\end{aligned}
$$

so that

$$
\left. \frac{di_L}{dt} \right|_{t=0^+} = -\frac{1}{5} C_1 + \frac{2}{5} C_2
$$

$$
= -\frac{2}{5}
$$

or

$$C_2 = -1$$

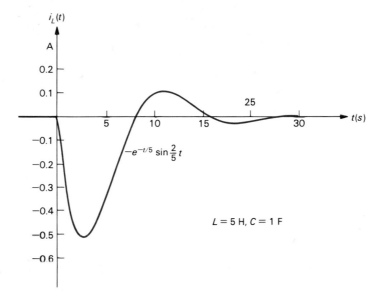

FIGURE 6.22
Example 6.7: Fig.
6.19 with $L = 5$ H,
$C = 1$ F.

The solution thus becomes

$$i_L(t) = -e^{-t/5} \sin \tfrac{2}{5}t \qquad t > 0$$

This is plotted in Fig. 6.22.

Example 6.8   For the parallel LC problem in Fig. 6.23a, solve for the current in the 1-Ω resistor $i_x(t)$ for $t > 0$.

Solution   The $t - 0^-$ circuit is shown in Fig. 6.23b. From it we obtain

$$i_L(0^-) = -\tfrac{3}{4} \text{ A}$$
$$= i_L(0^+)$$
$$v_C(0^-) = 0$$
$$= v_C(0^+)$$

The $t > 0$ circuit is shown in Fig. 6.23c. From it we obtain $R_{\text{TH}} = 3\ \Omega$ and

$$a = \frac{1}{R_{\text{TH}}C}$$
$$= \frac{4}{15}$$

$$b = \frac{1}{LC}$$
$$= \frac{4}{45}$$

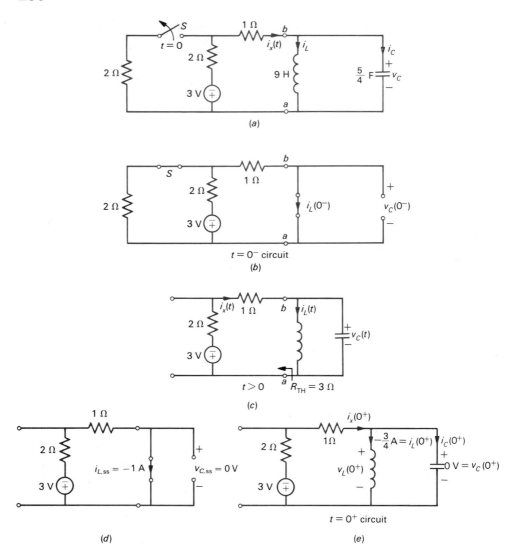

FIGURE 6.23
Example 6.8: a
parallel LC circuit.

So

$$p^2 + \tfrac{4}{15}p + \tfrac{4}{45} = 0$$

and the roots become

$$p_1, p_2 = -\tfrac{2}{15} \pm \tfrac{1}{2}\sqrt{(\tfrac{4}{15})^2 - \tfrac{16}{45}}$$
$$= -\tfrac{2}{15} \pm j\tfrac{4}{15}$$

The transient solution for the capacitor voltage is

$$v_{C,T}(t) = e^{-2t/15}(C_1 \cos \tfrac{4}{15}t + C_2 \sin \tfrac{4}{15})$$

The steady-state solution $v_{C,\mathrm{ss}}$ is found from Fig. 6.23d as

$$v_{C,\mathrm{ss}} = 0$$

so that

$$v_C(t) = e^{-2t/15}(C_1 \cos \tfrac{4}{15}t + C_2 \sin \tfrac{4}{15})$$

Since $v_C(0^+) = 0$, we find that

$$C_1 = 0$$

Now differentiating $v_C(t)$, we obtain

$$\frac{dv_C}{dt} = -\frac{2}{15}C_1 e^{-2t/15} \cos \frac{4}{15}t - \frac{4}{15}C_1 e^{-2t/15} \sin \frac{4}{15}t$$

$$-\frac{2}{15}C_2 e^{-2t/15} \sin \frac{4}{15}t + \frac{4}{15}C_2 e^{-2t/15} \cos \frac{4}{15}t$$

or

$$\left.\frac{dv_C}{dt}\right|_{t=0^+} = -\frac{2}{15}\cancel{C_1}^{\,0} + \frac{4}{15}C_2$$

or

$$C_2 = \left.\frac{15}{4}\frac{dv_C}{dt}\right|_{t=0^+}$$

How do we obtain $dv_C/dt|_{t=0^+}$? First draw the $t = 0^+$ circuit and observe that the only variables we know in this circuit are $i_L(0^+)$ and $v_C(0^+)$, as shown in Fig. 6.23e. Observe that

$$i_C = C\frac{dv_C}{dt}$$

so that

$$\left.\frac{dv_C}{dt}\right|_{t=0^+} = \frac{1}{C}i_C(0^+)$$

Therefore, we only need to find $i_C(0^+)$. Note that

$$i_x(0^+) = i_L(0^+) + i_C(0^+)$$

But by writing KVL around the outside loop we obtain

$$i_x(0^+) = \frac{-3\text{ V} - \cancel{v_C}^{\,0}(0^+)}{3\ \Omega}$$

$$= -1\text{ A}$$

Therefore,

$$i_C(0^+) = i_x(0^+) - i_L(0^+)$$
$$= -1 + \tfrac{3}{4}$$
$$= -\tfrac{1}{4}\text{ A}$$

and

$$\left.\frac{dv_C}{dt}\right|_{t=0^+} = \frac{4}{5}\left(-\frac{1}{4}\right)$$

$$= -\frac{1}{5}$$

Therefore,

$$C_2 = \frac{15}{4}\left.\frac{dv_C}{dt}\right|_{t=0^+}$$

$$= -\frac{3}{4}$$

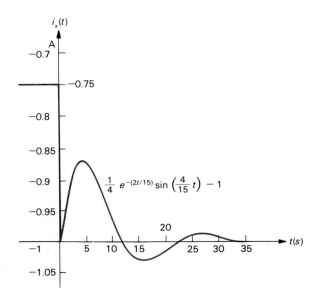

FIGURE 6.24
Example 6.8:
solution for the
circuit of Fig. 6.23
for $t \geq 0$.

and the solution for $v_C(t)$ is

$$v_C(t) = -\tfrac{3}{4}e^{-2t/15}\sin\tfrac{4}{15}t$$

Now we need to obtain the current $i_x(t)$. In the $t > 0$ circuit in Fig. 6.23c we obtain, by applying KVL around the outside loop,

$$i_x(t) = \frac{-3\,\mathrm{V} - v_C(t)}{3\,\Omega}$$

$$= -1 - \tfrac{1}{3}v_C(t)$$

$$= -1 + \tfrac{1}{4}e^{-2t/15}\sin\tfrac{4}{15}t$$

Note that as $t \to \infty$, $i_x \to -1$. Thus, $i_x = -1$ must be its steady-state value. This is seen to be true from Fig. 6.23d. The solution is plotted in Fig. 6.24.

PROBLEMS

6.1 Determine and sketch $i(t)$ in the circuit of Fig. P6.1.

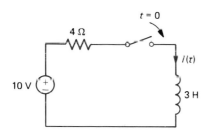

FIGURE P6.1

6.2 Determine and sketch $i(t)$ in the circuit of Fig. P6.2.

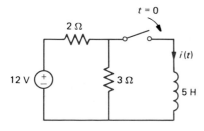

FIGURE P6.2

6.3 Determine and sketch $i(t)$ in the circuit of Fig. P6.3.

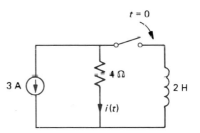

FIGURE P6.3

6.4 Determine and sketch $i(t)$ in the circuit of Fig. P6.4.

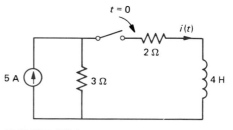

FIGURE P6.4

6.5 Determine and sketch $i(t)$ in the circuit of Fig. P6.5.

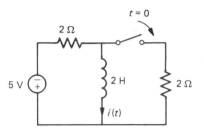

FIGURE P6.5

6.6 Determine and sketch $i(t)$ in the circuit of Fig. P6.6.

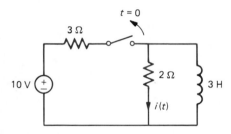

FIGURE P6.6

6.7 Determine and sketch $i(t)$ in the circuit of Fig. P6.7.

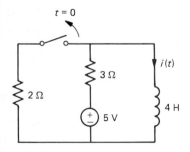

FIGURE P6.7

6.8 Determine and sketch $i(t)$ in the circuit of Fig. P6.8.

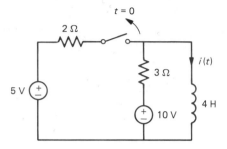

FIGURE P6.8

6.9 Determine and sketch $i(t)$ in the circuit of Fig. P6.9.

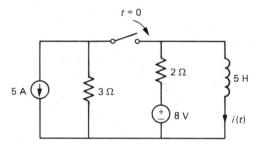

FIGURE P6.9

6.10 Determine and sketch $i(t)$ in the circuit of Fig. P6.10.

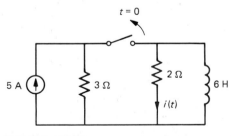

FIGURE P6.10

6.11 Determine and sketch $v(t)$ in the circuit of Fig. P6.11.

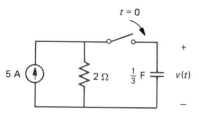

FIGURE P6.11

6.12 Determine and sketch $v(t)$ in the circuit of Fig. P6.12.

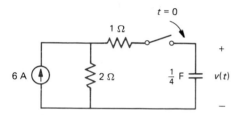

FIGURE P6.12

6.13 Determine and sketch $v(t)$ in the circuit of Fig. P6.13.

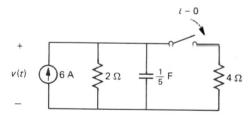

FIGURE P6.13

6.14 Determine and sketch $v(t)$ in the circuit of Fig. P6.14.

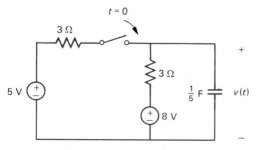

FIGURE P6.14

6.15 Determine and sketch $v(t)$ in the circuit of Fig. P6.15.

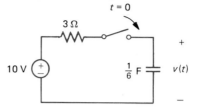

FIGURE P6.15

6.16 Determine and sketch $v(t)$ in the circuit of Fig. P6.16.

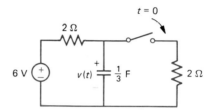

FIGURE P6.16

6.17 Determine and sketch $v(t)$ in the circuit of Fig. P6.17.

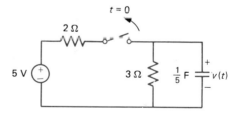

FIGURE P6.17

6.18 Determine and sketch $v(t)$ in the circuit of Fig. P6.18.

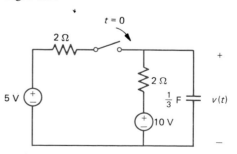

FIGURE P6.18

6.19 Determine and sketch $v(t)$ in the circuit of Fig. P6.19.

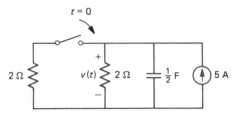

FIGURE P6.19

6.20 Determine and sketch $v(t)$ in the circuit of Fig. P6.20.

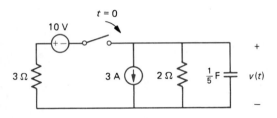

FIGURE P6.20

6.21 Determine and sketch $i_L(t)$ and $v_C(t)$ in the circuit of Fig. P6.21 for $v_S = 10$ V, $R = 4\,\Omega$, $L = 1$ H, and $C = \frac{1}{3}$ F.

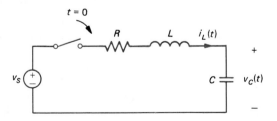

FIGURE P6.21

6.22 Determine and sketch $i_L(t)$ and $v_C(t)$ in the circuit of Fig. P6.21 for $v_S = 5$ V, $R = 4\,\Omega$, $L = 1$ H, and $C = \frac{1}{4}$ F.

6.23 Determine and sketch $i_L(t)$ and $v_C(t)$ in the circuit of Fig. P6.21 for $v_S = 10$ V, $R = 8\,\Omega$, $L = 2$ H, and $C = \frac{1}{26}$ F.

6.24 Determine and sketch $i_L(t)$ and $v_C(t)$ in the circuit of Fig. P6.24 for $i_S = 2$ A, $R = 2\,\Omega$, $L = \frac{3}{2}$ H and $C = \frac{1}{12}$ F.

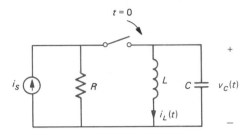

FIGURE P6.24

6.25 Determine and sketch $i_L(t)$ and $v_C(t)$ in the circuit of Fig. P6.24 for $i_S = 5$ A, $R = 2\,\Omega$, $L = \frac{4}{5}$ H, and $C = \frac{1}{20}$ F.

6.26 Determine and sketch $i_L(t)$ and $v_C(t)$ in the circuit of Fig. P6.24 for $i_S = 1$ A, $R = 2\,\Omega$, $L = \frac{8}{29}$ H, and $C = \frac{1}{8}$ F.

6.27 Determine and sketch $i_L(t)$ and $v_C(t)$ in the circuit of Fig. P6.27.

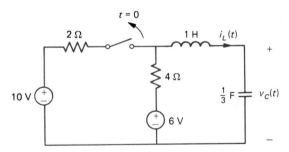

FIGURE P6.27

6.28 Determine and sketch $i_L(t)$ and $v_C(t)$ in the circuit of Fig. P6.28.

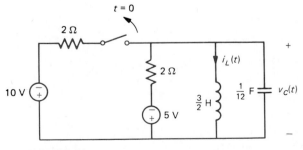

FIGURE P6.28

6.29 Determine and sketch $i_L(t)$ and $v_C(t)$ in the circuit of Fig. P6.29.

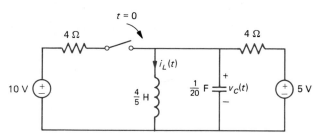

FIGURE P6.29

6.30 Determine and sketch $i_L(t)$ and $v_C(t)$ in the circuit of Fig. P6.30.

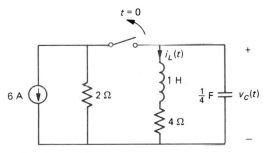

FIGURE P6.30

6.31 Determine and sketch $i_L(t)$ and $v_C(t)$ in the circuit of Fig. P6.31.

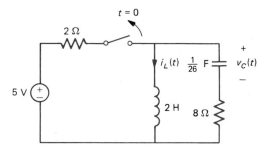

FIGURE P6.31

6.32   Determine and sketch $i_L(t)$ and $v_C(t)$ in the circuit of Fig. P6.32.

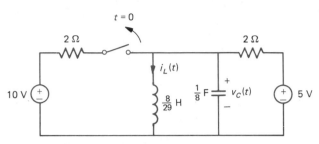

FIGURE P6.32

6.33   Determine and sketch $i(t)$ in the circuit of Fig. P6.33 for $L = \frac{1}{3}$ H, $C = \frac{3}{2}$ F.

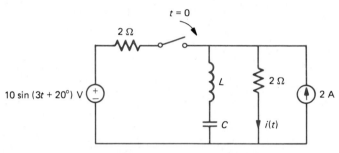

FIGURE P6.33

6.34   Determine and sketch $i_L(t)$ and $v_C(t)$ in the circuit of Fig. P6.34 when $R = 3\ \Omega$, $L = 1$ H, and $C = \frac{1}{2}$ F.

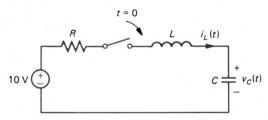

FIGURE P6.34

6.35   Determine and sketch $i_L(t)$ and $v_C(t)$ in the circuit of Fig. P6.34 when $R = 18\ \Omega$, $L = 3$ H, and $C = \frac{1}{27}$ F.

6.36   Determine and sketch $i_L(t)$ and $v_C(t)$ in the circuit of Fig. P6.34 when $R = 2\ \Omega$, $L = 1$ H, and $C = \frac{1}{5}$ F.

2

Electronic Circuits

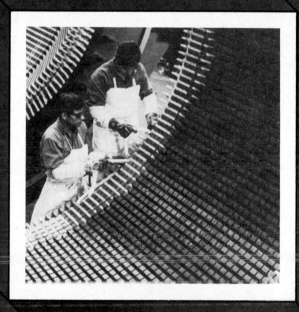

7

Diodes

In Part 1 we considered circuits composed of linear two-terminal elements: the resistor, the inductor, and the capacitor. In Part 2 we consider circuits which may contain—in addition to linear resistors, inductors, and capacitors—certain nonlinear resistors. The nonlinear resistors which we will consider are commonly referred to as *electronic elements*, and these nonlinear circuits are referred to as *electronic circuits*. Of course, the introduction of these nonlinear resistors into an otherwise linear circuit means that the resulting circuit becomes a nonlinear one. Consequently, the powerful circuit-analysis techniques for linear circuits such as superposition no longer hold, and the analysis of these nonlinear circuits becomes considerably more difficult than the analysis of linear circuits.

We study two such nonlinear resistors in this chapter—the semiconductor diode and the zener diode. An important use of these devices is in the construction of dc power supplies. A dc power supply converts an ac, or sinusoidal, waveform (typically the 60-Hz commercial signal) to a constant, or dc, voltage. We will find in later chapters of Part 2 that amplifiers which increase the magnitude of some information-bearing signal require a dc voltage source for proper operation; batteries are clumsy and require frequent maintenance or replacement. Most conventional radios (except portable ones) use a power supply to provide the required dc voltage, thus eliminating the need for batteries. There are many other uses for diodes which we also study.

7.1 THE SEMICONDUCTOR DIODE

The most common two-terminal nonlinear resistor is the semiconductor diode, whose symbol is shown in Fig. 7.1a. The terminal voltage and current are denoted as v_d and i_d, respectively.

The semiconductor diode is constructed by "growing" two pieces of semiconductor material—a p type and an n type—together to form a pn junction, as

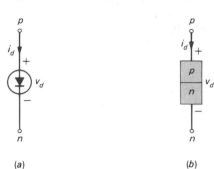

(a)

(b)

FIGURE 7.1 The semiconductor diode: (*a*) circuit symbol; (*b*) physical construction.

shown in Fig. 7.1*b*. Typical types of semiconductor materials used are silicon (Si) and germanium (Ge). Silicon has recently been the predominant material, especially in integrated circuits.

Both silicon and germanium have four valence electrons in their outer atomic orbits. The atoms of pure silicon form covalent bonds, as illustrated in Fig. 7.2; each neighboring atom shares an electron in the bond. Whereas the valence electrons in a metal are free to move about the material to participate in the conduction process, the valence electrons of a pure, or intrinsic, semiconductor (silicon) are so tightly bound that no free electrons are liberated to participate in conduction at a temperature of 0 kelvin (K). At room temperature (293 K), an insignificant number of the valence electrons are liberated.

The energy band description of a material is shown in Fig. 7.3, in which the bands of the atoms' allowed energy levels are separated by gaps or forbidden levels. For a metal, shown in Fig. 7.3*a*, the partially filled upper level is referred to as the conduction, or valence, band; the addition of small amounts of energy raises the energy levels of the electrons in this band quite easily. In the filled band, an increase in the gap energy E_g is required to raise an electron in a filled band to the next higher unfilled band. For a semiconductor material, shown in Fig. 7.3*b*, the conduction band is empty (at 0 K), and the next lower (valence) band is filled. In order for an electron

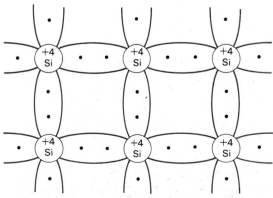

Pure (intrinsic) semiconductor (silicon)

FIGURE 7.2
Silicon and covalent bonds.

FIGURE 7.3
Energy band
characterization of
(*a*) metals, (*b*)
semiconductors,
and (*c*) insulators.

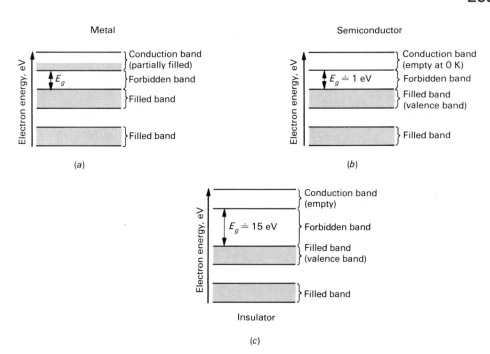

to be elevated into the unfilled conduction band, an energy of approximately 1 electronvolt (eV) must be added. At room temperature, an occasional electron is liberated by thermal energy to the conduction band; ordinarily, however, there are very few such electrons, and the semiconductor material behaves more as an insulator than a metal. The energy bands for an insulator are shown in Fig. 7.3*c*. To move an electron into the insulator's unfilled conduction band requires a much greater energy than for a semiconductor, on the order of 15 eV; hence, the insulator does not conduct except under the application of rather large voltages.

In order for an intrinsic semiconductor such as silicon to conduct current, certain additional sources of conduction electrons must be provided. Impurities are added to the intrinsic semiconductor for this purpose. Note that silicon and germanium are both tetravalent, which means that they have four electrons in the outer valence band; in a pure crystal, these valence electrons are shared with the other atoms in covalent bonds. Typical impurities are the pentavalent ones such as arsenic (As) and antimony (Sb), which have five valence electrons, or the trivalent ones such as gallium (Ga) and boron (B), which have only three valence electrons.

If a pentavalent impurity is added to an intrinsic semiconductor such as silicon, the atom replaces one of the semiconductor atoms and forms a covalent bond. The extra valence electron of the impurity does not participate in a bond and is available for conduction, as shown in Fig. 7.4*a*. The impurity is referred to as a *donor*, and the resulting material is referred to as *n-type material*, since electrons have been liberated. The addition of only very small amounts of these impurities causes a dramatic change in the conductivity of silicon. For example, when one atom in 10 million is replaced

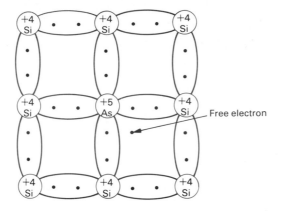 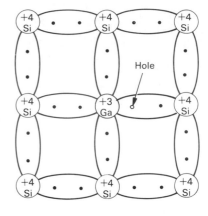

Free electron

Hole

n-type extrinsic semiconductor

(*a*)

p-type extrinsic semiconductor

(*b*)

FIGURE 7.4
Addition of
impurities to form
extrinsic
semiconductors.

by one of these donor impurity atoms, the conductivity of silicon at room temperature increases by a factor of 10^5!

If a trivalent impurity is added to an intrinsic semiconductor, the atom also replaces a host atom and forms a covalent bond. But there are only three valence electrons so that a hole or absence of an electron in a bond is created, as shown in Fig. 7.4*b*. This impurity atom may then accept an electron to complete the bond. When this happens, the hole appears to move to the position from which the electron came and can thus be thought of as a mobile charge carrier. A trivalent impurity is thus called an *acceptor*, and the resulting material is called a *p-type material* in reference to the holes.

Intrinsic semiconductors also have electron-hole pairs generated by thermal energy; however, at room temperature their numbers are quite small. The addition of an impurity to an intrinsic (pure) semiconductor results in an extrinsic semiconductor. Although hole-electron pairs are present via thermal agitation in both *n*-type and *p*-type materials, the dominant charge carrier (or majority carrier) is the one produced by the impurity. In *n*-type extrinsic semiconductors, for example, the electrons liberated by the impurity far exceed the holes produced by thermal agitation. Thus, for *n*-type materials, electrons are the majority carriers and holes are the minority carriers. For *p*-type materials, holes are the majority carriers and electrons are the minority carriers.

A semiconductor diode is formed as shown in Fig. 7.5 by growing a *p*-type material and an *n*-type material to form a *pn* junction diode. Leads (wires) attached to the ends of the material form the terminals of the device. The *n*-type material has a high density of electrons while the *p*-type has a high density of holes, and there is a tendency for these majority carriers to diffuse across the junction. This diffusion continues until a sufficient number of immobile donor and acceptor atoms have been exposed to provide an attractive force (electric field) or potential barrier V_b, which prevents any further diffusion, as shown in Fig. 7.5*b*. This region of immobile charges is called the *depletion region*, and the magnitude of the barrier voltage V_b is a few tenths of a volt.

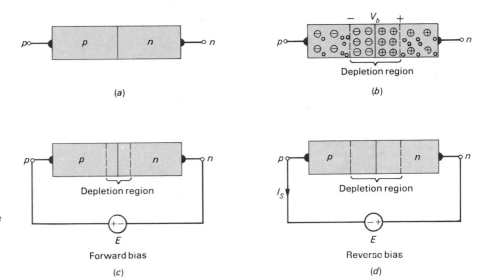

FIGURE 7.5 The formation of a semiconductor diode and its depletion region.

When a battery E is attached to the terminals of the diode with the positive terminal attached to the p-type material and the negative terminal attached to the n-type material, as shown in Fig. 7.5c, the diode is said to be *forward-biased*. With forward bias, more holes in the p region move into the n region. As these holes enter the n region, which has a larger number of electrons, recombination takes place. The electrons eliminated by recombination are replaced by electrons flowing from the negative terminal of the battery into the n region: a current flow. A current passes around the circuit through the battery and the diode. If the battery is reversed, as shown in Fig. 7.5d, the diode is said to be *reverse-biased*. More holes diffuse from the n region (where they are scarce) into the p region, and elecrons diffuse from the p region into the n region. Since a very small number of charge carriers are involved, the current is also quite small. This reverse current is called a *saturation current I_S*, and it eventually reaches a constant value which is quite small, on the order of 10^{-9} A. Thus, for practical purposes, we may say that the semiconductor diode conducts no current (i.e., is an open circuit) under reverse-biased conditions.

The symbol for a semiconductor diode is shown in Fig. 7.6a along with the device terminal characteristic. The direction of the arrow in the symbol serves as a convenient reminder of the direction of current flow through the device occurring for forward-biased conditions (p terminal positive and n terminal negative).

The behavior of the semiconductor diode is found to be characterized by the Shockley diode equation

$$i_d = I_S(e^{qv_d/(\eta kT)} - 1) \tag{7.1}$$

where q is the electron charge (1.6×10^{-19} C), k is Boltzmann's constant ($k = 1.38 \times 10^{-23}$ J/K), and T is the junction temperature in kelvins. The constant η depends on

FIGURE 7.6 The terminal characterization of a semiconductor diode: (a) circuit symbol; (b) device characteristic; (c) device characteristic for several scale factors of the axes.

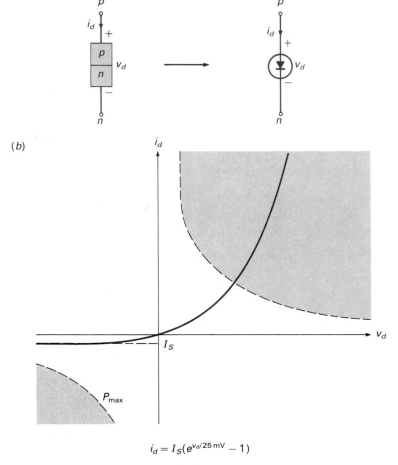

(b)

$$i_d = I_S(e^{v_d/25\,\text{mV}} - 1)$$

the semiconductor material ($\eta = 1$ for Ge and $\eta = 2$ for Si) but can be approximated for practical purposes as unity ($\eta \doteq 1$). At room temperature, or 293 K, we find that

$$\frac{kT}{q} = 0.025 \text{ V}$$

$$= 25 \text{ mV} \tag{7.2}$$

and Eq. (7.1) can be written as

$$i_d = I_S(e^{v_d/25\,\text{mV}} - 1) \tag{7.3}$$

where we have assumed $\eta = 1$ for simplicity. The diode equation in Eq. (7.3) is plotted in Fig. 7.6b. Note that for $v_d \gg 25$ mV, Eq. (7.3) simplifies to

$$i_d \approx I_S e^{v_d/25\,\text{mV}} \qquad v_d \gg 25 \text{ mV} \tag{7.4}$$

FIGURE 7.6
(*Continued*)

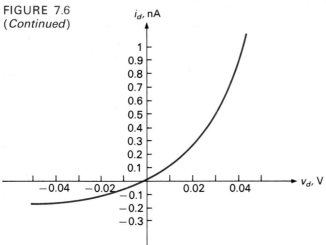

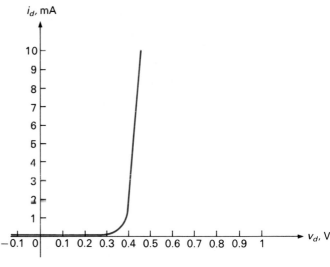

(*c*)

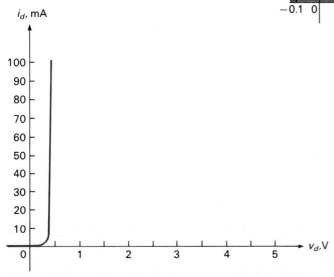

For $v_d \ll 25$ mV, it simplifies to

$$i_d \approx -I_S \qquad v_d \ll 25 \text{ mV} \qquad\qquad (7.5)$$

The diode characteristic is plotted in Fig. 7.6c for several scale factors of the horizontal v_d axis. Note that as the v_d scale is increased, the characteristic approaches that of a vertical line for $v_d > 0.5$ V. This "threshold voltage" where the curve transitions to a vertical line is, of course, not a precise number. It also depends on the material (0.3 for Ge and 0.7 for Si). We will use the easy-to-remember value of 0.5 V. This observation of the behavior for large v_d will be used in our modeling of the device.

Example 7.1

A semiconductor diode at room temperature is found to have a diode current of $i_d = 100$ mA when the diode voltage is $v_d = 0.5$ V. Determine the reverse saturation current I_S.

Solution   From Eq. (7.3) we have

$$I_S = \frac{i_d}{e^{v_d/25\,\text{mV}} - 1}$$

and

$$I_S = \frac{100 \text{ mA}}{e^{0.5/25\,\text{mV}} - 1}$$

$$= 0.206 \times 10^{-9} \text{ A}$$

As current passes through the diode, average power is dissipated and converted to heat. This heat must be dissipated into the surroundings, or it will accumulate and eventually destroy the diode. Semiconductor diodes are rated in terms of the maximum average power which they may safely dissipate; this is shown on the device characteristic in Fig. 7.6b as the hyperbola $P_{\max}$. Ordinarily, $P_{\max}$ is an average power quantity since it relates to the ability of the device to dissipate the accumulated heat. Consequently, it is a function of the waveshape of $v_d(t)$ and $i_d(t)$. For dc operation at some operating point $i_d = I_d$, $v_d = V_d$, the requirement for safe operation of the device is that $P_{\max} > V_d I_d$; thus, the operating point must not lie in the maximum average power dissipation region. For time-varying operation, the average power dissipated by the device is a function of the waveform of $v_d(t)$ and $i_d(t)$. For certain waveforms we may operate in the maximum power region such that $v_d i_d$ exceeds $P_{\max}$ if we do so only momentarily. A safe criterion is to ensure that at no time does the instantaneous product $v_d i_d$ exceed $P_{\max}$.

Heat sinks are metal plates or flanges which are attached to the diode to facilitate the heat dissipation. Heat sinks effectively increase the $P_{\max}$ rating. These heat sinks allow the heat generated in the diode to spread out rapidly, and they increase the surface area for heat dissipation.

7.2 LOAD LINES

The analysis of a circuit containing a diode is quite simple if we use the concept of a *load line*. For example, consider the circuit shown in Fig. 7.7a consisting of a diode, a resistor, and a battery. The diode characteristic in Fig. 7.6b relates the diode voltage v_d and diode current i_d. This may be thought of as an "equation" in the two unknowns v_d and i_d. Now if we are to solve for v_d and i_d, we need another equation containing these variables. This can be obtained by writing KVL around the loop in Fig. 7.7a to give

$$v_d = V_S - Ri_d \qquad (7.6)$$

Now we have two equations in the two unknowns v_d and i_d. If we plot Eq. (7.6) on the diode characteristic and locate the point of intersection of the two curves, we have found the simultaneous solution of these two equations. This point of intersection of the two curves is called the *operating point*. The values of the diode current and voltage are read off the graph and denoted I_d and V_d, respectively, as shown in Fig. 7.7b.

In plotting the load line from Eq. (7.6) on the graph, we must be careful to note that its slope is negative because of the negative sign in front of Ri_d in Eq. (7.6). Also, since the diode characteristic is plotted with i_d on the vertical axis and v_d on the

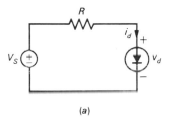

(a)

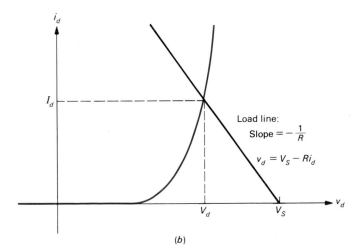

(b)

FIGURE 7.7
Using a load line
to solve a circuit
containing a
diode.

horizontal axis, the slope of the load line is $-1/R$, not $-R$. To fix the actual position of the load line, we note in Eq. (7.6) that when $i_d = 0$, $v_d = V_S$. The horizontal axis of the graph is the location of all points where $i_d = 0$. Therefore, the load line must cross the v_d axis at V_S.

This concept of load lines can be easily extended to more complicated circuits as long as they contain ideal sources, linear resistors, and only one diode. To do this, we reduce the circuit attached to the diode terminals to a Thévenin equivalent so that it appears in the form of Fig. 7.7a. In this case R becomes R_{TH} and V_S becomes V_{OC}. Then we draw the load line on the diode characteristic and solve for the diode voltage and current at the operating point, V_d and I_d. With a knowledge of the diode voltage and current, we work back into the circuit to obtain any other element voltage or current of interest.

Example 7.2

In the circuit of Fig. 7.8a, determine the current i.

Solution   First determine the Thévenin resistance seen by the diode, as shown in Fig. 7.8b:

$$R_{\text{TH}} = 4 \text{ k}\Omega$$

Next determine the open-circuit voltage at terminals a-b (by superposition), as shown in Fig. 7.8c:

$$V_{\text{OC}} = 3 \text{ V}$$

FIGURE 7.8
Example 7.2:
illustration of the
reduction of a
circuit by using a
Thévenin
equivalent and a
load line.

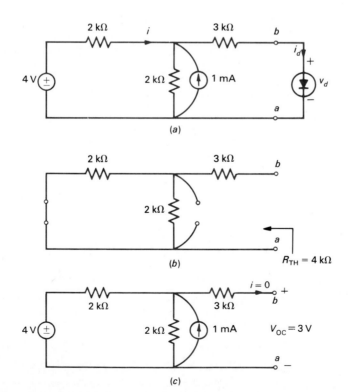

Attach the Thévenin equivalent circuit to the diode, as shown in Fig. 7.8*d*. Plot the load line on the diode characteristic, as in Fig. 7.8*e*, and obtain

$$V_d = 0.38 \text{ V}$$

$$I_d = 0.655 \text{ mA}$$

Now to determine *i*, we simply apply KCL at the node to which *i* enters to obtain

$$i = I_d - 1 \text{ mA} + \frac{V_d + 3kI_d}{2k}$$

$$= 0.8275 \text{ mA}$$

FIGURE 7.8
(*Continued*)

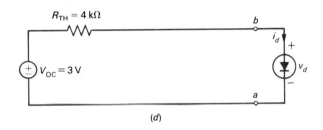

(*d*)

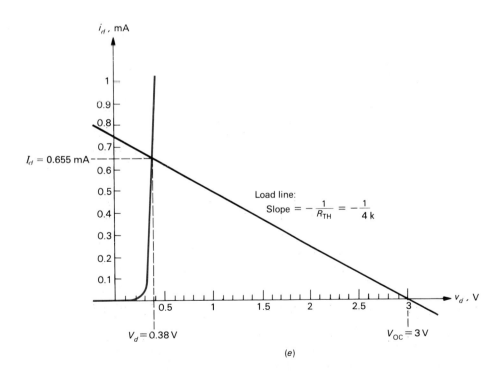

(*e*)

7.3 PIECEWISE-LINEAR APPROXIMATION AND SMALL-SIGNAL ANALYSIS

Although graphical analysis of resistive circuits containing one semiconductor diode is relatively straightforward when load lines are used, as we have seen in the previous section, there is nevertheless the problem of drawing the graphical characteristic of the diode, accurately plotting the load line, and reading off the values of the operating point. We will have numerous occasions to consider other approximate methods of analysis.

The primary approximate method which we employ is the *piecewise-linear approximation*, shown as a dashed line in Fig. 7.9a. The piecewise-linear approximation consists of two straight-line segments. At a particular operating point (I_d, V_d), we construct a tangent to the curve having slope $1/R_f$. The reciprocal of this slope, or R_f, is called the *forward resistance* of the diode *at this operating point* (I_d, V_d). The intersection of this line and the v_d axis $(i_d = 0)$ is denoted by E_f, which is referred to as the *forward voltage* of the diode for this operating point. For $v_d < E_f$, another straight-line segment with zero slope along the v_d axis completes the characterization. This latter segment corresponds to the condition that the diode is assumed to conduct no current for reverse-biased conditions $(v_d < 0)$. For $v_d > E_f$, the diode approximation appears as a resistor R_f, in series with a battery E_f, as shown in Fig. 7.9b.

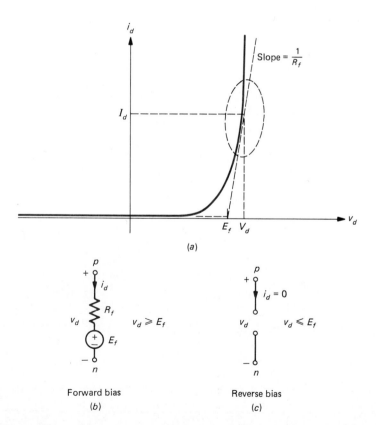

FIGURE 7.9 The piecewise-linear approximation of a diode.

For $v_d < E_f$, the diode is approximated as an open circuit conducting no current, as shown in Fig. 7.9c.

The value of the forward resistance R_f for a particular operating point (V_d, I_d) is found from Eq. (7.3):

$$\begin{aligned} R_f &= \left.\frac{dv_d}{di_d}\right|_{V_d, I_d} \\ &= \frac{25\,\text{mV}}{I_S e^{V_d/25\,\text{mV}}} \\ &= \frac{25\,\text{mV}}{I_d + I_S} \end{aligned} \tag{7.7}$$

Assuming I_S to be much smaller in magnitude than i_d, we have

$$R_f \approx \frac{25}{I_d\,(\text{mA})} \quad \Omega \tag{7.8}$$

For example, at an operating point of $I_d = 1$ mA, $R_f = 25\,\Omega$, and at $I_d = 10$ mA, $R_f = 2.5\,\Omega$. The forward voltage E_f is found from the equation of the tangent:

$$v_d = E_f + R_f i_d \tag{7.9}$$

For an operating point (I_d, V_d) we obtain

$$\begin{aligned} E_f &= V_d - R_f I_d \\ &= V_d - 25\,\text{mV} \end{aligned} \tag{7.10}$$

The representation of the diode with the piecewise-linear equivalent circuit in Fig. 7.9b only describes the characteristic in a small region about the operating point, as circled in Fig. 7.9a. This representation, however, is useful for what is known as small-signal analysis.

The use of this piecewise-linear approximation for small-signal analysis is illustrated in Fig. 7.10a. The ideal voltage source consists of a dc voltage E_S in series with a sinusoidal voltage $V_S \sin \omega t$. The load line, plotted on the characterisic in Fig. 7.10b, consists of a line of slope $-1/R_S$ intersecting the $i_d = 0$ axis at $v_d = E_S + V_S \sin \omega t$. As the sinusoid varies, the load line moves back and forth and the operating point moves along the characteristic. The actual values of i_d and v_d are plotted versus time as solid lines, whereas the approximate values using the tangent to the characteristic are plotted as dashed lines. Note that since the diode characteristic is nonlinear, the diode voltage and current only approximately resemble sinusoids; distortion of the input sinusoid $v_S(t)$ is said to have taken place. However, if we restrict the magnitude of the source sinusoid V_S to a small value, the nonlinearity of the device characteristic becomes less important and the diode voltage and current

FIGURE 7.10
Small-signal
operation.

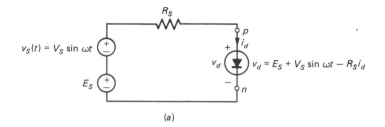

(a)

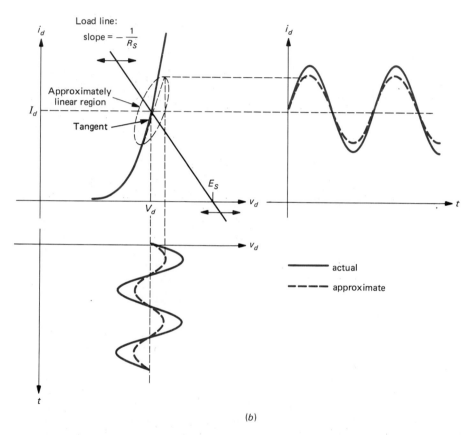

actual
--- approximate

(b)

more closely resemble sinusoids. In this case, we may simply replace the diode with its (linear) forward-biased circuit shown in Fig. 7.9b.

This is an important concept which is used often in Part 2. Note that so long as $v_S(t)$ is restricted to being "small enough" that excursions of the instantaneous operating point are restricted to an approximately linear region of the diode characteristic, we may replace the nonlinear diode with a linear circuit, as shown in Fig. 7.11a. So long as $v_S(t)$ is small enough, it doesn't matter that the diode characteristic is truly nonlinear; $v_S(t)$ never forces v_d and i_d outside of the approximately linear, circled region. Thus, the remainder of the characteristic may be approximated as shown. Under this restriction of small-signal approximately linear operation, we may replace the nonlinear diode with its small-signal linear equivalent circuit,

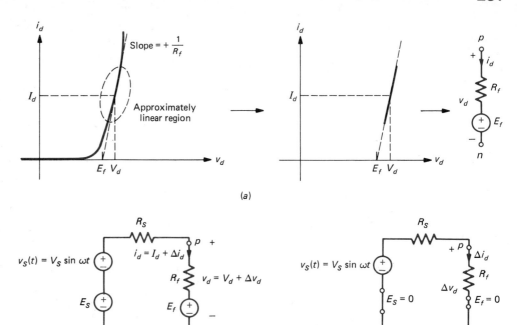

FIGURE 7.11
Small-signal
analysis.

as shown in Fig. 7.11b. Now we have a strictly *linear problem* to which we may apply all our previous linear resistive-circuit-analysis techniques, including superposition.

Granted, this is an approximation, but it is nevertheless a rather considerable simplification. We first find the operating point V_d and I_d by plotting the load line on the diode characteristic with $v_g(t) = 0$. We then find a small signal forward resistance R_f at the particular operating point from Eq. (7.8). The forward voltage E_f is found from Eqs. (7.8) and (7.10) (or is read off the characteristic) as

$$E_f = V_d - R_f I_d \tag{7.11}$$

Substituting Eq. (7.8) gives

$$E_f = V_d - 0.025 \tag{7.12}$$

Then we replace the diode with its small-signal linear equivalent circuit, as shown in Fig. 7.11b. Now, with simple linear circuit-analysis principles, we may solve for the diode voltage and current. Note that the circuit is, with this approximation, a linear one so that, by superposition, the diode current and voltage will consist of two pieces:

$$i_d = I_d + \Delta i_d \tag{7.13a}$$

$$v_d = V_d + \Delta v_d \tag{7.13b}$$

where

$$\left.\begin{matrix} I_d \\ V_d \end{matrix}\right\} \text{ due to } E_S \text{ and } E_f \qquad\qquad (7.14a)$$

$$\left.\begin{matrix} \Delta i_d \\ \Delta v_d \end{matrix}\right\} \text{ due to } v_S(t) \qquad\qquad (7.14b)$$

If we are only interested in the portion due to $v_S(t)$, we may set E_S and E_f equal to zero, as shown in Fig. 7.11c, and solve for Δi_d and Δv_d as

$$\Delta i_d = \frac{v_S(t)}{R_S + R_f} \qquad\qquad (7.15a)$$

$$\Delta v_d = \frac{R_f}{R_f + R_S} v_S(t) \qquad\qquad (7.15b)$$

This principle of small-signal approximately linear analysis is used on numerous occasions in Part 2 and should be studied carefully.

Quite often, for practical purposes, we may neglect R_f in the small-signal equivalent circuit of a diode. For typical semiconductor diodes, the value of R_f is quite small: between 1 and 100 Ω. Note that the diode characteristic for $v_d \gg 25\text{ mV}$

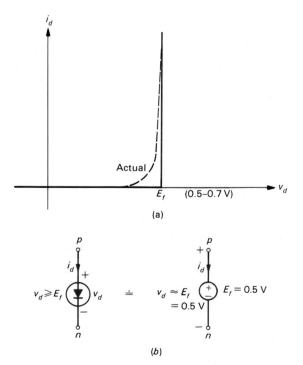

(a)

(b)

FIGURE 7.12 A practical approximation of a diode.

is exponential, as shown in Eq. (7.4). Consequently, for increasing v_d, the characteristic asymptotically approaches a vertical line quite rapidly, and a simpler approximate representation of the semiconductor diode is to neglect R_f, as shown in Fig. 7.12*a*. So long as the operating point is not in the "knee" of the characteristic, this approximation is usually sufficient. For silicon diodes, typical values of E_f range for 0.5 to 0.7 V. We use a value of 0.5 V in our analyses, as shown in Fig. 7.12*b*.

7.4 THE IDEAL DIODE

The semiconductor diode behaves in a fashion similar to a switch: Current flows in only one direction. For reverse-biased diodes ($v_d < 0$), the diode appears as an open circuit. This important feature of the diode permits the construction of several useful circuits.

To obtain preliminary estimates of the circuit performance, we replace the actual diode with an ideal diode, as shown in Fig. 7.13*a*. The ideal diode is effectively an ideal switch. When the diode is reverse-biased ($v_d < 0$), it appears as an open circuit, as is the case for an actual diode, as shown in Fig. 7.13*b*. Under forward-biased conditions ($i_d > 0$) it closes and appears as a short circuit, as shown in Fig. 7.13*c*.

The ideal diode is a further approximation to the piecewise-linear approximation in Fig. 7.9, in that for the ideal diode $R_f = 0$ and $E_f = 0$. For a semiconductor diode, R_f is quite small (typically only a few ohms), and E_f is on the order of 0.5 V.

Once a preliminary estimate of a circuit's performance is obtained by replacing the actual diodes with ideal diodes, a more refined estimate can be obtained by the

FIGURE 7.13
The ideal diode:
(*a*) device
characteristic;
(*b*) reverse bias;
(*c*) forward bias.

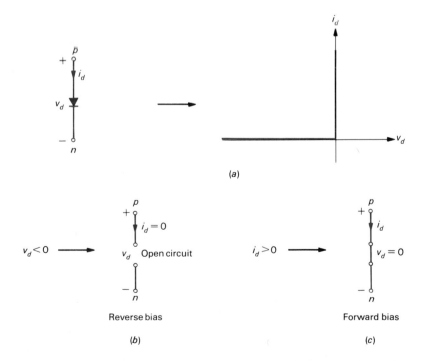

FIGURE 7.14
The piecewise-
linear model of a
diode, using an
ideal diode.

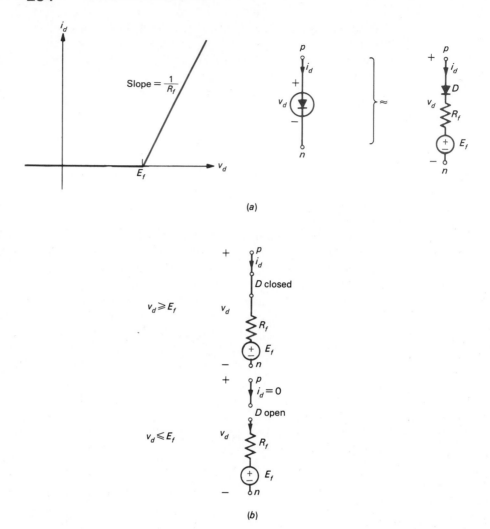

(a)

(b)

piecewise-linear models in series with the ideal diode, as shown in Fig. 7.14a. The ideal diode is closed in this model for $v_d > E_f$ and open for $v_d < E_f$, as shown in Fig. 7.14b.

Example 7.3

Nonlinear resistors with a wide range of characteristics can be obtained, approximately, with circuits containing diodes. For example, a square-law device is a two-terminal nonlinear resistor whose terminal voltage-current characteristic obeys

$$i = kv^2$$

where k is a normalization constant. The ideal characteristic is shown in Fig. 7.15a; this device may be used in modulators, e.g., to attach a voice signal to a higher-frequency carrier wave, as is done in amplitude modulation (AM) radio transmission. Design a square-law device to approximate the ideal characteristics for $0 \leq v \leq 5 \text{ V}$ with a normalization constant of $k = 0.001$.

FIGURE 7.15
Example 7.3:
synthesis of a
square-law
resistor, using
diodes.

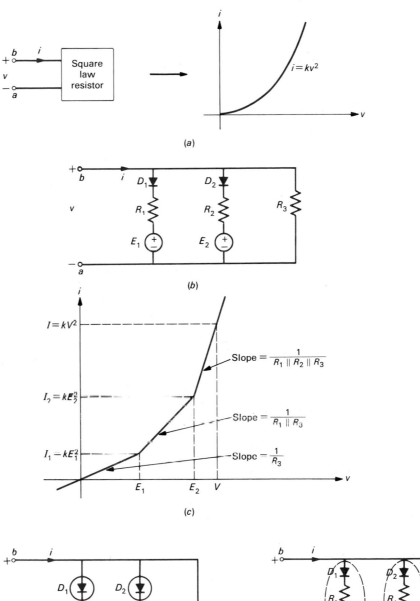

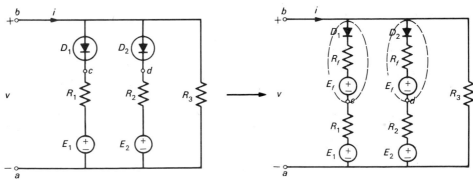

Solution A circuit using ideal diodes which will accomplish the task is shown in Fig. 7.15*b*. First we describe the operation of this circuit. Slowly increase v from zero and observe the behavior of i. Initially, diodes D_1 and D_2 will be reverse-biased and open. Once v reaches the smaller of the two battery voltages E_1 or E_2, the diode associated with that branch will close. Let us assume for illustration that $E_1 < E_2$. Thus, until v reaches E_1, both diodes will be open and only resistor R_3 will be in the circuit; for $0 \le v \le E_1$, the curve will thus have slope $1/R_3$. For $E_1 \le v \le E_2$, diode D_1 will be closed, but D_2 will be open. The input resistance at the terminals will be $R_1 \| R_3$. For $E_2 \le v \le V$, diode D_2 will also be closed, and the input resistance at the terminals will be $R_1 \| R_2 \| R_3$. Thus, the input characteristic will have the piecewise-linear characteristic shown in Fig. 7.15*c*. We would choose $V = 5$ V and choose the battery voltages E_1 and E_2 arbitrarily. Suppose we choose the battery voltages as

$$E_1 = 2.0 \text{ V}$$

$$E_2 = 3.5 \text{ V}$$

The resistors are chosen such that the breakpoints at E_1 and E_2 and the point at $V = 5$ V fall on the desired ideal curve. Thus, we compute

$$I_1 = kE_1^2$$
$$= 4 \text{ mA}$$

$$I_2 = kE_2^2$$
$$= 12.25 \text{ mA}$$

$$I = kV^2$$
$$= 25 \text{ mA}$$

Noting the slopes of each portion, we obtain

$$R_3 = \frac{E_1}{I_1}$$
$$= 500 \ \Omega$$

$$R_1 \| R_3 = \frac{E_2 - E_1}{I_2 - I_1}$$
$$= 182 \ \Omega$$

so that

$$R_1 = 286 \ \Omega$$

Finally,

$$R_1 \| R_2 \| R_3 = \frac{V - E_2}{I - I_2}$$
$$= 118 \ \Omega$$

so that

$$R_2 = 333 \ \Omega$$

Now that we have designed the circuit for ideal diodes, let us determine the actual values of R_1, R_2, R_3, E_1, and E_2 which should be used if we substitute actual diodes, as shown in Fig. 7.15d. Replacing the actual diodes with their piecewise-linear approximations, we see that we must subtract R_f from the above values of R_1 and R_2 and subtract E_f from the above values of E_1 and E_2. If we assume that both diodes have $R_f = 10\ \Omega$ and $E_f = 0.5$ V, then the actual circuit elements should be

$$R_1 = 276\ \Omega$$
$$R_2 = 323\ \Omega$$
$$R_3 = 500\ \Omega$$
$$E_1 = 1.5\ \text{V}$$
$$E_2 = 3.0\ \text{V}$$

In this case we may use ordinary flashlight batteries for E_1 and E_2 (a result of some preliminary planning).

In the following sections, we investigate some useful circuits containing semi-conductor diodes. These will be analyzed by replacing the actual diodes with ideal diodes. The modification of this behavior when piecewise-linear approximations are used will be indicated.

We will find it necessary to use the ideal transformers shown in Fig. 7.16; these were discussed in Chap. 4. Here we provide a brief review of their properties. It is sufficient to describe the ideal transformer with turns ratio $N_1 : N_2$ as a device such

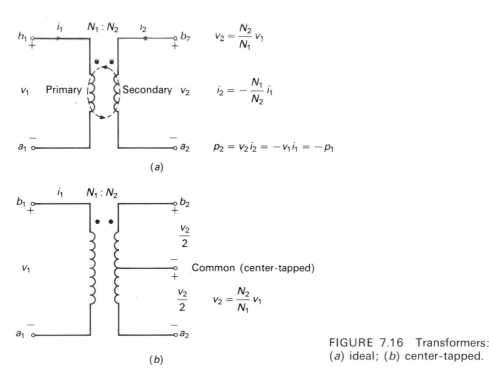

FIGURE 7.16   Transformers: (*a*) ideal; (*b*) center-tapped.

that the voltages v_2 and v_1 are related by the turns ratio:

$$v_2 = \frac{N_2}{N_1} v_1 \qquad (7.16)$$

If $N_2 > N_1$, the transformer can be used to step up or step down a voltage, depending on which side the voltage source is attached. The currents are related by

$$i_2 = -\frac{N_1}{N_2} i_1 \qquad (7.17)$$

Note that if the voltage v_2 is stepped up by the ratio of N_2/N_1, the current is stepped down by N_1/N_2. The power delivered to the transformer is

$$
\begin{aligned}
p &= p_2 + p_1 \\
&= \left(\frac{N_2}{N_1} v_1 \right)\left(-\frac{N_1}{N_2} i_1 \right) + v_1 i_1 \\
&= 0
\end{aligned}
\qquad (7.18)
$$

Thus, the ideal transformer consumes no power.

Practical transformers consist of turns of wire around a ferromagnetic metal. The magnetic flux generated in one winding is linked with the turns of the other winding, inducing a voltage in that winding. For this to occur, the voltages and currents must be varying with time so that transformers are commonly used to step up or step down ac (sinusoidal) voltages, such as commercial 60-Hz power.

If a terminal (common terminal) is attached to the center of the secondary winding, as shown in Fig. 7.16b, the transformer is said to be *center-tapped*. The total voltage across the secondary $v_2 = (N_2/N_1)v_1$ is now equally divided between the two halves. As terminal b_2 is increasing with respect to the common terminal, terminal a_2 is decreasing with respect to the common terminal. This arrangement is sometimes referred to as a *phase inverter* since it produces two voltages with respect to the common terminal which are 180° out of phase with each other.

7.5 RECTIFIERS

Rectifiers are circuits which produce a dc voltage or current from an ac source. The semiconductor diode is an essential component of these devices. As we will see in subsequent chapters, numerous electronic circuits require a dc voltage for operation. Batteries are cumbersome and require frequent maintenance and replacement. A more practical method of providing a dc voltage is to rectify an ac voltage; ac voltages are somewhat easier to generate, and transmission of ac power is, at present, more efficient than transmission of dc power, as evidenced by the numerous high-voltage ac transmission lines across the countryside. The generation and transmission of ac power are discussed in Part 3.

7.5.1 The Half-Wave Rectifier

A *half-wave rectifier* using an ideal diode is shown in Fig. 7.17a. The resistor R_L represents some load to which we would like to supply a dc voltage. A 1:1 ideal

FIGURE 7.17
The half-wave
rectifier.

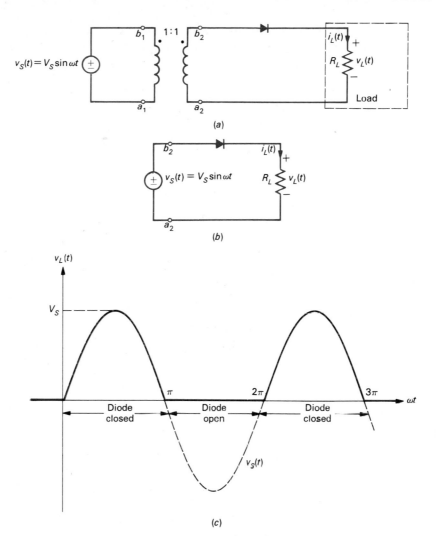

transformer isolates the load from the source. The transformer is useful in providing a larger or smaller apparent value of source voltage. If instead of a 1:1 turns ratio we had used a 1:2 turns ratio, the secondary voltage would be $2v_S(t)$. We may replace the source voltage and transformer with an equivalent source voltage, as shown in Fig. 7.17b.

During the positive half cycle of the source, the ideal diode is forward-biased and closed so that the source voltage is connected directly across the load. During the negative half cycle of the source, the ideal diode is reversed-biased so that the source voltage is disconnected from the load and the load voltage is zero. Thus, the load voltage is said to be *rectified* and is of one polarity, as shown in Fig. 7.17c. It may be described by

$$v_L(t) = V_S \sin \omega t \qquad 0 \le \omega t \le \pi \tag{7.19a}$$

$$v_L(t) = 0 \qquad \pi \le \omega t \le 2\pi \tag{7.19b}$$

Although the resulting load voltage is not truly dc as produced by a battery, it nevertheless has an average, or dc, component. The average (dc) value of v_L is

$$V_L = \frac{1}{2\pi} \int_0^\pi V_S \sin \omega t \, d(\omega t)$$

$$= \frac{V_S}{\pi} \tag{7.20}$$

Diodes are often rated in terms of their peak inverse voltage (PIV) and their maximum average power dissipation P_{max}, discussed previously. The PIV rating refers to the maximum reverse-biased voltage the diode can withstand. For the half-wave rectifier, this maximum reverse voltage occurs when $v_S(t)$ reaches its peak on the negative half cycle $-V_S$. Thus, for a 1:1 transformer, PIV = V_S. For a 1:5 transformer, PIV = $5V_S$. Ideal diodes obviously do not dissipate or consume power; thus, there is no power restriction P_{max}. Practical diodes do, and for this case we would have to find the diode voltage waveform $v_d(t)$ and diode current waveform $i_d(t)$ and determine whether

$$P_{AV} = \frac{1}{T} \int_0^T v_d(t) i_d(t) \, dt \tag{7.21}$$

is less than P_{max}.

7.5.2 Filtering the Half-Wave Rectifier Voltage

The pulsating dc voltage produced by the half-wave rectifier is periodic and thus has a Fourier series representation. (See App. B for a discussion of the Fourier series.) The average value was determined in the previous section. Denoting this representation of the load voltage as

$$\begin{aligned} v_L(t) = V_L &+ a_1 \sin \omega t + a_2 \sin 2\omega t + \cdots \\ &+ b_1 \cos \omega t + b_2 \cos 2\omega t + \cdots \end{aligned} \tag{7.22}$$

the Fourier coefficients are determined from App. B as

$$a_n = \frac{2}{T} \int_0^T v_L(t) \sin n\omega t \, dt \tag{7.23a}$$

$$b_n = \frac{2}{T} \int_0^T v_L(t) \cos n\omega t \, dt \tag{7.23b}$$

For the half-wave rectified voltage,

$$a_1 = \frac{2}{T} \int_0^T v_L(t) \sin \omega t \, dt$$

$$= \frac{1}{\pi} \int_0^\pi V_S \sin \omega t \sin \omega t \, d(\omega t)$$

$$= \frac{V_S}{\pi} \int_0^\pi \frac{1}{2} (1 - \cos 2\omega t) \, d(\omega t)$$

$$= \frac{V_S}{2}$$

Similarly, we may obtain

$$b_1 = 0 \qquad b_2 = -\frac{2V_S}{3\pi} \qquad b_3 = 0 \qquad b_4 = -\frac{2V_S}{15\pi} \qquad b_5 = 0$$

The coefficients of the sine terms a_n are zero for $n > 1$ since

$$a_n = \frac{2}{T} \int_0^T v_L(t) \sin n\omega t \, dt$$

$$= \frac{1}{\pi} \int_0^\pi V_S \sin \omega t \sin n\omega t \, d(\omega t)$$

$$= \frac{V_S}{\pi} \int_0^\pi \frac{1}{2} [\cos (n - 1)\omega t - \cos (n + 1)\omega t] \, d(\omega t)$$

$$= 0$$

but

$$a_1 = \frac{V_S}{2}$$

Thus, the Fourier series for the half-wave rectified signal is

$$v_L(t) = \frac{V_S}{\pi} + \frac{V_S}{2} \sin \omega t - \frac{2V_S}{3\pi} \cos 2\omega t - \frac{2V_S}{15\pi} \cos 4\omega t + \cdots \tag{7.24}$$

This shows that, in addition to the average (dc) value, higher-frequency harmonics are also present. This suggests the use of a filter to eliminate these unwanted sinusoidal components. One such satisfactory arrangement is obtained by placing a capacitor across the load resistor, as shown in Fig. 7.18a. We observe that the capacitor has a lower impedance to higher frequencies and is an open circuit to direct

FIGURE 7.18
Filtering of the
half-wave rectifier.

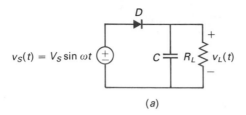

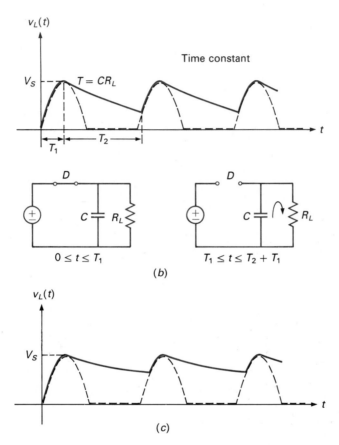

current; thus, the capacitor should tend to short-circuit the high-frequency components of the wave. As the source voltage initially increases positively, the diode is
forward-biased since the load voltage is zero and the source is directly connected to
the load. Once the source reaches its maximum value V_S and begins to decrease, the
load voltage and consequently the capacitor voltage are maintained momentarily at
V_S and the diode becomes reverse-biased and consequently open-circuited. The
capacitor then discharges over interval T_2 through R_L until the source voltage $v_S(t)$
has increased to a value equal to the load voltage. At this point, the source voltage
exceeds the capacitor voltage, and the diode is once again forward-biased and closed.
The capacitor once again charges to V_S. These points are illustrated in Fig. 7.18b.

Thus, the load voltage is smoothed somewhat and more closely resembles a true dc voltage. The smoothing of the filter can be improved by increasing the CR_L time constant so that the discharge rate is slowed. If R_L is fixed (as it usually is), this can be accomplished by using a larger capacitor, as shown in Fig. 7.18c.

7.5.3 Effects of Actual Diodes

If the piecewise-linear model of an actual diode replaces the ideal diode in the half-wave rectifier, the circuit becomes as shown in Fig. 7.19a. The ideal diode is open for $v_S(t) < E_f$ and is closed for $v_S(t) > E_f$, as shown in Fig. 7.19b. The phase

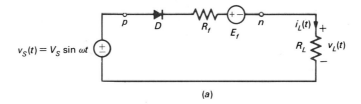

(a)

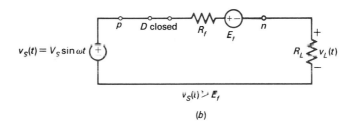

(b)

FIGURE 7.19
The half-wave
rectifier—modeling
the effects of
nonideal diodes
with piecewise-
linear equivalents.

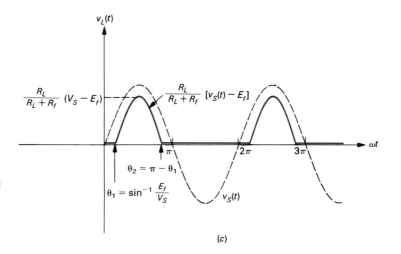

(c)

displacement $\omega t_1 = \theta_1$ at which the ideal diode closes is found by setting $v_S(t)$ equal to E_f:

$$V_S \sin \theta_1 = E_f \tag{7.25}$$

or

$$\theta_1 = \sin^{-1} \frac{E_f}{V_S} \tag{7.26}$$

The diode opens at $\theta_2 = \pi - \theta_1$. The load voltage when the ideal diode is closed is given by voltage division as

$$v_L(t) = \frac{R_L}{R_L + R_f} [v_S(t) - E_f] \tag{7.27}$$

The peak is reduced from the ideal case, as shown in Fig. 7.19c. One of the major effects is to lower the dc component of the voltage, since the average value of $v_L(t)$ has been lowered. This average value is determined from

$$V_L = \frac{1}{2\pi} \int_{\theta_1}^{\theta_2} \frac{R_L}{R_L + R_f} (V_S \sin \omega t - E_f)\, d(\omega t)$$

$$= \frac{1}{2\pi} \frac{R_L}{R_L + R_f} (-V_S \cos \theta_2 + V_S \cos \theta_1 - E_f \theta_2 + E_f \theta_1) \tag{7.28}$$

Since $\theta_2 = \pi - \theta_1$, we obtain

$$V_L = \frac{1}{2\pi} \frac{R_L}{R_L + R_f} (2V_S \cos \theta_1 - E_f \pi + 2E_f \theta_1) \tag{7.29}$$

by using the trigonometric identity $\cos (\pi - \theta) = -\cos \theta$.

7.5.4   The Full-Wave Rectifier

The half-wave rectifier produced a pulsating direct current yet appeared to use only the positive half cycle of the source voltage. The *full-wave rectifier* shown in Fig. 7.20a uses both half cycles of $v_S(t)$. We have used an ideal transformer with a center-tapped secondary having a $1:2$ turns ratio. Therefore, the voltage across each of the halves of the secondary is $v_S(t)$, but the voltages are out of phase with respect to the common, center-tapped terminal. Redrawing this circuit as shown in Fig. 7.20b, we see that for the positive half cycle of $v_S(t)$, diode D_1 is forward-biased and closed and diode D_2 is reverse-biased and open (as shown in Fig. 7.20c) and $v_L(t) = v_S(t)$. Over the negative half cycle of $v_S(t)$, D_1 is open and D_2 is closed and $v_L(t) = -v_S(t)$.

FIGURE 7.20
The full-wave
rectifier.

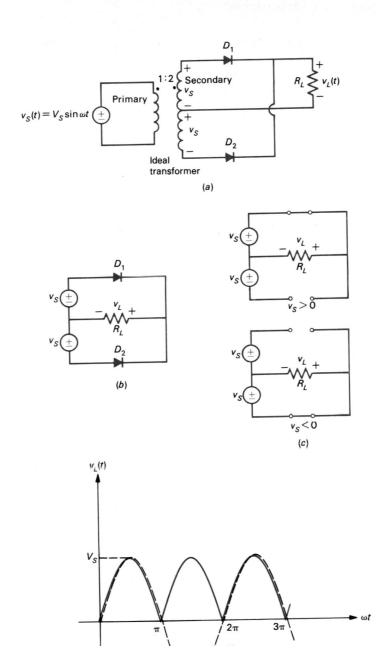

This has the net effect of reversing the polarity of the negative half cycle of $v_S(t)$ and applying this directly across the load.

The resulting load voltage waveform is shown in Fig. 7.20d. The average value is

$$V_L = \frac{1}{\pi} \int_0^{\pi} V_S \sin \omega t \, d(\omega t)$$

$$= \frac{2V_S}{\pi} \tag{7.30}$$

which is twice the dc component of the half-wave rectifier. The PIV of each diode is, from Fig. 7.20c, twice that of the half-wave rectifier: $2V_S$. The Fourier series of this waveform is

$$v_L(t) = \frac{2V_S}{\pi} - \frac{4V_S}{3\pi} \cos 2\omega t - \frac{4V_S}{15\pi} \cos 4\omega t + \cdots \tag{7.31}$$

Comparing Eqs. (7.31) and (7.24), we note an advantage of the full-wave rectifier. First, the fundamental harmonic ($\sin \omega t$) present in the half-wave rectifier result is

FIGURE 7.21
The full-wave
rectifier—modeling
the effects of
nonideal diodes
with piecewise-linear
equivalents.

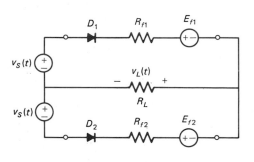

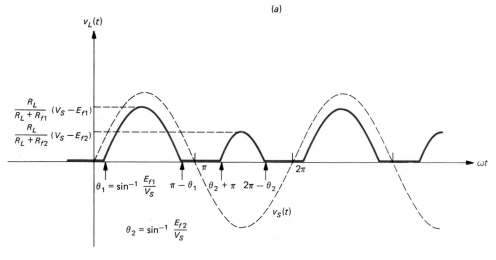

eliminated from the full-wave rectifier result. This indicates that the undesired (nonconstant) portion of $v_L(t)$ is reduced, as is clear from the waveform.

Replacing each diode with its piecewise-linear approximation, we obtain the circuit and resulting waveform shown in Fig. 7.21. We have shown the waveform for the case where the two diodes have different values of R_f and E_f. The different values of forward voltage E_f cause each ideal diode to delay its turn-on time by different amounts.

7.5.5 The Full-Wave Bridge Rectifier

The full-wave rectifier considered in the previous section must employ a center-tapped transformer for proper operation. A full-wave rectifier can be constructed without the need for a transformer, but we must use four diodes instead of two. This type of rectifier is called the *bridge rectifier* and is shown in Fig. 7.22a. When the source

FIGURE 7.22
The bridge
rectifier.

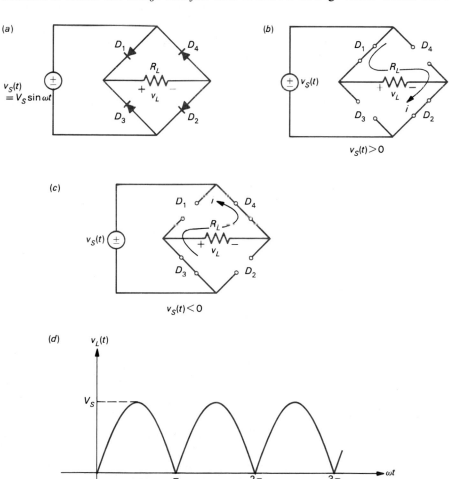

voltage is positive, diodes D_1 and D_2 are forward-biased and closed but D_3 and D_4 are reverse-biased and open, resulting in $v_L(t) = v_S(t)$, as shown in Fig. 7.22*b*. When the source voltage is negative, diodes D_1 and D_2 are reverse-biased and open and D_3 and D_4 forward-biased and closed, resulting in $v_L(t) = -v_S(t)$, as shown in Fig. 7.22*c*. The resulting waveform is a full-wave rectified wave having the same dc component as for the full-wave rectifier which used a 1:2 turns-ratio transformer. Of course, a transformer could be used to increase (or decrease) $v_S(t)$.

7.6   THE LIMITER

Often we need to restrict the variation of some voltage within certain limits to prevent an excessive voltage from appearing across the terminals of a device. A limiter, constructed from diodes, is a device which performs this function. For example, suppose we wish to restrict some voltage $v_L(t)$ across a load between the limits V_1 and $-V_2$, as shown in Fig. 7.23*a*. Inserting a limiter between the source and load will perform this function, as shown in Fig. 7.23*b*.

A simple limiter consisting of two ideal diodes and two batteries is shown in Fig. 7.24*a*. If the load voltage attempts to exceed the battery voltage V_1, diode D_1 closes at $v_L(t) = V_1$. Similarly, if $v_L(t)$ attempts to go more negative than $-V_2$, diode D_2 closes at $v_L(t) = -V_2$. For $-V_2 < v_L < V_1$ both diodes are open. These cases are illustrated in Fig. 7.24.

The analysis shows that the load voltage is limited between the extremes $-V_2$ and V_1. Using the resulting equivalent circuits in Fig. 7.24*b*, *c*, and *d* shows that the limiting occurs for the source voltage

$$-\frac{R_L + R_S}{R_L} V_2 < v_S(t) < \frac{R_L + R_S}{R_L} V_1 \tag{7.32}$$

which is obtained by using voltage division in the circuit of Fig. 7.24*d*. This gives the resulting "transfer characteristic" shown in Fig. 7.24*e*.

7.7   ZENER DIODES

It was pointed out in the discussion of semiconductor diodes that the diode conducts only a small reverse saturation current I_S under reverse-biased conditions. However, if the reverse-biased voltage is increased to a large value, the semiconductor diode will "break down" and conduct a significant current in the reverse direction. This phenomenon is referred to as *avalanche breakdown;* it results from the charge carriers acquiring sufficient energy between collisions to dislodge electrons from some of the covalent bonds. These dislodged electrons can cause other electrons to dislodge, which results in a cumulative, or avalanche, effect. A similar effect, the *zener effect*, occurs at lower reverse voltage. In this effect, the electric field becomes sufficiently intense to pull electrons from the covalent bonds, resulting in a large reverse current.

FIGURE 7.23
The ideal limiter.

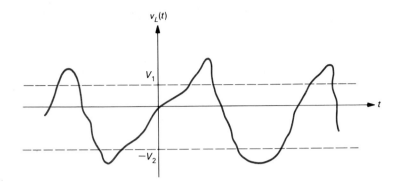

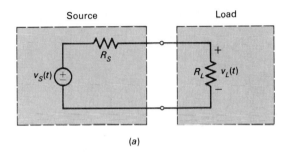

(a)

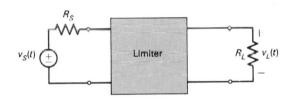

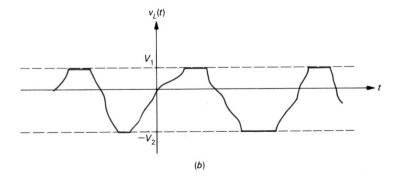

(b)

FIGURE 7.24
Synthesis of a
limiter using ideal
diodes.

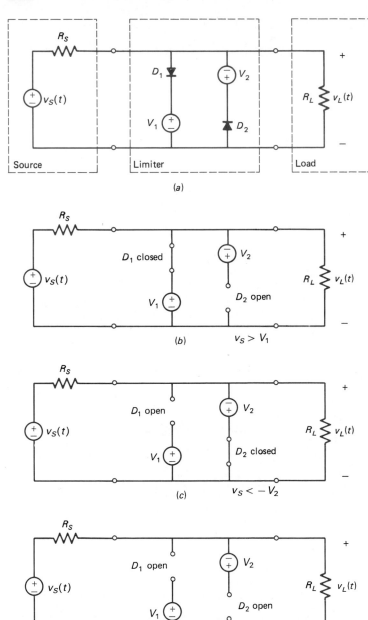

(a)

(b) $v_S > V_1$

(c) $v_S < -V_2$

(d) $-V_2 < v_L < V_1$

FIGURE 7.24
(*Continued*)

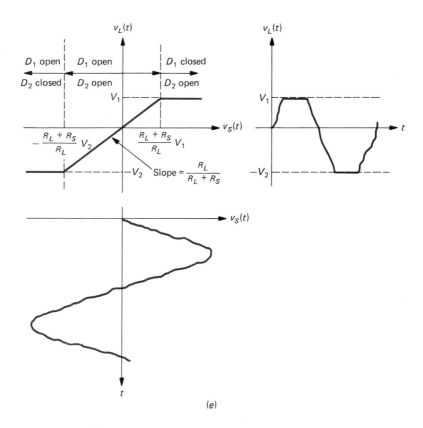

(e)

Certain semiconductor diodes are commercially available with precisely controlled zener breakdown voltages V_z, and these are called *zener diodes.*

A typical characteristic and a device symbol for a zener diode are shown in Fig. 7.25a. In the forward-biased direction ($v_d > 0$) the zener diode behaves in a manner which is very similar to the semiconductor diode. However, in the reverse-biased region the curve drops abruptly at $v_d = -V_z$ with a slope $1/R_z$. And R_z is called the *zener resistance;* typical values range from 0.1 to as much as 200 Ω, depending on the zener voltage V_z and the maximum average power the device can dissipate P_{max}. We assume in our analyses that $R_z = 0$. Thus, in the reverse-biased region the zener voltage is $v_d = -V_z$ for all $i_d < 0$. A piecewise-linear approximation of the characteristic is shown in Fig. 7.25b, and the equivalent circuit using two ideal diodes is shown in Fig. 7.25c.

Zener diodes as well as semiconductor diodes are rated in terms of the maximum average power which they may safely dissipate. This is shown on the device characteristic as a hyperbola: $P_{\mathrm{max}} = v_d i_d$. Ordinarily, this is an average power rating since it relates to the ability of the device to dissipate accumulated heat; consequently, it is a function of the waveshape of v_d and i_d. For certain waveforms we may operate outside the P_{max} hyperbola if we do so for a short time. A safe criterion, however, is to ensure that at no time does the product of the *instantaneous* values of v_d and i_d

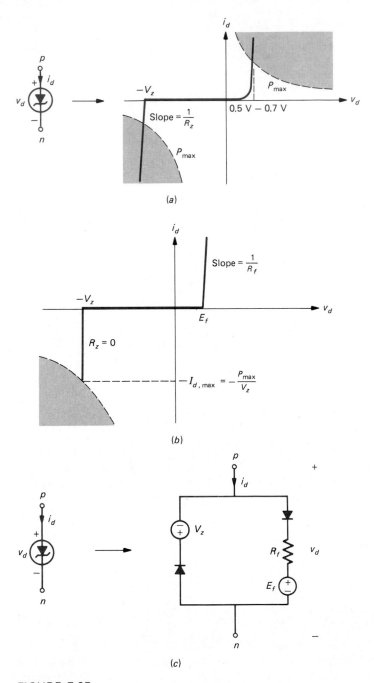

FIGURE 7.25
The zener diode: (*a*) device characteristic; (*b*) piecewise-linear characteristic; (*c*) piecewise-linear model.

exceed P_{max}. In the reverse-biased region, we assume that $v_d = -V_z$ ($R_z = 0$); thus, the maximum value of reverse current which the diode may conduct is

$$I_{d,max} = \frac{P_{max}}{V_z} \qquad v_d < 0 \tag{7.33}$$

as shown in Fig. 7.25*b*.

Zener diodes are commonly operated in the reverse region such that the operating point lies on the reverse breakdown region of the zener curve and the reverse current does not exceed $I_{d,max}$. In this mode of operation, the device functions as a regulator to maintain a constant voltage level of V_z across some load.

For example, typical dc voltage sources such as batteries are not ideal and have a source resistance R_S and an open-circuit voltage V_S, as shown in Fig. 7.26*a*. If this dc source is attached to a load R_L, the load voltage V_L will be dependent on the value of the source resistance since some of the open-circuit voltage, $I_L R_S$, is dropped across the source resistance. Changing values of R_L will result in changing values of load voltage since the load current will change.

To maintain a constant value of load voltage when R_L changes value, we may insert a regulator between the source and load, as shown in Fig. 7.26*b*. The regulator consists of a resistor R and a zener diode. Note the connection of the zener diode. The load voltage V_L is the negative of the diode voltage; $V_L = -v_d$. The battery voltage V_S is assumed known, as is the source resistance R_S. Given a range of possible values of R_L, we would like to determine a value of R such that the load voltage is maintained at the zener voltage $V_L = V_z$ for these various values of the load resistance.

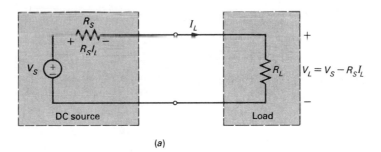

(a)

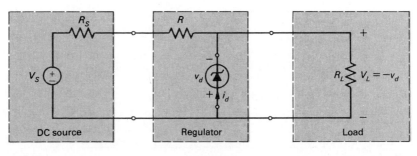

(b)

FIGURE 7.26
A zener diode-regulated power supply.

Another application is maintenance of a constant voltage across the load when the load resistance is fixed but the source voltage V_S varies. For example, the rectifiers discussed in a preceding section produce a pulsating dc voltage. However, we wish to obtain a constant value of voltage. Filters can be used to smooth this voltage somewhat (see Fig. 7.18, for example). A zener diode regulator may also be used for these purposes. For example, suppose the source voltage varies within the limits $V_{S,\text{min}} < v_S(t) < V_{S,\text{max}}$.

If we operate the zener diode in its reverse breakdown region, the voltage across it will be held constant at V_z and the load current will be held constant at V_z/R_L, as shown in Fig. 7.27a. In order to operate the zener in its reverse breakdown region, the current through it must always lie between (1) $i_d = 0$ and (2) $i_d = -P_{\text{max}}/V_z$. Therefore, the value of the regulator resistor R must be chosen to give a zener current that is between these two extremes, each of which is shown in Fig. 7.27b and c. The

FIGURE 7.27
Analysis of the regulated power supply.

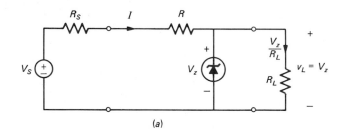

(a)

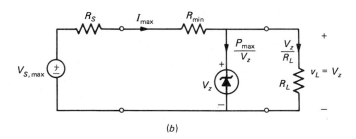

(b)

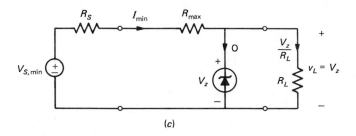

(c)

minimum regulator resistor gives the maximum current through it as shown in Fig. 7.27*b*:

$$I_{\text{max}} = \frac{V_{S,\text{max}} - V_z}{R_S + R_{\text{min}}}$$

$$= \frac{P_{\text{max}}}{V_z} + \frac{V_z}{R_L} \tag{7.34}$$

The maximum regulator resistor gives the minimum current through it, as shown in Fig. 7.27*c*:

$$I_{\text{min}} = \frac{V_{S,\text{min}} - V_z}{R_S + R_{\text{max}}}$$

$$= \frac{V_z}{R_L} \tag{7.35}$$

Example 7.4   A source voltage varies between 120 and 75 V. The source resistance is zero ($R_S = 0$) and the load resistance is 1000 Ω. It is desired to maintain the load voltage at 60 V. Determine a value of the regulator resistor R that will accomplish this and the required power rating of the zener.

Solution   We select a zener having a zener voltage of 60 V. The maximum value of the regulator resistance is determined from Fig. 7.27*c* as

$$R_{\text{max}} = \frac{V_{S,\text{min}} - V_z}{I_{\text{min}}} = 250 \, \Omega \qquad I_{\text{min}} = V_z/R_L$$

The required power rating of the zener is determined when $V_S = V_{S,\text{max}}$ and the zener draws the maximum current of P_{max}/V_z:

$$\frac{P_{\text{max}}}{V_z} = \frac{V_{S,\text{max}} - V_z}{R} - \frac{V_z}{R_L}$$

$$= 0.18 \, \text{A}$$

so that

$$P_{\text{max}} = 10.8 \, \text{W}$$

PROBLEMS

7.1   A silicon diode is operated at two temperatures, 15°C and 35°C. Sketch the v-i characteristic for these two temperatures assuming $\eta = 2$ and a saturation current of $I_S = 10$ nA. Repeat this for a germanium diode and assume $\eta = 1$ and $I_S = 1\ \mu A$.

7.2   Determine the diode voltage for a diode current of $I_d = 50$ mA for the diodes of Prob. 7.1.

7.3   A silicon diode ($\eta = 2$) has a voltage of 0.65 V and a current of 100 mA at room temperature (293 K). Determine the saturation current, I_S. Repeat this for a germanium diode ($\eta = 1$) that has a voltage of 0.25 V and a current of 50 mA at room temperature.

7.4   For the circuit shown in Fig. P7.4, determine the diode current and the average power dissipated by the diode.

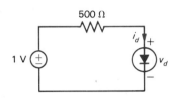

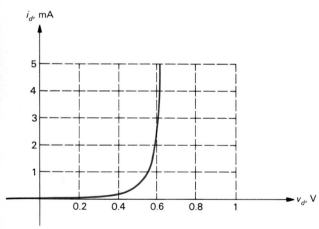

FIGURE P7.4

7.5   Repeat Prob. 7.4. when the resistor is changed from 500 to 250 Ω.

7.6   Suppose the voltage source and resistance in Fig. P7.4 are changed to 8 V and 2 kΩ. Determine the diode current and voltage and the power dissipated by the diode.

7.7   In the circuit shown in Fig. P7.7, determine the diode current and voltage and the power delivered by the source. The diode characteristic is given in Fig. P7.4.

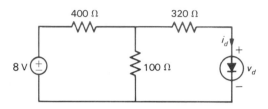

FIGURE P7.7

7.8   Repeat Prob. 7.7 if the voltage source is changed to 4 V.

7.9   For the circuit shown in Fig. P7.9, determine the diode current and voltage and the power delivered by the voltage source. The diode characteristic is given in Fig. P7.4.

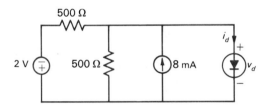

FIGURE P7.9

7.10   The diode in Fig. P7.4 is to be approximated by an ideal diode in series with a resistor R_f and a voltage source E_f. Determine R_f and E_f if the operating point of the diode is $V_d = 0.6$ V, $I_d = 2$ mA. Repeat this if the operating point is $V_d = 0.5$ V, $I_d = 0.275$ mA.

7.11   Repeat Prob. 7.10 using the diode equation where $I_S = 12.5$ nA, $\eta = 2$, and the temperature is 17°C.

7.12   For the circuit shown in Fig. 7.10a, with $V_S = 0.01$ V and $E_S = 5$ V, $R_S = 1$ kΩ and using the diode whose characteristic is given in Fig. P7.4, write an approximate equation for the diode current.

7.13   For the half-wave rectifier shown in Fig. 7.17a, assume an actual diode whose characteristic is given in Fig. P7.4 and sketch $v_L(t)$ for $V_S = 2$ V and $R_L = 500\ \Omega$.

7.14 For the circuit shown in Fig. 7.18a, assume an ideal diode and $V_S = 10$ V, $\omega = 2\pi \times 10^3$, $C = 10\ \mu F$, and $R_L = 1\ k\Omega$. Sketch $v_L(t)$. What is the minimum value of $v_L(t)$ at any time after steady-state operation has been obtained?

7.15 For the full-wave bridge rectifier shown in Fig. 7.22a, assume identical piecewise-linear diodes with $R_f = 20\ \Omega$, $E_f = 0.5$ V, $R_L = 100\ \Omega$, and $V_S = 100$ V. Sketch $v_L(t)$.

7.16 For the full-wave rectifier with ideal diodes shown in Fig. 7.20a, assume that diode D_2 has been inserted opposite the direction shown. Sketch $v_L(t)$.

7.17 For the limiter shown in Fig. 7.24a, assume identical piecewise-linear diodes with $R_f = 100\ \Omega$, $E_f = 0.5$ V, $V_1 = V_2 = 10$ V, $R_L = 100\ \Omega$, and $v_S(t) = 50 \sin \omega t$ V. Sketch $v_L(t)$.

7.18 For the zener diode regulator shown in Fig. 7.26b, assume that V_S varies between 40 and 60 V, $R_S = 1\ k\Omega$, $R_L = 1\ k\Omega$. Choose a zener diode and R such that V_L is maintained at 30 V.

7.19 For the zener diode regulator shown in Fig. 7.26b, assume that $V_S = 50$ V, $R_S = 100\ \Omega$, and $R_L = 100\ \Omega$. Suppose a 20-V 1-W zener is to be used so that $V_L = 20$ V with R_L attached (full load) and with R_L removed (no load). Determine a range of values for R which will accomplish this.

Transistors and Amplifiers

In the previous chapter we studied certain two-terminal nonlinear resistors—which were referred to as diodes. In this chapter we consider certain three-terminal nonlinear resistors—which are called *transistors*. The two transistors which we study are the field-effect transistor (FET) and the bipolar junction transistor (BJT). One of the main uses of these transistors is in the construction of amplifers. We study this use of these transistors in this chapter. Another important use of these devices is in the construction of digital circuits, such as computers. This use will also be studied in Chap. 10.

As was the case for circuits containing diodes, circuits containing transistors are nonlinear. Consequently, such powerful techniques of linear circuit analysis as superposition do not generally apply. Moreover, the only feasible way of specifying these nonlinear devices is with graphical techniques; the devices cannot generally be characterized with simple equations. Therefore, the analysis of circuits containing these nonlinear devices is generally more difficult than for linear circuits, and graphical techniques (such as load lines) must be employed.

Often we use certain approximations to simplify the analysis. The basis for these approximations is the small-signal linear operation of the device. Under this assumption of linear operation, the technique of superposition may be used to simplify the analysis. The use of these approximate techniques is important in gaining an understanding of the device performance. This approximate analysis technique is studied in this chapter.

8.1 PRINCIPLES OF LINEAR AMPLIFIERS

One of the primary uses of field-effect and bipolar junction transistors is in the construction of linear amplifiers. Essentially, a linear amplifier is a circuit which enlarges a signal (an amplifier) and preserves its shape (a linear one). For example, consider Fig. 8.1a. A sinusoidal voltage source

$$v_i(t) = V_i \sin \omega t \qquad\qquad (8.1)$$

is applied to the input terminals of a linear amplifier. We have shown this source to be, for illustration, a single-frequency sinusoid. However, no information is contained in a single-frequency sinusoid, so a more practical input signal would be some complicated waveform representing, e.g., music from a phonograph cartridge. Since any signal can be represented alternatively with Fourier methods as a sum of

FIGURE 8.1
Illustration of
general properties
of a linear
amplifier.

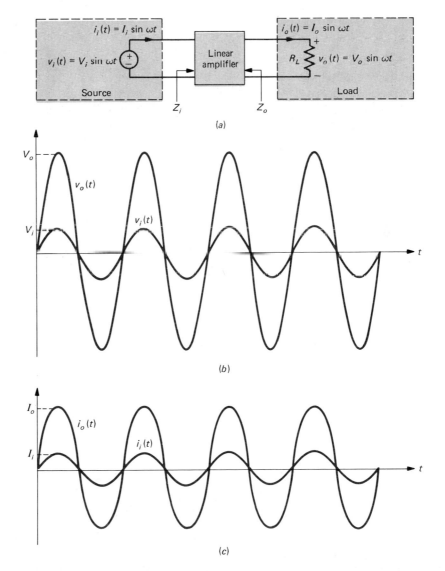

(a)

(b)

(c)

sinusoids at different frequencies and the amplifier is presumed to be linear, we may decompose the signal into its sinusoidal components and analyze the behavior of the amplifier for these sinusoidal components individually, and the sum of the responses to each will be, by superposition, the response to the more complicated signal. Thus, a single-frequency sinusoid can represent a rather general input signal.

Now suppose we wish to enlarge this sinusoidal signal and apply it to some load represented by R_L. This load may represent a loudspeaker. For the listener to hear the music, the signal applied to the loudspeaker

$$v_o(t) = V_o \sin \omega t \tag{8.2}$$

must be made much larger than the signal from the phonograph cartridge, as shown in Fig. 8.1b. If it is, then the amplifier is said to have *voltage gain*.

$$A_V = \frac{V_o}{V_i} > 1 \tag{8.3}$$

For example, if the phonograph cartridge produces a sinusoid with a peak voltage of 1 mV and the peak voltage across the load is 100 mV, then the amplifier has a voltage gain of $A_V = 100$.

We also do not want the amplifier to distort the signal as it is being amplified. If the input is a pure 1-kHz sinusoid, we want the output to be a pure 1-kHz sinusoid. This is what is meant by a *linear* amplifier. If the music is distorted by the amplifier, a large voltage gain is unimportant.

At this point, we might wonder if there is a simple way to produce a device which has voltage gain. The answer to this is simple: If we want a linear amplifier with a voltage gain of 100, we simply use an ideal transformer with a turns ratio of 1:100! So there must be other criteria for a circuit to be an amplifier other than simply possessing a voltage gain. Another "figure of merit" of an amplifier is its current gain. For example, if the input current to the amplifier is

$$i_i(t) = I_i \sin \omega t \tag{8.4}$$

and the output current delivered to the load is

$$i_o(t) = I_o \sin \omega t \tag{8.5}$$

as shown in Fig. 8.1c, then the amplifier is said to have a *current gain* of

$$A_I = \frac{I_o}{I_i} > 1 \tag{8.6}$$

An ideal transformer with a turns ratio of 1:100 has a voltage gain of $A_V = 100$ but a current gain of $A_I = \frac{1}{100}$ since if the voltage is stepped up by a factor of 100, the current is stepped down by a factor of $\frac{1}{100}$ and thus the ideal transformer has no power gain. Therefore, another figure of merit of an amplifier is its *power gain*:

$$
\begin{aligned}
A_P &= \frac{P_o}{P_i} \\
&= \frac{V_o I_o}{V_i I_i} \\
&= A_V A_I
\end{aligned}
$$

(8.7)

If the load represents a loudspeaker, it would be of little value to amplify the voltage if the power to drive the loudspeaker were not also amplified. Surely, the simple phonograph cartridge will not provide the power to drive a large loudspeaker. The additional power comes from the voltage sources within the amplifier which are used to "bias" it. This is discussed in later sections of this chapter.

Two final figures of merit of linear amplifiers should be discussed. First, the ratio of input voltage to input current for the linear amplifiers is known as the *input impedance* of the amplifier:

$$
Z_i = \frac{V_i}{I_i}
$$

(8.8)

It is an important quantity since if the source has a source resistance R_S, we would like to have $Z_i = R_S$ so that maximum power transfer will take place from the source to the amplifier; in other words, the source and the amplifier would be matched.

Second, the source and amplifier can be replaced, as far as the load is concerned, with a Thévenin equivalent. The Thévenin impedance seen by the load is called the *output impedance* of the amplifier and is denoted by Z_o. For maximum power transfer from the amplifier to the load, we should have $Z_o = R_L$. Matching of the amplifier to the load is also important in the overall efficiency of the system.

In this chapter we study linear amplifiers which are constructed from FETs and BJTs and analyze their performance. Our first method of analysis will be a graphical one so that we can understand how these devices operate. In this chapter we also investigate simpler ways of calculating A_V, A_I, A_P, Z_i, and Z_o for a particular amplifier.

8.2 THE JUNCTION FIELD-EFFECT TRANSISTOR (JFET)

There are several types of field-effect transistors (FETs). The primary ones are the junction field-effect transistor (JFET) and the metal-oxide-semiconductor field-effect transistor (MOSFET). The MOSFET is used primarily in integrated circuits and is

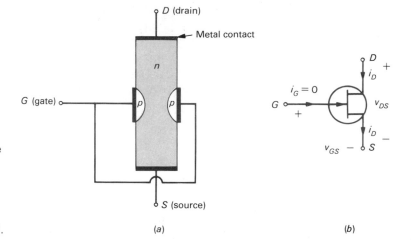

FIGURE 8.2 The junction field-effect transistor (JFET): (*a*) physical construction; (*b*) device symbol.

(*a*)   (*b*)

discussed in a later section. The analysis techniques for MOSFET amplifiers are virtually identical to those for JFET amplifiers.

The *n*-channel junction FET (JFET) is constructed of a piece of *n*-type material (called the *channel*) with two *p*-type regions embedded in it, as shown in Fig. 8.2*a*. Two terminals, *D* (drain) and *S* (source), are attached to the two ends of the channel. A third terminal, *G* (gate), is attached to the two *p* regions which are embedded in ("grown" into) the channel. The device symbol with terminal currents and voltages is shown in Fig. 8.2*b*. We will show why the gate current i_G is essentially zero [typically 0.01 microampere (μA)].

A *p*-channel JFET is also available, and it consists of a *p*-type channel with *n*-type regions embedded in it. The *n*-channel JFET requires positive dc power supplies (with respect to ground) in the amplifier biasing circuits and is thus more common than *p*-channel JFET amplifiers, which require negative power supplies. Our discussions therefore concentrate on the *n*-channel JFET. Once we design an amplifier with an *n*-channel JFET, we may substitute a *p*-channel JFET (of the same type number) if we simply reverse the polarity of the dc power supplies.

To determine the characteristic of the JFET, let us apply a voltage V_{DS} between the drain and the source with the positive terminal on the drain and the negative terminal on the source, as shown in Fig. 8.3*a*. Let us also attach the gate to the source so that $v_{GS} = 0$. If we increase the battery voltage v_{DS}, we observe a linear increase in drain current, as shown in Fig. 8.3*b*. This region is governed primarily by the resistance of the channel. As the battery voltage is increased further, we reach a point $v_{DS} = V_p$ where further increases in v_{DS} result in only very small increases in drain current. Increasing v_{DS} further, we reach a point $v_{DS} = BV_{DSS}$ where the drain current increases dramatically. The current I_{DSS} is normally specified for the device by the manufacturer and is the drain-source current i_D with a gate short-circuited to the source ($v_{GS} = 0$).

The operation and resulting terminal characteristics of an *n*-channel JFET can be easily understood if we reconsider the semiconductor diode shown in Fig. 8.4. A depletion layer devoid of mobile charges is established at the *pn* junction by diffusion

FIGURE 8.3
Terminal properties
of a JFET.

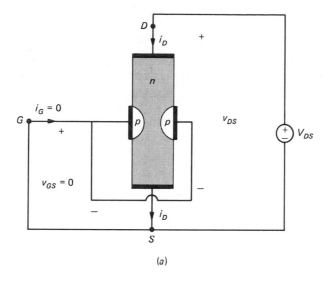

(a)

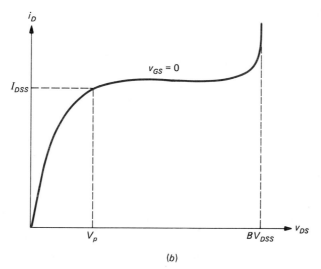

(b)

of the majority carriers of each material across the junction. This exposes the immobile donor and acceptor ions in a region of length d_o. If, as shown in Fig. 8.4b, a forward bias is applied by attaching the positive terminal of a battery V to the p region and the negative terminal to the n region, the depletion layer is reduced. If reverse bias is applied, as shown in Fig. 8.4c, the depletion layer is increased. Remember that the depletion layer is devoid of mobile charge carriers and thus may be thought of as having a large resistance.

This characteristic of an increased depletion region for a reverse-biased pn junction helps to explain the JFET characteristic in Fig. 8.3. Note that there are pn junctions formed between gate and source and between gate and drain. For $v_{GS} = 0$

FIGURE 8.4
Depletion regions
of a diode similar
to those formed in
a JFET.

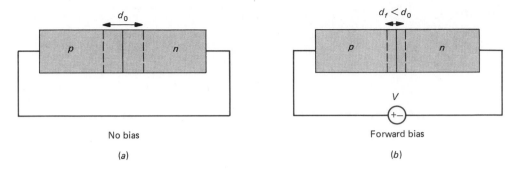

No bias

(a)

Forward bias

(b)

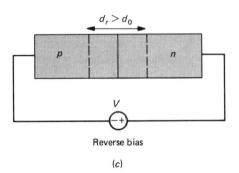

Reverse bias

(c)

in Fig. 8.3a, shown in Fig. 8.5a, a depletion region devoid of mobile charge carriers is established at the *pn* junctions. Since the gate is connected to the source, the gate-drain side of the *pn* junction will be more reverse-biased than the gate-source side since $v_{DG} > 0$ while $v_{GS} = 0$. This results in an asymmetric depletion region. In any case, negligible gate current flows since the *pn* junction is reverse-biased. This is why we assume that $i_G = 0$. If the gate to source is not reverse-biased ($v_{GS} > 0$), then gate current will flow.

The drain current flows through the *n* channel and out of the source terminal and is therefore determined by the resistance of the channel. As v_{DS} is increased, i_D increases, but the depletion region and the resulting resistance of the channel become larger since the gate-drain *pn* junction becomes more reverse-biased. We eventually reach a point where the two depletion regions merge and the channel is "pinched off," as shown in Fig. 8.5b. The value of drain-source voltage is denoted as $v_{DS} = V_p$ and is referred to as the *pinch-off voltage*. Further increases in v_{DS} greater than V_p cause a further increase in the depletion region, as shown in Fig. 8.5c, but the voltage of a point *p* in this depletion region remains essentially constant at the pinch-off voltage. Thus, further increases in v_{DS} bring only minor increases in drain current, as shown in Fig. 8.3b. If we further increase v_{DS}, we reach a point at which avalanche breakdown occurs, resulting in a rapid rise in drain current similar to that of semiconductor diodes. This voltage is denoted as BV_{DSS}, or breakdown voltage: drain to source with gate short-circuited (to source).

FIGURE 8.5
Illustration of
pinch off in a
JFET.

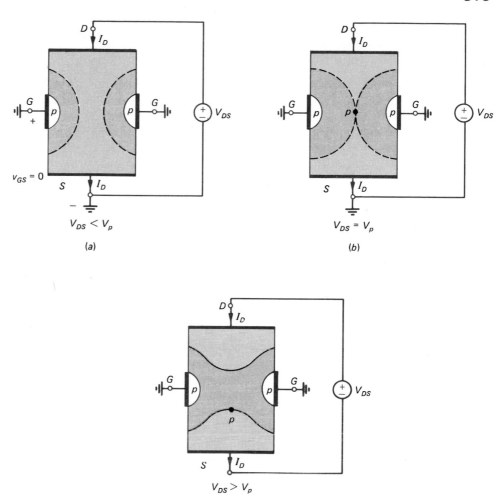

Now let us consider the resulting characteristic for other gate-source voltages. If we attach a battery V_{GS} between gate and source with the negative terminal attached to the gate, as shown in Fig. 8.6a, the *pn* junction will be reverse-biased and no gate current will flow. This is the typical mode of operation for the JFET. If we perform our previous experiment by increasing v_{DS} and recording the resulting drain current, we obtain the typical characteristic shown in Fig. 8.6b. For negative V_{GS}, the depletion region will be larger than for $V_{GS} = 0$ since the negative gate voltage adds a reverse bias to the *pn* junction in addition to the effect of v_{DS}. Thus, for the same value of v_{DS}, the drain current with $V_{GS} < 0$ will be less than with $V_{GS} = 0$, and the curves for increasingly negative V_{GS} will move downward. Note that the pinch-off and breakdown voltages become reduced for increasingly negative V_{GS}. Observe that there is a region where the curves appear as straight lines and are approximately horizontal. This is referred to as the linear region.

FIGURE 8.6 The terminal characteristics of a typical JFET.

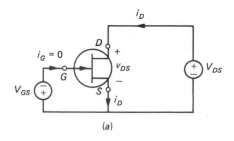

(a)

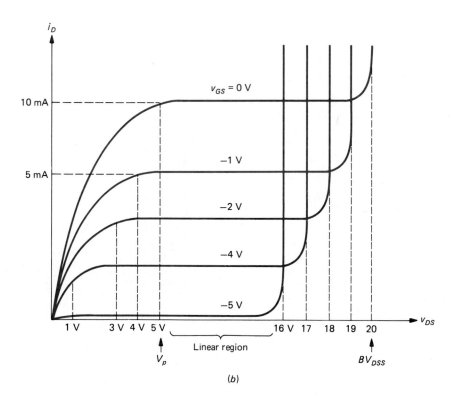

(b)

8.3 THE JFET AMPLIFIER

A simple JFET amplifier is shown in Fig. 8.7a. A battery V_{GG} is inserted in series with the voltage source $v_S(t) = V_S \sin \omega t$, with $V_{GG} > V_S$ so that the gate voltage always remains negative with respect to the source terminal. The output voltage is equal to the drain-source voltage $v_o = v_{DS}$. Here we have assumed that the load resistance in Fig. 8.1 is an open circuit. Writing KVL around the drain-source loop, we obtain

$$v_{DS} = V_{DD} - R_D i_D \qquad (8.9)$$

FIGURE 8.7 A simple JFET amplifier: (a) circuit diagram; (b) signal waveforms and the load line.

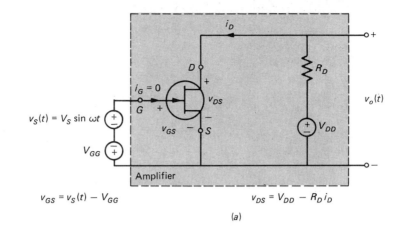

$$v_{GS} = v_S(t) - V_{GG}$$ $$v_{DS} = V_{DD} - R_D i_D$$

(a)

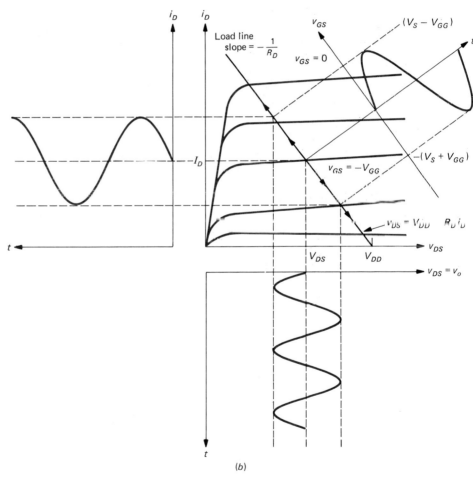

(b)

which is plotted on the characteristic as a load line in Fig. 8.7b. Writing KVL around the gate-source loop, we obtain

$$v_{GS} = -V_{GG} + V_S \sin \omega t \tag{8.10}$$

When $v_S(t) = 0$, $v_{GS} = -V_{GG}$, which locates the dc operating point (V_{DS}, I_D), as shown in Fig. 8.7b. As $v_S(t)$ varies, the gate-source voltage varies sinusoidally between $V_S - V_{GG}$ and $-(V_S + V_{GG})$, and the operating point moves along the load line. The resulting drain-source voltage v_{DS} is sketched versus time. Note that as $v_S(t)$ increases, $v_{DS} = v_o$ decreases; thus, the output voltage is 180° out of phase with the source voltage. If the source voltage is made larger in magnitude (V_S is increased), the excursions along the load line are larger and we observe a distortion of the output-voltage waveform, as shown in Fig. 8.8.

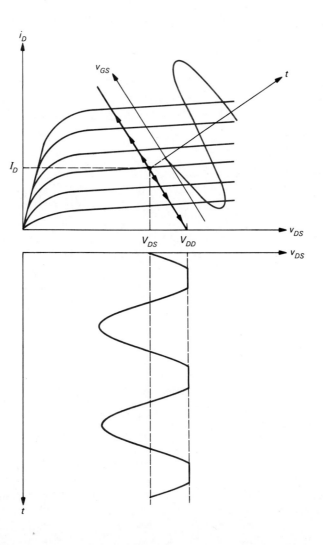

FIGURE 8.8
Illustration of distortion in a JFET amplifier caused by large input signals.

If $v_S(t)$ is kept small enough that distortion does not occur, we will have constructed a linear amplifier. Note that it is important for us to have chosen the battery V_{GG} such that the dc operating point with $v_S(t) = 0$ is in a reasonably linear region of the characteristic. If we have chosen $V_{GG} = 0$, a positive excursion would have caused the gate-source voltage to become positive and gate current would flow, which could damage the JFET. If V_{GG} has been chosen too large, the dc operating point would have been located near the bottom of the load line and distortion would occur with only a small $v_S(t)$, as shown in Fig. 8.8. Thus, we need to locate the dc operating point somewhere in the middle of the linear region in order to linearly amplify the largest possible $v_S(t)$.

Note also that the output voltage consists of a sinusoidal voltage superimposed on a dc level V_{DS}. It is only the ac portion of this output signal that we are interested in. If the output voltage swings from $V_{DS,\text{min}}$ to $V_{DS,\text{max}}$ symmetrically about V_{DS}, then the (ac) voltage gain is

$$A_V = \frac{V_{DS,\text{max}} - V_{DS,\text{min}}}{2V_S} \tag{8.11}$$

A factor of 2 is included in the denominator of (8.11) since $V_{DS,\text{max}} - V_{DS,\text{min}}$ is the peak-to-peak value, but V_S is the zero-to-peak value of the input sinusoid. Since the load is an open circuit, $i_o(t) = 0$ and current gain is meaningless. However, since the input current is i_G and $i_G = 0$ (for negative v_{GS}), the input impedance to the amplifier is infinite. This is one of the advantages of the FET over the BJT, as we will see.

There are a few problems with this simple amplifier. One is that it requires two batteries (or dc power supplies), V_{GG} and V_{DD}. Typical electronic systems have only one dc voltage available to be used by all the electronic subsystems, for example, $+5$ V, $+12$ V, etc. Another is that all semiconductor devices exhibit large unit-to-unit variations. For example, several 2N3819 JFETs purchased the same day from the same manufacturer will have different characteristic curves—the shapes will all be the same, as in Fig. 8.6b, but the numbers on these characteristics will differ between devices of the same type number. We must design our amplifiers with this variation in mind. If we wish to mass-produce a large number of amplifiers having approximately the same performance, the circuit should be relatively insensitive to these unit-to-unit variations. This is an important design criterion.

A more practical JFET amplifier which overcomes these problems is shown in Fig. 8.9a. Capacitors C_i and C_o are included to block any direct current from getting to the source and load, respectively. We will essentially assume that they have negligible impedance (for example, 1 Ω) at the frequency of $v_S(t)$ and that they can thus be asssumed to be short circuits to the ac signal.

The simple amplifier of Fig. 8.7a required two batteries, V_{GG} and V_{DD}. The function of the battery V_{GG} is to set a constant negative bias voltage of the gate to place the dc operating point in a linear region of the characteristic. This is performed by the resistor R_S and the combination of R_{G1} and R_{G2}. Capacitor C_S is typically placed across R_S so that ac voltages are not developed across R_S. The capacitor value is chosen such that its impedance at the frequency $v_S(t)$ is negligible. Thus, C_S

FIGURE 8.9 The
dc analysis of a
typical JFET
amplifier in
determining the
operating point.

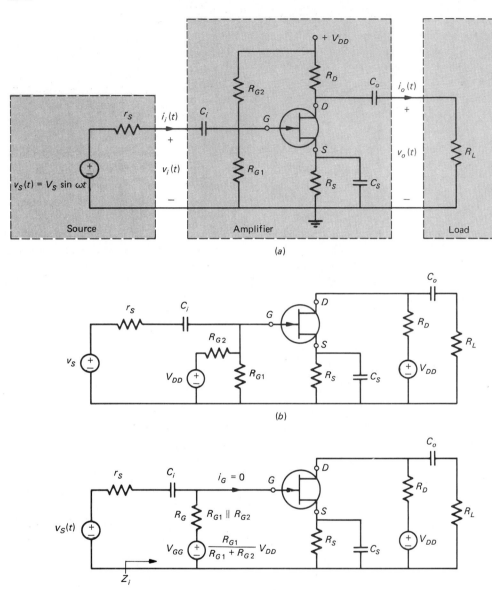

(a)

(b)

(c)

essentially short-circuits R_S for alternating current but direct current passes through R_S, developing a constant voltage across it. We designate the internal (Thévenin) impedance of the source [containing $v_S(t)$] as r_S to avoid confusion with the resistor in the source lead of the JFET, R_S.

Only one V_{DD} battery or dc power supply is used in this circuit connected between $+V_{DD}$ and ground ($\perp$). We may reduce the combination of R_{G1} and R_{G2} and its associated battery V_{DD} to a Thévenin equivalent, as shown in Fig. 8.9b and

c, where

$$R_G = R_{G1} \| R_{G2} \tag{8.12}$$

and

$$V_{GG} = \frac{R_{G1}}{R_{G1} + R_{G2}} V_{DD} \tag{8.13}$$

Note that since $i_G = 0$, the input impedance to this amplifer is

$$
\begin{aligned}
Z_i &= R_G \\
&= R_{G1} \| R_{G2}
\end{aligned}
\tag{8.14}
$$

The addition of the resistor R_S in the source lead of the JFET tends to stabilize against variations in JFET parameters. This is accomplished by the following. If we remove the ac signal source, we obtain the dc circuit shown in Fig. 8.10a. Since all

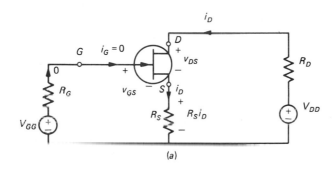

(a)

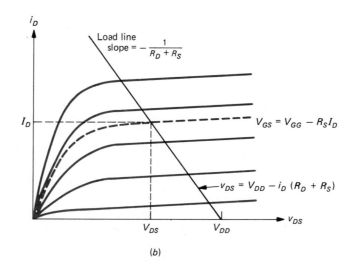

(b)

FIGURE 8.10
The dc load line and determination of the operating point.

currents and voltages are direct current in this circuit, all capacitors appear as open circuits and may be removed. Writing KVL around the right-hand loop gives

$$v_{DS} = V_{DD} - (R_D + R_S)i_D \qquad (8.15)$$

Plotting this equation on the JFET characteristic results in a dc load line with slope

$$-\frac{1}{R_{DC}} = -\frac{1}{R_D + R_S} \qquad (8.16)$$

which intersects the $i_D = 0$ axis at $v_{DS} = V_{DD}$. If the gate-to-source voltage V_{GS} is determined, this determines the operating point (I_D, V_{DS}). This can be found by interpolating between the limited number of V_{GS} curves shown on the characteristic.

However, the V_{GS} voltage is not completely determined by V_{GG}, as was the case for the simpler circuit shown in Fig. 8.7a. To illustrate this, write KVL around the left-hand loop in Fig. 8.10a to give

$$v_{GS} = V_{GG} - R_S i_D \qquad (8.17)$$

since $i_G \doteq 0$, so there is essentially no voltage drop across R_G. Observe that, according to (8.17), the operating-point gate-to-source voltage V_{GS} depends on the operating-point drain current I_D and vice versa. Given the JFET characteristic, we could select an operating point (V_{GS}, I_D, V_{DS}), choose R_S and V_{GG} from (8.17), and then determine R_D from (8.15).

Suppose we wish to construct many such supposedly identical amplifiers using the same type JFET. Although the JFETs to be used are of the same type, their characteristics will vary from unit to unit. This will result in different operating points for the amplifiers and different voltage, current, and power gains. The resistor R_S in the source lead of the JFET tends to stabilize against these variations. For example, suppose the operating-point drain current I_D is increased. From (8.17) the gate-to-source voltage will become more negative. (V_{GG} and $R_S I_D$ must be chosen to make V_{GS} negative in order to place the operating point in a linear portion of the characteristic and to avoid forward-biasing the gate-source junction and thereby drawing gate current.) According to the typical JFET characteristic of Fig. 8.6b, more negative v_{GS} will reduce I_D. Therefore, changes in the operating point due, perhaps, to variations in device characteristics are stabilized against.

There is one final calculation: determining R_{G1} and R_{G2}. These are related to V_{GG} by Eq. (8.13). Having selected V_{DD} and V_{GG}, we have

$$\frac{R_{G1}}{R_{G1} + R_{G2}} = \frac{V_{GG}}{V_{DD}} \qquad (8.18)$$

But we still haven't determined R_{G1} and R_{G2} explicitly. We need another relation among them. Note that the input impedance to the amplifier in Fig. 8.9c is

$$Z_i = R_G$$

$$= \frac{R_{G1}R_{G2}}{R_{G1} + R_{G2}} \tag{8.19}$$

Thus, we might specify the input impedance to the amplifier and then determine R_{G1} and R_{G2} from Eqs. (8.18) and (8.19).

Example 8.1   From the manufacturer's supplied characteristics, the JFET amplifier of Fig. 8.9 is to be designed for a dc operating point of $I_D = 2$ mA, $V_{DS} = 5$ V, and $V_{GS} = -2$ V. Determine R_{G1}, R_{G2}, R_D, and R_S to bias the device at this operating point with $V_{DD} = 20$ V and $Z_i = 100$ kilohms (kΩ).

Solution   If we choose $V_{GG} = 10$ V, then we may determine R_S from Eq. (8.17) as

$$R_S = \frac{V_{GG} - V_{GS}}{I_D}$$

$$= \frac{10 - (-2)}{2 \text{ mA}}$$

$$- 6 \text{ k}\Omega$$

From Eq. (8.15) we obtain

$$R_D = \frac{V_{DD} - V_{DS}}{I_D} - R_S$$

$$= \frac{20 - 5}{2 \text{ mA}} - 6 \text{ k}\Omega$$

$$= 1.5 \text{ k}\Omega$$

Finally, we must determine R_{G1} and R_{G2}. Now V_{GG} is related to R_{G1}, R_{G2}, and V_{DD} by

$$V_{GG} = \frac{R_{G1}}{R_{G1} + R_{G2}} V_{DD}$$

or

$$10 = \frac{R_{G1}}{R_{G1} + R_{G2}} 20$$

Therefore,

$$\frac{R_{G1}}{R_{G1} + R_{G2}} = \frac{1}{2}$$

Therefore, $R_{G1} = R_{G2}$, but we need another relation between R_{G1} and R_{G2} to determine them explicitly. This is provided by the input impedance specification:

$$Z_i = 100 \text{ k}\Omega$$

$$= R_{G1} \| R_{G2}$$

$$= \frac{R_{G1}}{R_{G1} + R_{G2}} R_{G2}$$

Thus

$$R_{G2} = 200 \text{ k}\Omega$$

and

$$R_{G1} = 200 \text{ k}\Omega$$

Note from Eq. (8.17) that if v_{GS} changes to -1 V (a 50 percent change in V_{GS}), this reduces I_D to 1.83 mA (an 8 percent change in I_D).

We always assume linear operation. In other words, we assume that the sinusoidal source is small enough that at no time does the operating point move outside the linear region of the characteristic. With this assumption, we essentially have a linear circuit to deal with. The device is linear over a large portion of its characteristic. Even though the characteristic is nonlinear in certain regions, we will never move into those regions under linear operation.

Under the assumption that $v_S(t)$ is "small enough" that the nonlinear device in the amplifier is being operated only in a linear region of its characteristic, we may apply superposition, and each circuit variable (branch voltage or branch current) will consist of two parts: one due to the dc source V_{DD} in the amplifier and one due to the sinusoidal source $v_S(t)$. We denote the dc component of a circuit voltage or current $x(t)$ with a capital letter X, and the sinusoidal component is denoted as a Δ quantity, Δx, so that

$$x(t) = X + \Delta x(t) \tag{8.20}$$

Remember that $\Delta x(t)$ is time-varying (sinusoidal) and that X is constant (direct current). For brevity, we will write Eq. (8.20) as

$$x = X + \Delta x \tag{8.21}$$

When we reattach the signal source and load, as shown in Fig. 8.11a, each voltage and current in the circuit will be composed of a direct current component due to V_{DD} (and V_{GG}) and an ac component due to $v_S(t)$. The dc components are governed by the dc load line determined previously with slope

$$-\frac{1}{R_{DC}} = -\frac{1}{R_D + R_S} \tag{8.22}$$

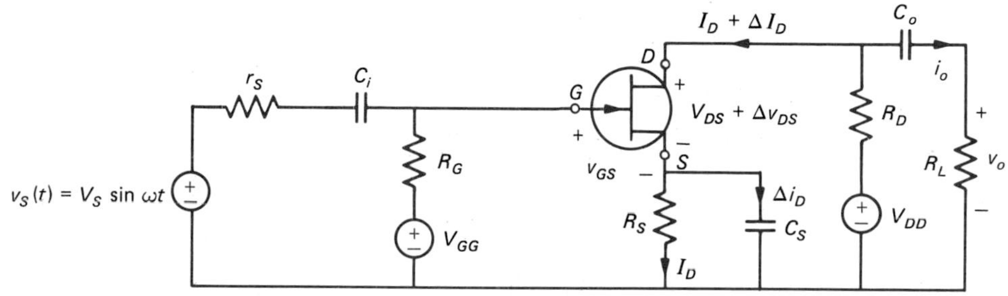

(a)

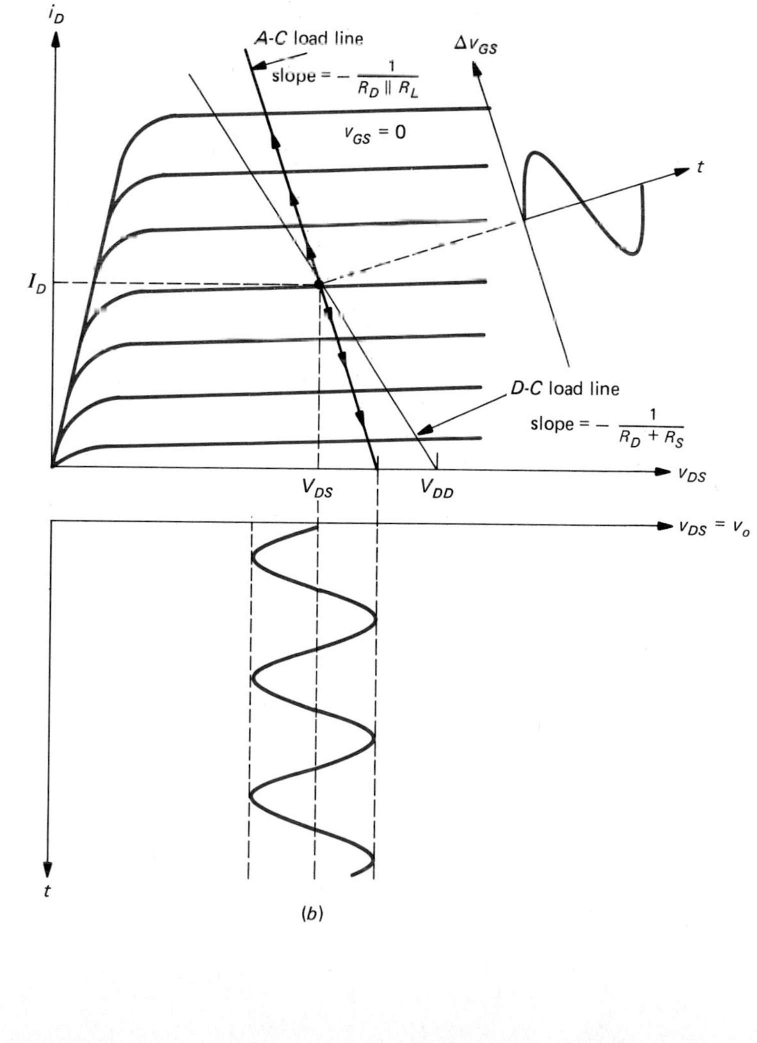

(b)

FIGURE 8.11
The ac load line
for determining
the circuit signals.

but the ac components are governed by an ac load line having a different slope. To determine the slope of this ac load line, we short-circuit all capacitors (assumed to have negligible impedance to the ac component) and deactivate the dc sources. This leaves only the delta (Δ), or ac, components of the voltages and currents, as shown in Fig. 8.12. Writing KVL around the drain-source loop for these conditions gives

$$\Delta v_{DS} = -(R_D || R_L)\Delta i_D \tag{8.23}$$

Thus, the ac load line has slope

$$-\frac{1}{R_{AC}} = -\frac{1}{R_D || R_L} \tag{8.24}$$

The ac variations must move along the ac load line. We reason that this ac load line must pass through the dc operating point since as we reduce the magnitude of $v_S(t)$, the variations along this ac load line must reduce to zero, which is the dc operating point. The ac variations of V_{GS} are seen from Fig. 8.12 (since $\Delta i_G = 0$) to be

$$\Delta v_{GS} = \frac{R_G}{R_G + r_S} v_S(t) \tag{8.25}$$

Thus, the gate voltage varies sinusoidally, and the resulting operating point moves along the ac load line, as shown in Fig. 8.11b.

The output voltage $v_o = v_{DS}$ also varies sinusoidally. If the magnitude of this output-voltage sinusoid is larger than the magnitude of the sinusoidal component of the input voltage to the amplifier, voltage gain will be achieved. The *voltage gain* is the ratio of the ac output voltage across R_L, Δv_o, to the ac input voltage to the amplifier Δv_i:

$$A_V = \frac{\Delta v_o}{\Delta v_i} \tag{8.26}$$

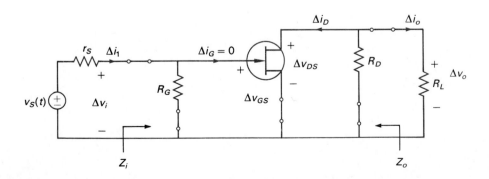

FIGURE 8.12
The small-signal
ac circuit.

The *current gain* is similarly defined:

$$A_I = \frac{\Delta i_o}{\Delta i_i} \tag{8.27}$$

Also the input and output impedances Z_i and Z_o of the amplifier are ac quantities and are determined from the ac circuit in Fig. 8.12. We now investigate the calculation of these ac figures of merit for an amplifier.

8.4 SMALL-SIGNAL ANALYSIS OF THE JFET AMPLIFIER

In the previous section, a typical JFET amplifier was investigated. To determine the important figures of merit for that amplifier—voltage gain A_V, current gain A_I, power gain A_P, input impedance Z_i, and output impedance Z_o—we follow the following procedure.

1 First we draw the dc circuit and determine the resistors and dc voltage sources in the biasing circuitry to locate the dc operating point at a desirable position in the linear portion of the device characteristic. The dc circuit is obtained by removing the ac source and load (caused by the open-circuit nature of the input and output capacitors to direct current).
2 Then we reattach the ac signal source, short-circuit all capacitors, and deactivate the dc sources to obtain the ac circuit—from which we calculate the ac quantities A_V, A_I, A_P, Z_i, and Z_o for the amplifier.

The justification for this procedure of treating dc and ac quantities separately is simply that we assume linear operation of the device so that superposition may be applied. Thus, each voltage and current in the circuit will consist of two parts. One part is direct current and is due to the dc batteries in the biasing circuitry; this is denoted with a capital letter X. The other part is alternating current and is due to the ac source $v_S(t)$; this is denoted as a delta (Δ) quantity, $\Delta x(t)$. The total value of the particular voltage or current is, by superposition, the sum of these two parts:

$$x(t) = X + \Delta x(t) \tag{8.28}$$

where $x(t)$ represents any voltage or current in the circuit.

Consider the JFET amplifier discussed in the previous section; it is shown in Fig. 8.9a. Once the dc analysis is completed and the operating point is determined, the signal variations move along the ac load line, as shown in Fig. 8.13a. Note from the ac circuit in Fig. 8.12 that

$$\Delta v_o = \Delta v_{DS} \tag{8.29}$$

and

$$\Delta i_o = -\frac{R_D}{R_D + R_L} \Delta i_D \tag{8.30}$$

FIGURE 8.13
Small-signal
linearization of
JFET amplifier
operation.

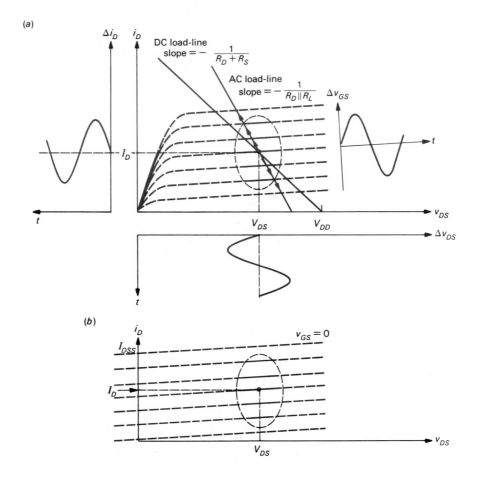

Thus, the ac output-voltage and output-current variations for the amplifier may be found from the characteristic, as shown in Fig. 8.13a.

Once again we observe an important point. If the excursions along the ac load line are restricted to a linear portion of the characteristic (circled in Fig. 8.13a), then the fact that the device characteristic is nonlinear over certain portions of the characteristic outside this linear region is immaterial; we never operate in those nonlinear regions. Therefore, for small-signal linear operation, we are free to redraw the device characteristic outside this linear region in any form we choose. Why not choose a simple representation. Simply extend the linear portion throughout the rest of the characteristic, as shown in Fig. 8.13b. We may write a linear equation describing this new characteristic. Note that three variables are related on the characteristic: i_D, v_{DS}, and v_{GS}. So our linear equation must relate these three variables. Let us presume a form of the equation to be

$$i_D = y_{fs}v_{GS} + y_{os}v_{DS} + I_{DSS} \tag{8.31}$$

where I_{DSS} is the drain-source current with the gate short-circuited to the source $v_{GS} = 0$. Obviously, this equation is a linear one, and once we determine the constants y_{fs}, y_{os}, and I_{DSS} from the characteristic at this dc operating point, we will have an alternative but equally valid representation of the characteristics.

The constant y_{fs} in Eq. (8.31) is calculated by taking a small change in v_{GS} (Δv_{GS}) and finding the resulting change in i_D (Δi_D) when no change in v_{DS} occurs ($\Delta v_{DS} = 0$), as shown in Fig. 8.14a:

$$y_{fs} = \left.\frac{\Delta i_D}{\Delta v_{GS}}\right|_{\Delta v_{DS}=0} \qquad \text{S} \qquad\qquad (8.32)$$

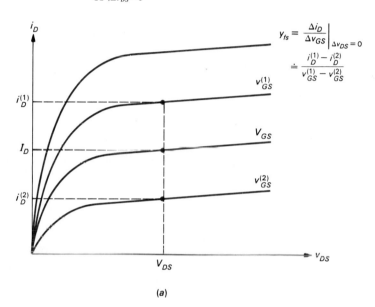

(a)

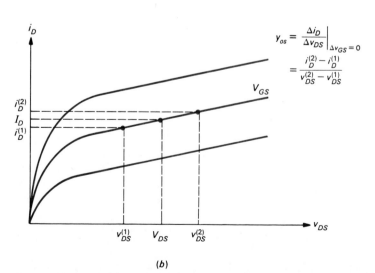

FIGURE 8.14
Calculation of
small-signal JFET
parameters.

(b)

The units of y_{fs} are the units of conductance (siemens) since y_{fs} is a ratio of a current to a voltage. Typical values range from $1000\ \mu S\ (10^{-3}\ S)$ to $6000\ \mu S\ (6 \times 10^{-3}\ S)$. The parameter y_{os} is the slope of the characteristic line passing through the dc operating point, since from Eq. (8.31).

$$y_{os} = \left. \frac{\Delta i_D}{\Delta v_{DS}} \right|_{\Delta v_{GS}=0} \qquad S \qquad (8.33)$$

Thus, the units of y_{os} are also siemens. This parameter may be calculated at the operating point from the JFET characteristic, as shown in Fig. 8.14b. Typical values of $1/y_{os}$ range from $10\ k\Omega$ to as much as 1 megohm ($M\Omega$). Of course, it cannot be stated that a JFET has unique values of y_{fs} and y_{os} since the values depend on where the parameters are being calculated on the characteristic—namely, the operating point. The parameter I_{DSS} is included in the equation since the approximation line in Fig. 8.13b for $v_{GS} = 0$ may not pass through $i_D = 0$ for $v_{DS} = 0$ (and generally doesn't).

Thus, the y symbols for these parameters indicate an admittance (or conductance). The origin of the subscripts is the following. The s subscript refers to the fact that the FET source terminal is common to the input and output; v_{GS} is input to one "port," and v_{DS} and i_D are output at the second port. The f subscript in y_{fs} symbolizes that this is a forward transfer conductance parameter since it yields the change in i_{DS} in the second port with a change in v_{GS} in the first port (with no change in v_{DS}). Similarly, the o subscript in y_{os} symbolizes that this is a conductance in the output port of the device.

An equivalent circuit for small-signal linear operation which represents Eq. (8.31) can be obtained as shown in Fig. 8.15. [The reader should verify that this circuit is equivalent to Eq. (8.31).] The diamond-shaped element is referred to as a *controlled source*. It has one of the characteristics of a current source since it maintains a current $y_{fs}v_{GS}$ through its terminals while its terminal voltage is not, as yet, known. It is quite different from an ideal current source for two reasons. First, the output current of this controlled source is not known but is determined by the value of some other branch variable in the circuit (in this case v_{GS}); thus, it is similar to a resistor in which the resistor current is "determined" by the resistor voltage $i = v/R$, and it is often

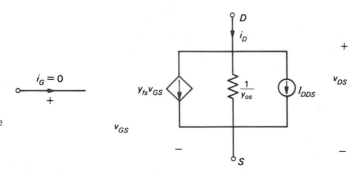

FIGURE 8.15 The small-signal linear equivalent circuit of the JFET.

referred to as a *controlled resistor*. Second, *superposition cannot be applied to controlled sources*. This should be clear since the output current of the controlled source is not fixed, as is the case for an ideal current source, but is dependent on another branch variable. Thus, its value will not appear on the right-hand side of the circuit equations but will instead be incorporated into a coefficient of v_{GS}.

Substituting this small-signal equivalent circuit for the JFET into the JFET amplifier circuit of Fig. 8.11a, we obtain the circuit in Fig. 8.16a. For ac analysis we set all dc sources equal to zero and replace (as approximations) the capacitors with short circuits, as shown in Fig. 8.16b. The resulting circuit is simplified in Fig. 8.16c. From this circuit we obtain the ac output voltage as

$$\Delta v_o = -y_{fs} \Delta v_{GS} \left(\frac{1}{y_{os}} || R_D || R_L \right) \tag{8.34}$$

and the ac input voltage to the amplifier is

$$\Delta v_i = \Delta v_{GS} \tag{8.35}$$

The ratio of Eqs. (8.34) and (8.35) yields the voltage gain of the amplifier:

$$A_V = \frac{\Delta v_o}{\Delta v_i}$$

$$= -y_{fs} \left(\frac{1}{y_{os}} || R_D || R_L \right) \tag{8.36}$$

The negative sign in Eq. (8.36) shows that the output voltage $\Delta v_o = \Delta v_{DS}$ and the input voltage $\Delta v_i = \Delta v_{GS}$ are 180° out of phase; as $\Delta v_i = \Delta v_{GS}$ increases, Δv_o decreases, which is also evident from Fig. 8.13a.

The input impedance is easily seen from Fig. 8.16c to be

$$Z_i = R_G \tag{8.37}$$

which can be made quite large (R_G is typically chosen to be 1 MΩ). The output (Thévenin) impedance seen by R_L is found by setting $v_S(t) = 0$ in the ac circuit, as shown in Fig. 8.16d. Since $\Delta v_{GS} = 0$, the controlled-source output current is zero and is therefore replaced by an open circuit. Note that we did *not* deactivate the controlled source; setting $v_S(t) = 0$ made $\Delta v_{GS} = 0$, which deactivated the controlled source. In this sense, a controlled source is treated no differently than a resistor in obtaining the Thévenin impedance. The output impedance then becomes

$$Z_o = \frac{1}{y_{os}} || R_D \tag{8.38}$$

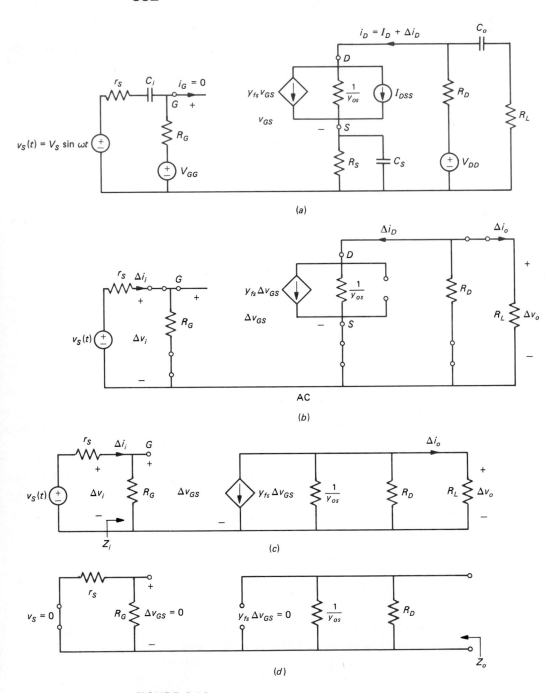

FIGURE 8.16
Small-signal
analysis of the
common-source
JFET amplifier.

The input current to the amplifier is

$$\Delta i_i = \frac{\Delta v_i}{R_G}$$

$$= \frac{\Delta v_i}{Z_i} \tag{8.39}$$

The output current is

$$\Delta i_o = \frac{\Delta v_o}{R_L} \tag{8.40}$$

and the current gain is

$$A_I = \frac{\Delta i_o}{\Delta i_i}$$

$$= \frac{\Delta v_o}{\Delta v_i} \frac{Z_i}{R_L}$$

$$= A_V \frac{R_G}{R_L} \tag{8.41}$$

The power gain is

$$A_p = A_V A_I \tag{8.42}$$

Example 8.2

The manufacturer lists nominal parameters for the 2N3819 JFET as $y_{os} = 50\ \mu S$, $y_{fs}(\text{min}) = 2000\ \mu S$, and $y_{fs}(\text{max}) = 6500\ \mu S$. Determine A_V, Z_i, and Z_o for this amplifier, assuming that the dc design yields $R_S = 5\ k\Omega$, $R_D = 1\ k\Omega$, $R_G = 100\ k\Omega$ and that the load is $R_L = 1\ k\Omega$. Assume a maximum JFET.

Solution   With $R_G = 100\ k\Omega$ and $R_L = 1\ k\Omega$, we find that

$$A_V = -y_{fs}\left(\frac{1}{y_{os}} || R_D || R_L\right)$$

$$= -3.17$$

$$Z_o = \frac{1}{y_{os}} || R_D$$

$$= 952.38\ \Omega$$

$$Z_i = R_G$$

$$= 100\ k\Omega$$

The current gain is

$$A_I = A_V \frac{Z_i}{R_L}$$

$$= -317$$

Thus, the power gain is

$$A_P = A_V A_I$$
$$= 1005$$

Note in the previous example that since $1/y_{os} = 20$ kΩ, it has little effect on any of the ac parameters. For example, in the voltage-gain expression in Eq. (8.36), $1/y_{os}$ is in parallel with $R_D||R_L$. But since $R_D||R_L = 500$ Ω, $1/(y_{os}||R_D||R_L) = 487.8$ Ω. Neglecting $1/y_{os}$ gives $A_V = -3.25$, which is reasonably close to the exact value of $A_V = -3.17$. Similarly, $1/y_{os}$ entered into the Z_o calculation in parallel with R_D, but since $R_D = 1$ kΩ, $1/y_{os}$ may be neglected, yielding $Z_o \doteq R_D = 1$ kΩ as compared to the exact value of $Z_o = 952.38$ Ω.

Therefore, for simplification of calculations with this model, we may generally omit $1/y_{os}$ from the small-signal linear equivalent circuit. Thus, the circuit which we will use is shown in Fig. 8.17a. Omitting $1/y_{os}$ means that we assume that the lines on the characteristic are flat with no slope (Fig. 8.17b). Remember that this will not

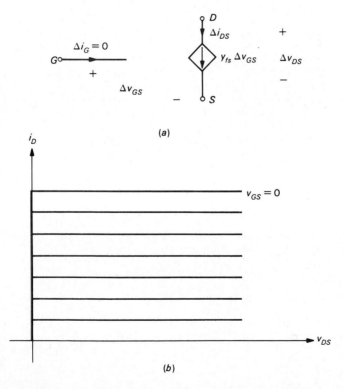

FIGURE 8.17 A simplified small-signal equivalent circuit of a JFET.

always be a valid approximation. For example, if R_D and R_L in the previous example had both been 100 kΩ, then $1/y_{os} = 20$ kΩ would have had an effect and could not be removed from the linear model. However, to simplify our calculations with this model, we always omit $1/y_{os}$ and use the model in Fig. 8.17a.

If the capacitor C_S which bypasses most of the ac signal around R_S had been removed, then R_S would appear in the ac circuit and would affect some of our ac figures of merit. To determine the effect of removing C_S, let us reexamine our calculations. The ac circuit is shown in Fig. 8.18a. We now replace the device symbol with the small-signal linear equivalent circuit of Fig. 8.17a, which results in the circuit in Fig. 8.18b. We must be careful to attach the correct terminals and identify the location of the controlling variable of the controlled source Δv_{GS}. Note that Δv_{GS} is now not equal to Δv_i. But with KVL we may obtain

$$\Delta v_i = \Delta v_{GS} + R_S(y_{fs}\Delta v_{GS}) \tag{8.43}$$

(since current through R_S is determined by the controlled current source: $y_{fs}\Delta v_{GS}$) or

$$\Delta v_i = (1 + R_S y_{fs})\Delta v_{GS} \tag{8.44}$$

The output voltage is, once again,

$$\Delta v_o = -y_{fs}\Delta v_{GS}(R_D||R_L) \tag{8.45}$$

Thus, the voltage gain is

$$A_V = \frac{\Delta v_o}{\Delta v_i}$$

$$= -\frac{y_{fs}(R_D||R_L)}{1 + R_S y_{fs}} \tag{8.46}$$

Comparing Eq. (8.46) to the result when C_S bypassed R_S (removing R_S from the ac circuit) in Eq. (8.34), we see that the voltage gain has been reduced by a factor of $1 + R_S y_{fs}$. This represents what is known as *feedback*; a portion of the effect of the current in the output circuit, $y_{fs}\Delta v_{GS}$, has been "fed back" to affect the input circuit via the voltage drop across R_S, $y_{fs}R_S\Delta v_{GS}$.

The input impedance is again

$$Z_i = \frac{\Delta v_i}{\Delta i_i}$$

$$= R_G \tag{8.47}$$

and R_S has no effect on this. The current gain is

$$A_I = A_V \frac{Z_i}{R_L} \tag{8.48}$$

FIGURE 8.18 AC analysis of the JFET amplifier, using the simplified small-signal equivalent circuit of Fig. 8.17.

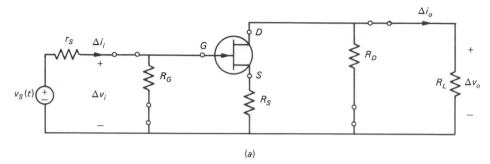

(a)

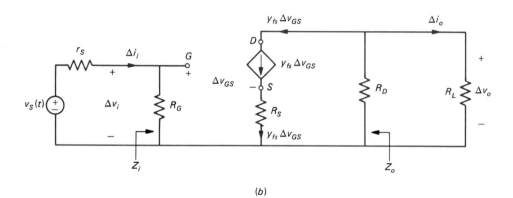

(b)

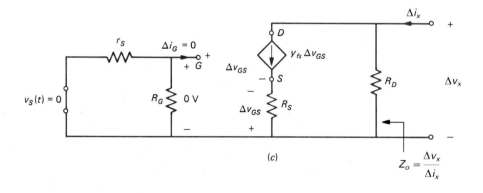

(c)

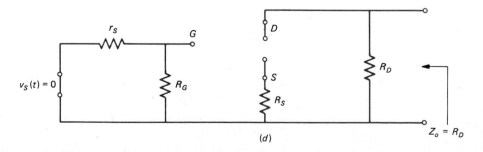

(d)

and since A_V has been reduced by the presence of R_S in the ac circuit, A_I is also reduced. The output impedance is found by setting $v_S(t) = 0$ and finding the Thévenin impedance seen by R_L, as shown in Fig. 8.18c. In this circuit

$$Z_o = \frac{\Delta v_x}{\Delta i_x} \tag{8.49}$$

The voltage across R_G is zero since $\Delta v_{GS} = 0$. Therefore, all the controlling voltage for the controlled source Δv_{GS} appears across R_S:

$$\Delta v_{GS} = -R_S y_{fs} \Delta v_{GS} \tag{8.50}$$

from which we conclude that $\Delta v_{GS} = 0$. Thus, the controlled current source is replaced with an open circuit, as shown in Fig. 8.18d, and

$$Z_o = R_D \tag{8.51}$$

Example 8.3

Compute A_V, A_I, Z_i, Z_o for the JFET amplifier of Example 8.2 with C_S removed.

Solution The circuit element values were given as $R_S = 5\text{ k}\Omega$, $R_D = 1\text{ k}\Omega$, $R_L = 1\text{ k}\Omega$, $R_G = 100\text{ k}\Omega$. Assume a maximum JFET with $y_{fs}(\text{max}) = 6500\ \mu\text{S}$. Thus

$$A_V = -\frac{y_{fs}(R_D\|R_L)}{1 + R_S y_{fs}}$$

$$= -0.097$$

$$Z_i = R_G$$

$$= 100\text{ k}\Omega$$

$$A_I = A_V \frac{Z_i}{R_L}$$

$$= -9.7$$

$$Z_o = R_D$$

$$= 1\text{ k}\Omega$$

Note that removing C_S dramatically reduces the voltage gain from 3.25 to 0.097 as a result of the feedback through the unbypassed R_S.

The JFET amplifier which we have considered is referred to as the *common-source amplifier* since the source terminal of the JFET is common to the amplifier input and output (except when C_S is removed). There are other possible configurations—the common-drain amplifier and the common-gate amplifier—which are used for special purposes such as impedance matching. The common-drain amplifier is similar to the common-source amplifier except that the output is taken off the JFET source terminal. The input is applied to the gate terminal, and the drain terminal of the JFET is common to the amplifier input and output. The common-drain JFET amplifier gives a voltage gain of approximately $A_V \cong 1$. Thus, the input and output voltages are approximately equal and in phase. Therefore, the common-drain amplifier is referred to as a *source follower* since the JFET source terminal voltage which is the output voltage "follows" the input voltage. The common-drain amplifier also has the unique characteristic of providing a very low output impedance. Values of Z_o on the order of 50 to 150 Ω can be attained. Note that the output impedance of the common-source amplifier considered previously is on the order of $Z_o = R_D$, which is usually not small. The other amplifier configuration, the common-gate JFET amplifier, has the input placed on the JFET source terminal, and the output is taken off the JFET drain terminal. Thus, the gate terminal is common to the amplifier input and output. The common-gate JFET amplifier is the dual to the common-drain JFET amplifier in the sense that the current gain is approximately unity (and positive), $A_I \cong 1$, and the input impedance can be made very small for matching to low-impedance signal sources. Typical values of Z_i that can be obtained are 50 to 150 Ω. Both these amplifiers can be analyzed to determine their ac figures of merit, such as the voltage gain, current gain, input impedance, and output impedance in the same fashion as for the common-source JFET amplifier which we considered in detail by simply substituting the ac equivalent circuit of Fig. 8.17a into the ac circuit.

8.5   FREQUENCY RESPONSE OF THE JFET AMPLIFIERS

The frequency response of the voltage gain of the common-source JFET amplifier is shown in Fig. 8.19. The low-frequency deterioriation of the voltage gain is due to the increasing impedance of the coupling capacitors C_o and C_i and of the bypass

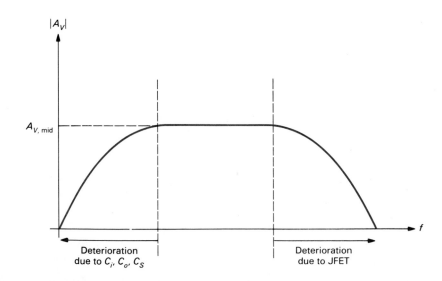

FIGURE 8.19 Frequency response of the voltage gain of a common-source JFET amplifier.

capacitor C_S at these lower frequencies. Recall that the coupling capacitors C_i and C_o of the amplifiers were chosen such that their impedances were small (negligible) at the operating frequency of the sinusoidal source $v_S(t)$. Similarly, the capacitor C_S which bypasses ac signals around the resistor R_S in the common-source amplifier shown in Fig. 8.9a was chosen such that its impedance was also negligible at this operating frequency. As the frequency is lowered, these capacitors present a larger impedance and tend to "open up" as the frequency is reduced to zero. If C_i and/or C_o is replaced by open circuits, then clearly the voltage gain is zero. Also, in Example 8.3 we found that removing the capacitor C_S causes the voltage gain of the common-source amplifier to drop dramatically. Consequently, increasing the impedances of either of these capacitances (as is the case when the frequency is reduced) will cause the voltage gain to drop.

The high-frequency deterioration of the voltage gain is due to the JFET. A high-frequency small-signal linear ac model of a JFET is shown in Fig. 8.20a. The gate-source pn junction is reverse-biased, as is the gate-drain junction. This results in depletion layers at the pn junctions which are a separation of mobile charge carriers. Thus, we would expect the capacitances C_{GS} and C_{GD} to represent this effect. The capacitance between drain and source C_{DS} is due to other, less predominant factors, such as the stray capacitance to the JFET case, and as such is smaller than C_{GS} or C_{GD} and is usually neglected. The depletion regions between gate and source and between gate and drain are not equal in extent but are somewhat asymmetric. However, this asymmetry does not cause these capacitances to be appreciably different. For example, the manufacturer of a 2N3819 JFET specifies that $C_{GS} = C_{GD} = 4$ picofarads (pF) and $C_{DS} = 0$. As the frequency is increased, these capacitances tend to "short-circuit" the JFET.

This high-frequency deterioration of the amplifier performance due to C_{GS}, C_{GD}, and C_{DS} has a more interesting and subtle effect. Substituting the equivalent circuit into a common-source amplifier of Fig. 8.12 and replacing C_o, C_i, and C_S with short circuits for the high-frequency analysis, we obtain the circuit in Fig. 8.20b. Capacitance C_{DS} is usually much smaller than C_{GS} or C_{GD} and is neglected here. This is redrawn in Fig. 8.20c by converting the controlled current source to a controlled voltage source with a Thévenin equivalent. Note that $\Delta v_i = \Delta v_{GS}$ and

$$\Delta i_x = \frac{\Delta v_{GS} + y_{fs}(R_D\|R_L)\,\Delta v_{GS}}{R_D\|R_L + 1/(j\omega C_{GD})} \tag{8.52}$$

The impedance Z_x indicated in Fig. 8.20c is

$$Z_x = \frac{\Delta v_{GS}}{\Delta i_x}$$

$$= \frac{R_D\|R_L + 1/(j\omega C_{GD})}{1 + y_{fs}(R_D\|R_L)} \tag{8.53}$$

Note that the midband voltage gain of the amplifier is

$$A_{V,\text{mid}} = -y_{fs}(R_D\|R_L) \tag{8.54}$$

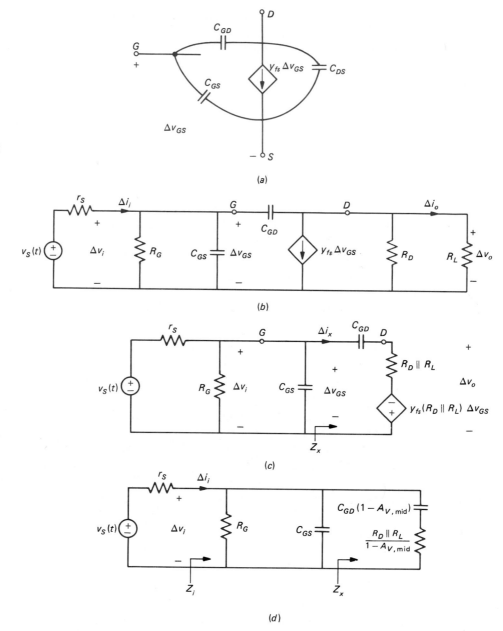

Writing Eq. (8.53) in terms of this quantity, we have

$$Z_x = \frac{R_D || R_L}{1 - A_{V,\mathrm{mid}}} + \frac{1}{j\omega C_{GD}(1 - A_{V,\mathrm{mid}})} \tag{8.55}$$

This leads to the equivalent circuit shown in Fig. 8.20d in which, according to Eq. (8.55), C_{GD} appears to the input circuit larger by a factor of $1 - A_{V,\mathrm{mid}}$. Similarly, the parallel combination of R_D and R_L appears to the input circuit smaller by a factor of $1 - A_{V,\text{mid}}$. This is known as the *Miller effect*. Therefore, the input impedance is reduced by these capacitances, and the interesting reflection of C_{GD} causes it to appear larger to the amplifier input than its physical value.

The voltage gain is similarly affected. For example, from Fig. 8.20c we obtain

$$\Delta v_o = -y_{fs}(R_D||R_L)\Delta v_{GS} + (R_D||R_L)\Delta i_x$$
$$= A_{V,\text{mid}}\Delta v_{GS} + (R_D||R_L)\Delta i_x \tag{8.56}$$

and

$$\Delta v_i = \Delta v_{GS} \tag{8.57}$$

Substituting Eqs. (8.52) and (8.57) into Eq. (8.56), we obtain

$$A_V = \frac{\Delta v_o}{\Delta v_i}$$

$$= A_{V,\text{mid}} + (R_D||R_L)\frac{1 - A_{V,\text{mid}}}{R_D||R_L + 1/(j\omega C_{GD})}$$

$$= A_{V,\text{mid}} + \frac{1 - A_{V,\text{mid}}}{1 + 1/[j\omega C_{GD}(R_D||R_L)]}$$

$$\doteq A_{V,\text{mid}}\frac{1}{1 + j\omega C_{GD}(R_D||R_L)} \tag{8.58}$$

Note that the high-frequency voltage gain rolls off like a low-pass filter with a time constant of $C_{GD}(R_D||R_L)$.

Example 8.4

For a common-source JFET amplifier with $R_L = R_D = R_S = 10\text{ k}\Omega$ and $R_G = 100\text{ k}\Omega$, compute the frequency at which the magnitude of the input impedance is reduced to 1 kΩ if a 2N3819 with $y_{fs} = 6500\ \mu\text{S}$, $C_{GS} = C_{GD} = 4\text{ pF}$, and $C_{DS} = 0$ is used.

Solution The midband voltage gain is

$$A_{V,\text{mid}} = y_{fs}(R_D||R_L)$$
$$= -32.5$$

so that C_{GD} appears in the input circuit as $(1 - A_{V,\text{mid}})C_{GD} = (33.5)(4)\text{ pF} = 134\text{ pF}$. Thus, the input impedance is

$$Z_i = R_G||\frac{1}{j\omega[C_{GS} + (1 - A_{V,\text{mid}})C_{GD}]}$$
$$= 1\text{ k}\Omega$$

and we have neglected $R_D \| R_L / (1 - A_{V,\text{mid}}) = 149.25 \ \Omega$. Since $C_{GS} + (1 - A_{V,\text{mid}}) C_{GD} = 138$ pF, for $Z_i = 1$ kΩ and $R_G = 100$ kΩ we have

$$1 \ \text{k}\Omega = \left| \frac{100 \ \text{k}\Omega [1/(j\omega 138 \times 10^{-12})]}{100 \ \text{k}\Omega + 1/(j\omega 138 \times 10^{-12})} \right|$$

$$= \left| \frac{100 \ \text{k}\Omega}{1 + j\omega 1.38 \times 10^{-5}} \right|$$

or

$$1 \ \text{k}\Omega = \frac{100 \ \text{k}\Omega}{\sqrt{1 + (\omega 1.38 \times 10^{-5})^2}}$$

and the frequency is 1.153 megahertz (MHz). This is a rather "low" frequency to have the input impedance deteriorate by such a large factor. It illustrates how dramatic the effect of these device capacitances can be.

8.6 THE BIPOLAR JUNCTION TRANSISTOR (BJT)

An *npn* BJT consists of two *n*-type semiconductor regions with a very thin *p*-type region sandwiched between them, as shown in Fig. 8.21*a*. There exist *pnp*-type BJTs where an *n*-type region is sandwiched between two *p*-type regions, but we consider only *npn* BJTs for the same reason that we considered only *n*-channel JFETs: amplifiers constructed from *npn* BJTs require dc power supplies which have a positive voltage with respect to ground, whereas *pnp* BJT amplifiers require negative-voltage power supplies. In fact, many types of BJTs (for example, a 2N718) exist in both *npn* and *pnp* construction. The only difference between these two is that the terminal

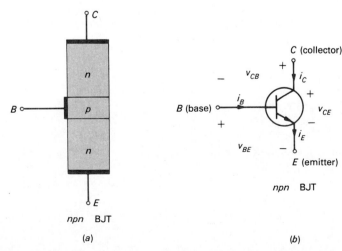

FIGURE 8.21
The bipolar junction transistor (BJT):
(*a*) physical construction;
(*b*) circuit symbol.

currents and voltages of the *pnp* BJT are opposite in sign to those of the *npn* BJT. Thus, if we design an *npn* BJT amplifier, we may simply substitute a *pnp* BJT (of the same type) if we reverse the polarity of the batteries (or dc power supplies) in the biasing circuitry.

Terminals are attached to each of the three regions as shown in Fig. 8.21*a*. These are designated *E* for emitter, *B* for base, and *C* for collector. The device symbol for an *npn* BJT is shown in Fig. 8.21*b*. Note that the base and collector currents i_B and i_C are defined as entering the terminals, whereas the emitter current i_E is defined as leaving the emitter terminal. Note also the (arbitrary) directions of the terminal voltages. There is a small arrow in the device symbol in the emitter lead which is in the direction of the (assumed) emitter current. We will find that this is the actual direction of the emitter direct current for an *npn* BJT under normal biasing conditions. A *pnp* BJT would have a similar device symbol, but the small arrow in the emitter lead would be reversed from that of the *npn* direction.

The operation and amplification ability of a BJT can be explained from Fig. 8.22*a*. We have supplied two batteries. One battery, V_{BE}, has its positive terminal connected to the *p*-type base region terminal and its negative terminal connected to the *n*-type emitter region terminal. The other battery, V_{CE}, has its positive terminal connected to the *n*-type collector region terminal and its negative terminal connected to the *n*-type emitter region terminal. This is the normal polarity for dc biasing, although we will use both batteries and series resistors with each battery, otherwise, the terminal voltages would be fixed at the battery voltages, and no variations (amplification) would be possible. However, note that

$$v_{BE} = V_{BE} \tag{8.59}$$

and that the base-emitter junction is forward-biased. Note also that

$$v_{CB} = V_{CE} - V_{BE} \tag{8.60}$$

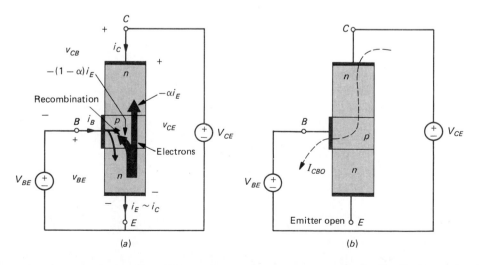

FIGURE 8.22 Charge flow in a BJT: (*a*) current amplification; (*b*) leakage current.

and that the collector-base junction is reverse-biased if $V_{CE} > V_{BE}$. In normal BJT operation, the junctions are always biased in this fashion. Each junction can be thought of as a *pn* semiconductor diode. The base-emitter junction being forward-biased permits electrons (majority carriers) in the *n*-type emitter region to be injected into the *p*-type base region. Consequently, the characteristic for this pair of terminals resembles that of a forward-biased *pn* semiconductor diode, as shown in Fig. 8.23*a*. For $V_{CE} > 1$ V, which is the usual case, the characteristic is independent of changes in $v_{CE} = V_{CE}$. For BJTs constructed of silicon, the curve breaks at approximately 0.5 to 0.7 V, as we saw for silicon *pn* diodes.

FIGURE 8.23
The terminal characteristics of a typical BJT.

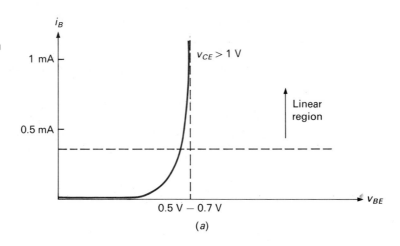

(a)

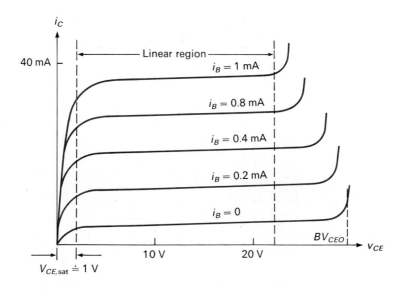

(b)

The BJT base region is very thin in order to minimize the recombination of the electrons injected from the emitter with holes (the majority carriers) in the *p*-type base region. Also the emitter region is much more heavily doped than the base region, so the hole current from base to emitter is much less than the electron current from emitter to base. Since the collector-base junction is reverse-biased, these injected electrons are swept into the collector region. However, a small amount of the electrons injected into the base region is "siphoned off" through the base terminal. Typically, only a small percentage is collected by the base so that the collector current ranges from 0.98 to 0.99 of the emitter current. Thus, we may write

$$i_C = \alpha i_E \tag{8.61}$$

where α (alpha) is called the *forward-current transfer ratio* whose value typically ranges from 0.98 to 0.99.

Now suppose that we open-circuit the emitter, as shown in Fig. 8.22*b*, thus reducing i_E to zero. Note that the base-collector junction resembles a reverse-biased *pn* diode; consequently we expect to find a leakage current which was designated as I_S for the diode. For the BJT, this is designated as I_{CBO}, where the subscript denotes collector to base with the emitter open-circuited. This reverse leakage current will always be present, regardless of whether the emitter is open, since the batteries reverse-bias the collector-base junction. Consequently, superimposing this current onto Eq. (8.61), we obtain

$$i_C = \alpha i_E + I_{CBO} \tag{8.62}$$

One important point about I_{CBO} should be mentioned—it exhibits a strong dependence on the temperature of the collector-base junction. If this temperature increases, I_{CBO} also increases. This can lead to an uncontrolled increase in the collector temperature (known as *thermal runaway*) unless the biasing circuit configuration is chosen to minimize this effect. For example, if an initial heating of the BJT occurs (owing, say, to heat generated by neighboring components or to other ambient temperature variations), this will result in an increase in I_{CBO}. However, this current increase causes an increase in junction temperature due to $I_{CBO}^2 R_{CB}$ losses, where R_{CB} is the junction resistance. These losses therefore increase the junction temperature, which increases I_{CBO}. The process may continue, with the result that the collector direct current $I_C = \alpha I_E + I_{CBO}$ increases. This may result in the operating point moving into a nonlinear region of the characteristic, as was the result for unit variations of the JFET. Consequently, our biasing circuitry should tend to minimize this effect.

We may rewrite Eq. (8.62) in terms of i_B and i_C, since by KCL

$$i_E = i_B + i_C \tag{8.63}$$

or

$$i_B = i_E - i_C \tag{8.64}$$

Equation (8.62) may be written as

$$i_E = \frac{1}{\alpha} i_C - \frac{1}{\alpha} I_{CBO} \qquad (8.65)$$

Substituting Eq. (8.65) into Eq. (8.64), we obtain

$$i_B = \frac{1}{\alpha} i_C - \frac{1}{\alpha} I_{CBO} - i_C$$

$$= \frac{1 - \alpha}{\alpha} i_C - \frac{1}{\alpha} I_{CBO} \qquad (8.66)$$

The symbol β (beta) is used to represented the ratio $\alpha/(1 - \alpha)$; thus

$$\beta = \frac{\alpha}{1 - \alpha} \qquad (8.67)$$

and

$$\alpha = \frac{\beta}{\beta + 1} \qquad (8.68)$$

Typical values of β range from $\beta = 49$ ($\alpha = 0.98$) to $\beta = 99$ ($\alpha = 0.99$). Writing Eq. (8.66) in terms of β, we obtain

$$i_C = \beta i_B + \frac{I_{CBO}}{1 - \alpha}$$

$$= \beta i_B + (\beta + 1)I_{CBO} \qquad (8.69)$$

The quantity $(\beta + 1)I_{CBO}$ is, from Eq. (8.69), the collector current when the base is open-circuited ($i_B = 0$) and is designated as

$$I_{CEO} = (\beta + 1)I_{CBO} \qquad (8.70)$$

where the subscript CEO denotes collector-emitter current with the base open-circuited. In terms of I_{CEO}, Eq. (8.69) becomes

$$i_C = \beta i_B + I_{CEO} \qquad (8.71)$$

The plot of v_{CE} and i_C versus i_B for a typical BJT is shown in Fig. 8.23b. For v_{CE} less than approximately 1 V, the collector-base junction is forward-biased [see Fig. 8.22a and Eq. (8.60)] and normal BJT operation does not occur. The collector current increases linearly with an increase in v_{CE} up to a point at which the v_{CE} is sufficiently large (>1 V) that the collector-base junction becomes reverse-biased and normal BJT operation occurs. In this case, Eq. (8.71) applies. Actually, Eq. (8.71) does not strictly apply since there is a slight increase in i_C with increasing v_{CE}, as evidenced by the small (but nonzero) slope of the curves in Fig. 8.23b. Further increases in v_{CE} result in only small increases in i_C, and the curves are relatively flat up to the point at which v_{CE} is sufficiently large that avalanche breakdown occurs. At this point, which is designated as BV_{CEO} for $i_B = 0$ (breakdown voltage collector to emitter with base open-circuited), the collector current increases rapidly with a small increase in v_{CE}.

8.7   THE BJT AMPLIFIER

Obviously, for linear operation we will design the biasing circuitry so that we operate in the linear regions of the characteristics. A dc-biasing network which will accomplish this is shown in Fig. 8.24a. Writing KVL around the base-emitter loop and the collector-emitter loop, we obtain

$$v_{BE} = V_{BB} - R_B i_B \tag{8.72}$$

and

$$v_{CE} = V_{CC} - R_C i_C \tag{8.73}$$

Plotting Eq. (8.72) as a dc load line on the base-emitter characteristic and plotting Eq. (8.73) on the collector-emitter characteristic, we find the dc operating points (I_B, $V_{BE} \doteq 0.5$ V, I_C, V_{CE}), as shown in Fig. 8.24b. Thus the dc load line has slope

$$-\frac{1}{R_{\text{DC}}} = -\frac{1}{R_C} \tag{8.74}$$

Now let us add an ac signal source and load, as shown in Fig. 8.25a. Once again we assume that $v_S(t)$ is small enough that linear operation is preserved. In this case each voltage and current in the circuit will consist, by superposition, of a dc quantity due to V_{BB} and V_{CC} and of a delta (Δ) ac quantity due to the sinusoidal source $v_S(t)$. Removing the ac source gives the dc circuit shown in Fig. 8.24a. The ac circuit is obtained by short-circuiting all capacitors and deactivating the two dc sources V_{BB} and V_{CC}, as shown in Fig. 8.25b. From this we see that the ac quantities Δv_{CE} and

FIGURE 8.24
DC load-line
analysis of a BJT
amplifier.

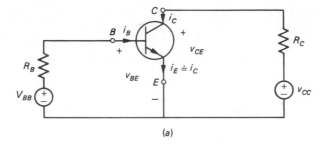

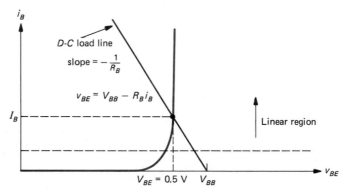

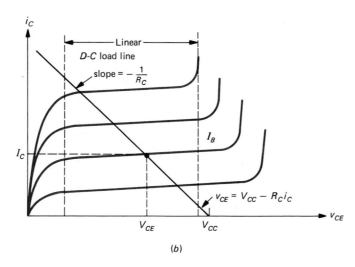

(b)

Δi_C are related by

$$\Delta v_{CE} = -(R_C||R_L)\Delta i_C \tag{8.75}$$

Thus, the ac signal variations follow an ac load line with slope

$$-\frac{1}{R_{AC}} = -\frac{1}{R_C||R_L} \tag{8.76}$$

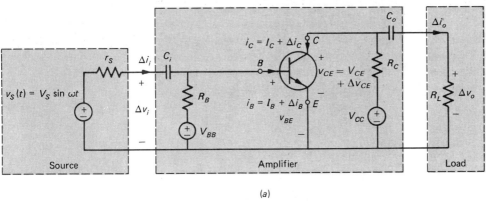

(a)

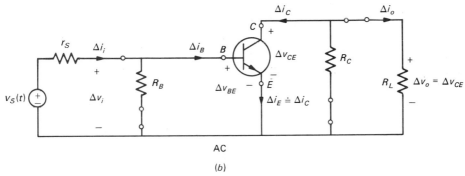

AC

(b)

FIGURE 8.25   AC small-signal analysis of a BJT amplifier.

as shown in Fig. 8.26. The ac load line again must pass through the dc operating point.

Once we have chosen an acceptable operating point in the linear portion of the i_C versus v_{CE} characteristic, (I_C, V_{CE}, I_B), where again we interpolate between the limited set of i_B curves shown, we can determine R_B and V_{BB} to force the desired value of I_B into the base of the BJT. Recall that the collector and base currents are related by Eq. (8.71) as (neglecting I_{CEO})

$$I_B \doteq \frac{I_C}{\beta} \tag{8.77}$$

Typical values of β are on the order of 100. Nevertheless, either reading I_B off the characteristic at the operating point or using Eq. (8.77) and an assumed value of β, we can determine the desired value of base current I_B. The values of R_B and V_{BB} can be chosen from Eq. (8.72), where we may assume $V_{BE} \doteq 0.5$ V. Once the operating point is set in this fashion, we next determine the ac quantities of interest—voltage and current gain as well as input and output impedance—for the design.

FIGURE 8.26
Load lines and
signals of a BJT
amplifier.

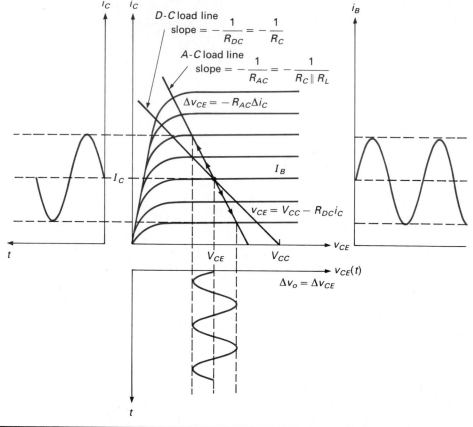

Example 8.5

Determine the value of R_B in the circuit of Fig. 8.27 such that the operating point is $I_C = 10$ mA. Assume $\beta = 100$. Determine the collector-emitter voltage V_{CE}.

Solution The required base current is

$$I_B = \frac{I_C}{\beta}$$

$$= \frac{10 \text{ mA}}{100}$$

$$= 0.1 \text{ mA}$$

Assuming $V_{BE} \doteq 0.5$ V,

$$R_B = \frac{V_{BB}(= V_{CC}) - V_{BE}(\cong 0.5 \text{ V})}{I_B}$$

$$= \frac{15 \text{ V} - 0.5 \text{ V}}{0.1 \text{ mA}}$$

$$= 145 \text{ k}\Omega$$

FIGURE 8.27
Example 8.5:
design of the
operating point.

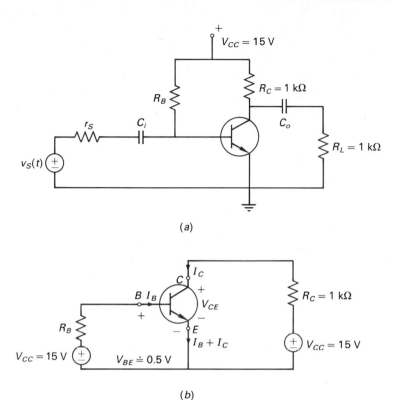

(a)

(b)

The dc circuit is drawn in Fig. 8.27*b*. From this we determine

$$V_{CE} = V_{CC} - R_C I_C$$
$$= 15\text{ V} - 1\text{ k}\Omega \times 10\text{ mA}$$
$$= 5\text{ V}$$

We have found that

$$i_C = \beta i_B + I_{CEO} \tag{8.78}$$

and that I_{CEO} is a leakage current which is strongly dependent on the base-collector junction temperature. It typically doubles for every $10°C$ increase in temperature. As pointed out previously, if the ambient temperature of the device increases, so does I_{CEO}, and this produces an additional increase in the junction temperature due to increased $i^2 R$ losses. Thus, I_{CEO} increases once again. The process may continue, with the result being an increase in collector current such that the operating point moves into a nonlinear region of the characteristic. This phenomenon is called *thermal runaway* and is clearly undesirable; once we design for a certain I_C, we want I_C to remain there.

Other parameters are also sensitive to temperature. For example, the base-emitter voltage V_{BE} decreases approximately 2.5 mV for every 1°C increase in temperature. Another parameter which varies with temperature is β; however, this variation is nonlinear with temperature and is difficult to estimate.

We design our biasing circuitry so that changes in the amplifier operating point due to changes in these parameters will be minimized. For silicon transistors, the major variations to be guarded against are changes in V_{BE} and β. For germanium transistors, the major variation to guard against is a temperature change resulting in an I_{CBO} change.

An amplifier circuit which has the ability to limit changes in the operating point due to these variations is shown in Fig. 8.28a. A resistor R_E bypassed for alternating current by a capacitor C_E is placed in the emitter lead. Two base-biasing resistors

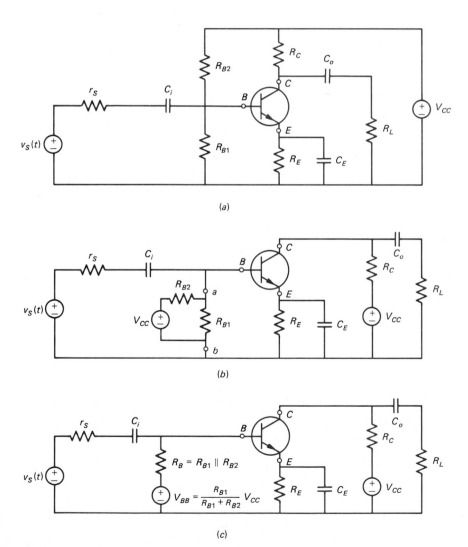

FIGURE 8.28 A practical BJT amplifier circuit to stabilize against operating-point changes.

R_{B1} and R_{B2} allow the use of one battery. These two resistors in combination with V_{CC} can be reduced to one resistor

$$R_B = R_{B1} \| R_{B2} \tag{8.79}$$

and a voltage source

$$V_{BB} = \frac{R_{B1}}{R_{B1} + R_{B2}} V_{CC} \tag{8.80}$$

with a Thévenin equivalent, as shown in Fig. 8.28b and c. With the exception of R_E and C_E, this is identical to the basic amplifier in Fig. 8.25a. In fact, this amplifier and the final JFET amplifier in Fig. 8.9 are identical in form.

This circuit tends to stabilize against changes in I_{CEO}, V_{BE}, and β due to temperature variations, and it also tends to stabilize against changes in operating point due to unit-to-unit variations in β, as we will see. The primary reason for this insensitivity is the addition of the resistor R_E in the emitter lead of the BJT. For example, the manufacturer of the 2N718 BJT specifies that $\beta_{min} = 50$, $\beta_{max} = 150$. Thus, a group of supposedly identical 2N718 BJTs can have β's (at the same operating points) which differ by as much as 3:1. For the original circuit which we considered (Fig. 8.25), note that I_B is fixed by V_{BB} and R_B:

$$I_B = \frac{V_{BB} - V_{BE} \text{ where } V_{BE} \doteq 0.5 \text{ V}}{R_B} \tag{8.81}$$

and

$$I_C = \beta I_B \tag{8.82}$$

Thus, for two supposedly identical amplifiers having 2N718 BJTs, the operating points I_C may differ by as much as 3:1.

The amplifier circuit of Fig. 8.28a does not share this extreme sensitivity due primarily to the resistor R_E in the emitter lead of the BJT. The dc circuit is shown in Fig. 8.29a. Writing KVL around the collector-emitter loop gives

$$\begin{aligned} V_{CE} &= V_{CC} - R_C I_C - R_E(I_C + I_B) \\ &\doteq V_{CC} - (R_C + R_E)I_C \end{aligned} \tag{8.83}$$

where we have approximated the emitter current as $I_E = I_C + I_B \doteq I_C$. Also writing KVL around the base-emitter loop gives

$$\begin{aligned} V_{BE} &= V_{BB} - R_B I_B - R_E(I_C + I_B) \\ &\doteq V_{BB} - R_B I_B - R_E I_C \end{aligned} \tag{8.84}$$

FIGURE 8.29
Analysis of the
circuit of Fig. 8.28
to determine
(a) dc load-line
slope and (b) ac
load-line slope.

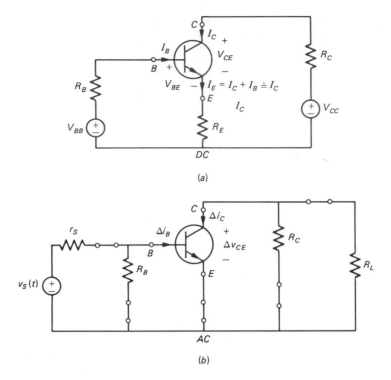

(a)

(b)

If the collector current attempts to increase due to β and/or temperature variations, a voltage is developed across R_E, $I_C R_E$, which tends to subtract from V_{BB} so that I_B is reduced, thereby reducing I_C.

From Eq. (8.83) we see that the dc load line has slope

$$-\frac{1}{R_{DC}} = -\frac{1}{R_C + R_E} \tag{8.85}$$

Similarly, from the ac circuit shown in Fig. 8.29b we obtain, by writing KVL around the collector-emitter loop,

$$\Delta v_{CE} = -(R_C || R_L)\Delta i_C \tag{8.86}$$

Therefore, the ac load line has slope

$$-\frac{1}{R_{AC}} = -\frac{1}{R_C || R_L} \tag{8.87}$$

Example 8.6

An *npn* BJT is to be used to construct the amplifier in Fig. 8.30. Assume an average value of $\beta = 100$, and determine R_E and R_C for an operating point of $I_C = 10$ mA and $V_{CE} = 7$ V. Determine the change in operating point for $\beta = 50$ and $\beta = 150$.

FIGURE 8.30
Example 8.6.

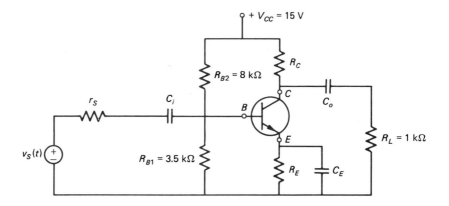

Solution   First we calculate

$$I_B = \frac{I_C}{\beta}$$

$$= \frac{10 \text{ mA}}{100}$$

$$= 0.1 \text{ mA}$$

and

$$V_{BB} = \frac{R_{B1}}{R_{B1} + R_{B2}} V_{CC}$$

$$- \frac{3.5 \text{ k}\Omega}{3.5 \text{ k}\Omega + 8 \text{ k}\Omega} 15 \text{ V}$$

$$= 4.565 \text{ V}$$

and

$$R_B = R_{B1} || R_{B2}$$
$$= 3.5 \text{ k}\Omega || 8 \text{ k}\Omega$$
$$= 2435 \ \Omega$$

From the dc circuit we obtain

$$V_{BB} = R_B I_B + V_{BE} + R_E I_C$$

so that

$$R_E = \frac{V_{BB} - R_B I_B - V_{BE}}{I_C}$$

$$= 382 \ \Omega$$

Similarly,

$$V_{CE} = V_{CC} - R_C I_C - R_E I_C$$

so that

$$R_C = \frac{V_{CC} - V_{CE} - R_E I_C}{I_C}$$

$$= \frac{15\text{ V} - 7\text{ V} - 382 \times 10\text{ mA}}{10\text{ mA}}$$

$$= 418\ \Omega$$

Using these values of R_E and R_C, we obtain the operating-point changes for changes in β by substituting Eq. (8.82) into Eq. (8.84) to give

$$I_C = \frac{V_{BB} - V_{BE}}{R_B/\beta + R_E}$$

Similarly,

$$V_{CE} = V_{CC} - (R_C + R_E)I_C$$

where $R_C + R_E = 800\ \Omega$. These equations show that the operating point changes to

$$I_{C,\beta=50} = 9.4\text{ mA} \qquad \text{and} \qquad V_{CE,\beta=50} = 7.45\text{ V}$$

and

$$I_{C,\beta=150} = 10.2\text{ mA} \qquad \text{and} \qquad V_{CE,\beta=150} = 6.83\text{ V}$$

Therefore large changes in β cause only small changes in the operating point—a very desirable design objective.

8.8 SMALL-SIGNAL ANALYSIS OF THE BJT AMPLIFIER

The analysis of the JFET amplifiers consisted of two parts: a dc analysis to place the dc operating point in a linear region of the characteristic and an ac analysis to determine the resulting voltage gain of the amplifier, as well as to determine the input and output impedances. For small-signal linear operation, we obtained a linear equivalent circuit representing small variations about the dc operating point. We assumed small-signal linear operation of this device such that the device terminal currents and voltages stayed in a locally linear region of the characteristic about the dc operating point. This assumption of small-signal linear operation allowed us to apply superposition even though the circuit contained a basically nonlinear element; since linear operation was preserved, it did not matter that the device was nonlinear outside the linear region.

Thus, we separated the analysis of JFET amplifiers into two parts. One part, the dc-biasing problem, is due only to dc sources in the circuit and uses the concept of selecting an operating point and determining the required values of the biasing resistors and batteries. The second part, the ac analysis problem, consists of setting all dc sources to zero, substituting the small-signal linear equivalent circuit of the device, and determinining the ac quantities of interest, such as the voltage gain and input and output impedances of the amplifier.

The analysis of amplifiers containing bipolar junction transistors follows much the same procedure. The only differences in the two procedures are that (1) the dc analysis is different since the JFET is basically a "voltage-controlled" device (v_{GS}), whereas the BJT is a "current-controlled" device (i_B), and (2) the small-signal linear ac model of the BJT is different from that for the JFET. But the general procedure remains the same.

Consider the typical set of characteristics of the BJT shown in Fig. 8.23. Note the location of the linear regions. These characteristics are linearized in Fig. 8.31a. Now since these characteristics are linear ones, we may write linear equations describing them. For example, for the base-emitter characteristic we may write

$$v_{BE} = h_{ie}i_B + V_{BEO} \qquad (8.88)$$

where h_{ie} is a constant to be determined; it represents the slope of the characteristic at the operating point, since from Eq. (8.88)

$$h_{ie} = \frac{\Delta v_{BE}}{\Delta i_B} \qquad \Omega \qquad (8.89)$$

where the Δ quantities represent small changes in the variables about the operating point. Since the base-emitter junction behaves as a forward-biased pn diode, we would expect h_{ie} to be approximately given by the small-signal incremental resistance of a pn diode, as determined in Chap. 7. For example, by analogy to the pn diode, we may write

$$i_B = I_{BEO}(e^{qv_{BE}/(kT)} - 1) \qquad (8.90)$$

where I_{BEO} is a reverse current for the base-emitter junction reverse-biased and the collector open-circuited. Once again, q is an electron charge, k is Boltzmann's constant, and T is the junction temperature in kelvins. At room temperature we have $kT/q \doteq 25$ mV. For $v_{BE} \gg 25$ mV $= 0.025$ V, Eq. (8.90) becomes

$$i_B \doteq I_{BEO}e^{v_{BE}/25\,\text{mV}} \qquad (8.91)$$

FIGURE 8.31
Small-signal
linearization of
the BJT:
(a) linearization of
device
characteristic;
(b) the small-signal
linear equivalent
circuit.

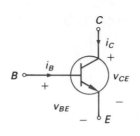

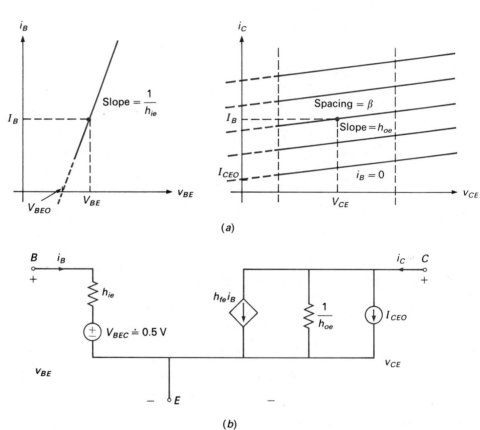

Differentiating this, we obtain

$$\frac{di_B}{dv_{BE}} = \frac{1}{25\ \text{mV}} I_{BEO} e^{v_{BE}/25\ \text{mV}} \tag{8.92}$$

so that

$$h_{ie} \doteq \frac{25\ \text{mV}}{I_{BEO} e^{v_{BE}/25\ \text{mV}}} \tag{8.93}$$

Substituting Eq. (8.91), we have

$$h_{ie} = \frac{25 \text{ mV}}{i_B} \tag{8.94}$$

But

$$i_B \doteq \frac{i_C}{\beta} \tag{8.95}$$

so that at the particular operating point $I_C \doteq \beta I_B$ we have

$$h_{ie} = \frac{25 \text{ mV}}{I_B}$$

$$\doteq \beta \frac{25 \text{ mV}}{I_C} \tag{8.96}$$

For a typical operating point of $I_C = 10$ mA and $\beta = 100$, we obtain $h_{ie} = 250 \ \Omega$.

Note that the intersection of the curve with the v_{BE} axis ($i_B = 0$) occurs typically in the range of 0.5 to 0.7 V for silicon semiconductor materials. This is denoted as V_{BEO}, where from Eq. (8.88)

$$V_{BEO} = v_{BE}|_{i_B = 0} \tag{8.97}$$

For simplification, we assume that

$$V_{BEO} \doteq 0.5 \text{ V} \tag{8.98}$$

Approximating the collector-emitter characteristic similarly, as shown in Fig. 8.31b, we may write

$$i_C = h_{fe} i_B + h_{oe} v_{CE} + I_{CEO} \tag{8.99}$$

which describes this characteristic. Note in Eq. (8.99) that

$$I_{CEO} = i_C \Big|_{\substack{i_B = 0 \\ v_{CE} = 0}} \tag{8.100}$$

Thus, I_{CEO} is the intersection of the $i_B = 0$ line with the i_C axis ($v_{CE} = 0$). Once again, this is the collector-emitter leakage current with the base open-circuited

$(i_B = 0)$, as was discussed previously. From Eq. (8.99) we obtain

$$h_{fe} = \frac{\Delta i_C}{\Delta i_B}\bigg|_{\Delta v_{CE} = 0} \qquad \text{dimensionless} \qquad (8.101)$$

$$h_{oe} = \frac{\Delta i_C}{\Delta v_{CE}}\bigg|_{\Delta i_B = 0} \qquad \text{S} \qquad (8.102)$$

By comparing Eq. (8.101) to Eq. (8.71), we see that†

$$h_{fe} \doteq \beta \qquad (8.103)$$

and is related to the spacing between the lines on the characteristic; on the collector-emitter characteristic for $v_{CE} = V_{CE}$, take a small change in i_B (Δi_B) and find the resulting change in i_C (Δi_C) for no change in v_{CE}. Similarly, Eq. (8.102) shows that h_{oe} is the slope of the $i_B = I_B$ curve through the operating point. Thus, the BJT may be approximated with the linear equivalent circuit shown in Fig. 8.31b, which represents Eqs. (8.88) and (8.99), as the reader should verify. Note that a current-controlled current source is used instead of the voltage-controlled current source used for FETs since, unlike the FET, the BJT second-port characteristic is a function of the first-port current i_B.

The origin of the symbols h_{ie}, h_{fe}, and h_{oe} is as follows. They are hybrid parameters in terms of their dimensions since h_{ie} has the dimensions of resistance, h_{fe} is dimensionless, and h_{oe} has the dimensions of conductance. The e subscript denotes that the emitter terminal is common to both ports. The subscript i on h_{ie} designates that the parameter appears as a resistance in the BJT input, as shown by Eq. (8.89). The f subscript on h_{fe} designates that this is a forward-current transfer ratio since it relates the change in a current in the output circuit Δi_C to a change in a current in the input circuit Δi_B, as shown by Eq. (8.101). The subscript o on h_{oe} designates that this parameter appears as a conductance in the BJT output, as shown by Eq. (8.102).

Now for small-signal linear operation, we may substitute the equivalent circuit in Fig. 8.31b for the BJT into the circuit shown in Fig. 8.28c, resulting in Fig. 8.32a. Observe that we have presumed that C_E short-circuits R_E and so both are absent from the ac circuit. Once again, since linear operation is assumed, each current and voltage in the circuit will be composed of two components. One component is due to the dc sources in the circuit [with $v_S(t) = 0$] and is therefore direct current. This dc component is denoted with a capital letter. The other component is due to $v_S(t)$ (with the dc sources set equal to zero) and is a sinusoidal quantity since $v_S(t)$ is sinusoidal. This sinusoidal component is denoted in the usual fashion as a Δ quantity. By superposition the total is the sum of these two contributions, e.g., $i_C = I_C + \Delta i_C$.

---

† The parameters h_{fe} and β are not identical; however, the symbols are often used interchangeably.

FIGURE 8.32
Use of the small-
signal equivalent
circuit in the ac
analysis of the
common-emitter
BJT amplifier:
(*a*) complete
circuit; (*b*) ac
circuit;
(*c*) redrawn ac
circuit;
(*d*) determining
the amplifier
output impedance.

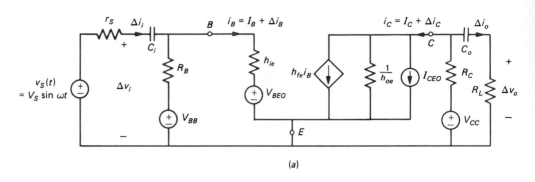

(*a*)

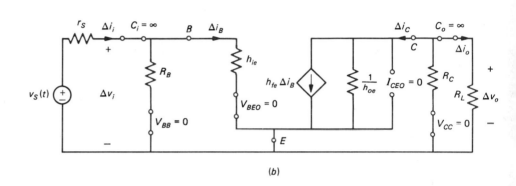

(*b*)

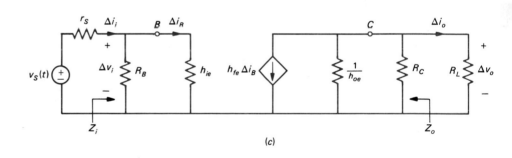

(*c*)

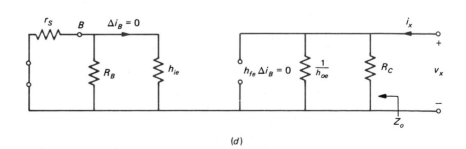

(*d*)

Now for computing the ac quantities of interest for the amplifier, such as

$$A_V = \frac{\Delta v_o}{\Delta v_i} \tag{8.104}$$

$$A_I = \frac{\Delta i_o}{\Delta i_i} \tag{8.105}$$

$$A_P = A_V A_I \tag{8.106}$$

and Z_i and Z_o, we may set all dc sources to zero, as shown in Fig. 8.32b, and all currents and voltages become only sinusoidal Δ quantities. To simplify the analysis, we assume that all capacitors are chosen such that their impedances are negligible at the frequency of $v_S(t)$ so that they may be replaced by short circuits. This circuit is further simplified in Fig. 8.32c.

From this resulting circuit we see that

$$\Delta v_o = -h_{fe} \Delta i_B \left(\frac{1}{h_{oe}} || R_C || R_L \right) \tag{8.107}$$

and

$$\Delta v_i = h_{ie} \Delta i_B \tag{8.108}$$

so that the voltage gain is

$$A_V = -\frac{h_{fe}}{h_{ie}} \left(\frac{1}{h_{oe}} || R_C || R_L \right) \tag{8.109}$$

The input impedance is, from Fig. 8.32c,

$$Z_i = \frac{\Delta v_i}{\Delta i_i}$$

$$= R_B || h_{ie} \tag{8.110}$$

From the circuit in Fig. 8.32c we see that, by current division,

$$\Delta i_o = -h_{fe} \Delta i_B \frac{R_C || 1/h_{oe}}{R_C || 1/h_{oe} + R_L} \tag{8.111}$$

and

$$\Delta i_B = \frac{R_B}{R_B + h_{ie}} \Delta i_i \tag{8.112}$$

Thus, the current gain becomes

$$A_I = \frac{\Delta i_o}{\Delta i_i}$$

$$= -\frac{R_B}{R_B + h_{ie}} h_{fe} \frac{R_C || 1/h_{oe}}{R_C || 1/h_{oe} + R_L} \tag{8.113}$$

This is equivalent (as the reader should verify) to

$$A_I = A_V \frac{Z_i}{R_L} \tag{8.114}$$

Note the presence of the two current-division quantities in Eq. (8.113) which are both less than unity; thus $|A_I| < h_{fe}$.

The output impedance of the amplifier Z_o can be found by finding the Thévenin impedance seen by R_L [set $v_S(t)$ to zero, of course], as shown in Fig. 8.32d. Since $\Delta i_B = 0$ in this situation, the controlled current source is zero, and it is replaced by an open circuit. Thus

$$Z_o = \frac{1}{h_{oe}} || R_C \tag{8.115}$$

Example 8.7   For the 2N718 common-emitter amplifier shown in Fig. 8.33a, with $\beta = 100 \doteq h_{fe}$, $h_{oe} = 10^{-5}$, compute the voltage gain, current gain, power gain, and input and output impedances.

Solution   The dc circuit becomes as shown in Fig. 8.33b. The operating point for the base current is found by summing KVL around the base-emitter loop as in Example 8.6 to yield

$$I_B = \frac{V_{BB} - V_{BE}}{R_B + \beta R_E}$$

$$= \frac{3.75 - 0.5 \text{ V}}{75 \text{ k}\Omega + 100 \times 500}$$

$$= 26 \ \mu A$$

Using $\beta = 100$, we obtain

$$I_C \doteq \beta I_B$$
$$= 100 I_B$$
$$= 2.6 \text{ mA}$$

Thus, V_{CE} can be found by writing KVL around the collector-emitter loop as

$$V_{CE} = V_{CC} - R_C I_C - R_E I_E$$
$$= 15 \text{ V} - (1 \text{ k}\Omega)(2.6 \text{ mA}) - (500)(2.6 \text{ mA})$$
$$= 11.1 \text{ V}$$

FIGURE 8.33
Example 8.7:
analysis of a
common-emitter
BJT: (*a*) complete
circuit; (*b*)
dc circuit.

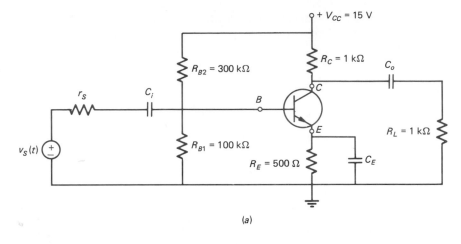

(a)

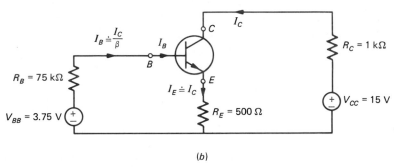

(b)

where we have again used the approximation that

$$I_E = I_C + I_B$$
$$\doteq I_C$$

The parameter h_{ie} is not specified but may be found from Eq. (8.96) as

$$h_{ie} = \beta \frac{25\,\text{mV}}{I_C}$$

$$= 100 \frac{25 \times 10^{-3}}{2.6 \times 10^{-3}}$$

$$= 961.5\,\Omega$$

Note that if I_C had been 1 mA, h_{ie} would have increased to 2500 Ω. The voltage gain is

$$A_V = -\frac{h_{fe}}{h_{ie}}\left(\frac{1}{h_{oe}}||R_C||R_L\right)$$

$$\doteq -52$$

since $1/h_{oe} = 100 \text{ k}\Omega$, which is much larger than $R_C||R_L = 500 \ \Omega$. The current gain is

$$A_I = -\underbrace{\frac{R_B}{R_B + h_{ie}}}_{1} \underbrace{h_{fe}}_{100} \underbrace{\frac{R_C||1/h_{oe}}{R_C||1/h_{oe} + R_L}}_{\frac{1}{2}}$$

$$\doteq -50$$

since $1/h_{oe} \gg R_C$; thus $R_C||1/h_{oe} \doteq R_C$ and $R_L = R_C$. The input impedance is

$$Z_i = R_B||h_{ie}$$
$$\doteq h_{ie}$$
$$= 961.5 \ \Omega$$

and the output impedance is

$$Z_o = \frac{1}{h_{oe}}||R_C$$
$$\doteq R_C$$
$$= 1 \text{ k}\Omega$$

Note that the current gain can also be computed more easily from

$$A_I = A_V \frac{Z_i}{R_L}$$
$$= -52 \frac{961.5 \ \Omega}{1 \text{ k}\Omega}$$
$$= -50$$

The power gain is

$$A_P = A_V A_I$$
$$= 2600$$

Note that for this rather typical range of parameters, the voltage gain is much larger than for the typical common-source JFET amplifiers. Also the input impedance of this common-emitter BJT amplifier is much lower than for typical common-source JFET amplifiers, which ranges from 100 kΩ to 1 MΩ, and it may easily be obtained simply by selecting R_G. For this BJT amplifier, however, Z_i is no larger than h_{ie} no matter what value of R_B occurs. Thus, the input impedance of this amplifier is limited by the value of h_{ie}.

Note in the previous example that the resistor $1/h_{oe}$ in the small-signal linear model which is in parallel with the controlled current source is $1/h_{oe} = 100 \text{ k}\Omega$ and is quite large. Consequently, in the previous example it has little effect on A_V since it is in parallel with $R_D||R_L = 500\ \Omega$. Similarly, it is in parallel with R_C in Z_o, and $R_C = 1\ \text{k}\Omega$. Thus, $1/h_{oe}$ has no significant effect and may be removed from the circuit.

Although this is not always possible (suppose $R_C = R_L = 100\ \text{k}\Omega$), we will omit $1/h_{oe}$ from the small-signal linear equivalent circuit of the BJT for the purposes of simplifying our calculations. This approximation has the effect of assuming that the curves in the second-port (collector-emitter) characteristic of the BJT are flat—which is again similar to the effect of neglecting $1/y_{os}$ in the JFET model. The resulting circuit is shown in Fig. 8.34 where we will assume that $h_{fe} \doteq \beta$. Compare this resulting circuit and the one for the JFET in Fig. 8.17a. We see two major differences. The controlled current source for the JFET depended on the gate-source voltage Δv_{GS}, whereas the controlled current source for the BJT depends on a current, the base current Δi_B. Also note that a resistance h_{ie} appears in the base lead of the BJT, whereas the gate lead of the JFET was an open circuit. This will have the effect of giving much smaller input impedances for BJT amplifiers than was the case for JFET amplifiers.

Removal of the capacitor C_E, which bypasses ac signals around R_E, will have an effect similar to the removal of C_S in the common-source JFET amplifier: voltage and current gain will drop. It also has another important effect: increasing the input impedance to the amplifier. Note in the previous example and in Fig. 8.32c that the input impedance is no larger than h_{ie}, and h_{ie} can be quite small. Removal of C_E will effectively increase Z_i, as we now show.

The ac circuit for the common-emitter amplifier in Fig. 8.28a with C_E removed will be similar to Fig. 8.29b, but with one major difference. Because C_E has been removed, R_E will appear in the ac circuit in the emitter lead. Substituting the small-signal linear equivalent circuit gives the circuit in Fig. 8.35a. Note that with KCL, the current through R_E is $(\beta + 1)\Delta i_B$. Thus

$$\Delta v_i = h_{ie}\,\Delta i_B + R_E(\beta + 1)\,\Delta i_B \tag{8.116}$$

Also

$$\Delta v_o = -\beta\,\Delta i_B(R_C||R_L) \tag{8.117}$$

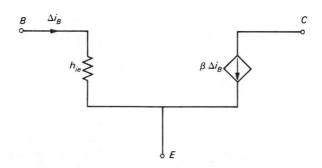

FIGURE 8.34
A simplified small-signal equivalent circuit of a BJT.

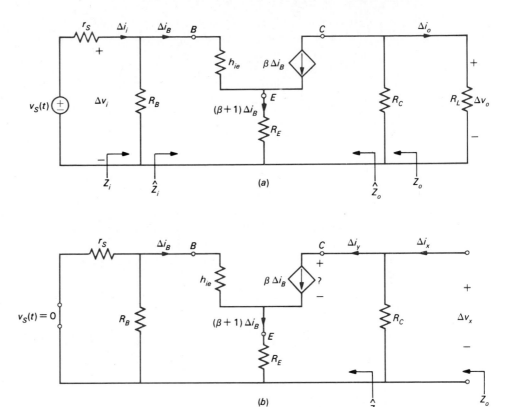

FIGURE 8.35
Analysis of the common-emitter BJT amplifier with C_F removed: (*a*) complete ac circuit; (*b*) determination of the amplifier output impedance.

so that the voltage gain is

$$A_V = \frac{\Delta v_o}{\Delta v_i}$$

$$= -\frac{\beta(R_C||R_L)}{h_{ie} + (\beta + 1)R_E} \tag{8.118}$$

Comparing Eq. (8.118) to the corresponding expression with C_E present, given by Eq. (8.109), we see that A_V is reduced when C_E is removed.

The input impedance is

$$Z_i = \hat{Z}_i || R_B \tag{8.119}$$

Note that

$$\hat{Z}_i = \frac{\Delta v_i}{\Delta i_B}$$

$$= h_{ie} + (\beta + 1)R_E \tag{8.120}$$

and

$$Z_i = R_B||[h_{ie} + (\beta + 1)R_E] \tag{8.121}$$

Comparing this to the corresponding expression with C_E present, given by Eq. (8.110), we see that $\hat{Z}_i$ can be much larger than h_{ie}. Thus, the overall input impedance to the amplifier Z_i is not "loaded down" as much by h_{ie}. Here we see that the effect of the controlled source is to make R_E appear larger to the input by a factor of $\beta + 1$. The output impedance is obtained from Fig. 8.35b as

$$Z_o = \frac{\Delta v_x}{\Delta i_x} \tag{8.122}$$

and

$$Z_o = \hat{Z}_o||R_C \tag{8.123}$$

To find $\hat{Z}_o = \Delta v_x/\Delta i_y$, we may constrain $\Delta i_y = 1$ A and find Δv_x. But if we do, we need to find the voltage across the controlled current source. We have no way of determining this. The dilemma is caused by our having neglected $1/h_{oe}$ in the model. If we reinsert $1/h_{oe}$, find $\hat{Z}_i$, and take the limit as $h_{oe} \to 0$, we find that

$$\hat{Z}_0 = \infty \tag{8.124}$$

so that

$$Z_o = R_C \tag{8.125}$$

The current gain is

$$A_I = A_V \frac{Z_i}{R_L} \tag{8.126}$$

Example 8.8   For the common-emitter BJT amplifier of Example 8.7, recompute A_V, A_I, Z_i, Z_o, and A_P with C_E removed.

Solution

$$A_V = -\frac{\beta(R_C||R_L)}{h_{ie} + (\beta + 1)R_E}$$

$$= -0.97$$

$$\hat{Z}_i = h_{ie} + (\beta + 1)R_E$$

$$= 51,462\ \Omega$$

$$Z_i = \hat{Z}_i||R_B$$

$$= 30,520\ \Omega$$

$$Z_o = R_C$$

$$= 1\ \text{k}\Omega$$

$$A_I = A_V \frac{Z_i}{R_L}$$

$$= -30$$

$$A_P = A_V A_I$$

$$= 29$$

Note that the voltage gain has been drastically reduced by the removal of C_E. The current gain has been reduced somewhat, while the input impedance has been increased by a factor of 32!

The amplifier we have been considering is referred to as a common-emitter amplifier since in the ac circuit the emitter is common to the amplifier input and output (except when C_E is removed). Other typical amplifier configurations are the common-base and common-collector amplifiers. These are special-purpose amplifiers often used for impedance matching. For example, the common-base amplifier has the input on the emitter lead and the output taken from the collector lead, so that the base is common to the input and output. For this amplifier, the current gain is less than unity, and the input impedance can be made quite large. This is the BJT equivalent to the common-gate JFET amplifier. Similarly, the common-collector amplifier has the input applied to the base terminal and the output taken from the emitter terminal, so that the collector terminal is common to the input and output. For this amplifier, the voltage gain is approximately unity, and the output impedance can be made quite small. This is often referred to as the *emitter-follower* amplifier since the emitter voltage (the output voltage) is in phase with and approximately equal to the base voltage (the input voltage). This is equivalent to the common-drain JFET amplifier which is referred to as the *source follower*. The analysis of these amplifiers for their dc and ac properties is virtually no different from that for the common-emitter amplifier which we examined in detail. First the dc operating point is set, and then the small-signal model is substituted into the ac circuit to determine the voltage and current gain and the input and output impedance.

8.9   FREQUENCY RESPONSE OF THE BJT AMPLIFIER

Consider the common-emitter BJT amplifiers in Fig. 8.28a. As the frequency of $v_S(t)$ is reduced, the impedances of the coupling capacitors C_i and C_o increase, resulting in a decrease in voltage gain. Also as the frequency is decreased, the impedance of C_E increases, thus no longer removing R_E from the ac circuit. This increasing presence of an impedance (R_E) in the emitter lead forms a "feedback" path between the input and output which also reduces the voltage gain for the JFET amplifier. Thus, the magnitude of A_V decreases with decreasing frequency, which results in a response similar to the JFET shown in Fig. 8.19.

If we increase the frequency above its midband value at which the capacitors were chosen to have negligible impedance, we find that the magnitude of A_V also decreases, as was the case for the JFET amplifier. This high-frequency deterioration is once again due to the deterioration of the device performance at these higher frequencies.

The small-signal linear equivalent circuit which we have been using to model the BJT for ac analysis is valid for midband frequencies and below. For higher frequencies, we must include the inherent capacitances of the device. The depletion layers at the base-emitter and base-collector junctions give rise to capacitances between the terminals. A useful high-frequency modification of the small-signal linear equivalent circuit is shown in Fig. 8.36a. Capacitances C_{BE} and C_{BC} have been added to the midband model. Note that Δi_B is still the current through h_{ie}. Ordinarily, C_{BC} is given by the manufacturer as C_{ob}. And C_{BE} can be calculated from

$$C_{BE} = \frac{\beta}{\omega_T h_{ie}} \tag{8.127}$$

where

$$\omega_T = 2\pi f_T \tag{8.128}$$

and f_T is given by the manufacturer on design specification sheets and is usually denoted as the current-gain–bandwidth product. It is the frequency at which the value of β drops to unity. Typical values for the 2N718 BJT are $C_{BC} = C_{ob} = 20$ pF and $f_T = 75$ MHz.

Replacing the BJT in the ac circuit for the amplifier circuit of Fig. 8.28a with the equivalent circuit of Fig. 8.36a, we obtain the circuit of Fig. 8.36b. (We are interested in high-frequency response, so we replace C_o, C_i, and C_E with short circuits.) For the purposes of calculating Δi_B as well as Z_i, we may replace the amplifier at the input terminals with the circuit of Fig. 8.36c. This may be determined from the following. Note in Fig. 8.36b that the voltage across h_{ie} as well as C_{BE} is Δv_i. Thus, if we find the current Δi_x entering C_{BC}, we can replace the circuit to the right

FIGURE 8.36
High-frequency
model of the BJT:
(*a*) circuit parasitic
capacitances;
(*b*) use in the
common-emitter
BJT amplifier;
(*c*) determining
the amplifier input
impedance and the
Miller effect.

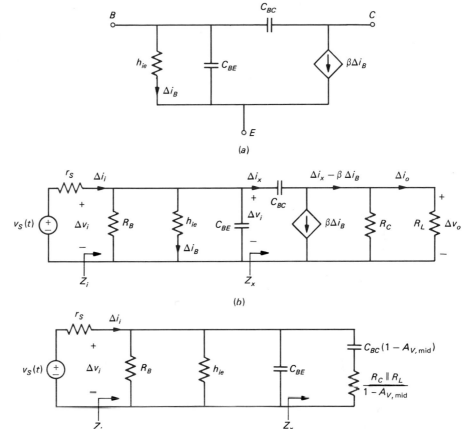

(*a*)

(*b*)

(*c*)

of C_{BE} with an equivalent impedance $Z_x = \Delta v_i/\Delta i_x$. From Fig. 8.36*b*, we obtain

$$\Delta i_B = \frac{\Delta v_i}{h_{ie}} \tag{8.129}$$

and

$$\Delta v_i = \Delta i_x \frac{1}{j\omega C_{BC}} + (\Delta i_x - \beta \,\Delta i_B)(R_C || R_L) \tag{8.130}$$

Substituting Eq. (8.129) into Eq. (8.130), we obtain

$$Z_x = \frac{\Delta v_i}{\Delta i_x}$$

$$= \frac{1}{j\omega C_{BC}(1 - A_{V,\text{mid}})} + \frac{R_C || R_L}{1 - A_{V,\text{mid}}} \tag{8.131}$$

where the midband voltage gain is used:

$$A_{V,\text{mid}} = -\frac{\beta}{h_{ie}}(R_C || R_L) \tag{8.132}$$

Thus, C_{BC} appears to the input as though it has been increased by a factor of $1 - A_{V,\text{mid}}$, and the parallel combination of R_C and R_L appears as though it has been reduced by a factor of $1 - A_{V,\text{mid}}$. This corresponds almost identically to the result for JFET amplifiers and is called the *Miller effect*. It has the effect of reducing the amplifier's input impedance (and its voltage and current gain) at high frequencies.

Example 8.9

For the common-emitter circuit of Example 8.7 (Fig. 8.33a), $I_C = 2.6$ mA, $\beta = 100$, $R_B = 75$ kΩ, $R_C = R_L = 1$ kΩ. A 2N718 with $C_{ob} = 20$ pF and $f_T = 75$ MHz is used. Compute the frequency at which the input impedance is reduced from its midband value of 961.5 Ω to 50 Ω.

Solution

$$C_{BC} = C_{ob}$$
$$= 20 \text{ pF}$$

$$C_{BE} = \frac{\beta}{\omega_T h_{ie}}$$

$$= \frac{100}{2\pi(75 \times 10^6)(961.5)}$$

$$= 220.7 \text{ pF}$$

and

$$A_{V,\text{mid}} = -52$$

From Fig. 8.36c

$$\frac{R_C || R_L}{1 - A_{V,\text{mid}}} = \frac{500}{53}$$

$$= 9.4 \text{ Ω}$$

which may be neglected in this analysis. Thus

$$Z_i = R_B || h_{ie} || \frac{1}{j\omega[C_{BE} + C_{BC}(1 - A_{V,\text{mid}})]}$$

$$= 949 || \frac{1}{j\omega(1281 \text{ pF})}$$

We wish to find f such that

$$50 = \left| \frac{949[1/(j2\pi f\, 1281 \text{ pF})]}{949 + 1/(j2\pi\, 1281 \text{ pF})} \right|$$

$$= \left| \frac{949}{1 + j7.63 \times 10^{-6} f} \right|$$

or

$$19 = \sqrt{1 + (7.63 \times 10^{-6} f)^2}$$

Solving, we obtain

$$f = 2.48 \text{ MHz}$$

8.10 INTEGRATED CIRCUITS

In this chapter and in Chap. 7, we have discussed various semiconductor devices—nonlinear resistors—along with useful circuits which utilize these devices. All these devices may be obtained in discrete form; i.e., each device may be obtained as an individual unit with the necessary device terminals protruding from the unit in the form of wires. A single (discrete) bipolar junction transistor is shown in Fig. 8.37a. Several such devices may also be manufactured as a single unit in the form of an integrated circuit. Integrated circuits (ICs) consisting of several hundred transistors, resistors, capacitors, and diodes can be fabricated on a single wafer which may be no larger than a pencil eraser. Thus, an entire amplifier or other special-purpose device can be obtained in an extremely miniaturized form. An operational amplifier, discussed in Chap. 9, is shown in Fig. 8.37b. It is commonly supplied in an 8-pin dual in-line package (DIP) for insertion into some larger circuit. A microprocessor consisting of a 40-pin DIP is shown in Fig. 8.37c.

There are currently two types of integrated circuits, monolithic and hybrid. The monolithic IC is one in which all the parts (transistors, resistors, capacitors, and diodes) necessary for a complete circuit, such as an amplifier, are constructed at the same time from one wafer. In the hybrid IC, the various circuit components constructed on individual chips are connected by wire bonds or other suitable techniques so that the hybrid IC resembles a discrete circuit packaged into a single, small case.

One obvious advantage of IC fabrication is a tremendous reduction in the space occupied by the circuit. A second advantage is an increase in the useful frequency range of a circuit. As frequency is increased so that the circuit dimensions become on the order of a wavelength, lumped-circuit concepts discussed previously no longer hold and distributed parameter effects become important. For example, in discussing the interconnection of lumped elements via wires, it was assumed that signal

FIGURE 8.37
Packaging of
transistors: (*a*)
discrete transistor;
(*b*) integrated
circuit; (*c*) dual
in-line package
(DIP).

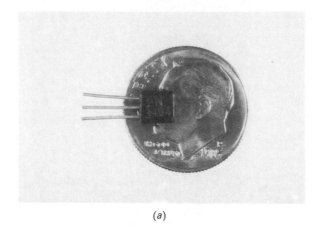

(*a*)

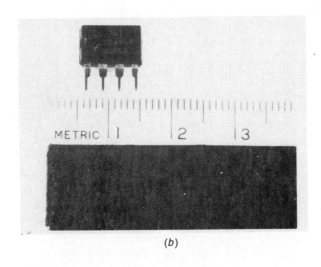

(*b*)

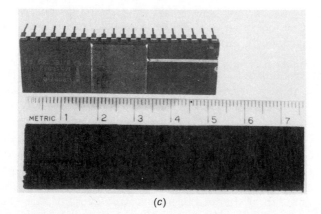

(*c*)

propagation between the lumped elements occurred instantaneously, but there is actually a certain amount of propagation delay introduced by the wire lengths. At frequencies where the circuit dimensions are electrically small, these propagation delays are inconsequential. As the frequency is increased, however, the connecting wires may introduce important propagation delays as well as other associated phenomena not present at the lower frequencies. The ability to place circuit elements closer on an IC chip thus acts to extend the frequency range of the devices.

A third and equally important advantage of IC technology is the increased reliability of circuits. All circuit elements, as well as their interconnections, can be fabricated in the initial manufacturing process. Thus, the reliability of all components in a circuit can be maintained to the same degree.

In this section we briefly discuss the construction of ICs. Our intent is not to give an exhaustive coverage, but instead to give the reader the essential "flavor" of the topic.

8.10.1 The General Fabrication Technique

An IC is commonly fabricated by "planar processing," which starts with a p-type silicon wafer that is typically 5 mils (0.005 in) thick. This silicon wafer is called the *substrate*. The wafer is placed in an oxygen atmosphere in an oven and heated to over 1000°C. This oxidizes the silicon surface and forms a silicon dioxide (SiO_2) layer. A photosensitive emulsion is then placed on the oxidized surface, and a mask having the required area to be etched as an opaque coating is placed over the emulsion, as shown in Fig. 8.38a. Ultraviolet light is shown on the mask. Those portions of the mask (to be etched) which are not exposed to the light are removed along with the underlying silicon dioxide layer, as shown in Fig. 8.38b.

The next step in the process is to place the etched wafer in an oven and expose it to an n-type dopant such as arsenic. The dopant diffuses into the silicon substrate, forming a heavily doped n^+ region, as shown in Fig. 8.39a. Passing a gas containing an n-type impurity over the wafer forms an n-type region over the n^+ region, as shown in Fig. 8.39b. This n-type layer will be the collector of a bipolar junction transistor. A p-type layer is diffused into this n-type layer to form the base, and an additional n^+-type layer is diffused into this p-type layer to form the emitter, as shown in Fig. 8.39c. A silicon dioxide coating is applied over the surface of the wafer, and metallic attachments are made to the various regions. Small n^+ regions are diffused to provide good electrical connections to each region.

A silicon wafer of about 1-in diameter is divided into squares called *chips*, which are on the order of 50 to 100 mils (0.05 to 0.1 in) wide. Each element of a circuit (including resistors, capacitors, and diodes) may be formed in the above manner on each chip. Thus, all components of a complete circuit may be fabricated at the same time from the same silicon wafer. This increases the reliability of the resulting circuit and tends to ensure that the values of the circuit elements will be relatively constant between similar elements. The values of two resistors constructed from the same wafer can be controlled to within a few percent of each other. The absolute values, however, may vary quite a bit from their desired values.

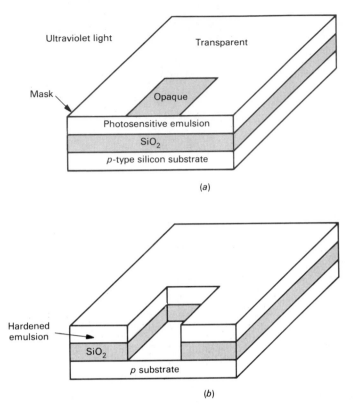

FIGURE 8.38
Fabrication of
integrated circuits—
initial etch of water.

8.10.2 The IC Bipolar Junction Transistor and Other Components

The basic process described in the preceding section has formed a bipolar junction transistor, whose equivalent circuit is given in Fig. 8.40. Resistor r_C represents the resistance of the relatively large n region connected to the collector. The n^+ heavily doped layer has a much lower resistance and is essentially in parallel with this n region, and it thus lowers r_C from what it would be without the n^+ region. A typical value for r_C is less than 1 Ω.

The diode shown in Fig. 8.40 represents the pn junction formed between the collector and the substrate. In most applications, the substrate is connected to the most negative part of the circuit or power supply to reverse-bias this diode.

An IC diode may be obtained by constructing a transistor in the above manner (Sec. 8.10.1) and using only the base-emitter leads. Recall that in normal transistor operation the base-emitter junction is essentially a forward-biased pn diode. Often the base and the collector leads are connected.

IC capacitors may also be fabricated as shown in Fig. 8.41. A metal plate forms one plate of the capacitor, and the n material forms the other plate.

Recall that under normal operating conditions the collector-base junction of a transistor appears as a reverse-biased pn junction. Also recall that a reverse-biased pn junction has a depletion region devoid of mobile charge which separates charges

FIGURE 8.39
The final steps in
the process of
forming a BJT in
integrated-circuit
form.

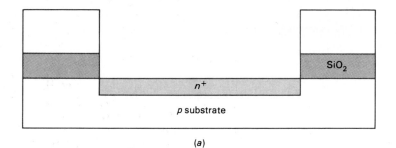

(a)

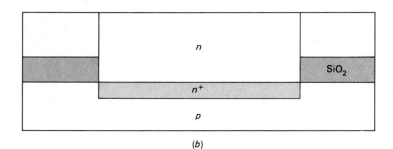

(b)

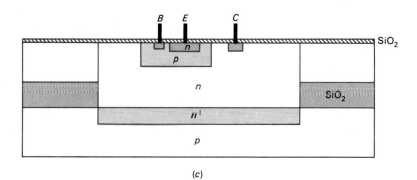

(c)

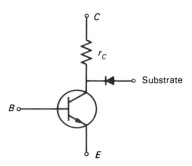

FIGURE 8.40
The IC BJT.

FIGURE 8.41
The IC capacitor.

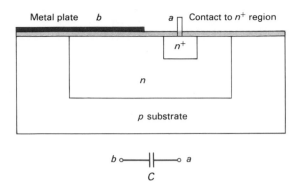

FIGURE 8.41
The IC capacitor.

on either side of the junction. Thus, a reverse-biased *pn* junction displays a capacitance. The collector-base terminals of the transistor can also be used for this component.

The IC resistor is obtained by stopping the diffusion process after the *p*-type base region has been deposited. The result is shown in Fig. 8.42. Both contacts are applied to the *p* region (via small n^+ regions), and the protective silicon dioxide layer is applied to the surface of the wafer.

A complete circuit can be fabricated in one step. Such an amplifier circuit and its cross-sectional realization are shown in Fig. 8.43. The usual n^+ contact regions are omitted.

8.10.3 The IC Field-Effect Transistor

The junction FET, or JFET, was considered in previous sections. Another type of FET which is used in ICs is the insulated-gate FET, or IGFET. Two types of IGFETs

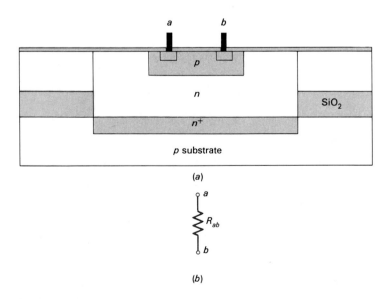

FIGURE 8.42
The IC resistor.

FIGURE 8.43
A complete
integrated-circuit
amplifier.

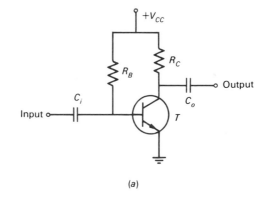

(a)

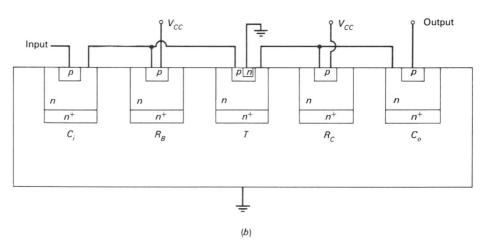

(b)

may be obtained, the depletion-mode IGFET and the enhancement-mode IGFET. Both are formed by insulating the metal gate from the rest of the device with a thin, silicon dioxide film; thus, these IGFETs are commonly referred to as metal-oxide-semiconductor FETs, or MOSFETs. In these types of FETs, the resistance of the oxide layer is so large ($10^{12}\ \Omega$) that the gate current is entirely negligible, regardless of the polarity of the gate-source voltage.

In the depletion-mode MOSFET, shown in Fig. 8.44a, two highly doped n^+ regions along with an n channel are diffused into the p substrate. Current flow is possible from drain to source for $v_{GS} = 0$, just as for the JFET. Applying a negative voltage to the gate such that $v_{GS} < 0$ produces an electric field which attracts holes toward the gate and pushes electrons away resulting in a depletion region being formed in the n channel near the gate. The width of the remaining portion of the channel determines its resistance and the resulting drain-source current, as with the JFET. Eventually, the channel is pinched at a value of $v_{GS} = -V_p$. The resulting characteristic in Fig. 8.44c is very similar to that of the JFET.

A transfer characteristic is shown in Fig. 8.45. Since the curves in the linear region of the characteristic in Fig. 8.44c are almost flat, the device is relatively independent of v_{DS} in the linear region. Thus, we may select a value of v_{GS} and

FIGURE 8.44
The depletion-
mode *n*-channel
MOSFET:
(*a*) physical
construction;
(*b*) device symbol;
(*c*) typical device
characteristic.

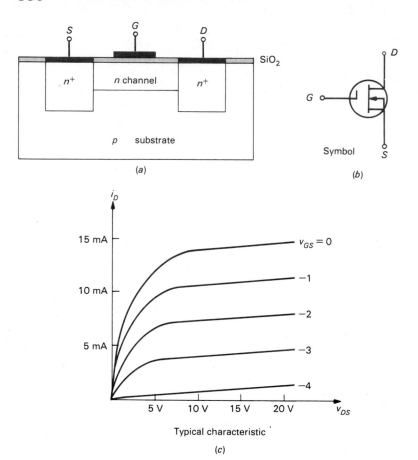

Typical characteristic

(*c*)

FIGURE 8.45
The transfer
characteristic of a
depletion-mode
n-channel
MOSFET.

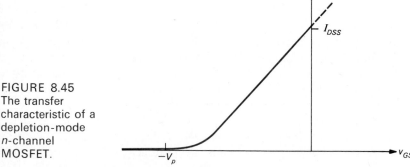

determine the corresponding value of i_D. Doing this for several values of v_{DS} gives the transfer characteristic in Fig. 8.45 (independent of v_{DS}). Note that for large negative values of v_{GS}, the transfer characteristic becomes nonlinear (around $v_{GS} = -V_p$). This represents a crowding of the curves on the characteristic for large negative values of v_{GS}. This relation between drain current and drain-source voltage is

$$ i_D = I_{DSS}\left(1 - \frac{v_{GS}}{V_p}\right)^2 \tag{8.133}$$

where I_{DSS} is the drain-source current with the gate short-circuited to the source $v_{GS} = 0$. Note that, as opposed to the JFET, the device may be operated with positive gate-source voltage. Since the gate is insulated, no gate current flows.

The n-channel enhancement-mode MOSFET is shown in Fig. 8.46. As opposed to the n-channel JFET and depletion-mode MOSFET, no n channel between drain and source is present. Thus, for $v_{DS} > 0$ the pn junction at the drain end is

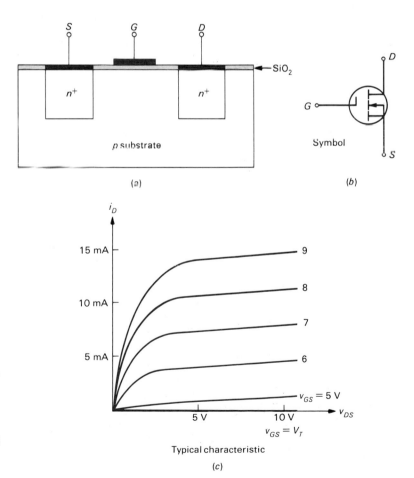

FIGURE 8.46
The enhancement-mode n-channel MOSFET:
(*a*) physical construction;
(*b*) device symbol;
(*c*) typical device characteristic.

FIGURE 8.47
The transfer
characteristic of an
enhancement-
mode *n*-channel
MOSFET.

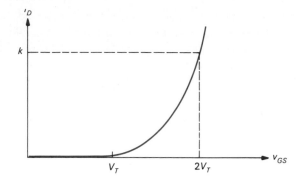

reverse-biased, and no drain current flows. This is indicated on the symbol in Fig. 8.46*b* by the broken lines. However, applying a positive gate-source voltage establishes an electric field which pulls electrons toward the gate and pushes holes away. Because of these mobile electrons being attracted to the gate, an *n*-type channel is formed adjacent to the gate which makes it similar to the *n*-channel depletion-mode MOSFET.

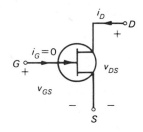

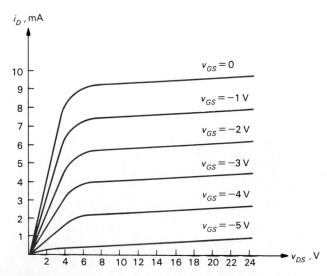

FIGURE 8.48

Above a certain value of v_{GS} known as the *threshold voltage* V_T, the channel forms along with a depletion region between this n channel and the p substrate. Thus, for positive v_{GS} the transfer characteristic of the *n*-channel enhancement-mode MOSFET shown in Fig. 8.46c is similar to that of the JFET and the depletion-mode MOSFET, with the exception that positive v_{GS} voltages are used.

The transfer characteristic is shown in Fig. 8.47. This curve obeys the relation

$$i_D = k\left(\frac{v_{GS}}{V_T} - 1\right)^2 \qquad (8.134)$$

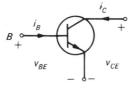

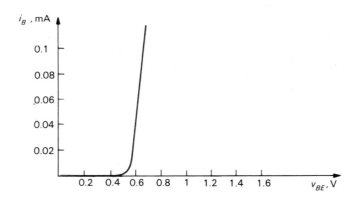

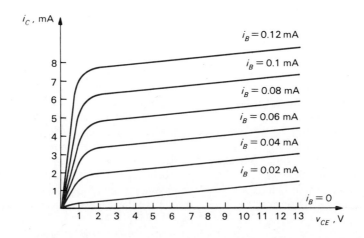

FIGURE 8.49

where V_T is the threshold voltage and $k = i_D$ when $v_{GS} = 2V_T$. The depletion-mode MOSFET discussed previously can also be operated in the enhancement mode by applying positive gate-source voltage.

Although not specifically discussed, p-channel MOSFETs in either the depletion-mode or enhancement-mode configuration (or both) may be obtained, as was the case for p-channel JFETs. The characteristics are similar to those for the n-channel MOSFETs except that the currents and voltages are reversed in polarity.

The majority of our design procedures developed in this chapter for the JFET hold true for the MOSFETs. The main exception is that amplifiers using enhancement-mode n-channel MOSFETs are designed to have a positive gate-source voltage.

PROBLEMS

8.1   For the linear amplifier shown in Fig. 8.1 suppose $V_i = 10$ mV, $I_o = 2$ mA, $Z_i = 5$ kΩ, and $R_L = 1$ kΩ. Determine A_V and A_I. Can you determine Z_o?

8.2   If the load resistor in the linear amplifier of Fig. 8.1 is changed from $R_L = 1$ kΩ to $R_L = 2$ kΩ, the output voltage changes from $V_o = 1$ V to $V_o = 1.2$ V. Determine the output impedance of the amplifier. (*Hint*: Model the output of the amplifier as a Thévenin equivalent circuit.)

8.3   For the JFET in Fig. 8.6a with the battery voltages $V_{DS} > V_{GS}$, the depletion regions at the gate-channel junction will be asymmetric rather than symmetric, as shown in Fig. 8.5. Explain why this will be the case.

8.4   Severe distortion of the output-voltage waveform for the JFET amplifier of Fig. 8.7a occurs if $v_S(t)$ has too large a magnitude, as shown in Fig. 8.8. Consider the JFET characteristic shown in Fig. 8.48. Assume $V_{DD} = 20$ V, $R_D = 2$ kΩ, and $V_{GG} = 3$ V, and determine the operating point for the amplifier (I_D, V_{DS}). Determine the largest output voltage swing such that distortion of the output waveform does not occur. Determine the voltage gain A_V. Repeat these calculations if R_D is changed to 500 Ω.

8.5   For the JFET amplifier of Fig. 8.7 assume $V_{GG} = 2$ V and $V_{DD} = 24$ V. Determine R_D to give an operating-point voltage of $V_{DS} = 16$ V. Assume that the JFET has the characteristic given in Fig. 8.48. Determine I_D. How large can the output voltage sinusoid be before distortion occurs? Determine the voltage gain A_V.

8.6   For the JFET amplifier shown in Fig. 8.9a, assume that $V_{DD} = 24$ V, $R_D = 2$ kΩ, $R_S = 1$ kΩ, $R_{G1} = 10$ kΩ, and $R_{G2} = 230$ kΩ. Determine the operating point (V_{DS}, I_D, V_{GS}) if the JFET has the characteristic given in Fig. 8.48. If $R_L = 2$ kΩ, determine the largest output voltage swing before distortion of the output-voltage waveform occurs if $r_S = 20$ kΩ. Determine the voltage gain of the amplifier.

8.7   For the JFET amplifier shown in Fig. 8.9a, determine R_{G1}, R_{G2}, R_D, and R_S to attain an operating point of $V_{DS} = 12$ V, $I_D = 4$ mA, $V_{GS} = -3$ V, and an input impedance of $Z_i = 75$ kΩ if $V_{DD} = 20$ V. Compute the voltage gain of your design.

8.8   Determine y_{fs} and y_{os} for the JFET whose characteristic is shown in Fig. 8.48 at an operating point of $V_{DS} = 12$ V, $I_D = 4$ mA, and $V_{GS} = -3$ V.

8.9   For the amplifier design in Prob. 8.4, determine y_{fs} and y_{os} at the operating point and recompute the voltage gain, using the small-signal model of Fig. 8.15.

8.10   For the amplifier design in Prob. 8.5, determine y_{fs} and y_{os} at the operating point and recompute the voltage gain, using the small-signal model of Fig. 8.15.

8.11   For the amplifier design in Prob. 8.6, determine y_{fs} and y_{os} at the operating point and recompute the voltage gain, using the small-signal model of Fig. 8.15.

8.12   For the amplifier design in Prob. 8.7, determine y_{fs} and y_{os} at the operating point and recompute the voltage gain using the small-signal model of Fig. 8.15.

8.13 For the amplifier shown in Fig. 8.9, assume $R_D = 5$ kΩ, $R_L = 10$ kΩ, $R_S = 3$ kΩ, $R_{G1} = 100$ kΩ, $R_{G2} = 200$ kΩ. Determine the voltage gain, input impedance, current gain, and output impedance if the JFET has $y_{fs} = 1000$ μS and $y_{os} = 30$ μS. Repeat, using the simplified small-signal model of Fig. 8.17 which neglects y_{os}. Is y_{os} important for this amplifier?

8.14 Repeat the calculations of Prob. 8.13 with C_S removed.

8.15 For the amplifier shown in Fig. 8.9, assume a value of $R_D = 10$ kΩ to match to a load of $R_L = 10$ kΩ. Determine the voltage gain and power gain if $y_{fs} = 1200$ μS and $y_{os} = 0$ and $R_{G1} = 50$ kΩ and $R_{G2} = 50$ kΩ. If $R_D = 5$ kΩ so that the amplifier is not matched, determine the voltage and power gain.

8.16 For the amplifiers of Prob. 8.15, determine the voltage gain at 10 MHz if the capacitors C_i, C_o, and C_S are chosen such that they have 1-Ω impedance at 1 MHz and the JFET is specified as having $C_{GS} = C_{GD} = 2$ pF and $C_{DS} = 0$.

8.17 Determine the voltage gain of the amplifiers of Prob. 8.15 at an input frequency of 1 kHz if the capacitors are chosen to have values of $C_i = C_o = C_S = 0.01$ μF and $R_S = 2$ kΩ. The capacitances of the JFET can be ignored (why?).

8.18 A common-drain JFET amplifier is shown in Fig. P8.18. The input is at the gate, and the output is taken off the source terminal of the JFET. If $y_{fs} = 3000$ μS, $v_{os} = 0$, $R_g = 10$ kΩ, $R_L - 10$ kΩ,

$R_D = 10$ kΩ, $R_{G1} = 100$ kΩ, and $R_{G2} = 200$ kΩ, determine the voltage and current gain as well as the input and output impedances for the amplifier.

8.19 A common-gate JFET amplifier is shown in Fig. P8.19. The input is on the source terminal of the JFET, and the output is taken off the drain terminal of the JFET. If $y_{fs} = 3000$ μS, $y_{os} = 0$, $R_S = 10$ kΩ, $R_L = 10$ kΩ, $R_D = 10$ kΩ, $R_{G1} = 100$ kΩ, and $R_{G2} = 200$ kΩ, determine the voltage and current gain as well as the input and output impedances for the amplifier.

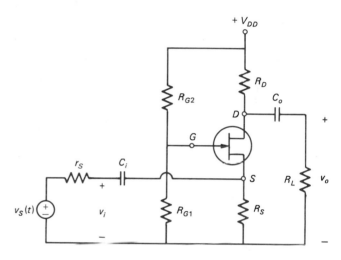

FIGURE P8.19

8.20 For the BJT amplifier of Fig. 8.25a determine the operating point (I_C, V_{CE}, V_{BE}, I_B) if $V_{CC} = 12$ V, $R_C = 1$ kΩ, $V_{BB} = 1.6$ V, $R_B = 17$ kΩ, and the BJT has the characteristic shown in Fig. 8.49. Determine the largest magnitude of the output voltage such that distortion of the output-voltage waveform does not occur if $R_L = 1$ kΩ.

8.21 For the BJT amplifier of Fig. 8.25a, determine R_B to bias the BJT at $I_C = 3.5$ mA and $V_{CE} = 5$ V if $V_{BB} = V_{CC} = 10$ V and the BJT has the characteristic shown in Fig. 8.49.

8.22 Repeat Prob. 8.20, using the approximate relation $I_C = \beta I_B$ and assuming $\beta = 60$. How does this compare to the values obtained by using the BJT characteristic?

8.23 Repeat Prob. 8.21, using the approximate relation $I_C = \beta I_B$ and assuming $\beta = 60$. How well does this compare to the values obtained by using the BJT characteristic?

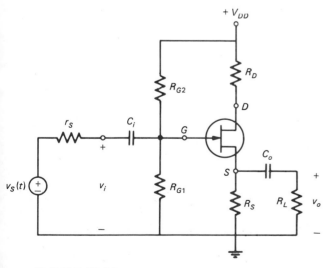

FIGURE P8.18

8.24 For the BJT amplifier shown in Fig. 8.28a where the BJT has the characteristic shown in Fig. 8.49, determine the operating point if $V_{CC} = 12$ V, $R_E = 400\ \Omega$, $R_C = 1\ \text{k}\Omega$, $R_{B1} = R_{B2} = 135\ \text{k}\Omega$. Determine the largest magnitude of the output-voltage swing such that distortion does not occur if $R_L = 2\ \text{k}\Omega$.

8.25 Repeat Prob. 8.24, using the approximate relationship $I_C = \beta I_B$ with $\beta = 63$. How well does this compare to the values obtained by using the BJT characteristic.

8.26 Determine h_{fe} and h_{oe} at $I_C = 5$ mA, $V_{CE} = 5$ V for the BJT whose characteristic is given in Fig. 8.49.

8.27 Determine the voltage gain, current gain, input impedance, output impedance, and power gain of the amplifier shown in Fig. 8.27 where $R_B = 145\ \text{k}\Omega$, $\beta = 60$, $h_{fe} = 90$, $h_{oe} = 5 \times 10^{-5}$ S.

8.28 Repeat Prob. 8.27 for the circuit of Fig. 8.30 where $R_E = 300\ \Omega$ and $R_C = 2\ \text{k}\Omega$.

8.29 Recompute the results of Prob. 8.28 if C_E is removed.

8.30 The BJT amplifier circuit of Fig. 8.30 has $R_C = 2\ \text{k}\Omega$, $R_E = 300\ \Omega$. The BJT has $\beta = 60$, $h_{fe} = 90$, $h_{oe} = 5 \times 10^{-5}$ S, and $C_{ob} = 10$ pF and $f_T = 100$ MHz.The capacitors C_i, C_o, and C_E are designed to have impedances of $1\ \Omega$ at 1 MHz. Determine the input impedance at 50 MHz.

8.31 The circuit of Fig. P8.31 is referred to as a common-base BJT amplifier. The input is on the emitter, and the output is on the collector. Assume that the BJT has $\beta \doteq h_{fe} = 50$, $h_{oe} = 0$, and an operating point of $I_C = 1$ mA. Also assume that $R_L = R_C = 10\ \text{k}\Omega$, $R_E = 1\ \text{k}\Omega$, $r_S = 600\ \Omega$. Determine the voltage gain, current gain, input impedance, output impedance, and power gain for this amplifier.

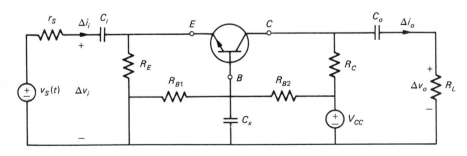

FIGURE P8.31

8.32   The circuit of Fig. P8.32 is referred to as a common-collector BJT amplifier. The input is on the base, and the output is on the emitter. Assume that the BJT has $\beta \doteq h_{fe} = 100$, $h_{oe} = 0$, and an operating point of $I_C = 10$ mA. Also assume that $R_C = R_E = 1\ k\Omega$, $R_L = 50\ \Omega$, $r_S = 600\ \Omega$, and $R_{B1} \| R_{B2} = 5\ k\Omega$. Determine the voltage gain, current gain, input impedance, output impedance, and power gain for this amplifier.

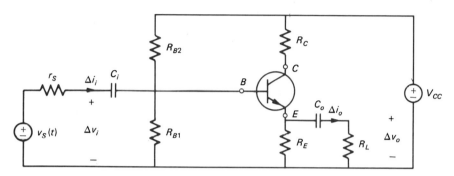

FIGURE P8.32

8.33   For the amplifier shown in Fig. P8.33, the operating point is $V_{CE} = 6$ V, $I_C = 2.3$ mA, and the BJT whose characteristic is given in Fig. 8.49 is used. If $V_{CC} - 10$ V, determine R_B and R_C. Suppose $h_{fe} = 90$, $h_{oe} = 0$, $h_{ie} = 650\ \Omega$, $R_C = 500\ \Omega$, $R_L = 1\ k\Omega$, and $R_B = 50\ k\Omega$, determine the voltage gain, current gain, input impedance, output impedance, and power gain of the amplifier.

8.34   The Darlington compound transistor configuration shown in Fig. P8.34 uses identical BJTs having $\beta = 100$ and $I_{CEO} = 1$ nA. Determine an expression relating i_o to i_i. From this can you deduce the β of the overall structure?

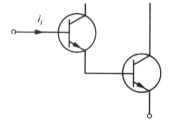

FIGURE P8.34

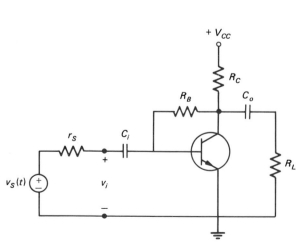

FIGURE P8.33

8.35 Two common-source JFET amplifiers are cascaded as shown in Fig. P8.35 to provide a larger overall voltage gain than could be provided by one stage. Calculate the overall voltage gain for the amplifier, assuming that both JFETs have $y_{fs} = 10^{-3}$ S and $y_{os} = 0$.

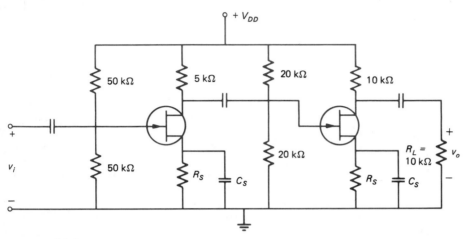

FIGURE P8.35

8.36 A two-stage BJT amplifier consisting of a common-emitter amplifier followed by a common-collector amplifier is used to match to a 50-Ω load and to provide voltage gain, as shown in Fig. P8.36. Calculate the overall voltage and current gains and the input and output impedances, assuming both BJTs have $h_{ie} = 500$ Ω, $h_{fe} = 100$, and $h_{oe} = 0$.

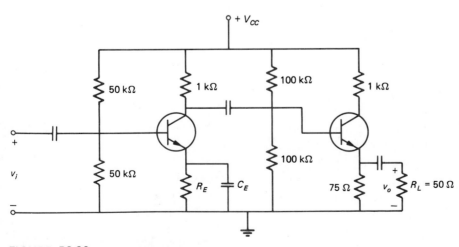

FIGURE P8.36

The Operational Amplifier

In Chap. 8 we discussed the construction of linear signal amplifiers. The complete design consisted of determining a biasing circuit to place the operating point in a linear region of the nonlinear device characteristic and then of determining the ac properties of interest, such as the voltage gain, current gain, power gain, input impedance, and output impedance. All these stages of the design required considerable analytical effort. Furthermore, it was not known until the dc-biasing design was completed what values of the ac parameters would result.

In this chapter we discuss an important type of amplifier that essentially removes all these steps so that the design of an amplifier becomes an almost trivial process. This is the operational amplifier, which we refer to as the *op amp*. The op amp consists of several transistors (BJTs and/or FETs), diodes, capacitors, and resistors and is available in integrated-circuit form in a small DIP with external leads, as shown in Fig. 8.37*b*. These op amps can presently be purchased for less than one U.S. dollar!

9.1   THE IDEAL OPERATIONAL AMPLIFIER

First we discuss the ideal op amp shown in Fig. 9.1, along with some useful circuits, and in the next section we discuss practical op amps. A useful preliminary analysis of circuits containing op amps can be made by replacing the actual op amps with ideal ones. Once the general characteristics of the circuit are obtained, the limitations of practical op amps can be factored into the analysis.

The symbol for the op amp is shown in Fig. 9.1. Two terminals labeled + and − are available for inputs, as in an output terminal. The assumed voltages of these terminals are labeled (as shown in Fig. 9.1) with respect to the common terminal, denoted with a ground symbol ($\perp$). The output voltage is related to the difference of the two input voltages as

$$v_o = A(v_p - v_n) \tag{9.1}$$

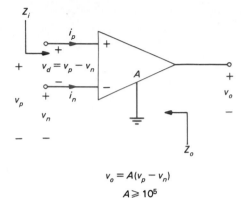

FIGURE 9.1
The operational
amplifier.

$$v_o = A(v_p - v_n)$$
$$A \geqslant 10^5$$

Thus, the op amp is basically a form of differential, or difference, amplifier. Practical op amps have gains A which are on the order of 10^5; thus, for practical op amps where the output voltage may be practically restricted to something on the order of 12 V, the difference voltage $v_d = v_p - v_n$ must be less than 0.12 mV, or 120 μV. The input impedance Z_i between the $+$ and $-$ terminals is on the order of 1 MΩ, while the output impedance Z_o is on the order of 100 Ω to 1 kΩ.

These practical op-amp characteristics are approximated in the *ideal op amp* shown in Fig. 9.2. By virtue of the high input impedance, the very large gain, and the resulting small difference voltage v_d in practical op amps, the ideal op amp is approximated by the following two important properties:

1   The input currents i_p and i_n are zero:

$$i_p = i_n = 0 \tag{9.2a}$$

2   The difference voltage v_d is zero:

$$v_d = 0 \tag{9.2b}$$

Note that although no current flows into the $+$ and $-$ terminals, the voltage between them v_d is assumed to be zero. This is referred to as the principle of *virtual short circuit* and will be crucial to our ability to analyze circuits containing ideal op amps.

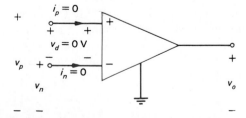

FIGURE 9.2
The ideal
operational
amplifier.

As an example, consider the circuit shown in Fig. 9.3. A resistor R_i is placed in the − lead while the + lead is connected to the common terminal (grounded). Also a resistor R_f connects the − lead and the output lead; this feedback resistor R_f is always connected to the − terminal for stability purposes, which is discussed later. Note that because $i_n = 0$, $i_i = i_f$. Summing KVL around the outside loop, we obtain

$$v_i = R_i i_i + R_f i_f + v_o \tag{9.3}$$

Since $v_d = 0$, the input voltage and current are related by (sum KVL through R_i and v_d)

$$i_i = \frac{v_i}{R_i} \tag{9.4}$$

Substituting Eq. (9.4) and $i_i = i_f$ into Eq. (9.3) yields

$$v_i = (R_i + R_f)\frac{v_i}{R_i} + v_o \tag{9.5}$$

Solving Eq. (9.5) for the ratio of v_o/v_i yields

$$\frac{v_o}{v_i} = -\frac{R_f}{R_i} \tag{9.6}$$

Thus, the voltage gain of this overall amplifier (containing R_i, R_f, and the op amp) is simply the ratio of R_f to R_i! The negative sign symbolizes that this is an "inverting amplifier"; i.e., the output voltage is 180° out of phase with the input voltage. Nevertheless, constructing a linear amplifier with a prescribed voltage gain is exceedingly simple: Select the resistors to achieve the desired ratio.

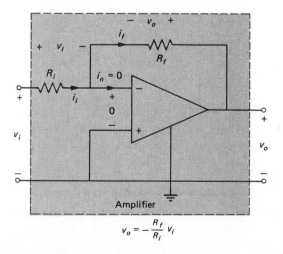

FIGURE 9.3
The inverting amplifier.

Assuming that the ideal op amp has zero output impedance, we can show that the amplifier in Fig. 9.3 also has $Z_o = 0$. However, the input impedance is the ratio $Z_i = v_i/i_i$. From Eq. (9.4) we see that $Z_i = R_i$. Thus, to construct an amplifier with a voltage gain of 10 and an input impedance of 100 kΩ, we select $R_i = 100$ kΩ and $R_f = 1$ MΩ.

Another useful amplifier is shown in Fig. 9.4. Here the input is directly connected to the + terminal. Resistors R_i and R_f are connected as in the previous amplifier. Note that since the input current i_p is zero, the input impedance to this amplifier is infinite. Once again, because $i_n = 0$, we see that $i_i = i_f$. Summing KVL around the loop containing R_i and R_f, we obtain

$$v_o = -(R_i + R_f)i_i \qquad (9.7)$$

But because of the virtual short circuit

$$v_i = -R_i i_i \qquad (9.8)$$

so that

$$\frac{v_o}{v_i} = 1 + \frac{R_f}{R_i} \qquad (9.9)$$

Thus, we have a simple noninverting amplifier.

The op amp can be used to construct numerous other useful devices. Some of these are discussed later in this chapter.

A *summer* is a device which produces as an output the sum of several input signals. A simple summer using only resistors is shown in Fig. 9.5. With $v_2 = v_3 = \cdots = v_n = 0$, we obtain

$$v_o = \frac{R||R_2||\cdots||R_n}{R_1 + R||R_2||\cdots||R_n} v_1 \qquad (9.10)$$

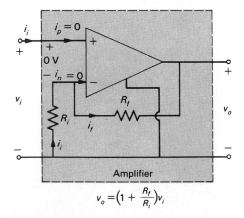

FIGURE 9.4
The noninverting amplifier.

$$v_o = \left(1 + \frac{R_f}{R_i}\right)v_i$$

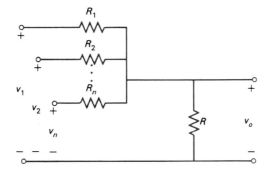

FIGURE 9.5
The resistor
summer.

If we choose R to be much smaller than $R_1, R_2, \ldots, R_n$, then

$$v_o \doteq \frac{R}{R_1 + R} v_1 \tag{9.11}$$

Similarly, we obtain by superposition

$$v_o \doteq \frac{R}{R_1 + R} v_1 + \frac{R}{R_2 + R} v_2 + \cdots + \frac{R}{R_n + R} v_n \tag{9.12}$$

so that v_o is the weighted sum of the n input signals.

A much more useful summer can be constructed with an op amp, as shown in Fig. 9.6. With a simple extension of the analysis for Fig. 9.3, we obtain

$$v_o = -\left(\frac{R_f}{R_1} v_1 + \frac{R_f}{R_2} v_2 + \cdots + \frac{R_f}{R_n} v_n \right) \tag{9.13}$$

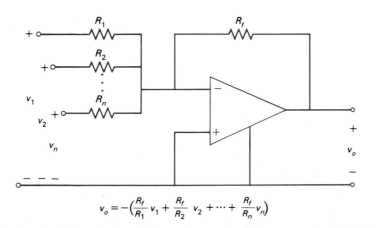

FIGURE 9.6
The operational-
amplifier summer.

$$v_o = -\left(\frac{R_f}{R_1} v_1 + \frac{R_f}{R_2} v_2 + \cdots + \frac{R_f}{R_n} v_n \right)$$

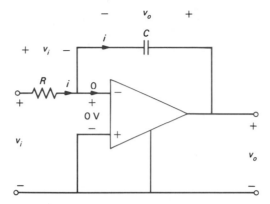

FIGURE 9.7
The integrator.

Another useful circuit is the integrator, shown in Fig. 9.7. Here a capacitor is used in place of the usual feedback resistor. Note that, once again, due to the principle of virtual short circuit,

$$i = \frac{v_i}{R} \tag{9.14}$$

and v_o is directly across C. Thus

$$i = -C\frac{dv_o}{dt} \tag{9.15}$$

Equating Eqs. (9.14) and (9.15), we obtain

$$\frac{dv_o}{dt} = -\frac{1}{RC}v_i \tag{9.16}$$

or

$$v_o = -\frac{1}{RC}\int v_i(\tau)\,d\tau \tag{9.17}$$

Example 9.1   Determine the input impedance to the circuit in Fig. 9.8.

Solution   The input impedance to the circuit is the ratio of v_i/i_i. Utilizing the principle of virtual short circuit, we label the various voltages and currents as shown. Summing KVL around the loop containing v_i, Z, and the two resistors, we obtain

$$v_i = Zi_i + 2v_i \tag{9.18}$$

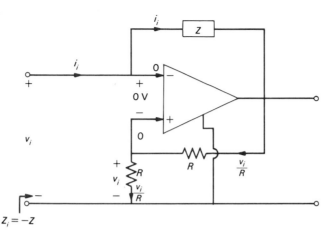

FIGURE 9.8 The negative impedance converter.

$$Z_i = -Z$$

or

$$\frac{v_i}{i_i} = -Z \tag{9.19}$$

Thus, the circuit is sometimes called a *negative impedance converter*.

9.2 THE PRACTICAL OPERATIONAL AMPLIFIER

The ideal op amp in the previous section was simplified in its analysis primarily by virtue of the principle of virtual short circuit; i.e., the currents into the $+$ and $-$ input terminals were assumed to be zero, and the voltage between these two terminals was assumed to be zero. Practical op amps, even though they do not have these ideal characteristics, approximate them very closely.

A practical op amp has the equivalent circuit shown in Fig. 9.9a. Typical values of Z_i and Z_o are 1 MΩ and 100 to 1 kΩ, respectively, while A is typically 10^5 or larger. Also shown are the usual $+V_{CC}$ and $-V_{CC}$ dc power supplies required to bias the transistors in the op amp. The transfer characteristic for open-circuited output is shown in Fig. 9.9b; it exhibits a linear region with slope A. Also shown is a saturation region so that v_o cannot exceed V_{CC}, the values of the power supplies; typically, V_{CC} is on the order of 10 V. Thus, for $A = 10^5$ we find a maximum differential voltage which may be applied such that linear operation is maintained as

$$v_{d,\,max} = \frac{V_{CC}}{A}$$

$$= \frac{10}{10^5}$$

$$= 0.1 \text{ mV}$$

$$= 100 \ \mu\text{V} \tag{9.20}$$

FIGURE 9.9 A practical operational amplifier. (*a*) The equivalent circuit, and (*b*) the input-output characteristic.

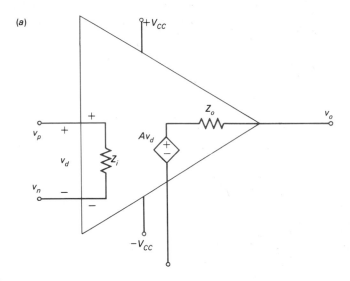

(*a*)

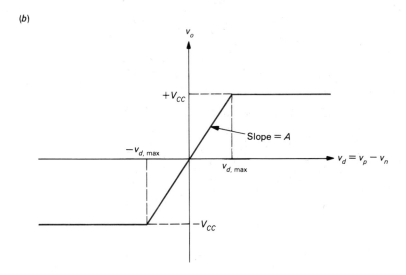

(*b*)

Even though the input currents are not zero, if $Z_i = 1\ \text{M}\Omega$, they are on the order of $i = v_{d,\text{max}}/Z_i = 10^{-10}$ amp and are essentially zero.

The restriction imposed by Eq. (9.20) that v_d must not exceed 0.1 mV to maintain linear operation seems to rule out the use of this device as a practical amplifier. But recall that in the circuits considered previously, the difference voltage v_d is *not* the input voltage to the circuit. For example, consider the circuit in Fig. 9.3. Inserting the equivalent circuit of Fig. 9.9*a*, we obtain Fig. 9.10. Let us presume that $Z_i = \infty$ and $Z_o = 0$ to simplify the analysis. In this case we obtain

$$v_o = Av_d \tag{9.21}$$

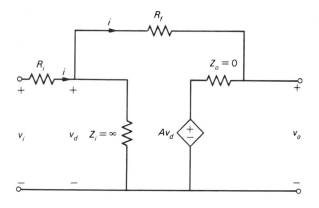

FIGURE 9.10
The inverting
amplifier with operational-
amplifier circuit substituted.

but by voltage division

$$v_d = \frac{R_f}{R_f + R_i}(v_i - v_o) + v_o \tag{9.22}$$

Thus

$$v_o = \frac{AR_f}{R_f + R_i}(v_i - v_o) + Av_o \tag{9.23}$$

or

$$\frac{v_o}{v_i} = \frac{AR_f}{R_f + (1 - A)R_i} \tag{9.24}$$

For A very large, this simplifies to

$$\frac{v_o}{v_i} = -\frac{R_f}{R_i} \tag{9.25}$$

which was obtained by using the principle of the virtual short circuit, which
essentially assumed that $A \to \infty$. The ratio of v_d to v_i can then be found from

$$\frac{v_d}{v_i} = \frac{v_d}{v_o}\frac{v_o}{v_i} \tag{9.26}$$

Substituting Eqs. (9.21) and (9.24), we obtain

$$\frac{v_d}{v_i} = \frac{1}{A}\frac{AR_f}{R_f + (1 - A)R_i}$$

$$= \frac{1}{1 + (1 - A)R_i/R_f} \tag{9.27}$$

Since A is very large, the ratio in Eq. (9.27) is quite small. (For example, if $A = 10^5$ and $R_f/R_i = 10$, then $v_d/v_i \doteq -10^{-4}$.) Thus, v_d can be very small and yet v_i can be considerably larger. The feedback configuration thus allows the construction of practical amplifiers which are not limited to small input voltages.

Example 9.2   Construct an amplifier with the configuration shown in Fig. 9.3 to amplify a 200-mV (peak) sinusoid. Assume that $A = 10^5$ and $V_{CC} = 15$ V.

Solution   For linear operation, v_o must not exceed 15 V. Thus, the maximum overall gain is $15/0.2 = 75$. Rather than working so close to this margin, we might select an overall gain of 68. Selecting $R_i = 10 \text{ k}\Omega$ and $R_f = 680 \text{ k}\Omega$, we see that the output-voltage sinusoid will have a peak value of $68 \times 0.2 = 13.6$ V, and linear operation will be maintained.

Another practical aspect of op amps is their frequency response. The open-loop gain is the gain

$$A_o = \frac{v_o}{v_d} \tag{9.28}$$

of the op amp without any external feedback circuitry. A plot of this gain as a function of frequency for a typical op amp is shown in Fig. 9.11a. The frequency axis is plotted logarithmically, and the gain is plotted in decibels (dB), where

$$A_{\text{dB}} = 20 \log A \tag{9.29}$$

If $A_o = 10^5$, then in decibels $A_o = 100$ dB. Note that this large open-loop gain is constant only up to around 10 Hz and drops at a rate of 20 dB per decade to a value of 0 dB (or absolute value of unity gain) at 1 MHz. Although these numbers differ among op amps, they are fairly typical. The important point here is that the bandwidth is very narrow, approximately 10 Hz. The frequency at which the open-loop gain (in decibels) drops to zero is the gain-bandwidth product. (A gain of 0 dB corresponds to unity gain.)

When the op amp is inserted into the circuit of Fig. 9.3, the gain-versus-frequency curve becomes as shown in Fig. 9.11b. Note that the maximum (dc) closed-loop gain is $-R_f/R_i$, so for a gain of 10 or 20 dB the frequency response is reduced in magnitude but the bandwidth is increased dramatically to 10^5 Hz. What we observe is that the gain-bandwidth product is the same for both amplifiers, a property which is true for other feedback amplifiers.

FIGURE 9.11
Frequency
response of
operational
amplifiers:
(*a*) device
frequency
response;
(*b*) compensated
amplifier frequency
response.

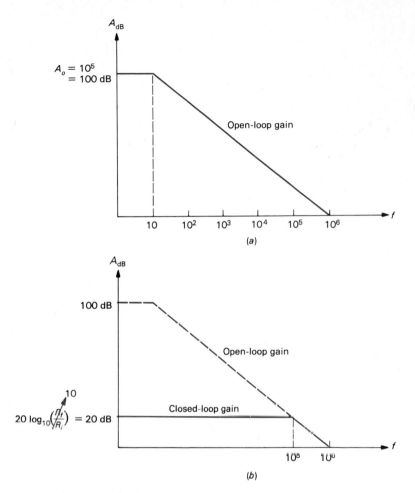

9.3 INDUCTORLESS (ACTIVE) FILTERS

In Chap. 5 we considered filters—low-pass, high-pass, bandpass, and band-reject—which were used to pass or eliminate certain frequency components of a signal. Some of these required the use of inductors. For example, the bandpass and band-reject filters required an inductor in series or in parallel with a capacitor. The resonance phenomenon associated with the series or parallel connection of these two elements allowed the elimination or passage of certain frequency components.

In Chap. 8 we discussed the small-scale integrated-circuit fabrication of various transistors, resistors, diodes, and capacitors. Inductors are obviously not suited to IC fabrication. Thus, to construct filters in the form of an integrated circuit, we need some alternative to the inductor.

Filters which are suitable for IC fabrication but which do not contain inductors are referred to as *active filters*. One common component of active filters is the operational amplifier.

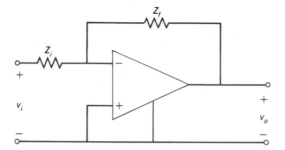

FIGURE 9.12
The operational-
amplifier active
filter.

Consider the basic op-amp circuit with frequency-dependent impedances shown in Fig. 9.12. The voltage-gain or voltage transfer function of this device is

$$\frac{v_o}{v_i} = -\frac{Z_f}{Z_i} \tag{9.30}$$

A low-pass filter can be constructed as shown in Fig. 9.13a. In this case

$$Z_i = R_i \tag{9.31a}$$

$$Z_f = R_f \| \frac{1}{j\omega C_f}$$

$$= \frac{R_f}{1 + j\omega R_f C_f} \tag{9.31b}$$

Substituting into Eq. (9.30), we obtain

$$\frac{v_o}{v_i} = -\frac{R_f}{R_i} \frac{1}{1 + j\omega R_f C_f} \tag{9.32}$$

Denoting

$$f_l = \frac{1}{2\pi R_f C_f} \tag{9.33}$$

Eq. (9.32) may be written as

$$\frac{v_o}{v_i} = -\frac{R_f}{R_i} \frac{1}{1 + jf/f_l} \tag{9.34}$$

The magnitude is

$$\left| \frac{v_o}{v_i} \right| = \frac{R_f}{R_i} \frac{1}{\sqrt{1 + (f/f_l)^2}} \tag{9.35}$$

FIGURE 9.13 A low-pass active filter.

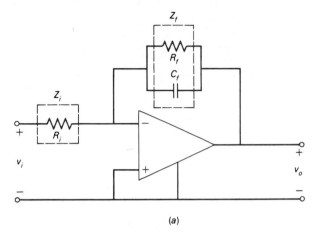

(a)

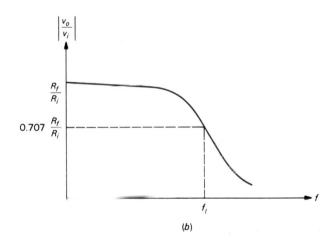

(b)

which is sketched in Fig. 9.13*b*. The dc value is R_f/R_i, which is down by a factor of $1/\sqrt{2}$ at the half-power, or 3-dB, point of f_l.

Similarly, a high-pass filter is shown in Fig. 9.14*a*. Here

$$Z_i = R_i + \frac{1}{j\omega C_i} \tag{9.36a}$$

$$Z_f = R_f \tag{9.36b}$$

Substituting into Eq. (9.30) and defining

$$f_h = \frac{1}{2\pi R_i C_i} \tag{9.37}$$

FIGURE 9.14 A high-pass active filter.

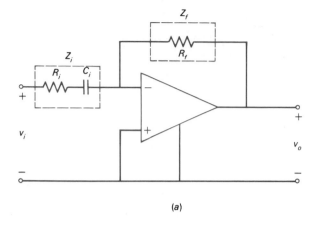

(a)

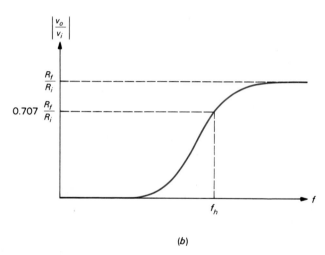

(b)

we obtain

$$\frac{v_o}{v_i} = -\frac{R_f}{R_i}\frac{jf/f_h}{1+jf/f_h} \tag{9.38}$$

The magnitude is

$$\left|\frac{v_o}{v_i}\right| = \frac{R_f}{R_i}\frac{1}{\sqrt{1+(f_h/f)^2}} \tag{9.39}$$

which is sketched in Fig. 9.14b. Note that the high-frequency gain is again R_f/R_i and that the half-power, or 3-dB, point is f_h.

To construct a bandpass filter, we might try using a low-pass and a high-pass filter and overlapping their transfer functions, as shown in Fig. 9.15. To do this, we

FIGURE 9.15 The band-pass filter as a combination of low-pass and high-pass characteristics.

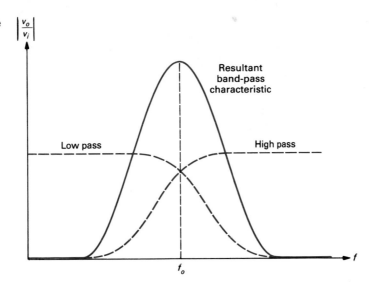

need to isolate the two filters, as shown in Fig. 9.16a. Here

$$Z_f = R_f || \frac{1}{j\omega C_f}$$

$$= \frac{R_f}{1 + j\omega R_f C_f} \tag{9.40a}$$

$$Z_i = R_i + \frac{1}{j\omega C_i} \tag{9.40b}$$

and

$$\frac{v_o}{v_i} = -\frac{Z_f}{Z_i}$$

$$= -\frac{R_f}{R_i} \frac{1}{(1 + j\omega R_f C_f)[1 + 1/(j\omega R_i C_i)]} \tag{9.41}$$

Denoting

$$f_l = \frac{1}{2\pi R_i C_i} \tag{9.42a}$$

$$f_h = \frac{1}{2\pi R_f C_f} \tag{9.42b}$$

FIGURE 9.16 A bandpass active filter.

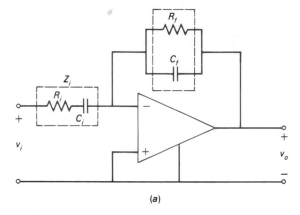

(a)

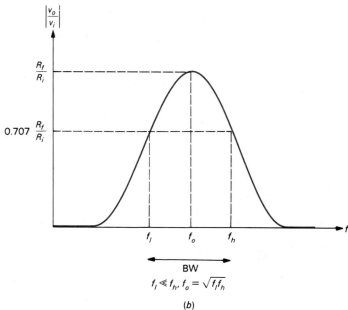

$$f_l \ll f_h, \quad f_o = \sqrt{f_l f_h}$$

(b)

Eq. (9.41) can be written as

$$\frac{v_o}{v_i} = -\frac{R_f}{R_i} \frac{1}{(1 + jf/f_h)(1 - jf_l/f)}$$

$$= -\frac{R_f}{R_i} \frac{1}{(1 + f_l/f_h) + j(f/f_h - f_l/f)}$$

$$= -\frac{R_f}{R_i} \frac{1}{(f_h + f_l)/f_h + j(f^2 - f_l f_h)/(ff_h)}$$

$$= -\frac{R_f}{R_i} \frac{f_h/(f_h + f_l)}{1 + j(f^2 - f_l f_h)/[f(f_l + f_h)]}$$

(9.43)

Note that the response is a maximum at

$$f_o = \sqrt{f_l f_h} \qquad (9.44)$$

If $f_l \ll f_h$, then the magnitude of Eq. (9.43) becomes $0.707 R_f / R_i$ at f_l and f_h as shown in Fig. 9.16b. In this case we may define the bandwidth (BW) as

$$\text{BW} = f_h - f_l \qquad f_l \ll f_h \qquad (9.45)$$

9.4 ANALOG COMPUTERS

Presently (and in the conceivable future) the digital computer forms the basic numerical analysis tool for the solution of equations (algebraic as well as differential). In the past, another type of computer—the analog computer—played an equally important role. Although not used as much as the digital computer, the analog computer still retains some important advantages over the digital computer.

The majority of the basic elements of an analog computer, shown in Fig. 9.17, are constructed with operational amplifiers. The basic amplifier in Fig. 9.17a is used to provide integer multiplication (often by a factor of 10), whereas the potentiometer in Fig. 9.17b is used for providing noninteger gains <1. The summer in Fig. 9.17c and the integrator in Fig. 9.17d were discussed previously. In the integrator, C_f is charged to an initial voltage $V_C(0)$ to provide the initial condition on y. In the summer of Fig. 9.17c, if we choose $R_f / R_1 = R_f / R_2 = \cdots = R_f / R_n = 1$, then $y = x_1 + x_2 + \cdots + x_n$. Similarly, in Fig. 9.17d, if we choose $R_i C_f = 1$, then $y = -\int_0^t x(\tau)\, d\tau + y(0)$. These cases will be indicated by 1s in the unit boxes.

For example, suppose we wish to solve the ordinary differential equation

$$\frac{d^2 y(t)}{dt^2} + a_1 \frac{dy(t)}{dt} + a_2 y(t) = f(t) \qquad (9.46)$$

subject to the initial conditions

$$y(0) = y_0$$

$$\left. \frac{dy}{dt} \right|_{t=0} = y_1 \qquad (9.47)$$

The connection shown in Fig. 9.18 will accomplish the *simulation* of the solution. We write Eq. (9.46) by isolating the highest derivative as

$$\ddot{y} = -a_1 \dot{y} - a_2 y + f \qquad (9.48)$$

where the dots are used to denote the various derivatives with respect to time t. Thus, we need to sum $-a_1 \dot{y}$, $-a_2 y$, and f. The summer does this. The first integrator

FIGURE 9.17
Analog computer
components:
(a) basic amplifier;
(b) potentiometer;
(c) summer;
(d) integrator.

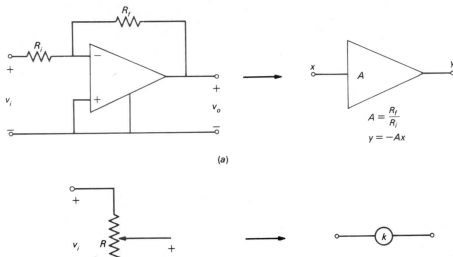

$$A = \frac{R_f}{R_i}$$

$$y = -Ax$$

(a)

(b)

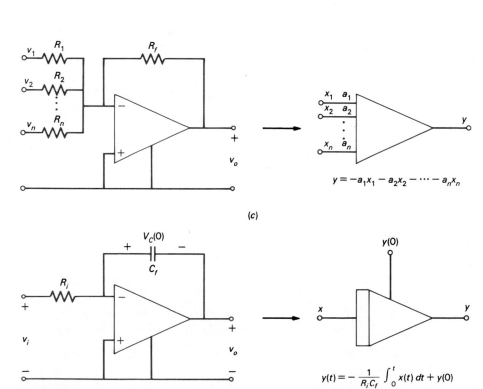

$$y = -a_1 x_1 - a_2 x_2 - \cdots - a_n x_n$$

(c)

$$y(t) = -\frac{1}{R_i C_f} \int_0^t x(t)\, dt + y(0)$$

(d)

FIGURE 9.18
Analog computer
implementation of
a second-order
differential
equation.

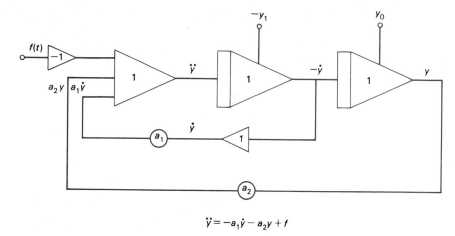

$$\ddot{y} = -a_1\dot{y} - a_2 y + f$$

integrates $\ddot{y}$ to give $\dot{y}$:

$$\dot{y} = -\int_0^t \ddot{y}\, d\tau - \underbrace{\dot{y}(0)}_{y_1} \tag{9.49}$$

The second integrator integrates this result:

$$y = -\int_0^t \dot{y}\, d\tau - \underbrace{y(0)}_{y_0} \tag{9.50}$$

These variables are multiplied by constants a_1 and a_2 and added, along with $-f$, at the summer.

One problem with this arrangement is that if we need to investigate the actual solution for a very long time, say 1 hour (h), we must observe the output of this analog computer for that actual time interval. Similarly, if the important part of the response lasts a very short time, say 1 microsecond (μs), then the observation time may be too short for a recording instrument such as a chart recorder. These problems may be overcome by time scaling. Redefine t as

$$t = \alpha\tau \tag{9.51}$$

Thus

$$\frac{dy}{dt} = \frac{dy}{d\tau}\frac{d\tau}{dt}$$

$$= \frac{1}{\alpha}\frac{dy}{d\tau} \tag{9.52}$$

Thus, the differential equation (9.46) becomes, in terms of the new time variable,

$$\frac{1}{\alpha^2}\frac{d^2 y(\tau)}{d\tau^2} + \frac{a_1}{\alpha}\frac{dy(\tau)}{d\tau} + a_2 y(\tau) = f(\tau) \tag{9.53}$$

or

$$\frac{d^2 y(\tau)}{d\tau^2} = -\alpha a_1 \frac{dy(\tau)}{d\tau} - \alpha^2 a_2 y(\tau) + \alpha^2 f(\tau) \tag{9.54}$$

Consequently, the coefficients may simply be scaled by the appropriate factors to change the solution time of the analog computer. For example, if $\alpha = 10$, then $t = 10$ s corresponds to $\tau = 1$ s, and thus the solution time may be reduced.

PROBLEMS

9.1 For the circuit shown in Fig. P9.1, assume an ideal op amp and calculate v_o.

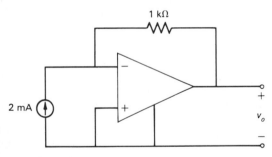

FIGURE P9.1

9.2 For the circuit shown in Fig. P9.2, assume an ideal op amp and calculate v_o.

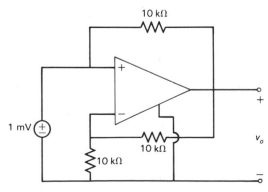

FIGURE P9.2

9.3 For the circuit shown in Fig. P9.3, assume an ideal op amp and calculate v_o.

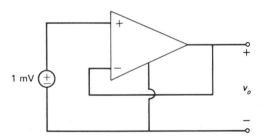

FIGURE P9.3

9.4 The circuit shown in Fig. P9.4 is used as an electronic ohmmeter for measuring an unknown resistance. Assume an ideal op amp, and determine the voltmeter reading V in terms of the unknown value of R.

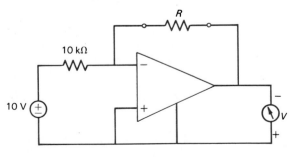

FIGURE P9.4

9.5 The circuit shown in Fig. P9.5 will serve as a "difference amplifier" to measure the difference between two voltages v_1 and v_2. Write v_o in terms of v_1 and v_2, assuming an ideal op amp.

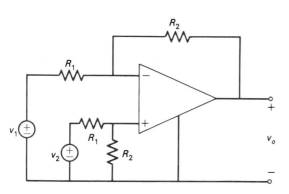

FIGURE P9.5

9.6 For the circuit shown in Fig. P9.6, calculate v_o/v_i. Assume an ideal op amp. What effect do R_L and R have? Compare this circuit with that of Fig. 9.4.

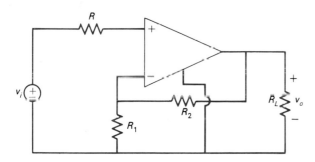

FIGURE P9.6

9.7 For the op-amp circuit shown in Fig. 9.4, calculate v_o/v_i. Assume that the op amp has finite A, $Z_i = \infty$, and $Z_o = 0$. Show that as $A \to \infty$, $v_o/v_i \to (R_f + R_i)/R_i$.

9.8 For the circuit shown in Fig. P9.8, assume an ideal op amp and calculate v_o/v_i. Determine the input impedance seen by v_i.

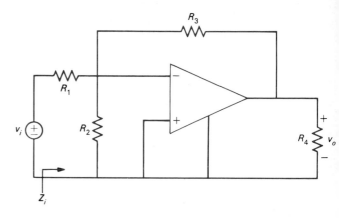

FIGURE P9.8

9.9 For the circuit shown in Fig. P9.2, calculate the input impedance seen by the 1-mV source.

9.10 For the low-pass filter shown in Fig. 9.13, determine C_f such that the 3-dB point is at 1 kHz. Assume $R_i = R_f = 1\ M\Omega$.

9.11 For the high-pass filter shown in Fig. 9.14, calculate R_f and R_i to give a 3-dB point at 1 MHz. Assume $v_o/v_i = 2$ at 10 MHz and $C_i = 100\ pF$.

9.12 Construct an analog computer diagram to solve the differential equation

$$\frac{d^2y(t)}{dt^2} + 10\frac{dy(t)}{dt} + 4y(t) = 5$$

with $y(0) = 2$, $\dot{y}(0) = 0$.

9.13 If the solution to the differential equation in Prob. 9.12 is to be obtained over $0 \le t \le 1$ ms but we wish to expand this over an interval of 1 s, redraw the analog computer diagram.

9.14 An integrator shown in Fig. 9.17d is to be constructed to solve the differential equation

$$\frac{dy(t)}{dt} + 10^3 y(t) = 0$$

with $y(0) = 2$. If $R_i = 10\ k\Omega$, determine C_i.

9.15 Determine v_x and i_x in the circuit of Fig. P9.15.

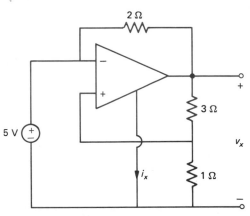

FIGURE P9.15

9.16 Determine I in the circuit of Fig. P9.16.

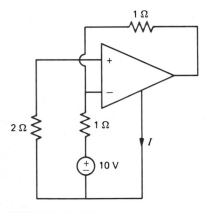

FIGURE P9.16

9.17 Determine v_o, i_x, and the power absorbed by the 4-Ω resistor in the circuit of Fig. P9.17.

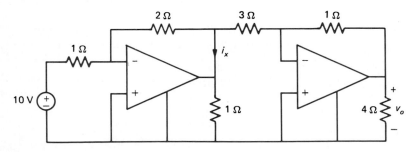

FIGURE P9.17

9.18 Design the circuit of Fig. P9.18 to provide a low-pass filter having a cutoff frequency of 10 kHz and a dc gain of 2.

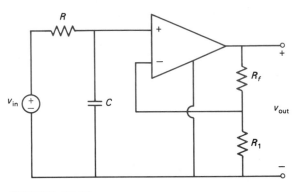

FIGURE P9.18

9.19 Design the circuit of Fig. P9.19 to provide a high-pass filter having a cutoff frequency of 1 MHz and a gain of 2.

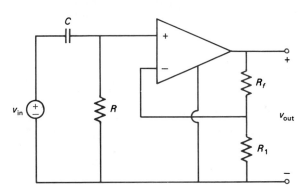

FIGURE P9.19

Digital Electronic Circuits

In previous chapters we studied the use of nonlinear devices (the BJT and the FET) in constructing linear amplifiers. Although these devices are inherently nonlinear, we confined the operation to the linear portions of the characteristics in order to produce linear amplification of a signal.

In this chapter we are more interested in the operation of these devices in the nonlinear regions of their characteristics. The primary use of these devices lies in constructing electronic switches for use in computers and other digital devices. Digital electronic circuits are becoming of increasing importance for several reasons. Current integrated-circuit (IC) technology allows the construction of an enormous number of transistors and diodes, as well as resistors and capacitors, on a very small chip no larger than a pencil eraser. The present technology with regard to density of components on a chip is

1 Small-scale integration (SSI) contains fewer than 100 components.
2 Medium-scale integration (MSI) contains 100 to 1000 components.
3 Large-scale integration (LSI) contains 1000 to 10,000 components.
4 Very large-scale integration (VLSI) contains over 10,000 components on a single chip.

The large number of switches (gates) required by digital systems can now be placed on a chip of very small size. Prior to the revolutionary advances in IC technology, such systems would occupy a much larger space and analog systems tended to be predominant.

Another important advantage of digital systems over analog ones is their inherent noise and interference immunity. Digital systems operate on discrete voltage levels; for typical BJT gates, these levels are 0 and 5 V. (The logic levels are assumed, for illustration, to be 0 and 5 V. In digital circuits these are actually defined to be

0.8 and 2 V but we will use the levels of 0 and 5 V to simplify our discussion.) The data bits are logical 0s (0 V) and 1s (5 V). Strings of these logical 0s and 1s can be used to represent numbers, as discussed in Part 4. These numbers can be manipulated—such as in addition, subtraction, multiplication, and division—by operating on the bits. Since the bit levels are widely separated (on the order of 5 V for BJT devices), any noise or other random voltage introduced to the system will have to be of sufficient magnitude to cause a "bit error" to result. The noise tolerance level tends to be larger for digital systems than for the comparable analog versions.

In this chapter we study the construction of the basic digital circuit elements. These are composed of gates (AND, OR, NAND, NOR, NOT) for bit stream manipulation, memory circuits (flip-flops) for storage of data, and timing circuits for synchronization of the data manipulation. We take a brief journey through this field and highlight the important points. There is much more to study in the design of these circuits, and this is reserved for later courses.

10.1 DIODE GATES

The diode AND gate is shown in Fig. 10.1a. The symbol is shown in Fig. 10.1b. Suppose that the input voltages to the gate V_A and V_B consist of pulses which are either 0 or $+5$ V, with 0 V representing a logical 0 and $+5$ V representing a logical 1. If either V_A or V_B is 0 V, the associated diode is forward-biased (by the $+5$-V battery) and is closed. Thus, the output voltage V_o is zero. If both V_A and V_B are $+5$ V, the output is also $+5$ V. (If V_A and V_B are both $+5$ V, then the diode is either open or closed; the precise state is hard to determine, for when the voltage across an ideal diode is 0 V, is it open or closed? Nevertheless, for either condition $V_o = +5$ V.) Thus, this gate performs the logical AND function:

A	B	$A \cdot B$
0	0	0
0	1	0
1	0	0
1	1	1

The AND operation is denoted with a dot ($\cdot$): $A \cdot B$. Typical bit stream inputs to the AND gate are shown in Fig. 10.1c along with the resulting output stream.

The diode OR gate is shown in Fig. 10.2a, and the symbol is shown in Fig. 10.2b. If either V_A or V_B is $+5$ V, the associated diode will be forward-biased and closed and the output will be $V_o = 5$ V. If both V_A and V_B are 0 V, both diodes are reverse-biased and open and the output voltage will be 0 V. This gate performs the logical OR function:

A	B	$A + B$
0	0	0
0	1	1
1	0	1
1	1	1

FIGURE 10.1
The AND gate.
(*a*) Diode
implementation;
(*b*) symbol;
(*c*) digital
operation.

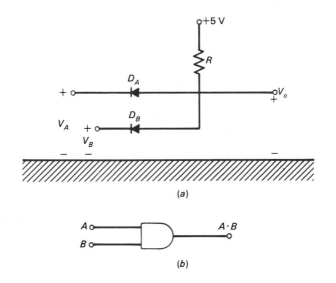

(*a*)

(*b*)

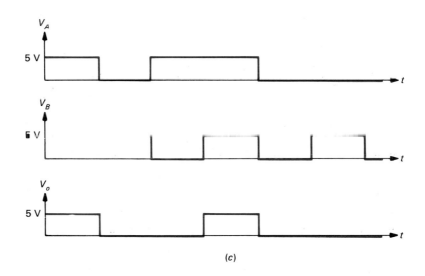

(*c*)

The OR function is denoted with a plus $(+)$: $A + B$. Typical input bit streams to the OR gate are shown in Fig. 10.2*c* along with the resulting output.

Although these gates are conceptually simple, they have an important problem. In the illustration we used ideal diodes, but when we construct these gates with actual diodes, a drop of approximately 0.5 V (actually closer to 0.7 V) will appear across a diode when it is closed. Thus, when the output of the OR gate in Fig. 10.2*c* is supposed to be 5 V, it will instead be 4.5 V, as shown in Fig. 10.3. When these gates are cascaded, the error eventually reaches a point where the output level has dropped such that a logical 1 is interpreted as a logical 0. The AND gate has the reverse problem; namely,

FIGURE 10.2
The OR gate.
(*a*) Diode
implementation;
(*b*) symbol;
(*c*) digital
operation.

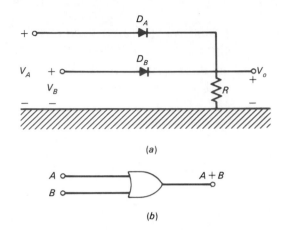

(*a*)

(*b*)

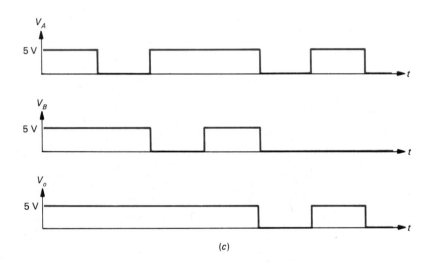

(*c*)

0.5 V is added to the input voltage, so that a 0-V logical 0 at the input becomes a 0.5-V level at the output of the gate. In a later section we will see a way of overcoming this propagating error by using transistors to construct these gates.

A NOT gate can be constructed by using the inherent inversion property of a transistor, as shown in Fig. 10.4*a*; the symbol is shown in Fig. 10.4*b*. If V_A is 0 V (*A* is grounded), the resistors R_{B1} and R_{B2} and battery V_{BB} produce a negative base voltage on the transistor, so that the base-emitter junction is reverse-biased (for this *npn* BJT). Thus, the base current is approximately zero, as is the collector current ($i_C = \beta i_B$), and thus $V_o = 5$ V. If $V_A = 5$ V and R_{B1}, R_{B2}, and V_{BB} are properly chosen, the base voltage will be greater than the turn-on voltage of the base-emitter junction (approximately 0.5 V) and base current will flow. Thus, collector current will flow and $V_o = 5$ V $- R_C i_C$. If enough collector current is caused to flow, $V_o = 0$ V and

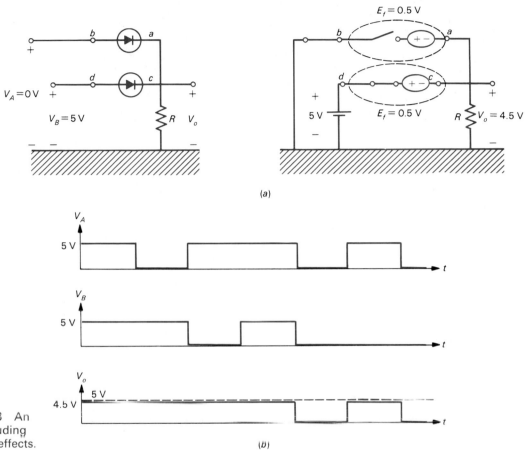

FIGURE 10.3 An OR gate including actual diode effects.

the NOT operation results. The NOT function is denoted with a bar over the operation. Typical bit streams are shown in Fig. 10.4c.

The NAND gate

A	B	$\overline{A \cdot B}$
0	0	1
0	1	1
1	0	1
1	1	0

and the NOR gate

A	B	$\overline{A + B}$
0	0	1
0	1	0
1	0	0
1	1	0

FIGURE 10.4
The NOT gate.
(a) BJT
implementation;
(b) symbol;
(c) digital
operation.

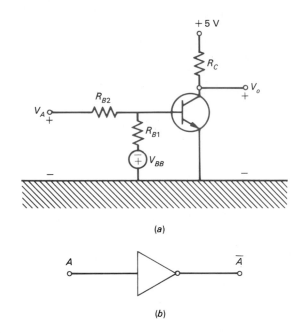

(a)

(b)

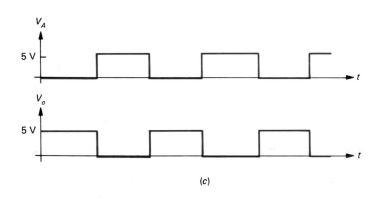

(c)

can be constructed by feeding the output of a diode AND or OR gate to the transistor NOT gate. The symbol for these gates is similar to the AND and OR gate symbols except that a small circle is placed on the output, as was the case for the NOT gate.

Example 10.1

The diode AND gate can be used as a pulse-code modulator. In this application the signal is sampled over equally spaced intervals of time, and the samples are then processed by a digital circuit. For the AND gate, suppose that V_A is a 3-V 1-kHz sinusoid

$$V_A = 3 \sin 2000\pi t \quad \text{V}$$

and that V_B is a sequence of 6-V pulses with duration 0.1 ms and separated by 0.1 ms. Sketch the output of the AND gate

FIGURE 10.5 An AND gate, pulse-code modulator.

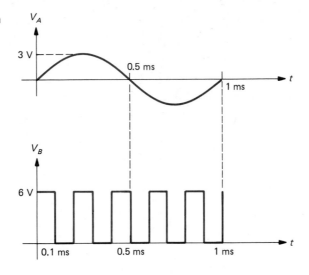

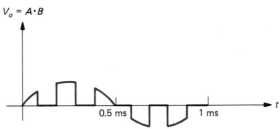

Solution If either input to the AND gate in Fig. 10.1a is less than the battery voltage 5 V, then the associated diode will be forward-biased and closed. Thus, the output will be connected to that input. When a 6-V pulse is present at B, diode D_B is open, and when the 6-V pulse is absent at B, diode D_B is closed. By comparing the two signals, we obtain the sampled output shown in Fig. 10.5.

Suppose the sinusoid has a peak value of 10 V. What changes would occur?

10.2 THE BJT SWITCH

Transistors (BJT or FET) can also be used to construct gates. They have the ability to restore the proper logic levels in a cascade of diode gates which has been degraded by the accumulation of 0.5-V drops across those diodes which are closed Cascading a set of diode gates has another problem—impedance loading. If a diode gate is to drive successive diode gates, the input impedances to those successive gates must be much larger than R in this first gate, so that the next gate will not affect the operation of the first. Transistor gates also tend to remedy this loading problem to some degree.

The operation of a BJT switch is summarized in Fig. 10.6. Writing KVL around the collector-emitter loop yields $v_{CE} = V_{CC} - R_c i_C$, from which the load line may be

FIGURE 10.6
The BJT switch.
(*a*) Circuit;
(*b*) typical
operation using
load lines.

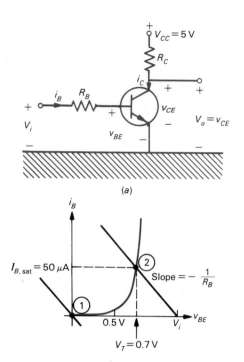

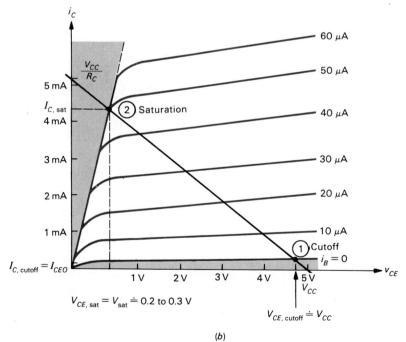

drawn on the collector-emitter characteristic. Similarly, writing KVL around the base-emitter loop yields $v_{BE} = V_i - R_B i_B$, and the corresponding load line may be drawn on the base-emitter characteristic. Suppose the input is at 0 V (logical 0); then the base current is zero, and the operating point is at ①. At this point the transistor is said to be *cut off*, or simply *off*, and only a small value of collector current flows ($I_{C,\text{cutoff}} \doteq I_{CEO}$). Thus, in the cutoff state

$$
\begin{aligned}
V_o &= V_{CE,\text{cutoff}} \\
&\doteq V_{CC} - R_C I_{CEO} \\
&\doteq V_{CC} \\
&= 5 \text{ V}
\end{aligned}
\tag{10.1}
$$

and the output is at the logical 1 level.

Now suppose that V_i changes to $+5$ V (logical 1 level). Base current flows which is determined from

$$
I_{B,\text{sat}} = \frac{V_i - v_{BE}}{R_B}
\tag{10.2}
$$

where $v_{BE} \doteq 0.7$ V. Note that although the base-emitter characteristic breaks at about 0.5 V, the actual value of v_{BE} is more like 0.7 V. We refer to this as the *threshold voltage* $V_T - 0.7$ V. Previously, we ignored this practicality since this base-emitter voltage turned out to be negligible in our dc amplifier calculations. But here the voltage levels are much lower, and the true value of v_{BE} (0.7 V) should be used. This applies to the diode gates, too. Now suppose that R_B is chosen such that $I_{B,\text{sat}} = 50$ μA. The operating point "switches" to point ② on the collector-emitter characteristic, and the transistor is said to be *saturated*, or simply *on*. In this saturated state $V_{CE,\text{sat}} = V_{\text{sat}}$, which is typically 0.2 to 0.3 V, depending on i_B. Note that increasing i_B further by reducing R_B causes no additional change. The collector current in saturation is

$$
\begin{aligned}
I_{C,\text{sat}} &= \frac{V_{CC} - V_{\text{sat}}}{R_C} \\
&\doteq \frac{V_{CC}}{R_C}
\end{aligned}
\tag{10.3}
$$

Thus, the transistor behaves as an ideal switch, as shown in Fig. 10.7.

In the above analysis, we assumed that R_B was small enough that $i_B \geq 50$ μA so as to drive the transistor into saturation. Increasing i_B further gains nothing; the BJT operating point remains at ②. But we must be sure that i_B is sufficiently large. Since $i_C \doteq \beta i_B + I_{CEO}$, we thus must have (neglecting I_{CEO}) $I_{B,\text{sat}} > I_{C,\text{sat}}/\beta$. But i_C at

FIGURE 10.7
Modeling the BJT
switch: (*a*) cutoff
(off);
(*b*) saturation
(on).

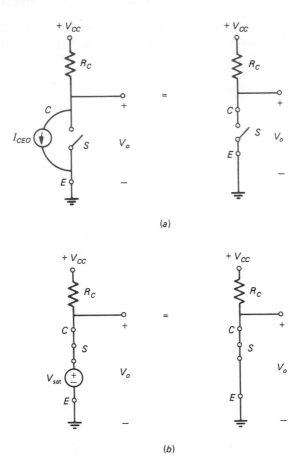

saturation is $I_{C,\text{sat}} = (V_{CC} - V_{\text{sat}})/R_C$. Thus, $I_{B,\text{sat}} > (V_{CC} - V_{\text{sat}})/(\beta R_C)$. Also $I_{B,\text{sat}} = (V_i - V_T)/R_B$. Therefore, saturation will occur when

$$\frac{V_i - V_T}{R_B} > \frac{V_{CC} - V_{\text{sat}}}{\beta R_C} \tag{10.4}$$

or

$$V_i > (V_{CC} - V_{\text{sat}})\frac{R_B}{\beta R_C} + V_T \tag{10.5}$$

Note that the power dissipated in the transistor $p = v_{CE}i_C + v_{BE}i_B \doteq v_{CE}i_C$ is approximately zero (or at least very small) in either cutoff or saturation. However, power is expended in switching through the linear, or active, region from one state to the other.

Example 10.2   The transistor switch in Fig. 10.6*a* is to be designed to operate in saturation and in cutoff when a pulse signal which varies between 0 and 5 V is applied to the input. The width of the pulses is 5 μs, and 5 μs separates the pulses. The supply voltage is $V_{CC} = 5$ V and $R_C = 500$ Ω. Determine the minimum value of R_B, and sketch the output voltage waveform. Assume an ideal transistor with $β = 100$, $V_T = 0.7$ V, $V_{sat} = 0.2$ V, and $I_{CEO} = 0.1$ mA.

Solution   When the signal is zero, the transistor is cut off and $I_{C,\text{cutoff}} = I_{CEO} = 0.1$ mA. Thus, $V_o = V_{CC} - R_C I_{CEO} = 4.95$ V. From Fig. 10.6 and with the transistor in saturation

$$I_{C,\text{sat}} = \frac{V_{CC} - V_{\text{sat}}}{R_C}$$

$$= \frac{5 - 0.2}{500}$$

$$= 9.6 \text{ mA}$$

But

$$i_C \doteq β i_B + I_{CEO}$$

so that

$$I_{B,\text{sat}} \doteq \frac{I_{C,\text{sat}} - I_{CEO}}{β}$$

$$= 95 \text{ μA}$$

For the transistor to be in saturation, we must have $I_B > I_{B,\text{sat}}$. R_B can be found from

$$I_B = \frac{V_i - V_T}{R_B}$$

$$\gtrless I_{B,\text{sat}}$$

or

$$R_B \le \frac{V_i - V_T}{I_{B,\text{sat}}}$$

$$= 45.26 \text{ kΩ}$$

The resulting output voltage is sketched in Fig. 10.8.

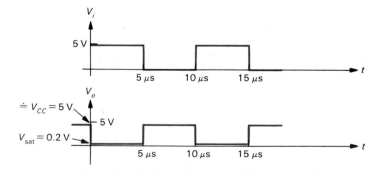

FIGURE 10.8
Example 10.2: design of a BJT switch.

The transfer characteristic relating V_i and V_o is shown in Fig. 10.9. A plot of a typical input waveform and the resulting output-voltage waveform is shown in Fig. 10.10. Note the inherent inversion property of the switch. Also note the important distortion of the waveform. The collector-current waveform is shown with important time parameters defined on it. With $V_i = 0$ V initially, the transistor is off, $V_o \doteq 5$ V, and $I_{C,\text{cutoff}} = I_{CEO} \doteq 0$. When V_i abruptly changes from 0 to 5 V, the output voltage and collector current do not initially react; there is a certain amount of delay t_d for a change to occur. This delay is referred to as *propagation delay* and is the time to change from 0 to 10 percent of the final value. (Sometimes propagation delay is defined as the time between the 50 percent level of V_i and the 50 percent level of V_o.) The rise time t_r is the time required to change from 10 to 90 percent of the final level. The various capacitances inherent in the transistor, as well as other stray capacitances, influence these times.

Once the transistor is established in saturation and the input pulse returns to 0 V (logic level 0), there is again a propagation delay t_s as well as a fall time t_f required for the output voltage and collector current to change state. The propagation delay t_s is a result of the time required to remove charge stored in the base region before the transistor begins to switch out of saturation and is usually longer than t_d. The fall time is the time required to switch through the active region from saturation to cutoff; this, too, is influenced by the transistor capacitances.

These propagation delays and the rise and fall times are important parameters influencing the design of a digital circuit. They are grouped under the category of switching speed.

Another important performance parameter is *fan-out*, which refers to the maximum number of switches that may be *driven* by a switch. The importance of this fan-out restriction is illustrated in Fig. 10.11. When the input is low ($V_i = V_L$) at the logic level 0, the output is high ($V_o = V_H$); in this state the transistor is supplying, or sourcing, current to the next switch. When the state is reversed so that

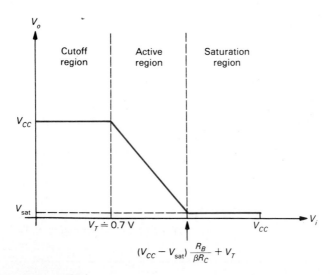

FIGURE 10.9
The transfer characteristic of a BJT switch.

FIGURE 10.10
Illustration of
propagation
delays, rise and
fall times of a BJT
switch.

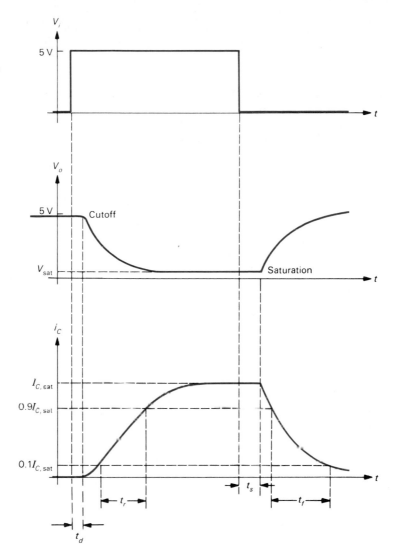

the output is in the low state, $V_o = 0$ and the switch is drawing, or sinking, current from the next switch. The maximum number of switches which can *precede* a switch is referred to as *fan-in*. Other factors affecting fan-out are the input capacitances of the driven stages, as shown in Fig. 10.11c. If one stage has an input capacitance of 8 pF, for instance, then 10 stages connected in parallel will have a net input capacitance of 80 pF, and this will affect the rise and fall times of the driving stage.

The third important performance parameter is the noise margin. This is illustrated in Fig. 10.12. The logic levels are not defined precisely at 0 and 5 V but occupy acceptable ranges. For example, the manufacturer may specify that the output of a switch which is not driving others will not fall below 2.4 V in the high state or rise above 0.4 V in the low state. Similarly, the driven stage is specified as interpreting a

FIGURE 10.11
Effect of loads on
a BJT switch.
(*a*) Sourcing
current;
(*b*) sinking current;
(*c*) capacitive
loads.

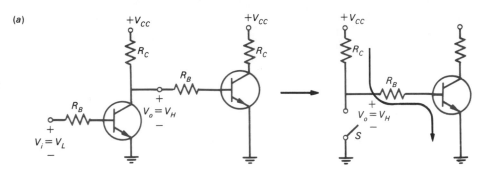

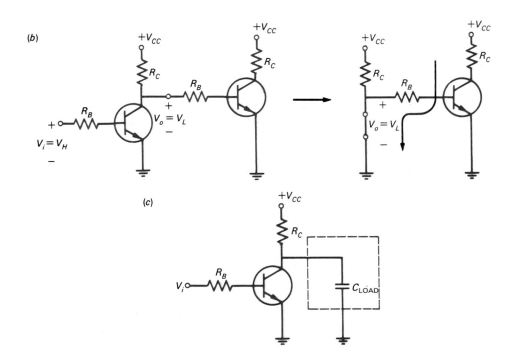

level as a logic 0 if the input voltage is below 0.8 V and as a logic 1 if the input
voltage is above 2 V. Thus, if the output of one stage is at 0.4 V, the next stage would
interpret that as low, since it is less than 0.8 V. Suppose that noise added 0.4 V to
the output of the first stage; then the next stage would still interpret this as low, but
the addition of any more noise could result in an incorrect interpretation of the level
by the second stage. These differences in manufacturer-specified logic levels

$$\Delta 1 = V_{oH} - V_{iH}$$

$$\Delta 0 = V_{iL} - V_{oL}$$

are called *noise margins*.

FIGURE 10.12
Illustration of
noise margins.

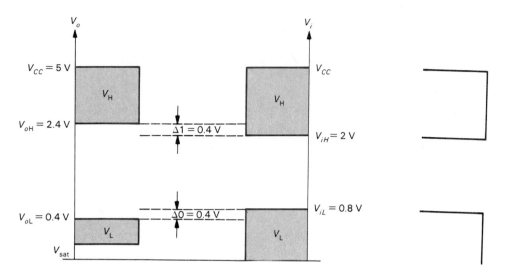

10.3   LOGIC FAMILIES

Logic gates to perform the basic logic functions (AND, OR, NAND, NOR, NOT) can be constructed with the transistor switch. To some degree, these gates overcome the disadvantage of the simple diode gates. In designing a digital system it is generally not necessary to "design" the individual gates; these have already been designed and are available in the form of small packages, such as the dual in-line package, or DIP, shown in Fig. 10.13*a*.

A typical schematic of a unit containing several gates is shown in Fig. 10.13*b*. The "designer" merely needs to connect the proper supply voltage to the unit; observe fan-out restrictions (the maximum number of gates which this gate may drive), fan-in restrictions (the maximum number of gates which may drive this device), and propagation delays; and ensure that the gates are properly connected so that the intended logic function will be performed—a topic covered in Part 4. Gates of the same logic family can be interconnected in this fashion since they have the same logic voltage levels, impedance characteristics, and switching times. Several logic families are discussed in this section.

A gate using resistor-transistor logic (RTL) is shown in Fig. 10.14. Suppose V_A is high but V_B and V_C are low. In this case, T_1 is on and T_2 and T_3 are off, so that V_o is low. If both V_A and V_B are high, V_o is also low. And V_o is high only if V_A, V_B, and V_C are all low. Thus, this gate performs the NOR function. The noise margins as well as the switching speeds of the RTL family are low. The fan-out is usually limited to about five gates.

Diode transistor logic (DTL) NOR and NAND gates are shown in Fig. 10.15. These are obtained by connecting the appropriate OR or AND diode gate to a transistor switch (inverter). The noise margins and fan-out of DTL are generally better than those of RTL, but the switching speeds are about the same.

FIGURE 10.13
Packaging of
digital circuits:
(a) dual in-line
package (DIP);
(b) schematic of
quad NAND gate
package.

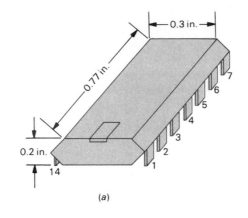

(a)

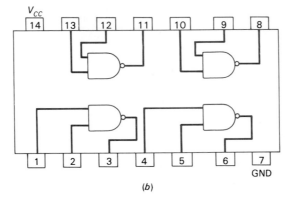

(b)

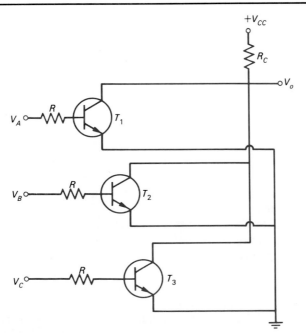

FIGURE 10.14   A
resistor transistor
logic (RTL) gate.

FIGURE 10.15
Diode transistor
logic (DTL) gates:
(*a*) NOR:
(*b*) NAND.

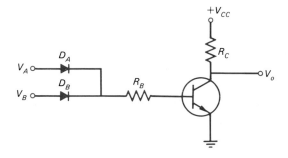

DTL NOR gate
(a)

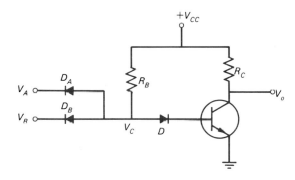

DTL NAND gate
(b)

Example 10.3

For the DTL NOR gate in Fig. 10.15*a*, assume diodes with $V_T = 0.7$ V, $V_{CC} = 5$ V, $R_B = 12$ kΩ, and $R_C = 500$ Ω. For the transistor, assume $\beta = 35$, $V_T = 0.7$ V, $I_{CEO} = 0$, and $V_{sat} = 0.2$ V. If $V_A = 5$ V and $V_B = 0$ V, determine V_o. Repeat these calculations if $V_A = V_B = 0$ V.

Solution   With $V_A = 5$ V and $V_B = 0$ V, diode D_A will be closed and D_B will be open. Thus, the base current is

$$i_B = \frac{V_A - 0.7 - 0.7}{R_B}$$

$$= 0.3 \text{ mA}$$

For the transistor to be in saturation, i_C must be greater than

$$I_{C,\,sat} = \frac{V_{CC} - V_{sat}}{R_C}$$

$$= 9.6 \text{ mA}$$

But $i_C \doteq \beta i_B$, so

$$i_C = 35 i_B$$
$$= 10.5 \text{ mA}$$

and the transistor is in saturation, with

$$V_o = V_{\text{sat}}$$
$$= 0.2 \text{ V}$$

If $V_A = V_B = 0$, both diodes will be open and $i_B = 0$. Thus, the transistor is cut off and $V_o \doteq V_{CC}$. Thus, this functions as a NOR gate.

In recent years, the most popular BJT logic family has been transistor-transistor logic (TTL). A basic TTL NAND gate is shown in Fig. 10.16a. Transistor T_1 is a multiple-emitter *npn* BJT, which acts as an AND gate. Replacing the base-emitter and base-collector junctions with diodes, we arrive at Fig. 10.16b. With V_A, V_B, and V_C high, all three diodes D_A, D_B, and D_C are reverse-biased and open. Current flows through R_B and D to saturate T_2, which saturates T_3. By adding the base-emitter drops of T_2 and T_3, the base voltage of T_2 becomes $2V_T = 1.4$ V. Adding the 0.7-V drop of T_1 across the base-collector diode D gives the base current to T_2 of $(V_{CC} - 2.1 \text{ V})/R_B$. With T_2 on, the base voltage of T_4 is too low to cause it to saturate, and T_4 is off. Thus, with all inputs in the high state, V_o will be in the low state.

Suppose that at least one input, say V_A, is in the low state. The associated base-emitter diode will be closed, and the base voltage of T_1 will be $V_A + 0.7$ V = 0.7 V. Combined with the 0.7-V drop across D, the base voltage of T_2 will be $-0.7 + 0.7 = 0$ V, so that T_2 is cut off. This serves to cut off T_3, and the output is high. Since T_2 is cut off, its collector voltage rises, turning on T_4.

The combination of R_4 and T_4 acts as a variable resistance. When T_3 is on, T_4 is off, lowering the power consumption; but when T_3 is off, T_4 is on, which provides a low resistance seen by the succeeding gate, thus improving the switching time (see Fig. 10.11). The diode between T_4 and T_3 serves to add a voltage drop to help ensure that T_4 is off when T_3 is on.

TTL is the most popular of the members of the bipolar logic families. The fan-out is quite large (on the order of 10 or more gates may be driven by one TTL NAND gate), the propagation delays are quite small [on the order of 2 to 10 nanoseconds (ns)], and the power consumption is typically 2 mW. By contrast, DTL has a typical fan-out of 8 to 10, a propagation delay on the order of 30 to 90 ns, and a power consumption of around 15 mW.

One of the reasons for the popularity of TTL over DTL is its higher speed. A primary reason for the speed restrictions in DTL and TTL is that the transistors are switched into saturation and time is required to remove the stored charge. (Driving the transistor into saturation with more than the minimum required collector current $I_{C,\text{sat}}$ accomplishes nothing, but it increases the required time to switch back out of saturation.) Part of this problem of switching speed could be eliminated if, in

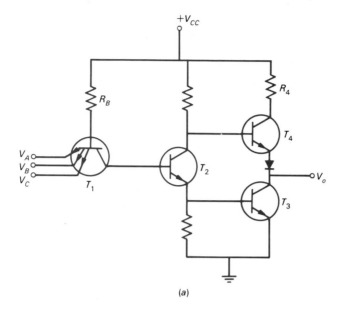

(a)

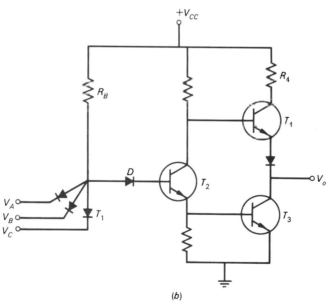

(b)

FIGURE 10.16 The transistor-transistor logic (TTL) gate. (a) Physical schematic; (b) diode replacement of three-input transistor.

switching, we stayed only in the active region. DTL and TTL could not be reliably operated in this mode since the range of base-emitter voltage required to switch from cutoff to saturation is only a few tenths of a volt (from about 0.4 to 0.7 V). Temperature variations, manufacturer variability, etc., would not allow us to ensure that we would not roam out of cutoff or saturation with an "ideal" design. Furthermore, driving the transistors completely into saturation or cutoff gives reliable

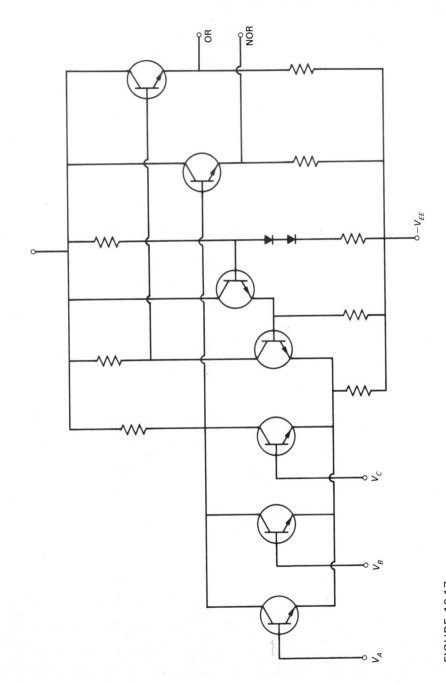

FIGURE 10.17
The emitter-
coupled logic
(ECL) gate.

logic voltage levels ($V_{CC} = 5$ V or $V_{sat} \doteq 0.2$ V). If we tried to switch in the active region only, minor variations could cause rather large changes in these logic levels.

The emitter-coupled logic (ECL) gate shown in Fig. 10.17 provides a way of reliably doing this and thus increasing switching times. The name arises from the common attachment of the emitters of the input transistors. The propagation delays are on the order of 1 ns, but the power consumption is quite high (on the order of 25 mW per gate). This latter disadvantage of ECL, combined with relatively small noise margins (<0.3 V), has tended to make TTL—and particularly the high-speed Schottky diode TTL gates—the popular choice.

10.4 COMPLEMENTARY METAL-OXIDE SEMICONDUCTOR (CMOS) GATES

Switches can also be constructed with FETs. A typical FET switch constructed from an n-channel JFET (or a depletion-mode MOSFET) is shown in Fig. 10.18. Voltage V_T is referred to as the threshold voltage (previously called the pinch-off voltage). For $v_{GS} < -V_T$, no drain current flows and the device is cut off. With $V_i = 0$ V, the FET is in saturation, and with V_i more negative than the threshold voltage V_T, the FET is cut off. Here, $V_i = 0$ V may represent a logical 1 and $V_i = -V_p = -V_T$ may represent a logical 0. Thus, the most positive of the two levels represents a logical 1. This is referred to as *positive logic*. If the most positive level represented a logical 0, this would be referred to as *negative logic*. The depletion-mode MOSFET in Fig. 10.18 can also be operated in the enhancement mode in which case V_i may go positive.

FIGURE 10.18
The FET switch.
(*a*) *n*-channel JFET
or depletion-mode
MOSFET switch;
(*b*) terminal
characteristic
operation;
(*c*) digital operation.

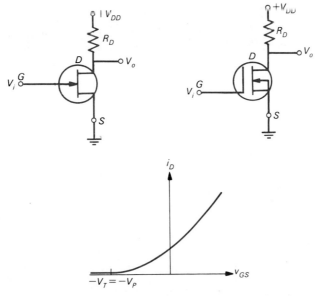

(a)

FIGURE 10.18
(*Continued*)

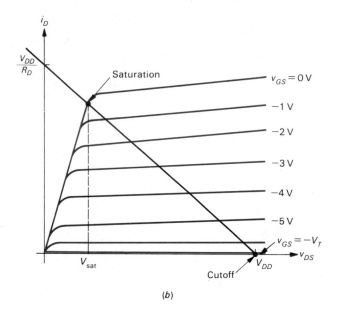

(*b*)

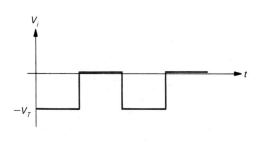

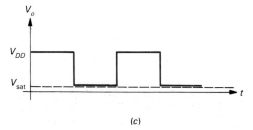

(*c*)

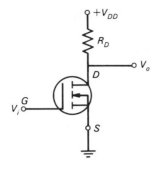

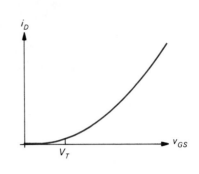

(a)

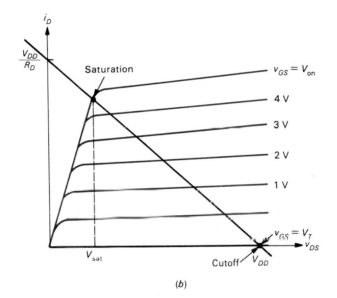

(b)

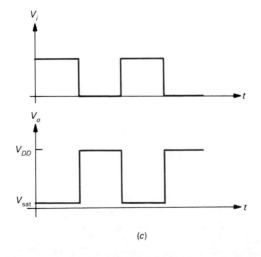

(c)

FIGURE 10.19
The n-channel enhancement-mode MOSFET switch.
(a) Circuit diagram and transfer characteristic;
(b) terminal characteristic and load line;
(c) digital operation.

FET logic circuits can also be used with positive pulses by using enhancement-mode n-channel MOSFETs as switches, as shown in Fig. 10.19. A transfer characteristic for this n-channel enhancement-mode MOSFET is shown; note that the threshold voltage V_T is positive and is typically on the order of a few volts (how many depends on the particular device). A p-channel enhancement-mode MOSFET has the characteristics shown in Fig. 10.20. Note that the characteristics of the p-channel and

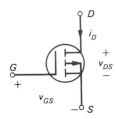

FIGURE 10.20
The p-channel enhancement-mode MOSFET switch.

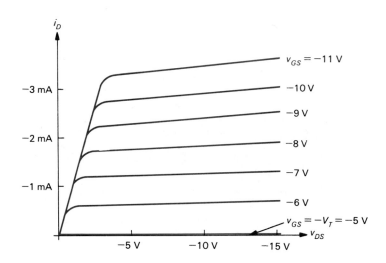

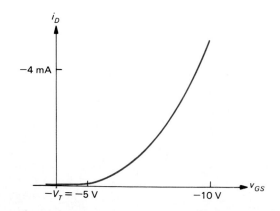

n-channel devices differ only in that the directions of i_D and the polarities of v_{DS} and v_{GS} are reversed. For the p-channel MOSFET, we must make the gate more *negative* than the source terminal by V_T in order to turn on the device. Making the gate more negative causes drain current to flow *from* the source *to* the drain.

FET switches offer important advantages over BJT switches. Since the gate current of a FET is, for all practical purposes, zero, a FET switch does not draw current from a previous stage; consequently, no significant power-loading effects are present. In contrast, BJTs do load down previous stages, and this must be taken into account in the design. Another advantage of FETs is that their logic voltage levels tend to be somewhat higher (typically, $V_{DD} = 15$ V) than with BJTs (typically, $V_{CC} = 5$ V); thus, logic circuits constructed from FETs tend to tolerate more noise than do comparable circuits using BJTs. Because of the larger inherent capacitances of FETs, however, their switching speeds tend to be somewhat slower than those for BJTs.

MOSFETs are preferred over JFETs for digital integrated circuits. They can be constructed as either a p-channel metal-oxide semiconductor (pMOS) or an n-channel metal-oxide semiconductor (nMOS). MOSFETs offer tremendous packing densities, since a typical MOSFET requires about 15 percent of the chip area of a BJT.

The most popular of the MOSFET family (for reasons soon to become apparent) is the complementary MOS, or CMOS (pronounced "see-moss"). A basic CMOS inverter is shown in Fig. 10.21. A pMOS and an nMOS (both are enhancement-mode) are used. When V_i is low, the gate-source voltage of the nMOS is less than the threshold voltage (see Fig. 10.19) and is cut off. The voltage from gate to source of the pMOS, however, is $-V_{SS}$, where $V_{SS} > V_T$ is the supply voltage. Thus, the pMOS is on (see Fig. 10.20), and the supply voltage appears at the output. When V_i goes to the high state ($V_i = V_{SS}$), the pMOS turns off and the nMOS turns on, whereupon V_{SS} appears across the drain-source terminals of the pMOS and V_o drops to zero.

Thus, the circuit functions as an inverter with an important property. Note that when the output is in the lower state, the nMOS is on but the pMOS is off and virtually no current is drawn from the power supply. On the other hand, when the output is in the high state, the pMOS is on but the nMOS is off and once again no power-supply current is drawn. This property of virtually no power consumption,

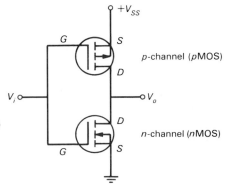

FIGURE 10.21 The complementary symmetry MOSFET (CMOS) switch.

coupled with the small consumption of chip area, makes the CMOS very attractive for such miniature, low-power applications as wristwatches and calculators. The poor switching speeds (relative to TTL), however, relegate it to low- to medium-speed devices. The power-supply voltage and the logic level in the high state (5 to 15 V) can provide noise margins larger than those for comparable TTL gates.

Two input NAND and NOR CMOS gates are shown in Fig. 10.22. For the NOR gate in Fig. 10.22b if V_A and V_B are low, both nMOS devices are off but both

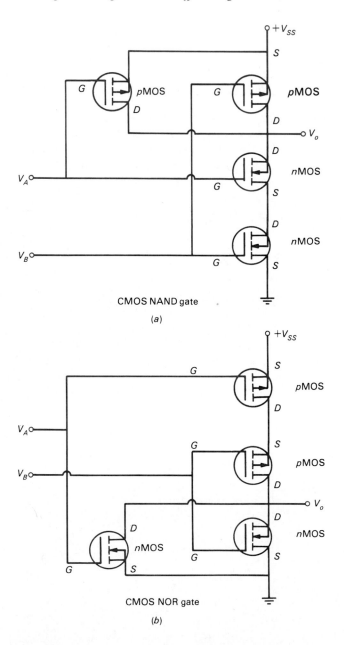

FIGURE 10.22
CMOS gates;
(a) NAND;
(b) NOR.

$pMOS$ devices are on, and V_o is around V_{SS}. If one of the inputs goes high, the associated $pMOS$ device turns off and the corresponding $nMOS$ device turns on, and thus V_o drops to the low state. Thus, the function of a NOR gate is produced. For the NAND gate in Fig. 10.22a, if at least one input is in the low state, the associated $pMOS$ device will be on and the $nMOS$ device will be off, giving a high output state. If both V_A and V_B are high, both $pMOS$ devices will be off while both $nMOS$ devices will be on, and V_o will be low. Thus, the device functions as a NAND gate. Neither gate draws virtually any power-supply current, so there is virtually no power consumption.

10.5 MULTIVIBRATORS

The above gates can be used to construct *combinational* logic circuits, in which a gate output at a particular time depends only on the gate inputs at that time. *Sequential* logic circuits have memory, in that an output at a particular time depends not only on the inputs at that time but also on inputs at previous times. Both types of logic circuits are discussed in Part 4. Here we discuss the basic construction of these types of logic devices.

The output of the BJT and FET switches discussed previously has two stable states: With the input in one state (0 or 1), the output is in a well-defined state (1 or 0). Multivibrators are devices in which, as illustrated in Fig. 10.23, (a) the output may appear in either of two states for the same input (bistable), (b) only one state is stable (monostable), or (c) no output state is stable (astable). For the bistable flip-flop, a spike at t_1 causes the output to switch from its present stable state to the other stable state. The device remains in that stable state until another spike is applied at t_2, and the device switches states again. Note that we do not need a long-duration pulse to switch the device; a short-duration spike is sufficient. The monostable multivibrator switches from its present stable state with the application of a spike, but after Δt (an adjustable parameter) it switches states again. The astable multivibrator switches continuously from one state to the other at well-defined times without the application of an input spike. Astables are often called *oscillators*.

A BJT flip-flop is shown in Fig. 10.24a. Resistor R_C is usually chosen much less that R_A. Assume for the moment that T_1 is off and T_2 is on, as shown in Fig. 10.24b. The collector C_2 of T_2 is at zero potential (neglecting $V_{sat} \doteq 0.2$ V for T_2). The current entering the base of T_2

$$i_{B2} = \frac{V_{CC} - 0.7 \text{ V}}{R_C + R_A}$$

is such that T_2 is kept on. The collector current for T_2 is

$$i_{C2} = \frac{V_{CC}}{R_C}$$

FIGURE 10.23
Multivibrators.
(*a*) Bistable (flip-
flop);
(*b*) monostable;
(*c*) astable.

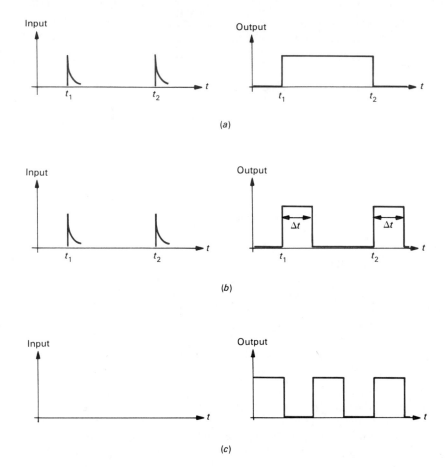

Since C_2 is low, virtually no base current flows into T_1, ensuring its off condition. Assuming T_1 on and T_2 off, one can show that this device can indeed be designed to have two stable spikes at 0 (and their opposite at $\bar{0}$). Now to change states, we simply apply a spike to S (the SET input); this causes base current to flow into T_1, turning it on, and C_1 thus drops, turning T_2 off. Note that additional spikes to S cause no change. To change states, we apply a spike to R (the RESET input), causing T_2 (which was turned off by S) to turn on, which causes T_1 to turn off. Thus, the SET (S) input causes the device to change from its present state ($\bar{0}$ high, 0 low) to its other state ($\bar{0}$ low, 0 high). A pulse must be applied to the RESET (R) input to change the device from its present state ($\bar{0}$ low, 0 high) to the other state ($\bar{0}$ high, 0 low). A typical pulse sequence to the S and R inputs and the resulting outputs are shown in Fig. 10.24*c*. Note that the flip-flop has memory, since it remembers the state into which it has been triggered.

The monostable multivibrators generate a single pulse of adjustable duration when a trigger is applied, as shown in Fig. 10.23*b*. These are used to generate a control, or gating, pulse of proper magnitude and duration when an event has taken place or to provide intervals of elapsed time—a timer. As shown in Fig. 10.23*c*, the

FIGURE 10.24
Operation of the
BJT flip-flop.
(*a*) Circuit;
(*b*) equivalent
circuit; (*c*) digital
operation.

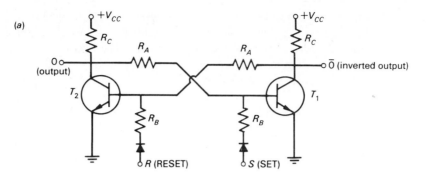

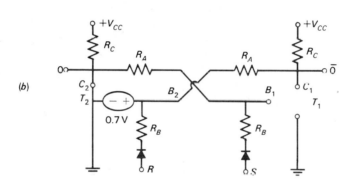

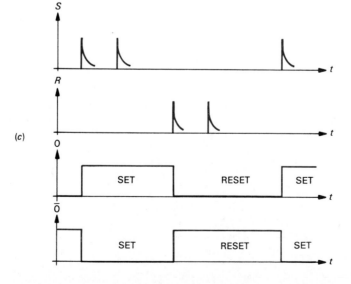

astable multivibrators generate a periodic waveform which is used to synchronize the sequence of operations of the system; this is usually referred to as the system *clock*. There is currently a standard IC package known as the 555 timer which can be used to perform either of these functions with the simple addition of a capacitor and one or two resistors to its external terminals. The pulse duration for the monostable output can be easily adjusted by proper selection of these external elements; in fact, rules for selection of the values of these external elements are given in the handbooks provided by the manufacturer. The 555 timer can also produce an astable multivibrator with a variable frequency (period of oscillation) adjustable by proper selection of the external resistors and capacitor. A functional block diagram of the 555 timer is given in Fig. 10.25. C_1 and C_2 are comparators (to be described) which provide SET and RESET signals to a flip-flop. The output of the flip-flop drives a buffer, or power, stage which may drive devices requiring larger power, such as electromechanical relays.

Before describing how the 555 timer functions as a monostable or astable multivibrator, we first describe the comparator. Comparators are essentially high-gain difference amplifiers, much like the op amp considered in the previous chapter. The characteristic of an ideal comparator is shown in Fig. 10.26a. When $V_i = 0$, $V_o = V_H$. As V_i increases, V_o remains at V_H until V_i reaches V_2, at which point V_o drops to V_L. (Note the arrows on the characteristic.) The output remains at V_L until V_i is reduced to V_1, at which point the output switches back to V_H. Input-output sample waveforms are shown in Fig. 10.26b. The comparator shown in Fig. 10.26a is said to have the property of *hysteresis* in that the switching levels V_1 and V_2 depend upon whether the input is going from low to high or high to low. A comparator with

FIGURE 10.25
The 555 timer.

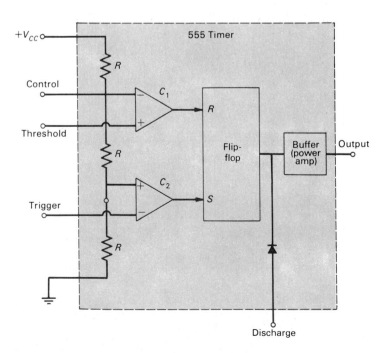

FIGURE 10.26
The ideal
comparator.
(*a*) Device symbol
and transfer
characteristic; (*b*)
device operation.

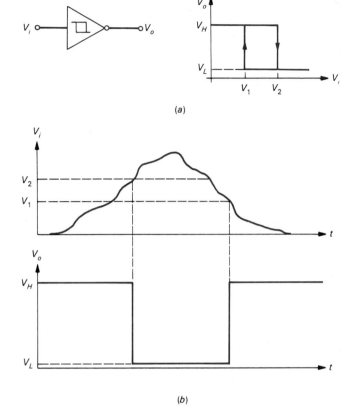

(*a*)

(*b*)

hysteresis is referred to as a *Schmitt trigger*. Hysteresis reduces the possibility of random switching due to noise or other voltage variations.

A monostable multivibrator can be constructed by adding a capacitor and resistors to the input of a comparator, as shown in Fig. 10.27*a*. For the input waveform shown in Fig. 10.27*b*, when $V_i = 0$, the capacitor voltage is zero and the voltage across R, or v, is zero. When the input abruptly rises to $V_{CC} = 15$ V, the voltage across the capacitor remains zero and v jumps to $V_{CC} = 15$ V. Eventually, the capacitor charges up, and v drops to zero. When v passes $V_2 = 5$ V, the comparator output switches from V_H to V_L and remains there until the RC circuit charges up sufficiently that v decays to $V_1 = 3$ V. At this point the comparator switches back to V_H. The time required for this to happen is governed by the charge time of the RC circuit. This circuit can be solved (neglecting any loading of the comparator, which is typically true for CMOS comparators) to yield

$$T = RC \ln \frac{V_H}{V_1} \tag{10.6}$$

Thus, the duration of the output pulse can be adjusted by selecting R and C. The polarity of the output can be inverted by feeding it to a NOT gate.

FIGURE 10.27   A monostable multivibrator using a comparator.

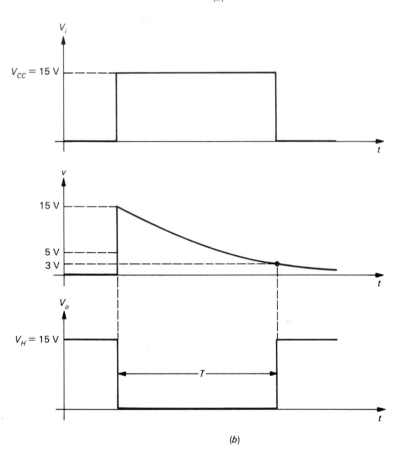

$V_H = 15$ V, $V_L = 0$ V
$V_1 = 3$ V, $V_2 = 5$ V

(a)

(b)

Example 10.4   Show that Eq. (10.6) is true.

Solution   Considering V_i, R, and C alone, we may write the equation for v as

$$v(t) = Ae^{-t/(RC)} + v_{SS}$$

For V_i constant at V_H, $v_{SS} = 0$; thus

$$v(t) = Ae^{-t/(RC)}$$

The initial condition on $v(t)$ is $v(0^+) = V_i - v_C(0^+) = V_i$; thus

$$v(t) = V_H e^{-t/(RC)}$$

The time required for v to decay to V_i is

$$V_1 = V_H e^{-T/(RC)}$$

or

$$\ln \frac{V_1}{V_H} = -\frac{T}{RC}$$

Thus

$$T = RC \ln \frac{V_H}{V_1}$$

The astable multivibrator, too, can be constructed from comparators, as shown in Fig. 10.28a. Suppose that both comparators have identical characteristics, as shown

FIGURE 10.28
An astable
multivibrator using
comparators.
(a) Circuit;
(b) transfer
characteristic;
(c) circuit
operation.

(a)

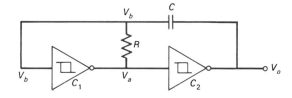

(b)

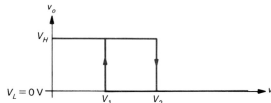

(c)

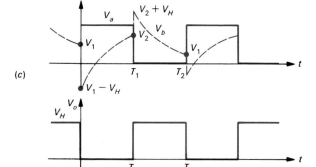

in Fig. 10.28b. To begin, suppose that V_o is high and V_a is low. Since V_a (the output of C_1) is low, V_b (the input to C_1) must be greater than V_1. The capacitor is in the process of charging up, and V_b is decreasing with time constant $T = RC$ (see Fig. 10.27). When V_b decreases to V_1, capacitor C_1 switches to the high state and V_a goes high, and C_2 switches to the low state and V_o goes low. Immediately prior to this change, the voltage across C is $V_b - V_o = V_1 - V_H$, negative at b. This voltage cannot charge the capacitor instantaneously, so when V_o drops to low (0 V), V_b drops to $V_1 - V_H$. Since $V_o = 0$ V and $V_a = V_H$, the voltage across the RC circuit is $V_a - V_o = V_H$ and the capacitor charges up toward V_H. When V_b reaches V_2, capacitor C_1 switches to the low state, causing V_a to be zero, which causes C_2 to switch to the high state—and $V_o = V_H$, and the process repeats.

Example 10.5

For the astable multivibrator in Fig. 10.28, assume that $R = 500$ Ω, $C = 100$ pF, $V_L = 0$ V, $V_H = 15$ V, $V_1 = 3$ V, and $V_2 = 5$ V. Calculate the frequency of this clock.

Solution The time required for the waveform to increase from $V_1 - V_H$ to V_2 at T_1 is determined by writing the solution of the voltage across the capacitor. See Fig. 10.29a. Essentially, a dc voltage of $V_a - V_o = V_H - 0 = V_H$ is applied across the RC combination for $0 < t < T_1$. The initial capacitor voltage is $V_b - V_o = V_1 - V_H$, so the form of the solution for $V_b - V_o = V_b$ is

$$V_b = (V_1 - 2V_H)e^{-t(RC)} + V_H$$

FIGURE 10.29
Example 10.5.

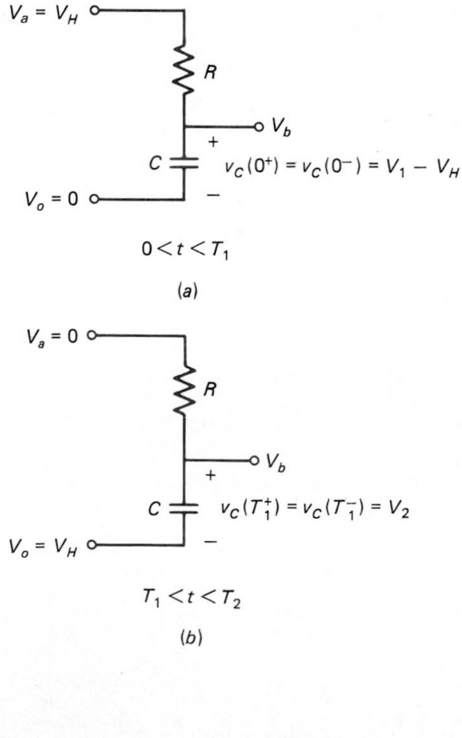

(Check it at $t = 0$ and steady state.) The time T_1 required to increase to V_2 is found from

$$V_2 = (V_1 - 2V_H)e^{-T_1/(RC)} + V_H$$

or

$$T_1 = RC \ln \frac{V_1 - 2V_H}{V_2 - V_H}$$

This is repeated for the interval $T_1 < t < T_2$ from the circuit shown in Fig. 10.29b with the result

$$V_b = (V_2 + V_H)e^{-(t - T_1)/(RC)}$$

so that at T_2

$$V_1 = (V_2 + V_H)e^{-(T_2 - T_1)/(RC)}$$

or

$$T_2 = RC \ln \frac{V_2 + V_H}{V_1} + T_1$$

Thus, the period of the waveform is $T_1 + T_2$, or

$$T \quad RC\left(2 \ln \frac{V_1 - 2V_H}{V_2 - V_H} + \ln \frac{V_2 + V_H}{V_1}\right)$$

with the prescribed values

$$T = 1.45 \times 10^{-7}\,\text{s}$$

and the clock frequency is

$$f = \frac{1}{T}$$

$$= 6.92\ \text{MHz}$$

Comparators can be constructed from op amps, as illustrated in Fig. 10.30. Recall from Chap. 9 that for negative v_d (between the $+$ and $-$ terminals of the op amp) the op-amp output will be saturated at its negative supply voltage $-V_{CC}$. Similarly,

FIGURE 10.30
Op-amp
comparators.

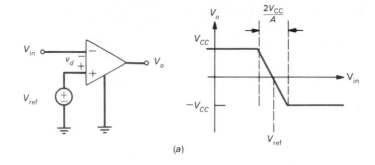

(a)

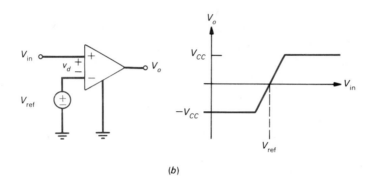

(b)

for positive v_d, the op-amp output voltage will be at the positive supply voltage V_{CC}. (See Fig. 9.9.) With these properties in mind, we see that v_d in Fig. 10.30a will be $v_d = V_{ref} - V_{in}$. So long as $V_{in} < V_{ref}$, voltage v_d will be positive and the op-amp output will be at $+V_{CC}$. Recall that a small linear region determined by the gain of the amplifier A exists. For realistic op-amp gains, this region is quite narrow, for example, $V_{CC} = 15$ V, $A = 10^5$, $2V_{CC}/A = 300$ μV. Hysteresis can be incorporated into this basic comparator with the addition of resistors in a feedback arrangement.

The 555 timer can be used to construct an astable multivibrator with the addition of resistors R_1 and R_2 and capacitor C, as shown in Fig. 10.31a. The three resistors R internal to the timer divide the supply voltage such that the voltage at the negative terminal of C_1 is $\frac{2}{3}V_{CC}$ and the voltage at the positive terminal of C_2 is $\frac{1}{3}V_{CC}$. The threshold and trigger terminals are connected to the upper terminal of the capacitor so that their voltages are the capacitor voltage v_C. The capacitor charges through R_1 and R_2 until $v_C = \frac{2}{3}V_{CC}$, at which time C_1 resets the flip-flop output state to zero. The capacitor discharges through R_2 until its voltage is reduced to $\frac{1}{3}V_{CC}$. At this point C_2 sets the flip-flop. The cycle continues over and over to produce the periodic waveform shown in Fig. 10.31b.

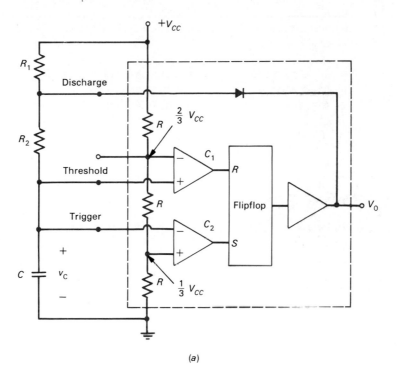

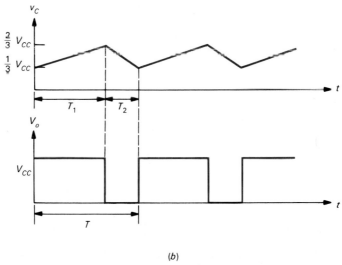

FIGURE 10.31
The 555 timer
connected as an
astable multivibrator.

Example 10.6 Determine equations for times T_1 and T_2 in Fig. 10.31b. Determine values of R_1, R_2, and C to produce a 100-kHz square wave.

Solution The various times can be calculated with the appropriate RC time constants of the charge-discharge circuits. Once v_C has charged to a value of $\frac{2}{3}V_{CC}$ through $R_1 + R_2$, the time required to discharge through R_2 to a value of $\frac{1}{3}V_{CC}$ is calculated from

$$\tfrac{2}{3}V_{CC}e^{-T_2/(R_2C)} = \tfrac{1}{3}V_{CC}$$

giving

$$T_2 = R_2 C \ln 2$$

The time required to charge up to $v_C = \frac{2}{3}V_{CC}$ can similarly be obtained as

$$T_1 = (R_1 + R_2)C \ln 2$$

The frequency of the output waveform is

$$f = \frac{1}{T_1 + T_2}$$
$$= \frac{1.44}{(R_1 + 2R_2)C}$$

To obtain an oscillator at a frequency of 100 kHz, we may choose

$$R_1 = R_2$$
$$= 480\ \Omega$$
$$C = 0.01\ \mu F$$

Choosing standard resistor values of 470 Ω would yield a frequency of 102.128 kHz.

10.6 A PRACTICAL NOTE

Although we have considered in some detail the design and operation of a large class of digital circuit "building blocks," it is not necessary to perform these laborious and detailed analysis tasks when interconnecting such devices to build a digital system, described in Part 4. The numerous handbooks furnished by the device manufacturers contain more than enough design aids, simple equations, nomographs, etc., to interconnect these devices properly. In fact, many of these handbooks could (and often do) serve as textbooks for courses in logic circuit design and make enjoyable and—now that we have discussed the fundamentals—easy reading.

PROBLEMS

10.1   For the diode AND gate shown in Fig.
10.1, determine V_o if the two signals shown in Fig.
P10.1 are applied. Assume ideal diodes.

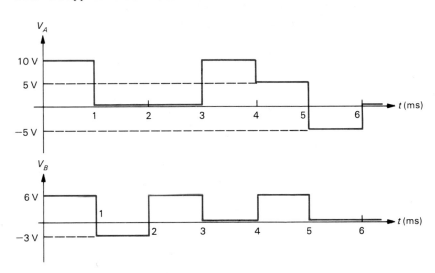

FIGURE P10.1

10.2   Repeat Prob. 10.1 if piecewise-linear
diodes having $E_f = 0.5$ V and $R_f = 0$ are used.

10.3   For the diode OR gate shown in Fig.
10.2a, sketch V_o for the two input signals shown in
Fig. P10.1

10.4   Repeat Prob. 10.3 if piecewise-linear
diodes having $E_f = 0.5$ V and $R_f = 0$ are used.

10.5   The NOT gate shown in Fig. 10.4 uses a
BJT having characteristics given in Fig. 8.49. For the
input signal shown in Fig. P10.5, sketch V_o if
$R_{B1} = R_{B2} = 40$ kΩ, $V_{BB} = 6$ V, $R_C = 1$ kΩ.

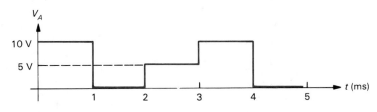

FIGURE P10.5

10.6   Change V_{BB} to 2 V and repeat Prob. 10.5.

10.7 For the BJT switch shown in Fig. 10.6a, $V_{CC} = 5$ V, $V_T = 0.7$ V, $V_{sat} = 0.2$ V, and $\beta = 20$. If V_i switches between 0 and 5 V and $i_B \leq 0.1$ mA, determine the minimum values of R_B and R_C for proper operation.

10.8 Repeat Prob. 10.7 if the input signal switches between 0 and 1 V.

10.9 Plot the transfer characteristic for the BJT switch in Fig. 10.6a with $V_{CC} = 5$ V, $V_{sat} = 0.2$ V, $V_T = 0.7$ V, $R_C = 1$ kΩ, $R_B = 10$ kΩ, $\beta = 50$.

10.10 Repeat Prob. 10.9 with R_C changed to 500 Ω.

10.11 The transistor switch shown in Fig. 10.6 uses a BJT which has the characteristics shown in Fig. 8.49. For the input signal shown in Fig. P10.11, sketch V_o if $R_B = 10$ kΩ, $R_C = 750$ Ω.

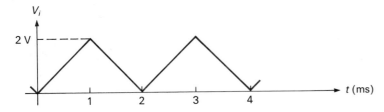

FIGURE P10.11

10.12 For the DTL NOR gate shown in Fig. 10.15a, assume ideal diodes, $V_T = 0.7$ V, $V_{sat} = 0.2$ V, $R_B = 10$ kΩ, $R_C = 500$ Ω, and $V_{CC} = 5$ V. Sketch V_o for the input signals shown in Fig. P10.1. Use reasonable approximations and assume $\beta = 20$ for the BJT.

10.13 For the TTL gate shown in Fig. 10.16, assume that the inputs vary between 0 and 5 V and that $V_{CC} = 5$ V. Determine the maximum value of R_B to saturate T_2 if $i_{C,sat} = 3.8$ mA.

10.14 The switch shown in Fig. 10.18, with $R_D = 3$ kΩ and $V_{DD} = 12$ V, uses a JFET whose characteristic is given in Fig. 8.48. If the input voltage is as shown in Fig. P10.14, sketch the output voltage.

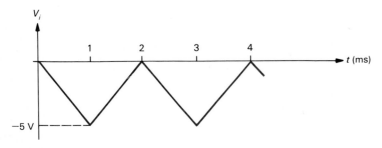

FIGURE P10.14

10.15 The MOSFETs in the CMOS gate shown in Fig. 10.21 have $V_T = 5$ V. If $V_{SS} = 20$ V and V_i is as shown in Fig. P10.15, sketch V_o. Assume that $V_{sat} = 1$ V for both MOSFETs.

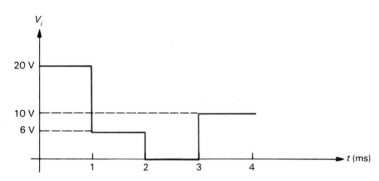

FIGURE P10.15

10.16 The monostable multivibrator show in Fig. 10.27a has $V_H = 10$ V, $V_L = 1$ V, $V_1 = 3$ V, and $V_2 = 5$ V. For the input voltage shown in Fig. P10.16, sketch the output voltage if $R = 1$ kΩ and $C = 1$ μF.

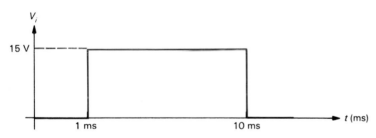

FIGURE P10.16

10.17 Repeat Prob. 10.16 if R is changed to 100 kΩ.

3

Electric Machines and Transformers

Polyphase Circuits

Polyphase circuits—three-phase circuits in particular—are used in the majority of transmission, distribution, and energy conversion systems where the power levels are above 10 to 20 kilowatts (kW). The transmission and distribution of electric power to residences, industries, and business is accomplished almost entirely by means of three-phase networks consisting of transmission and distribution lines, transformers, and associated protective devices such as circuit breakers, fuses, relays, and lightning protectors.

The basic reasons for using polyphase systems are related to *power density*, which is defined as the ratio of either power to weight or power to volume of the device. Thus, an electric machine of a given weight is capable of delivering more power in polyphase than in single-phase designs. In some electrical applications the voltampere rating is often substituted for the power rating. For example, the specific weight of an ac motor is generally less for a three-phase motor than for a single-phase motor at output power ratings above 1 horsepower (hp). Because three-phase systems are by far the most common, we restrict our discussion to them.

11.1   THREE-PHASE SYSTEMS

Suppose we have a system of three ac voltages of a certain frequency such that their amplitudes are equal but these voltages are displaced from one another by 120° in time. We may mathematically express this system† of voltages as

$$v_{a'a} = V_m \sin \omega t \tag{11.1}$$

$$v_{b'b} = V_m \sin (\omega t - 120°) \tag{11.2}$$

$$v_{c'c} = V_m \sin (\omega t - 240°) \tag{11.3}$$

---

† In this double-subscript notation, we consider that the voltage rises from terminal a' to terminal a.

These voltages are graphically depicted in Fig. 11.1. In terms of their RMS values, these voltages may be written in phasor notation as

$$V_{a'a} = V\underline{/0°}$$
$$= V(1 + j0) \tag{11.4}$$

$$V_{b'b} = V\underline{/-120°}$$
$$= V(-0.5 - j0.866) \tag{11.5}$$

$$V_{c'c} = V\underline{/-240°}$$
$$= V(-0.5 + j0.866) \tag{11.6}$$

Figure 11.2a shows the phasor representation of Eqs. (11.4) to (11.6). Notice that in Fig. 11.2b we have also shown (hypothetically) three voltage sources corresponding to Eqs. (11.4) to (11.6). Consequently, we may define a three-phase (voltage) source having three equal voltages which are 120° out of phase with one another. In particular, we call this system a *three-phase balanced system*—in contrast to an unbalanced system, in which the magnitudes may be unequal and/or the phase displacements may not be 120°. For a balanced three-phase system, it follows from Eqs. (11.1) to (11.3), as well as from Eqs. (11.4) to (11.6), that the phasor sum of the three voltages is zero.

We abbreviate $v_{a'a}$, $v_{b'b}$, and $v_{c'c}$ as v_a, v_b, and v_c, respectively. Now, referring to Figs. 11.1 and 11.2a, we observe that the voltages attain their maximum values in the order v_a, v_b, and v_c. This order is known as the phase *sequence abc*. A reverse phase sequence will be *acb*, in which case the voltages v_c and v_b lag v_a by 120° and 240°, respectively.

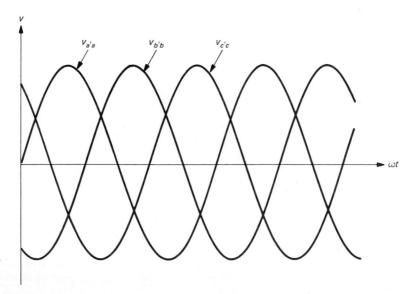

FIGURE 11.1 A system of three voltages of equal magnitude but displaced from each other by 120° in phase.

FIGURE 11.2
(*a*) Balanced three-phase phasor representation.
(*b*) Three-phase voltage source.

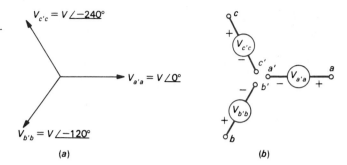

(*a*)

(*b*)

11.2 THREE-PHASE CONNECTIONS

There are two common and practical ways to interconnect the three voltage sources shown in Fig. 11.2*b*. These two forms of interconnection are illustrated in Fig. 11.3*a* and *b*, respectively labeled the *wye connection* and the *delta connection*. The result, for either type of connection, is a three-terminal (*ABC*) ac source of power supplying a balanced set of three voltages to a load. In the wye connection, terminals *a'*, *b'*, and *c'*, are joined to form the *neutral* point *o*. If a lead is brought out from the neutral

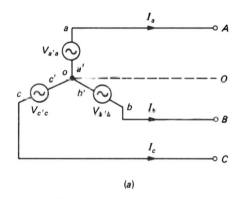

(*a*)

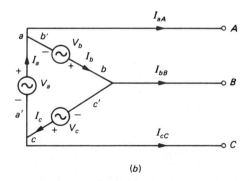

FIGURE 11.3
(*a*) Wye connection.
(*b*) Delta connection.

(*b*)

point o, the system becomes a *four-wire three-phase* system. Terminals a and b', b and c', and c and a' are joined individually to form the delta connection.

Notice from Fig. 11.3a that we can identify two types of voltages: voltages $V_{a'a}$, $V_{b'b}$, and $V_{c'c}$ across the three individual phases, known as *phase voltages*, and voltages V_{ab}, V_{bc}, and V_{ca} across lines a, b, and c (or A, B, and C), known as *line voltages*. Line voltages are related to phase voltages (using Kirchhoff's voltage law) such that

$$V_{oa} + V_{ab} = V_{ob}$$

or

$$V_{ab} = V_{ob} - V_{oa} \tag{11.7}$$

Similarly,

$$V_{bc} = V_{oc} - V_{ob} \tag{11.8}$$

and

$$V_{ca} = V_{oa} - V_{oc} \tag{11.9}$$

These relationships of the phase voltages and line voltages are illustrated in the phasor diagram of Fig. 11.4. Furthermore, we may combine Eqs. (11.4), (11.5), and (11.7) to obtain

$$V_{ab} = V(-0.5 - j0.866) - V(1 + j0)$$
$$= V(-1.5 - j0.866)$$

or

$$V_{ab} = \sqrt{3}\, V\underline{/-120^\circ} \tag{11.10}$$

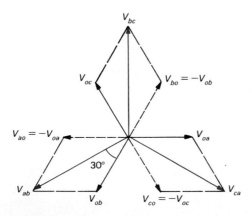

FIGURE 11.4
Voltage phasors
for Y connection.

FIGURE 11.5
Current phasors
for Y connection.

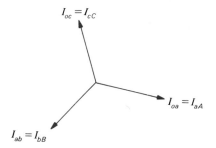

which is consistent with the phasor V_{ab} of Fig. 11.4. Relationships similar to Eq. (11.10) are valid for phasors V_{bc} and V_{ca}. Because V_{ab} is the voltage across lines a and b and V is the magnitude of the voltage across the phase, we may generalize Eq. (11.10) to

$$V_l = \sqrt{3}V_p \qquad (11.11)$$

where V_l is the voltage across any two lines and V_p is the phase voltage.

For the wye connection, it is clear from Fig. 11.3a that line currents I_l and phase currents I_p are the same. Thus, we may write

$$I_l = I_p \qquad (11.12)$$

The mutual phase relationships of the currents are given in Fig. 11.5.

Turning now to the delta connection, we can verify from Fig. 11.3b that the line voltages V_l are the same as the phase voltages V_p. Hence,

$$V_l = V_p \qquad (11.13)$$

The phase relationships of the three phase (and line) voltages are shown in Fig. 11.6. Next we show in Fig. 11.7 the phase currents and line currents for the

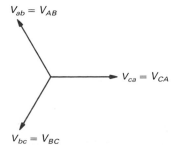

FIGURE 11.6
Voltage phasors
for Δ connection.

FIGURE 11.7
Current phasors
for Δ connection.

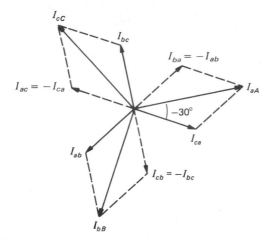

delta-connected system of Fig. 11.3b. The phase currents and line currents are related to each other (using Kirchhoff's current law) by

$$I_{aA} = I_{ca} - I_{ab}$$
$$= \sqrt{3}I_{ca}\underline{/30^\circ} \tag{11.14}$$

$$I_{bB} = \sqrt{3}I_{ab}\underline{/30^\circ} \tag{11.15}$$

and

$$I_{cC} = \sqrt{3}I_{bc}\underline{/30^\circ} \tag{11.16}$$

where $\cos 2\omega t$ is the pulsating component of double frequency. Of course, the time-average value of the power is $P = (V_m/\sqrt{2})(I_m/\sqrt{2}) = VI$, where V and I are the RMS values.

$$I_l = \sqrt{3}I_p \tag{11.17}$$

11.3 POWER IN THREE-PHASE SYSTEMS

We know from the study of single-phase ac circuits that the instantaneous power in the circuit is pulsating in nature. For instance, if $v = V_m \sin \omega t$ and $i = I_m \sin \omega t$ are the voltage across and the current through a resistor, then the instantaneous power p is given by

$$p = V_m I_m \sin^2 \omega t = \tfrac{1}{2} V_m I_m (1 - \cos 2\omega t)$$

where $\cos 2\omega t$ is the pulsating component of double frequency. Of course, the time-average value of the power is $P = (V_m/\sqrt{2})(I_m/\sqrt{2}) = VI$, where V and I are the RMS values.

It is interesting to note that the instantaneous power in a balanced three-phase circuit is a constant. This fact can be demonstrated as follows, where we consider a purely resistive circuit for the sake of simplicity. The instantaneous power may be written as

$$
\begin{aligned}
p &= v_a i_a + v_b i_b + v_c i_c \\
&= (V_m \sin \omega t)(I_m \sin \omega t) \\
&\quad + [V_m \sin (\omega t - 120°)][I_m \sin (\omega t - 120°)] \\
&\quad + [V_m \sin (\omega t + 120°)][I_m \sin (\omega t + 120°)] \\
&= V_m I_m (\sin^2 \omega t + \tfrac{1}{4} \sin^2 \omega t + \tfrac{3}{4} \cos^2 \omega t \\
&\quad + \tfrac{1}{4} \sin^2 \omega t + \tfrac{3}{4} \cos^2 \omega t) \\
&= \tfrac{3}{2} V_m I_m (\sin^2 \omega t + \cos^2 \omega t) \\
&= \tfrac{3}{2} V_m I_m \\
&= 3 V_p I_p
\end{aligned}
\tag{11.18}
$$

where V_p and I_p are RMS values of phase voltage and current. Since the instantaneous power is constant, the average power is the same as the total instantaneous power; i.e.,

$$
P_T = 3 V_p I_p
\tag{11.19}
$$

If the circuit is not purely resistive and has a power-factor angle θ_p, then the average power delivered by each phase is given by $V_p I_p \cos \theta_p$. So the total power delivered to the load is given (for this *balanced* system) by

$$
P_T = 3(V_p I_p \cos \theta_p)
\tag{11.20}
$$

But for the wye connection, the phase current is the same as the current in line I_l, and we showed earlier that $V_l = \sqrt{3} V_p$; thus, we can express the total power in terms of line voltages and currents as follows:

$$
P_T = 3 \frac{V_l}{\sqrt{3}} I_l \cos \theta_p = \sqrt{3} V_l I_l \cos \theta_p
\tag{11.21a}
$$

Note that the angle θ_p is still the angle between the phase voltage and phase current.

The expression for the total power in a delta-connected system is the very same. Whereas the line voltage equals the phase voltage for a delta-connected system, the line current is $\sqrt{3}$ times greater than the phase current, as we could easily show. Hence, if Eq. (11.20) is converted to line quantities for the delta-connected system, we obtain

$$
P_T = 3 V_l \frac{I_l}{\sqrt{3}} \cos \theta_p = \sqrt{3} V_l I_l \cos \theta_p
\tag{11.21b}
$$

FIGURE 11.8
Power in three-
phase system.

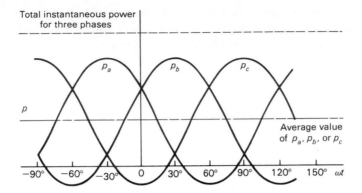

Whereas Eqs. (11.21a) and (11.21b) yield the true power, or active power (in watts), the reactive power is given by:

$$Q_T = \sqrt{3}\,V_l I_l \sin \theta_p \qquad (11.22)$$

And $\sqrt{3}\,V_l I_l$ is the total voltampere, which is called the apparent power.

A graphical representation of the instantaneous power in a three-phase system is given in Fig. 11.8. The total instantaneous power is constant and is equal to 3 times the average power of each phase. This feature is of great value in the operation of three-phase motors where the constant instantaneous power implies an absence of torque pulsations and consequent vibrations.

Example 11.1

A three-phase delta-connected load having a $(3 + j4)$-Ω impedance per phase is connected across a 220-V three-phase source. Calculate the magnitude of the line current and the total power supplied to the load.

Solution

$$V_{\text{phase}} = 220 \text{ V}$$
$$Z_{\text{phase}} = 3 + j4$$
$$= 5\underline{/53.2}\ \Omega$$
$$I_{\text{phase}} = \tfrac{220}{5}$$
$$= 44 \text{ A}$$
$$I_{\text{line}} = \sqrt{3} \times 44$$
$$= 76.21 \text{ A}$$
$$\text{Power} = \sqrt{3} \times 220 \times 76.21 \times \cos 53.2$$
$$= 17.4 \text{ kW}$$

Example 11.2

A 220-V three-phase source supplies a three-phase wye-connected load having an impedance of $(3 + j4)\Omega$ per phase. Calculate the magnitude of the phase voltage across each phase of the load and the total power consumed by the load.

Solution

$$V_{\text{phase}} = \frac{220}{\sqrt{3}}$$

$$= 127 \text{ V}$$

$$Z_{\text{phase}} = 3 + j4$$

$$= 5\underline{/53.2} \ \Omega$$

$$I_{\text{phase}} = \frac{127}{5}$$

$$= 25.4 \text{ A} = I_{\text{line}}$$

$$\text{Power} = \sqrt{3} \times 220 \times 25.4 \cos 53.2$$

$$= 5.8 \text{ kW}$$

Example 11.3   A three-phase balanced load has a 10-Ω resistance in each of its phases. The load is supplied by a 220-V three-phase source. Calculate the power absorbed by the load if it is connected in wye; calculate the same if it is connected in delta.

Solution   In the wye connection

$$V_p = \frac{220}{\sqrt{3}}$$

$$= 127 \text{ V}$$

$$I_p = \frac{127}{10}$$

$$= 12.7 \text{ A} = I_l$$

$$\cos \theta_p = 1 \qquad \text{load purely resistive}$$

Hence,

$$P = \sqrt{3} V_l I_l \cos \theta_p$$

$$= \sqrt{3} \times 220 \times 12.7 \times 1$$

$$= 4.84 \text{ kW}$$

In the delta connection

$$V_l = 220 \text{ V}$$

$$I_p = \frac{220}{10}$$

$$= 22 \text{ A}$$

$$I_l = \sqrt{3} \times 22 = 38.1 \text{ A}$$

Hence,

$$P = \sqrt{3} V_l I_l \cos \theta_p$$

$$= \sqrt{3} \times 220 \times 38.1 \times 1$$

$$= 14.52 \text{ kW}$$

Notice that the power consumed in the delta connection is 3 times that of the wye connection.

Example 11.4   For the balanced load shown in Fig. 11.9a, draw a phasor diagram showing all currents and voltages and determine the power consumed in the load.

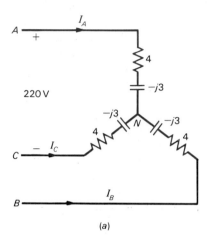

FIGURE 11.9
Example 11.4.

(a) (b)

Solution   The phasor diagram is shown in Fig. 11.9b.

$$V_p = \frac{220}{\sqrt{3}}$$
$$= 127 \text{ V}$$
$$Z_p = 4 - j3$$
$$= 5\underline{/-36.87}\ \Omega$$
$$I_l = I_p$$
$$= \frac{127}{5\underline{/-36.87}}$$
$$= 25.4\underline{/36.87}\ \text{A}$$

Hence,

$$P = \sqrt{3} \times 220 \times 25.4 \cos 36.87$$
$$= 7.74 \text{ kW}$$

11.4 WYE-DELTA EQUIVALENCE

In working with *balanced* three-phase systems, computations are usually made on a *per-phase* basis, and then the total results for the entire circuit are obtained on the basis of the symmetry and other factors which must apply. Thus, if a set of three

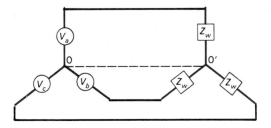

FIGURE 11.10
Reduction of a three-phase system to three single-phase systems.

identical wye-connected impedances is connected to a wye-connected three-phase source (as shown in Fig. 11.10), then because of the symmetry, point O' will prove to have the same potential as point O and could therefore be connected to it. Thus, each generator seems to supply only its own phase, and computations on a per-phase basis are therefore legitimate.

Sometimes it is necessary, or desirable, to mathematically convert a set of identical impedances connected in wye to an equivalent set connected in delta, or vice versa (Fig. 11.11). For this transformation to be valid, the impedances seen between any two of the three terminals must be the same. Thus, getting the impedance Z_{ab} for both the wye and the delta, and equating them, we get

$$2Z_w = \frac{Z_d(Z_d + Z_d)}{Z_d + Z_d + Z_d}$$

$$= \tfrac{2}{3}Z_d \tag{11.23}$$

From this we can solve for wye impedances in terms of delta impedances, or vice versa, such that

$$Z_w = \frac{Z_d}{3} \tag{11.24}$$

and

$$Z_d = 3Z_w \tag{11.25}$$

Hence, we can switch back and forth between balanced wye and delta as desired.

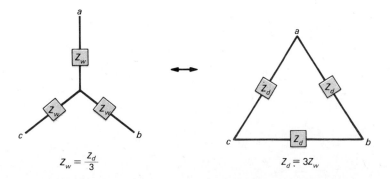

FIGURE 11.11
Equivalence of YΔ impedances.

$$Z_w = \frac{Z_d}{3}$$

$$Z_d = 3Z_w$$

Example 11.5   The load on a three-phase wye-connected 220-V system consists of three 6-Ω resistors connected in wye and in parallel with three 9-Ω resistors connected in delta. Calculate the magnitude of the line current.

Solution   Converting the delta load to a wye, from Eq. (11.24) we have (Fig. 11.12)

$$R_w = \tfrac{1}{3} \times 9$$
$$= 3\,\Omega$$

This resistance combined with the wye-connected load of 6 Ω per phase gives a per-phase resistance R_p as

$$R_p = \frac{6 \times 3}{6 + 3}$$
$$= 2\,\Omega$$

The phase voltage is

$$\frac{220}{\sqrt{3}} = 127\text{ V}$$

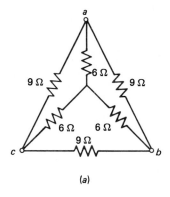

(a)

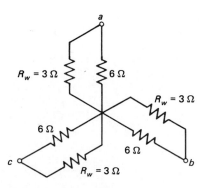

FIGURE 11.12   (b)

Hence,

$$I_p = I_l$$
$$= \frac{127}{2}$$
$$= 63.5 \text{ A}$$

Example 11.6

The voltage across the lines of a three-phase wye-connected generator is 11 kV. The generator supplies a 6-MW load at 0.8 lagging power factor. Calculate the active and reactive components of the current in each phase of the generator.

Solution In terms of line values from Eq. (11.21) we have

$$\sqrt{3} \times 11 \times 10^3 I_l \times 0.8 = 6 \times 10^6$$

or

$$I_l = \frac{6 \times 10^6}{\sqrt{3} \times 11 \times 10^3 \times 0.8} = 393.66 \text{ A} = I_p$$

$$\text{Active current} = I_p \cos \theta_p$$
$$= 393.66 \times 0.8 = 314.93 \text{ A}$$

$$\text{Reactive current} = I_p \sin \theta_p$$
$$- 393.66 \times 0.6 = 236.20 \text{ A}$$

Example 11.7

In the generator of Example 11.6, the load is changed such that the power factor becomes 0.9 and the line current decreases by 10 percent. What is the power supplied to the load?

Solution The power is given by Eq. (11.21), where we substitute $\cos \theta = 0.9$, $V_l = 11$ kV, and $I_l = 0.9(393.66)$, from Example 11.6. Thus

$$P = \sqrt{3} \times 11 \times 10^3 \times 0.9 \times 393.66 \times 0.9 = 6.075 \text{ MW}$$

PROBLEMS

11.1 Three identical impedances, each having a value $\mathbf{Z} = 12\underline{/-15°}$ Ω, are connected in delta to a balanced three-phase source of 100 V (line to line). Determine the line and phase currents, three-phase power, and three-phase voltamperage; draw the phasor diagram.

11.2 Three identical impedances, each having a value $\mathbf{Z} = (3 + j4)$ Ω, are connected in wye and supplied by a 400-V three-phase source. Determine the

quantities listed in Prob. 11.1, and draw the phasor diagram.

11.3 Three impedances $\mathbf{Z}_A = 20\underline{/0°}$ Ω, $\mathbf{Z}_B = 10\underline{/30°}$ Ω, and $\mathbf{Z}_C = 25\underline{/0°}$ Ω are connected in delta to a 100-V source. Calculate the line currents.

11.4 Repeat Prob. 11.3 if the impedances are connected in wye and the supply voltage remains unchanged.

11.5 Repeat Prob. 11.3 if the sequence of the source voltage is reversed.

11.6 Determine the power supplied to the load of Probs. 11.3 and 11.5.

11.7 The power factor of each phase of a balanced wye-connected 5-Ω impedance is 0.6 lagging. If this impedance is connected to a source having 100-V phase voltage, calculate the reactive power drawn by the circuit.

11.8 A balanced delta-connected load draws a 10-A lagging line current and 3-kW power at 220 V. Calculate the complex impedance of each phase of the load.

11.9 The apparent power input to a balanced wye-connected inductive load is 30 kVA, and the corresponding true power is 15 kW at 50 A. Calculate (*a*) the reactive power drawn by the load and (*b*) the phase and line voltages.

11.10 Calculate the input power to a three-phase load formed by a delta connection of the load of Prob. 11.9, at the value of the line voltage determined in Prob. 11.9.

11.11 A neutral is provided to the three-phase system of Prob. 11.4. What is the neutral current?

11.12 A balanced three-phase wye-connected load having an *RC* parallel circuit in each phase is supplied by a 220-V 60-Hz source. If $R = 80\ \Omega$ and $C = 50\ \mu\text{F}$, determine the total power and voltamperes taken by the load.

11.13 The load on a 400-V three-phase system consists of a delta-connected load which has a 10-Ω resistance in each phase and operates in parallel with a wye-connected load also having a 10-Ω resistance in each phase. Calculate the line current and the power supplied by the source.

11.14 A balanced delta-connected inductive load takes 4 kW of power at 220 V while drawing a line current of 15 A. What is the impedance in each phase?

11.15 If the impedances determined in Prob. 11.14 are connected in wye and supplied by a 220-V

source, determine the power and voltamperes supplied to the load.

11.16 A 460-V three-phase 60-Hz source supplies energy to the following three-phase balanced loads: a 200-hp motor load operating at 94 percent efficiency and 0.88 power factor lagging; a 50-kW resistance heating load; and a combination of miscellaneous loads totaling 40 kW at 0.70 lagging power factor. Calculate (*a*) the total kilowatts supplied, (*b*) the total reactive kilovoltamperage supplied, (*c*) the total apparent power, and (*d*) the total line current.

11.17 A balanced wye-connected load and a balanced delta-connected load are supplied by a three-phase 480-V 50-Hz generator. The branch impedances of the wye and delta loads are $10\underline{/30°}\ \Omega$ and $20\underline{/-50°}\ \Omega$, respectively. Determine the active and reactive powers drawn by each three-phase load. Determine the phasor voltage and phasor current for *any* one branch of each three-phase load, and then substitute into the power equation for balanced three-phase loads.

11.18 Determine the total apparent power and the overall power factor of the circuit of Prob. 11.17.

11.19 A four-wire three-phase 450-V 60-Hz system supplies power to the following loads: three impedances $13\underline{/20°}\ \Omega$ each connected in wye, an unbalanced delta load consisting of $15\underline{/45°}\ \Omega$ between lines *a* and *b*, $3\underline{/30°}\ \Omega$ between lines *a* and *c*, and $10\underline{/60°}\ \Omega$ between lines *b* and *c*; a single-phase load of $(3 + j4)\ \Omega$ between line *c* and the neutral line. Draw the corresponding circuit, and calculate the current in line *a*.

11.20 A wye-connected load and a delta-connected load are supplied by a four-wire 400-V three-phase 50-Hz system. The connections for the wye load are $2\underline{/20°}\ \Omega$ to line *a*, $30\underline{/50°}\ \Omega$ to line *c*, and $6\underline{/75°}\ \Omega$ to line *b*. The connections for the delta load are $50\underline{/30°}\ \Omega$ between lines *a* and *b*, $25\underline{/-60°}\ \Omega$ between lines *b* and *c*, and $17.3\underline{/90}\ \Omega$ between lines *a* and *c*. Sketch the circuit and calculate the current $\mathbf{I}_b$.

REFERENCES

1 C. I. Hubert, *Electric Circuits AC-DC: An Integrated Approach*, McGraw-Hill, New York, 1982.

2 J. D. Irwin, *Basic Engineering Circuit Analysis*, Macmillan, New York, 1983.

3 D. F. Mix and N. M. Schmitt, *Circuit Analysis for Engineers*, Wiley, New York, 1985.

Transformers

The transformer is extremely important as a component in many different types of electric circuits, from small-signal electronic circuits to high-voltage power transmission systems. A knowledge of the theory, design relationships, and performance capabilities of transformers is essential for understanding the operation of many electronic control and power systems. Therefore, both as a vehicle for understanding some basic electromagnetic principles and as an important component of electrical systems, the transformer deserves serious study.

The most common functions of transformers are (1) changing the voltage and current levels in an electrical system, (2) impedance matching, and (3) electrical isolation. The first of these functions is probably best known to the reader and is typified by the distribution lines from, say, 2300 V, to the household voltage of 115/230 V. The second function is found in many communication circuits and is used, for example, to match a load to a line for improved power transfer and minimization of standing waves. The third feature is used to eliminate electromagnetic noise in many types of circuits, blocking dc signals, and user safety in electrical instruments and appliances.

Because the construction and operation of transformers involve magnetic circuits, in the following we give a brief introduction to magnetic circuits before we present a discussion of transformers. An understanding of magnetic circuit concepts is very useful in the study of transformers and other electromagnetic devices.

12.1 MAGNETIC CIRCUITS

A magnetic circuit provides a path for magnetic flux, just as an electric circuit provides a path for the flow of electric current. Magnetic circuits are an integral part of transformers and electric machines. The analysis of magnetic circuits depends on the following basic concepts.

We define the *magnetic flux density B* by the force equation

$$F = BlI \tag{12.1}$$

where F is the force [in newtons (N)] experienced by a straight conductor of length l [in meters (m)] carrying a current I [in amperes (A)] and oriented at right angles to a magnetic field of flux density B [in teslas (T)]. In other words, if a conductor is 1 m long, carries a 1-A current, and experiences a 1-N force when located at right angles to certain magnetic flux lines, then the flux density is 1 T. The force, the conductor, and the flux density are mutually perpendicular.

The magnetic flux ϕ through a given surface is defined by

$$\phi = \int_s \mathbf{B} \cdot d\mathbf{s} \tag{12.2}$$

If B is uniform over an area A and is perpendicular to A, then

$$\phi = BA \tag{12.3}$$

from which

$$B = \frac{\phi}{A} \tag{12.4}$$

The unit of magnetic flux is the weber (Wb), and 1 Wb/m² = 1 T.

The source of magnetic flux is either a permanent magnet or an electric current. To measure the effectiveness of electric current in producing a magnetic field (or flux), we introduce the concept of *magnetomotive force* (or mmf) $\mathscr{F}$, defined as

$$\mathscr{F} \equiv NI \tag{12.5}$$

where I is the current in amperes (A) flowing in an N-turn coil. The unit of mmf is the ampere-turn (A·turn). Schematically, a magnetic circuit with an mmf and magnetic flux are shown in Fig. 12.1.

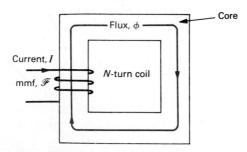

FIGURE 12.1
A magnetic circuit showing mmf and flux.

Although we have defined the mmf by Eq. (12.5), the mutual relationship between an electric current I and the corresponding magnetic field intensity H is given by *Ampere's circuit law*, expressed as

$$\oint \mathbf{H} \cdot d\mathbf{l} = I \qquad (12.6)$$

When the closed path is threaded by the current N times, as in Fig. 12.1, Eq. (12.6) becomes

$$\oint \mathbf{H} \cdot d\mathbf{l} = NI \equiv \mathscr{F} \qquad (12.7)$$

12.2 CORE MATERIAL

The core of a magnetic circuit (constituting a transformer or an electric machine) is generally of such a material having the variation of B with H as depicted by the saturation curve of Fig. 12.2. The slope of the curve depends upon the operating flux density, as classified in regions a, b, and c. For region b, which has a constant slope, we may write

$$B = \mu H \qquad (12.8)$$

where μ is defined as the *permeability* of the material and is measured in henrys per meter (H/m). For free space (or air), we have $\mu = \mu_o = 4\pi \times 10^{-7}$ H/m. In terms of μ_o, Eq. (12.8) is sometimes written as

$$B = \mu_r \mu_o H \qquad (12.9)$$

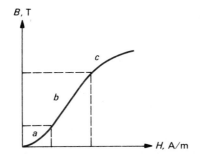

FIGURE 12.2
A *B-H* curve.

where $\mu_r = \mu/\mu_o$ and is called *relative permeability*. For ferromagnetic materials, $\mu_r \gg 1$.

12.3 ANALOGY WITH DC RESISTIVE CIRCUITS

Based on an analogy between a magnetic circuit and a dc resistive circuit, Table 12.1 summarizes the corresponding quantities. In this table, l is the length and A is the cross-sectional area of the path for the current in the electric circuit or for the flux in the magnetic circuit. Based on the above analogy, the laws of resistances in series or parallel also hold for reluctances. Note from Table 12.1 that a reluctance of a magnetic circuit is similar to the resistance in the dc resistive circuit analog.

TABLE 12.1

DC Resistive Circuit	Magnetic Circuit
Current I	Flux, ϕ Wb
Voltage V	Magnetomotive force, $\mathscr{F}$ A · turns
Conductivity σ	Permeability μ H/m
Ohm's law: $I = V/R$	$\phi = \mathscr{F}/\mathscr{R}$
Resistance $R = l/(\sigma A)$	Reluctance $\mathscr{R} = l/(\mu A)$ H^{-1}
Conductance $G = 1/R$	Permeance $\mathscr{P} = 1/\mathscr{R}$ H

The differences between a dc resistive circuit and a magnetic circuit are that (1) we have an I^2R loss in a resistance but do not have a $\phi^2\mathscr{R}$ loss in a reluctance, (2) magnetic fluxes take *leakage* paths (as ϕ_l in Fig. 12.3) but electric currents (flowing through resistances) do not, and (3) in magnetic circuits with air gaps we encounter *fringing* of flux lines (Fig. 12.3) but do not have fringing of currents in electric circuits. Fringing increases with the length of the air gap and increases the effective area of the air gap.

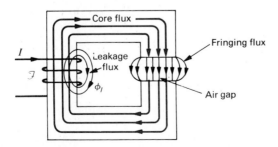

FIGURE 12.3
Leakage flux and
fringing flux.

12.4 CORE LOSSES

Consider the magnetic circuit of Fig. 12.1. If the mmf acting in a magnetic circuit is alternating current, then the *B-H* curve takes the form shown in Fig. 12.4. The loop shown is known as a hysteresis loop, and the area within the loop is proportional to the energy loss (as heat) per cycle. This energy loss is known as *hysteresis loss.* *Eddy-current loss*, the loss due to the eddy currents induced in the core material of a magnetic circuit excited by an ac mmf, is another feature of an ac-operated magnetic circuit. The power losses due to hysteresis and eddy currents—collectively known as *core losses* or *iron losses*—are approximately given by

eddy-current loss:

$$P_e = k_e f^2 B_m^2 \qquad \text{W/m}^3 \tag{12.10}$$

hysteresis loss

$$P_h = k_h f B_m^{1.5 \text{ to } 2.5} \qquad \text{W/m}^3 \tag{12.11}$$

where k_e is a constant depending on the material conductivity and thickness, k_h is another constant depending on the hysteresis loop of the material, B_m is the maximum core flux density, and f is the frequency of excitation.

The hysteresis loss component of the core loss in a magnetic circuit is reduced by using "good-quality" electrical steel (having a narrow hysteresis loop) for the core material.

The eddy-current loss is reduced by making the core of laminations, or thin sheets, with very thin layers of insulation alternating with the laminations. The laminations are oriented parallel to the direction of flux (Fig. 12.5). Laminating a core increases its cross-sectional area and hence its volume. The ratio of the volume

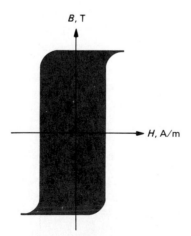

FIGURE 12.4
A hysteresis loop
of a core material.

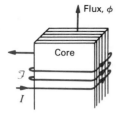

FIGURE 12.5
A laminated core.

actually occupied by the magnetic material to the total volume of the core is called the *stacking factor*. Table 12.2 gives some values of stacking factors.

TABLE 12.2

Lamination Thickness, mm	Stacking Factor
0.0127	0.50
0.0254	0.75
0.0508	0.85
0.10 to 0.25	0.90
0.27 to 0.36	0.95

Typical magnetic characteristics of certain core materials are given in Fig. 12.6.

The magnetic circuit concepts developed so far are now illustrated by the following examples.

FIGURE 12.6
Characteristics
of certain
core materials.

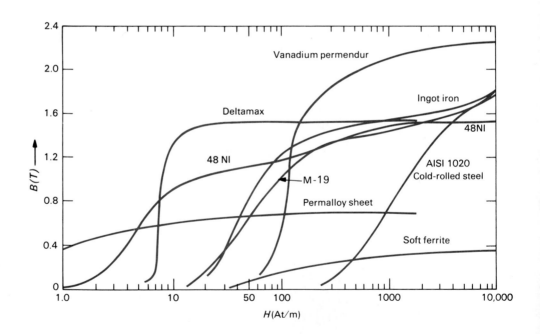

Example 12.1   The magnetic circuit shown in Fig. 12.7 carries a 50-turn coil on its core; the total core is made of 0.15-mm-thick laminations of M-19 (Fig. 12.6); the total core length is 10 cm, and the air-gap length is 0.1 mm. Calculate the coil current to establish an air-gap flux density of 1 T. Thus, what is the core flux if the core's cross section is 2.5 by 2.5 cm? Neglect fringing and leakage.

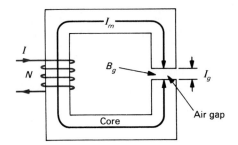

FIGURE 12.7
Magnetic circuit.

Solution For the air gap,

$$H_g = \frac{B_g}{\mu_0}$$

$$= \frac{1.0}{4\pi \times 10^{-7}}$$

$$= 7.95 \times 10^5 \text{ A/m}$$

$$\mathscr{F}_g = H_g l_g$$
$$= (7.95 \times 10^5)(10^{-4})$$
$$= 79.5 \text{ A} \cdot \text{turns}$$

For the core, given from Table 12.2 that the stacking factor for a 0.15-mm-thick lamination is 0.9,

$$B_c = \frac{B_g}{0.9}$$

$$= \frac{1}{0.9}$$

$$= 1.11 \text{ T}$$

From Fig. 12.6, for M-19 at 1.11 T we have

$$H_c = 130 \text{ A/m}$$

$$\mathscr{F}_c = H_c l_c$$
$$= 130 \times 0.1$$
$$= 13 \text{ A} \cdot \text{turns}$$

For the entire magnetic circuit,

$$NI = \mathscr{F}_{\text{total}}$$
$$= \mathscr{F}_g + \mathscr{F}_c$$
$$= 79.5 + 13$$
$$= 92.5 \text{ A} \cdot \text{turns}$$

from which

$$I = \frac{92.5}{50}$$

$$= 1.85 \text{ A}$$

The core flux is thus

$$B_c A_c = 1.11 \times (2.5)^2 \times 10^{-4}$$

$$= 0.694 \text{ milliweber (mWb)}$$

Example 12.2

Define flux linkage λ by $\lambda = N\phi$, where ϕ is the magnetic flux threading an N-turn coil; also define *inductance L* by $L = \lambda/i$ (or flux linkage per ampere), i being the current in the coil to produce the flux ϕ. Hence, determine the inductance of the coil of Example 12.1.

Solution   From Example 12.1, $\phi = 6.94 \times 10^{-4}$ Wb and $N = 50$. Thus, for the inductance of the coil we have

$$\lambda = N\phi$$

$$= 50 \times 6.94 \times 10^{-4}$$

$$= 3.47 \times 10^{-2} \text{ Wb} \cdot \text{turns}$$

$$L = \frac{\lambda}{i}$$

$$= \frac{3.47}{1.85} \times 10^{-2}$$

$$= 18.76 \text{ mH}$$

12.5   MAGNETIC AND ELECTRIC CIRCUITS OF A TRANSFORMER

The electromagnetic structure of a transformer consists of one or more electrical windings linked together magnetically by a magnetic circuit or core. The magnetic circuit of most transformers is constructed of a magnetic material, but nonmagnetic materials—often called "air cores"—are found in some applications. When there are more than two windings on a transformer, two of the windings are usually performing the identical functions. Therefore, in terms of understanding the theory and performance of a multiwinding transformer, only the relationships in two of the windings need be considered. The two basic windings are often called primary and secondary. The meaning usually attached to this nomenclature is that the input or source energy is applied to the primary windings and the output energy is taken from the secondary winding. However, since a transformer is a bilateral device and is often operated bilaterally, this meaning is not very significant and these words are used more as a way to distinguish the two windings. Depending on the voltage rating, the windings are also designated as high-voltage (HV) and low-voltage windings, respectively.

12.6 TRANSFORMER CONSTRUCTION

The construction of transformers varies greatly, depending on their applications, winding voltage and current ratings, and operating frequencies. The electromagnetic structure of a transformer is contained within a housing or case for safety and protection. In several types of transformers the space surrounding the electromagnetic structure is filled with an electrically insulating material, such as transformer oil, to prevent damage to the windings or core and to prevent their movement or to facilitate heat transfer between the electromagnetic structure and the case.

Transformer oil serves an added function of improving the insulation characteristics of the transformer, since it has a higher dielectric strength than air. In most oil-filled transformers the oil is permitted to circulate through cooling fins or tubes on the outside of the case to improve further the heat-transfer characteristics. The fins or tubes are often cooled by forced air. The magnetic core of a transformer must be constructed in a manner to minimize the magnetic losses. Power transformer cores are generally constructed from soft magnetic materials in the form of punched laminations or wound tapes. The most common lamination materials are silicon-iron, nickel-iron, and cobalt-iron alloys.

Transformer windings are constructed of solid or stranded copper or aluminum conductors. The conductor used in electronics transformers—as well as in many small and medium-size motors and generators—is magnet wire. The windings of large power transformers generally use conductors with heavier insulation than magnet wire insulation. The windings are assembled with much greater mechanical support, and windings layers are insulated from each other. Larger high-power windings are often preformed, and the transformer is assembled by stacking the laminations within the preformed coils.

12.7 PRINCIPLE OF OPERATION OF A TRANSFORMER

A transformer is an electromagnetic device having two or more mutually coupled windings. Figure 12.8 shows a two-winding ideal transformer. The transformer is *ideal* in the sense that its core is lossless, is infinitely permeable, and has no leakage fluxes and that its windings have no losses.

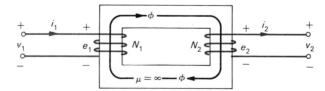

FIGURE 12.8
Schematic of a
two-winding
transformer.

In Fig. 12.8 the basic components are the core, primary winding N_1, and the secondary winding N_2. If ϕ is the mutual (or core) flux linking N_1 and N_2, then, according to Faraday's law of electromagnetic induction, emf's e_1 and e_2 are induced in N_1 and N_2 owing to a finite rate of change of ϕ such that

$$e_1 = N_1 \frac{d\phi}{dt} \tag{12.12}$$

and

$$e_2 = N_2 \frac{d\phi}{dt} \tag{12.13}$$

The direction of e_1 is such as to produce a current which opposes the flux change, according to Lenz's law. The transformer being ideal, $e_1 = v_1$ (Fig. 12.8). From Eqs. (12.12) and (12.13)

$$\frac{e_1}{e_2} = \frac{N_1}{N_2}$$

which may also be written in terms of RMS values as

$$\frac{E_1}{E_2} = \frac{N_1}{N_2} = a \tag{12.14}$$

where a is known as the turns ratio.

Since $e_1 = v_1$ (and $e_2 = v_2$), the flux and voltage are related by

$$\phi = \frac{1}{N_1} \int v_1 \, dt = \frac{1}{N_2} \int v_2 \, dt$$

If the flux varies sinusoidally such that

$$\phi = \phi_m \sin \omega t$$

then the corresponding induced voltage e linking an N-turn winding is given by

$$e = \omega N \phi_m \cos \omega t \tag{12.15}$$

From Eq. (12.15), the RMS value of the induced voltage is

$$E = \frac{\omega N \phi_m}{\sqrt{2}}$$

$$= 4.44 f N \phi_m \tag{12.16}$$

which is known as the emf equation. In Eq. (12.16), $f = \omega/2\pi$ is the frequency in hertz.

12.8 VOLTAGE, CURRENT, AND IMPEDANCE TRANSFORMATIONS†

Major applications of transformers are in voltage, current, and impedance transformations and in providing isolation (that is, eliminating direct connections between electric circuits). The voltage transformation property of an ideal transformer (mentioned in Sec. 12.7) is expressed as

$$\frac{V_1}{V_2} = \frac{E_1}{E_2} = a = \frac{N_1}{N_2} \tag{12.17}$$

where the subscripts 1 and 2 correspond to the primary and secondary sides, respectively. This property of a transformer enables us to interconnect transmission and distribution systems of different voltage levels in an electric power system.

For an ideal transformer, the net mmf around its magnetic circuit must be zero, implying that

$$N_1 I_1 - N_2 I_2 = 0 \tag{12.18}$$

where I_1 and I_2 are the primary and secondary currents, respectively. From Eqs. (12.17) and (12.18) we get

$$\frac{I_2}{I_1} = \frac{N_1}{N_2} = a \tag{12.19}$$

From Eqs. (12.17) and (12.19) it can be shown that if an impedance Z_2 is connected to the secondary, the impedance Z_1 seen at the primary satisfies

$$\frac{Z_1}{Z_2} = \left(\frac{N_1}{N_2}\right)^2 \equiv a^2 \tag{12.20}$$

Example 12.3   How many turns must the primary and secondary windings of a 220/110-V 60-Hz ideal transformer have if the core flux is not allowed to exceed 5 mWb?

Solution   From the emf equation, Eq. (12.16), we have

$$N = \frac{E}{4.44 f \phi_m}$$

---

† See also Section 4.4.

Consequently,

$$N_1 = \frac{220}{4.44 \times 60 \times 5 \times 10^{-3}}$$

$$\cong 166 \text{ turns}$$

$$N_2 = \tfrac{1}{2}N_1$$
$$= 83 \text{ turns}$$

Example 12.4 A 220/110-V 10-kVA transformer has a primary winding resistance of 0.25 Ω and a secondary winding resistance of 0.06 Ω. Determine the primary and secondary currents at rated load, the total winding resistance referred to the primary, and the total winding resistance referred to the secondary.

Solution Given the transformation ratio $a = 220/110 = 2$, the primary current

$$I_1 = \frac{10 \times 10^3}{220}$$

$$= 45.45 \text{ A}$$

and the secondary current

$$I_2 = aI_1$$
$$= 2 \times 45.45$$
$$= 90.9 \text{ A}$$

From Eq. (12.20), the secondary winding resistance referred to the primary $= a^2R_2 = 2^2 \times 0.06 = 0.24$ Ω. The total resistance referred to the primary is thus

$$R'_e = R_1 + a^2R_2$$
$$= 0.25 + 0.24$$
$$= 0.49 \text{ Ω}$$

The primary winding resistance referred to the secondary $= R_1/a^2 = 0.25/4 = 0.0625$ Ω. The total resistance referred to the secondary is thus

$$R''_e = \frac{R_1}{a^2} + R_2$$
$$= 0.0625 + 0.06$$
$$= 0.1225 \text{ Ω}$$

Example 12.5 Determine the I^2R loss in each winding of the transformer of Example 12.4 and thus find the total I^2R loss in the two windings. Verify that the same result can be obtained by using the equivalent resistance referred to the primary winding.

Solution $I_1^2 R_1 \text{ loss} = (45.45)^2 \times 0.25$

$$= 516.425 \text{ W}$$

$$I_2^2 R_2 \text{ loss} = (90.9)^2 \times 0.06$$

$$= 495.768 \text{ W}$$

$$\text{Total } I^2 R \text{ loss} = 1012.19 \text{ W}$$

For the equivalent resistance,

$$I_1^2 R_e' \text{ loss} = (45.45)^2 \times 0.49$$

$$= 1012.19 \text{ W}$$

which is consistent with the preceding result.

12.9   THE NONIDEAL TRANSFORMER AND ITS EQUIVALENT CIRCUITS

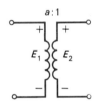

FIGURE 12.9   An ideal transformer.

In contrast to an ideal transformer, a nonideal (or actual) transformer has hysteresis and eddy-current losses (core losses) and has resistive $(I^2 R)$ losses in its primary and secondary windings. Furthermore, the core of a nonideal transformer is not perfectly permeable and thus requires a finite mmf for its magnetization. Also, because of leakage, not all fluxes link with the primary and the secondary windings simultaneously in a nonideal transformer.

An equivalent circuit of an ideal transformer is shown in Fig. 12.9. To include the nonideal effects of winding resistances, leakage reactances, magnetizing reactance, and core losses, we proceed as follows. First, we consider a simple magnetic circuit such as that shown in Fig. 12.1, excited with an ac mmf. We wish to represent it by an equivalent circuit. We will develop the equivalent circuit step by step as follows. First, we assume that the core has a finite and constant permeability but no losses. We also assume that the coil has no resistance. In such an idealized case the magnetic circuit and the coil can be represented just by an inductance, L_m. If the coil is excited by a sinusoidal ac voltage $v = V_m \sin \omega t$, it is conventional to express L_m as an inductive reactance $X_m = \omega L_m$. This reactance is known as the *magnetizing reactance*. Thus in Fig. 12.10a we have X_m across which we show the terminal voltage V_1, equal

FIGURE 12.10
Equivalent circuits
of an iron core
excited by an ac
mmf.

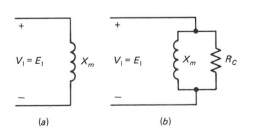

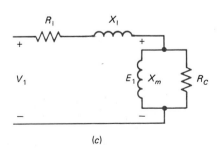

to the induced voltage E_1. Note that V_1 and E_1 are root-mean-square (RMS) values. Next, we include the hysteresis and eddy-current losses by the resistance R_c in parallel with X_m, such that the voltage V_1 appears across R_c also, the core losses being directly dependent on V_1. We thus obtain Fig. 12.10b, where again $V_1 = E_1$. Finally, we include the series resistance R_1, the resistance of the coil, and X_1, its *leakage reactance*, which arises from leakage fluxes. Hence we obtain the electrical equivalent shown in Fig. 12.10c, which represents an iron core excited by an ac mmf. We will use this basic equivalent circuit to obtain the equivalent circuit of a transformer. Accounting for the imperfections, the circuit of Fig. 12.9 is modified to that of Fig. 12.11, where the primary and the secondary are coupled by an ideal transformer. By using Eqs. (12.17), (12.19), and (12.20), we may remove the ideal transformer from Fig. 12.11 and refer the entire equivalent circuit either to the primary, as shown in Fig. 12.12, or to the secondary, as shown in Fig. 12.13.

The various symbols used in Figs. 12.11 through 12.13 are

$$a \equiv \text{turns ratio}$$
$$E_1 \equiv \text{primary induced voltage}$$
$$E_2 \equiv \text{secondary induced voltage}$$
$$V_1 \equiv \text{primary terminal voltage}$$
$$V_2 \equiv \text{secondary terminal voltage}$$
$$I_1 \equiv \text{primary current}$$
$$I_2 \equiv \text{secondary current}$$
$$I_0 \equiv \text{no-load (primary) current}$$
$$R_1 \equiv \text{resistance of the primary winding}$$
$$R_2 \equiv \text{resistance of the secondary winding}$$
$$X_1 \equiv \text{primary leakage reactance}$$
$$X_2 \equiv \text{secondary leakage reactance}$$
$$I_m, X_m \equiv \text{magnetizing current and reactance}$$
$$I_c, R_c \equiv \text{current and resistance accounting for the core losses}$$

The constants of the equivalent circuit can be found from certain tests. The major use of the equivalent circuit of a transformer is for determining its characteristics. The characteristics of most interest to power engineers are voltage regulation and efficiency. *Voltage regulation* is a measure of the change in the terminal voltage

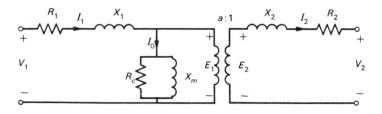

FIGURE 12.11
A nonideal
transformer.

FIGURE 12.12
Equivalent circuit
referred to primary.

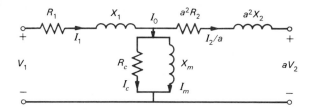

of the transformer with load. Specifically, we define voltage regulation as

$$\text{Percent regulation} = \frac{V_{\text{no-load}} - V_{\text{load}}}{V_{\text{load}}} \times 100 \qquad (12.21)$$

With reference to Fig. 12.12, we may rewrite Eq. (12.21) as

$$\text{Percent regulation} = \frac{V_1 - aV_2}{aV_2} \qquad (12.22)$$

for a given load.

There are two kinds of transformer efficiencies of interest to us; they are known as *power efficiency* and *energy efficiency*. These are defined as follows:

$$\text{Power efficiency} = \frac{\text{output power}}{\text{input power}} \qquad (12.23)$$

$$\text{Energy efficiency} = \frac{\text{output energy for a given period}}{\text{input energy for the same period}} \qquad (12.24)$$

Generally, energy efficiency is taken over a 24-h period and is called *all-day efficiency*. In such a case, Eq. (12.24) becomes

$$\text{All-day efficiency} = \frac{\text{output for 24 h}}{\text{input for 24 h}} \qquad (12.25)$$

We will use the term *efficiency* to mean power efficiency from now on. The output power is less than the input power because of losses: the I^2R losses in the windings and the hysteresis and eddy-current losses in the core. Thus, in terms of

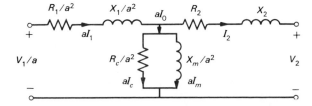

FIGURE 12.13
Equivalent circuit
referred to
secondary.

these losses, Eq. (12.23) may be more meaningfully expressed as

$$\text{Efficiency} = \frac{\text{input power} - \text{losses}}{\text{input power}}$$

$$= \frac{\text{output power}}{\text{output power} + \text{losses}}$$

$$= \frac{\text{output power}}{\text{output power} + I^2 R \text{ loss} + \text{core loss}} \qquad (12.26)$$

Obviously, $I^2 R$ loss is load-dependent, whereas the core loss is constant and independent of the load on the transformer. The next three examples show the voltage regulation and efficiency calculations for a transformer.

Example 12.6

A 150-kVA 2400/240-V transformer has the following parameters: $R_1 = 0.2 \, \Omega$, $R_2 = 0.002 \, \Omega$, $X_1 = 0.45 \, \Omega$, $X_2 = 0.0045 \, \Omega$, $R_c = 10,000 \, \Omega$, and $X_m = 1550 \, \Omega$, where the symbols are shown in Fig. 12.12. Refer the circuit to the primary. From this circuit, calculate the voltage regulation of the transformer at rated load with 0.8 lagging power factor.

Solution   The circuit referred to the primary is shown in Fig. 12.12. From the given data we have $V_2 = 240$ V, $a = 10$, and $\theta_2 = -\cos^{-1} 0.8 = -36.8°$.

$$I_2 = \frac{150 \times 10^3}{240}$$

$$= 625 \text{ A}$$

$$\frac{\mathbf{I}_2}{a} = \frac{I_2}{a} \underline{/\theta_2}$$

$$= 62.5\underline{/-36.8°}$$

$$= 50 - j37.5 \quad \text{A}$$

$$a\mathbf{V}_2 = 2400\underline{/0°}$$

$$= 2400 + j0 \quad \text{V}$$

$$a^2 R_2 = 0.2 \, \Omega \quad \text{and} \quad a^2 X_2 = 0.45 \, \Omega$$

Hence

$$\mathbf{E}_1 = (2400 + j0) + (50 - j37.5)(0.2 + j0.45)$$

$$= 2427 + j15$$

$$= 2427\underline{/0.35°} \text{ V}$$

$$\mathbf{I}_m = \frac{2427\underline{/0.35°}}{1550\underline{/90°}}$$

$$= 1.56\underline{/-89.65°}$$

$$= 0.0096 - j1.56 \quad \text{A}$$

$$I_c = \frac{2427 + j15}{10,000}$$

$$\cong 0.2427 + j0 \quad \text{A}$$

$$\mathbf{I}_0 = \mathbf{I}_c + \mathbf{I}_m$$
$$= 0.25 - j1.56 \quad \text{A}$$

$$\mathbf{I}_1 = \mathbf{I}_0 + \frac{\mathbf{I}_2}{a}$$

$$= 50.25 - j39.06$$
$$= 63.65\underline{/-37.85^\circ} \quad \text{A}$$

$$\mathbf{V}_1 = (2427 + j15) + (50.25 - j39.06)(0.2 + j0.45)$$
$$= 2455 + j30$$
$$= 2455\underline{/0.7^\circ} \quad \text{V}$$

$$\text{Percentage regulation} = \frac{V_1 - aV_2}{aV_2} \times 100$$

$$= \frac{2455 - 2400}{2400} \times 100$$

$$= 2.3\%$$

Example 12.7   Determine the efficiency of the transformer of Example 12.6 operating on rated load and 0.8 lagging power factor.

Solution

$$\text{Output} = 150 \times 0.8 = 120 \text{ kW}$$

$$\text{Losses} = I_1^2 R_1 + I_c^2 R_c + I_2^2 R_2$$
$$= (63.65)^2 \times 0.2 + (0.2427)^2 \times 10,000 + (625)^2 \times 0.002$$
$$= 2.18 \text{ kW}$$

$$\text{Input} = 120 + 2.18$$
$$= 122.18 \text{ kW}$$

$$\text{Efficiency} = \frac{120}{122.18}$$

$$= 98.2\%$$

Example 12.8   The transformer of Example 12.6 operates on full-load 0.8 lagging power factor for 12 h, on no-load for 4 h, and on half-full-load unity power factor for 8 h. Calculate the all-day efficiency.

Solution   The output for 24 h is

$$(150 \times 0.8 \times 12) + (0 \times 4) + (150 \times \tfrac{1}{2} \times 1.0 \times 8) = 2040 \text{ kWh}$$

The losses for 24 h are as follows.

The core loss is

$$(0.2427)^2 \times 10{,}000 \times 24 = 14.14 \text{ kWh}$$

The I^2R loss on full load for 12 h is

$$12[(63.65)^2 \times 0.2 + (625)^2 \times 0.002] = 19.1 \text{ kWh}$$

The I^2R loss on half-full load for 8 h is

$$8\left[\left(\frac{63.65}{2}\right)^2 \times 0.2 + \left(\frac{625}{2}\right)^2 \times 0.002\right] = 3.18 \text{ kWh}$$

The total losses for 24 h are

$$14.14 + 19.1 + 3.18 = 36.42 \text{ kWh}$$

The input for 24 h is

$$2040 + 36.42 = 2076.42 \text{ kWh}$$

The all-day efficiency is

$$\frac{2040}{2076.42} = 98.2\%$$

12.10   TRANSFORMER POLARITY

Polarities of a two-winding transformer identify the relative directions of induced voltages in the two windings. The polarities result from the relative directions in which the two windings are wound on the core. For operating transformers in parallel, it is necessary that we know the relative polarities. Polarities can be checked by a

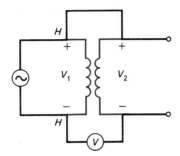

FIGURE 12.14
Polarity test on a
transformer.

simple test requiring only voltage measurements with the transformer on no-load. In this test, rated voltage is applied to one winding, and an electrical connection is made between one terminal from one winding and one from the other, as shown in Fig. 12.14. The voltage across the two remaining terminals (one from each winding) is then measured. If this measured voltage is larger than the input test voltage, the polarity is *additive;* if smaller, the polarity is *subtractive.*

A standard method of marking transformer terminals is as follows: The high-voltage terminals are marked $H_1, H_2, II_3, \ldots$, with H_1 being on the right-hand side of the case when facing the high-voltage side. The low-voltage terminals are designated $X_1, X_2, X_3, \ldots$, and X_1 may be on either side, adjacent to H_1 or diagonally opposite. The two possible locations of X_1 with respect to H_1 for additive and subtractive polarities are shown in Fig. 12.15. The numbers must be so arranged that the voltage difference between any two leads of the same set, taken in order from smaller to larger numbers, must be of the same sign as that between any other pair of the set taken in the same order. Furthermore, when the voltage is directed from H_1 to H_2, it must simultaneously be directed from X_1 to X_2. Additive polarities are required by the American National Standards Institute (ANSI) in large (>200-kVA) high-voltage (>8660-V) power transformers. Small transformers have subtractive polarities (which reduce voltage stress between adjacent leads).

Knowing the polarities of a transformer aids in obtaining a combination of several voltages from the transformer. Certain single-phase connections of transformers are illustrated in the next example.

FIGURE 12.15
(*a*) Subtractive
and (*b*) additive
polarities of a
transformer.

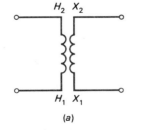

(*a*)

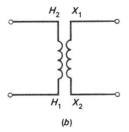

(*b*)

Example 12.9

Two 1150/115-V transformers are given. Using appropriate polarity markings, show the interconnections of these transformers for 2300/230-V operation and for 1150/230-V operation.

Solution   We mark the high-voltage terminals by H_1 and H_2 and the low-voltage terminals by X_1 and X_2 (as in Figs. 12.14 and 12.15). According to the ANSI convention, H_1 is marked on the terminal on the right-hand side of the transformer case (or housing) when facing the high-voltage side. With this nomenclature, the desired connections are shown in Fig. 12.16a and b.

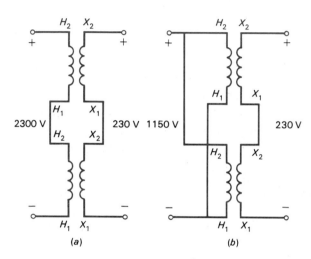

FIGURE 12.16
Example 12.9.

12.11   AUTOTRANSFORMERS

In contrast to the two-winding transformers considered so far, the autotransformer is a single-winding transformer having a tap brought out at an intermediate point. Thus, as shown in Fig. 12.17, ac is the single winding (wound on a laminated core) and b is the intermediate point where the tap is brought out. The autotransformer may be used either as a step-up or a step-down operation, as with a two-winding transformer. Considering a step-down arrangement, let the primary applied (terminal) voltage be V_1, resulting in a magnetizing current and a core flux ϕ_m. Let the secondary

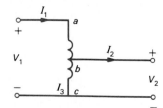

FIGURE 12.17
A step-down
autotransformer.

FIGURE 12.18
A three-winding
transformer.

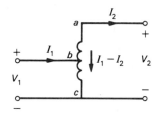

be open-circuited. Then the primary and secondary voltages obey the same rules as in a two-winding transformer, and we have

$$\frac{V_1}{V_2} = \frac{E_1}{E_2} = \frac{N_1}{N_2} = a$$

with $a > 1$ for step-down.

Finally, Fig. 12.18 shows a step-up autotransformer, including the various currents and voltages.

PROBLEMS

12.1 From the magnetic material characteristics shown in Fig. 12.6, determine the relative amplitude permeability at a flux density of 0.1 T for the following materials: M-19 and AISI 1020.

12.2 A certain air gap has an area of 1 in² and a length of 1 mm. What are the reluctance and permeance of this gap in SI units?

12.3 A flux of 0.02 Wb exists in the air gap of Prob. 12.2. Assuming no fringing effects, what is the magnetic potential (mmf) across the gap?

12.4 For the magnetic circuit shown in Fig. P12.4, determine the mmf of the exciting coil required to produce a flux density of 1.6 T in the air gap. Neglect leakage and fringing. In Fig. P12.4, the mean lengths of portions of the magnetic circuit of constant cross-sectional area are $l_{m1} = 45$ cm, $A_{m1} = 24$ cm²; $l_{m2} = 8$ cm, $A_{m2} = 16$ cm²; $l_g = 0.08$ cm, and $A_g = 16$ cm². The material is M-19.

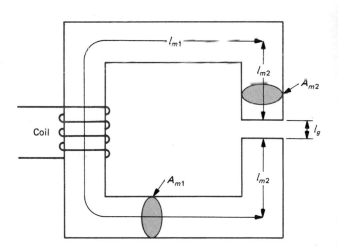

FIGURE P12.4

12.5 For the magnetic circuit shown in Fig. P12.5, determine the current in a 100-turn coil to establish a flux of 10^{-3} Wb in the air gap. Assume a fringing factor of 1.09, and neglect the leakage flux.

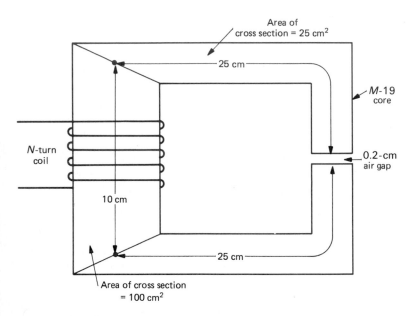

FIGURE P12.5

12.6 A 1-turn coil carries a current of 5 A. Determine the inductance of this coil if the total flux linked by the coil is 0.004 Wb.

12.7 A transformer is to be assembled using a magnetic core constructed of M-19 transformer steel laminations (see Fig. 12.6). The *net* cross-sectional area of the core is 10 cm^2. The transformer will be excited from a sinusoidal voltage source of 100 V (RMS) at 400 Hz. Determine the number of primary winding turns required so that the maximum core flux density at rated applied voltage is at the knee of the saturation curve for M-19 steel (1.2 T). Neglect winding resistance and core loss.

12.8 The mean magnetic length of the core of Prob. 12.7 is 40 cm. Determine the root-mean-square magnetizing current with rated voltage applied using the turns calculated in Prob. 12.7.

12.9 The sinusoidal flux in the core of a transformer is given by the expression $\phi = 0.0012 \sin 377t$ Wb; this flux links a primary winding of 150 turns. Determine the root-mean-square induced voltage in the primary winding.

12.10 In a certain transformer tested with open secondary (no-load), a core loss of 150 W is measured when the no-load current is 2.5 A and the induced voltage is 400 V (voltage and current are root-mean-square values). Neglect the winding resistance and leakage flux. Determine (*a*) the no-load power factor, (*b*) the root-mean-square magnetizing current, and (*c*) the root-mean-square-core-loss component of current.

12.11 A transformer is rated 10 kVA (kilovoltamperes), 400-V primary and 200-V secondary voltages. Neglecting the magnetizing and core-loss voltamperes, determine the turns ratio and the winding current ratings.

12.12 A transformer is rated at 3 kVA and 100/400 V. A 50-Ω load resistance is connected across the 400-V winding. With this load, determine the current in the two windings and the equivalent load resistance referred to the 100-V winding; assume that the transformer is ideal.

12.13 A transmission line is to be terminated with a resistance of 100 Ω. The actual load resistance is 500 Ω. Determine the required turns ratio of the transformer to match the actual load to the transmission line.

12.14 A transformer rated 500 kVA, 2400/120 V, and 60 Hz has no-load loss (at rated voltage) of 1600 W and a short-circuit loss (at rated current) of 7500 W. Determine the efficiency of this transformer under the following conditions of load: (a) Rated current at 0.8 power factor lagging. (b) Three hundred kilowatts at 0.8 power factor leading. (c) One hundred kilowatts at 0.8 power factor lagging.

12.15 A load of 12 kVA at 0.7 power factor lagging is to be supplied at 110 V from a 120-V supply by an autotransformer. Specify the voltage and current ratings of each section of the autotransformer.

12.16 A transformer with additive polarity is rated 15 kVA, 2300/115 V, 60 Hz. Under rated conditions, the transformer has an excitation loss of 75 W and short-circuit loss of 250 W. The transformer is to be connected as an autotransformer to transform 2300 to 2415 V. With a load of 0.8 power factor lagging, what volt-ampere load can be supplied without exceeding the current rating of any winding? Determine the efficiency at this load.

12.17 An ideal transformer is rated 2400/240 V. A certain load of 50 A, unity power factor is to be connected to the low-voltage winding and must have exactly 200 V across it. With 2400 V applied to the high-voltage winding, what resistance must be added in series with the transformer if it is (a) located in the low-voltage winding, (b) located in the high-voltage winding?

12.18 Two 1150/115-V transformers are given. Show the interconnections of these transformers for (a) 2300/115-V operation and (b) 1150/115- V operation. Indicate the polarity markings.

12.19 A 25-kVA 440/220-V 60-Hz transformer has a constant core loss of 710 W. The equivalent resistance and reactance referred to the primary are each 0.16 Ω. Plot the following curves: (a) core loss versus primary current, (b) I^2R loss versus primary current, and (c) efficiency versus primary current—the primary current to range from 0 to 100 A. What is the I^2R loss when the efficiency of the transformer is maximum? (*Note*: This is a special case of the general rule that the efficiency of a transformer is maximum when the total I^2R losses in the two windings equal the core losses.)

12.20 A 75-kVA transformer has an iron loss of 1 kW and a full-load copper loss of 1 kW. Calculate the transformer efficiency at unity power factor at (a) full load and (b) half-full load.

12.21 The transformer of Prob. 12.20 operates on full load at unity power factor for 8 h, on no-load for 8 h, and on one-half load at unity power factor for 8 h, during 1 day. Determine the all-day efficiency.

12.22 The maximum efficiency of a 100-kVA transformer is 98.4 percent and occurs at 90 percent of full load. Calculate the efficiency of the transformer at unity power factor at (a) full load and (b) 50 percent of full load.

REFERENCES

1 S. A. Nasar, *Schaum's Outline of Theory and Problems in Electric Machines and Electromechanics*, McGraw-Hill, New York, 1981.

2 J. J. Cathey and S. A. Nasar, *Schaum's Outline of Theory and Problems in Basic Electrical Engineering*, McGraw-Hill, New York, 1983.

3 S. A. Nasar and L. E. Unnewehr, *Electromechanics and Electric Machines*, 2d ed., Wiley, New York, 1983.

4 S. A. Nasar, *Electric Machines and Transformers*, Macmillan, New York, 1983.

DC Machines

In Chap. 12 we studied the transformer, which is in a sense an *energy transfer* device: energy is transferred from the primary to the secondary. During the energy transfer process, the form of energy remains unchanged; i.e., the energy at the input (primary) and output (secondary) terminals is electrical. In a rotating electric machine, on the other hand, electrical energy is converted to mechanical form and vice versa, depending on the mode of operation of the machine. An electric *motor* converts electrical energy to mechanical energy, whereas an electric *generator* converts mechanical energy to electrical energy. For this reason, electric machines are also called *electromechanical energy converters*.

In essence, the process of electromechanical energy conversion can be expressed as

$$\text{Electrical energy} \quad \underset{\text{generator}}{\overset{\text{motor}}{\rightleftarrows}} \quad \text{mechanical energy}$$

Figure 13.1 schematically represents an ideal electric machine, for which we have, over a certain time interval Δt,

$$vi\,\Delta t = T_e \omega_m \,\Delta t$$
$$vi = T_e \omega_m \tag{13.1}$$

FIGURE 13.1
A general
representation of
an electric
machine.

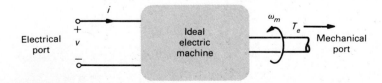

where, respectively, v and i are the voltage and the current at the electrical port and T_e and ω_m are the torque [in newton-meters (N·m)] and the angular rotational velocity [in radians per second (rad/s)] at the mechanical port. We wish to reiterate that Eq. (13.1) is valid for an ideal machine, in that the machine is lossless.

The first machine devised for electromechanical energy conversion was the dc machine—in the form of the Faraday disk.

13.1 THE FARADAY DISK AND FARADAY'S LAW

Applying his law of electromagnetic induction, Michael Faraday demonstrated in 1832 that if a copper disk with sliding contacts, or brushes, mounted at the rim and at the center of the disk is rotated in an axially directed magnetic field (produced by a permanent magnet), then a voltage will be available at the brushes. Such a machine is shown in Fig. 13.2 and is commonly known as the *Faraday disk*, or *homopolar generator* (another name for it is the *acyclic* machine); this device is also capable of operating as a motor. It is called a *homopolar* machine because the conductors (which may be thought of as spokes in the disk) are under the influence of the magnetic field of one polarity at all times. In contrast, the moving conductors in *heteropolar* machines are under the influence of magnetic fields of opposite polarities in an alternating fashion, as we shall soon see.

Faraday's law of electromagnetic induction states that an emf is induced in a circuit placed in a magnetic field if either (1) the magnetic flux linking the circuit is time-varying or (2) there is a relative motion between the circuit and the magnetic field such that the conductors comprising the circuit cut across the magnetic flux lines. The first form of the law, stated as (1) above, is the basis of operation of transformers. The second form, stated as (2) above, forms the basic principle of operation of electric generators, including the homopolar generator.

Consider a conductor of length l located at right angles to a uniform magnetic field of flux density B, as shown in Fig. 13.3. Let the conductor be connected with

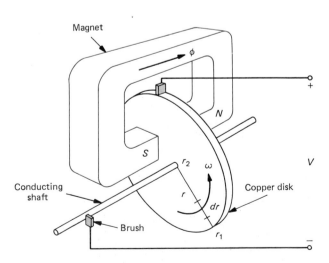

FIGURE 13.2
Faraday disk, or homopolar machine.

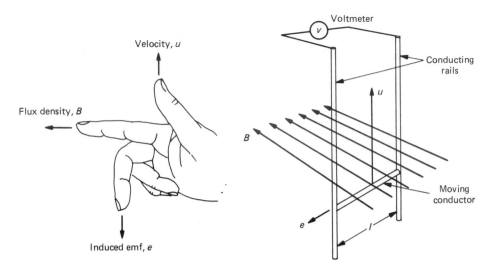

FIGURE 13.3
The right-hand
rule generator
action.

fixed external connections to form a closed circuit. These external connections are
shown in the form of conducting rails in Fig. 13.3. The conductor can slide on the
rails, and thereby the flux linking the circuit changes. Hence, according to Faraday's
law, an emf will be induced in the circuit, and a voltage v will be measured by the
voltmeter. If the conductor moves with a velocity u [in meters per second (m/s)] in
a direction at right angles to B and l both, then the area swept by the conductor in
1 s is lu. The flux per unit time in this area is Blu, which is also the flux linkage (since,
in effect, we have a single-turn coil formed by the conductor, the rails, and the
voltmeter). In other words, flux linkage per unit time is Blu, which is thus the induced
emf e. We write this in equation form as

$$e = Blu \tag{13.2}$$

This form of Faraday's law is also known as the *flux-cutting rule*. Stated in words,
an emf e, as given by Eq. (13.2), is induced in a conductor of length l if it "cuts"
magnetic flux lines of density B by moving at right angles to B at a velocity u (at
right angles to l). The mutual relationships between e, B, l, and u are given by the
right-hand rule, as shown in Fig. 13.3.

Example 13.1
A conducting disk of 0.5-m radius rotates at 1200 revolutions per minute (r/min) in a uniform
magnetic field of 0.4 T. Calculate the voltage available between the rim and the center of the
disk.

Solution   Referring to Eq. (13.2), in this problem we have $B = 0.4$ T, $l = 0.5$ m, and $u = r\omega$,
where $\omega = 2\pi n/60$ radians per second (rad/s) is the angular velocity of the disk. Substituting
$n = 1200$ r/min and $r = 0.5$ m, we obtain the linear velocity of the rim as

$$u_{rim} = 0.5 \times 2\pi \times \frac{1200}{60} = 20\pi \text{ m/s}$$

But the linear velocity of the center of the disk is zero. Hence the average velocity of the disk (or a radial conductor) is given by

$$u = \tfrac{1}{2}(20\pi + 0) = 10\pi \text{ m/s}$$

Hence, from Eq. (13.2)

$$e = 0.4 \times 0.5 \times 10\pi$$
$$= 6.28 \text{ V}$$

Alternatively, we may solve this problem by integration. Referring to Fig. 13.2, the voltage, de, induced in the element dr is given by

$$de = B\, dr(r\omega)$$

Or, total emf becomes

$$e = B\omega \int_{r_2}^{r_1} r\, dr = \tfrac{1}{2}B\omega(r_1^2 - r_2^2)$$

Substituting numerical values yields (with $\omega = 2\pi n/60$)

$$e = \tfrac{1}{2} \times 0.4 \times \frac{2\pi \times 1200}{60}(0.5^2 - 0) = 6.28 \text{ V}$$

Let us now summarize the salient points of this section:

1 The Faraday disk operates on the principle of electromagnetic induction enunciated by Faraday.

2 There are two forms of expression of Faraday's law—time rate of change of flux linkage (with a circuit) or flux cut (by a conductor) results in an induced emf, but the latter is contained in the former.

3 Electric generators operate on the flux-cutting principle.

4 Example 13.1 shows that a rather small voltage (6.28 V) is induced in a disk of large diameter (1 m) rotating at a reasonably high speed (1200 r/min).

Thus, the Faraday disk in its primitive form is not a practical type of dc generator. Indeed, present-day dc generators have little resemblance to the Faraday disk.

13.2 THE HETEROPOLAR, OR CONVENTIONAL, DC MACHINE

Consider an N-turn coil rotating at a constant angular velocity ω in a uniform magnetic field of flux density B, as shown in Fig. 13.4a. Let l be the axial length of

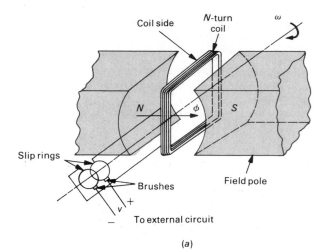

(a)

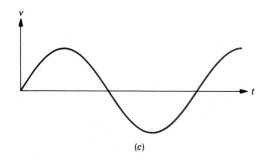

(b)

(c)

FIGURE 13.4 (a) An elementary generator.
(b) Resolution of velocities into parallel and
perpendicular components. (c) Voltage waveform at the
brushes.

the coil and r its radius. The emf induced in the coil can be found by an application of Eq. (13.2). However, we must be careful in determining the u of Eq. (13.2). Recall from the last section that u is the velocity perpendicular to B. In the system under consideration, we have a rotating coil. Thus, u in Eq. (13.2) corresponds to that component of velocity which is perpendicular to B. We illustrate the components of velocities in Fig. 13.4b, where the tangential velocity $u_t = r\omega$ has been resolved into a component u_1 along the direction of the flux and another component u_2 across (or perpendicular to) the flux. This latter component cuts the magnetic flux and is the component responsible for the emf induced in the coil. The u in Eq. (13.2) should thus be replaced by u_2. The next term in Eq. (13.2) that needs careful consideration is l, the effective length of the conductor. For the N-turn coil of Fig. 13.4a, we effectively have $2N$ conductors in series (since each coil side has N conductors and there are two coil sides). If l_1 is the length of each conductor, then the total effective length of the N-turn coil is $2Nl_1$, which should be substituted for l in Eq. (13.2). The form of Eq. (13.2) for the N-turn coil thus becomes

$$e = B(2Nl_1)u_2 \tag{13.3}$$

From Fig. 13.4b we have

$$u_2 = u_t \sin \theta = r\omega \sin \theta \tag{13.4}$$

If the coil rotates at a constant angular velocity ω, then

$$\theta - \omega t \tag{13.5}$$

Consequently, Eqs. (13.3) to (13.5) yield

$$e = 2BNl_1 r\omega \sin \omega t = E_m \sin \omega t \tag{13.6}$$

where $E_m = 2BNl_1 r\omega$. A plot of Eq. (13.6) is shown in Fig. 13.4c. The conclusion is that a sinusoidally varying voltage will be available at the slip rings, or bushes, of the rotating coil shown in Fig. 13.4a. The brushes reverse polarities periodically.

To obtain a unidirectional voltage at the coil terminals, we replace the slip rings of Fig. 13.4a with the commutator segments shown in Fig. 13.5a. It can be readily verified, by applying the right-hand rule, that in the arrangement shown in Fig. 13.5a the brushes will maintain their polarities regardless of the position of the coil. In other words, brush a will always be positive, and brush b always negative, for the given relative polarities of the flux and the direction of rotation. Thus, the arrangement shown in Fig. 13.5 forms an elementary heteropolar dc machine. The main characteristic of a heteropolar dc machine is that the emf induced in a conductor, or the current flowing through it, has its direction reversed as it passes from a north-pole to a south-pole region. This reversal process is known as *commutation* and is accomplished by the commutator-brush mechanism, which also serves as a connection to the external circuit. The voltage available at the brushes will be of the form shown in Fig. 13.5b.

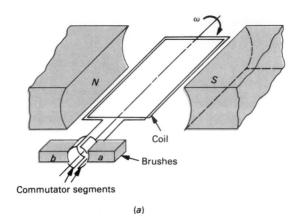

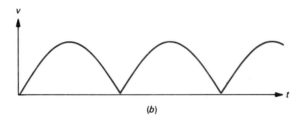

FIGURE 13.5
(*a*) An elementary dc generator.
(*b*) Output voltage at the brushes.

The natural question to ask at this point is: What have we gained, compared to the Faraday disk, by introducing the complications of a commutator? The answer lies in the fact that a full range of ratings can be realized from a heteropolar cylindrical configuration. Recall from Example 13.1 that we could get only about 6 V at the terminals of a homopolar machine. Obtaining much higher voltages is no problem with a heteropolar machine, as we will presently see. As a matter of convention, we will call the dc heteropolar commutator machine simply a dc machine (unless otherwise stated).

Before leaving the subject of the homopolar and elementary heteropolar machines, let us recall Eq. (13.1), according to which these machines must also be able to operate as motors. The production of a unidirectional torque—and hence a machine's operation as a dc motor—can be verified by referring to Fig. 13.6.

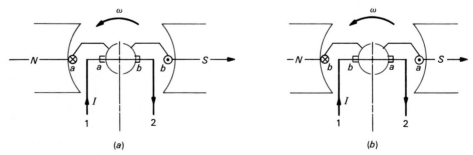

FIGURE 13.6   Production of a unidirectional torque and operation of an elementary dc motor: (*a*) position of conductor *a* under *N* pole; (*b*) position of conductor *a* under *S* pole.

FIGURE 13.7
The left-hand
rule.

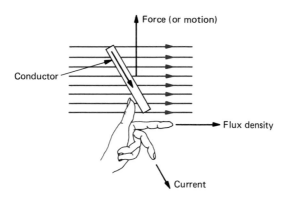

Conductors *a* and *b* are the two sides of a coil. Thus *a* and *b* are connected in series. Notice from Fig. 13.6*a* that the current enters into conductor *a* through brush *a* and leaves conductor *b* through brush *b*. By applying the left-hand rule (Fig. 13.7), the conductors will experience a force to produce a counterclockwise rotation. After half a rotation, the conductors interchange their respective positions, as shown in Fig. 13.6*b*. Now the current in conductor *b* is again in such a direction that the force on the coil will tend to rotate it in a counterclockwise direction. In other words, the torque is unidirectional and is independent of conductor position.

13.3 CONSTRUCTIONAL DETAILS

In order that we may subsequently consider some conventional dc machines, it is best that we briefly study the constructional features of their various parts and discuss the usefulness of these parts.

We observe from the last section that the basic elements of a dc machine are the rotating coil, a means for the production of flux, and the commutator-brush arrangement. In a practical dc machine the coil is replaced by the *armature winding* mounted on a cylindrical magnetic structure. The flux is provided by the *field winding* wound on field poles. Generally, the armature winding is placed on the rotating member, the rotor, and the field winding is on the stationary member, the *stator*, of the dc machine. The schematic of Fig. 13.8 shows most of the important parts of a dc machine.

The field poles, mounted on the stator, carry the field windings. Some machines carry more than one separate field winding on the same core. The cores and faces of the poles are built of sheet-steel laminations. The pole faces have to be laminated because of their proximity to the armature windings. Because the field windings carry direct current, it is not necessary to have the pole cores laminated; nevertheless, the use of laminations for the cores as well as for the pole faces facilitates assembly.

The rotor (or armature) core, which carries the rotor or armature windings, is generally made of sheet-steel laminations. These laminations are stacked together to form a cylindrical structure. On its outer periphery, the armature (rotor) has the *slots* in which the armature coils that make up the armature winding are located. For mechanical support, protection from abrasion, and greater electrical insulation,

FIGURE 13.8
Parts of a dc
machine.

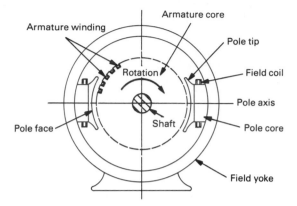

nonconducting slot liners are often wedged in between the coils and the slot walls. The projections of magnetic material between the slots are called *teeth*. A typical slot/teeth geometry for a large dc machine is shown in Fig. 13.9.

The commutator is made of hard-drawn copper segments insulated from one another by mica. The details of the commutator assembly are given in Fig. 13.10. The armature windings are connected to the commutator segments or bars, over which the carbon brushes slide and serve as leads for electrical connection.

The armature winding may be a lap winding (Fig. 13.11*a*) or a wave winding (Fig. 13.11*b*), and the various coils forming the armature winding may be connected in a series-parallel combination. In practice, the armature winding is housed as two layers in the slots of the armature core. In large machines the coils are preformed in the shapes shown in Fig. 13.12 and are interconnected to form an armature winding.

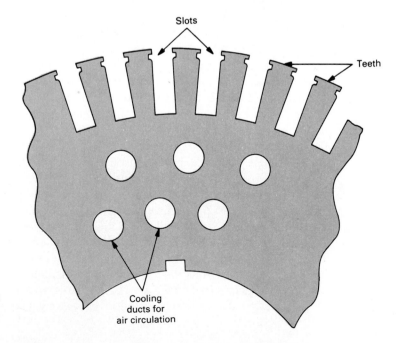

FIGURE 13.9
Portion of an
armature
lamination of a dc
machine showing
slots and teeth.

FIGURE 13.10
Details of
commutator
assembly.

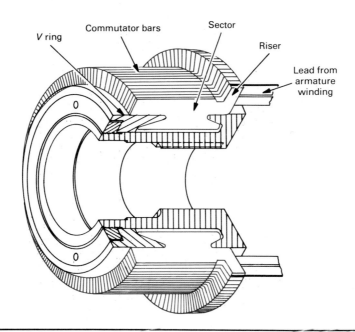

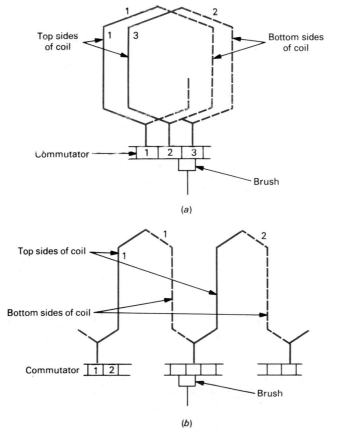

FIGURE 13.11
Elements
of (*a*) a lap winding;
(*b*) a wave winding. Odd-
numbered conductors are
at the top and even-
numbered conductors are
at the bottom of the slots.

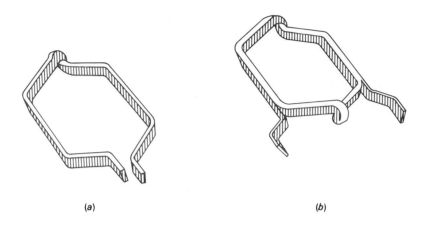

(a)   (b)

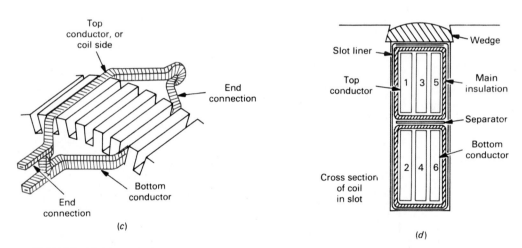

(c)   (d)

FIGURE 13.12   (a) Coil for a lap winding, made of a single bar. (b) Multiturn coil for a wave winding. (c) A coil in a slot. (d) Slot details, showing several coils arranged in two layers.

The coils span approximately a pole pitch—the distance between two consecutive poles.

13.4   CLASSIFICATION ACCORDING TO FORMS OF EXCITATION

Conventional dc machines having a set of field windings and armature windings can be classified, on the basis of mutual electrical connections between the field and

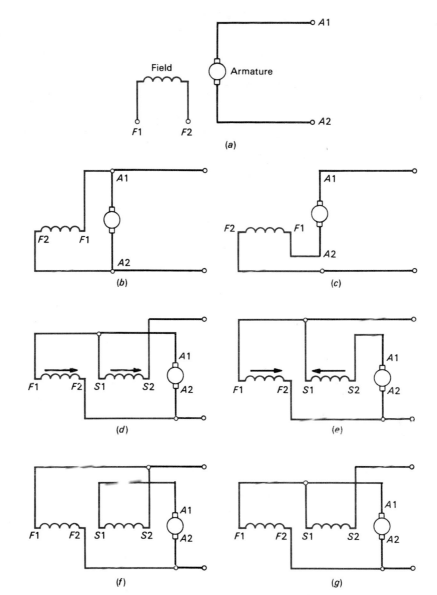

FIGURE 13.13
Classification of
dc machines:
(a) separately
excited;
(b) shunt;
(c) series;
(d) cumulative
compound;
(e) differential
compound;
(f) long-shunt;
(g) short-shunt.

armature windings, as shown in Fig. 13.13. These interconnections of field and armature windings essentially determine the machine's operating characteristics.

13.5 PERFORMANCE EQUATIONS

The three quantities of greatest interest in evaluating the performance of a dc machine are (1) the induced emf, (2) the electromagnetic torque developed by the machine, and (3) the speed corresponding to (1) and/or (2). We now derive the equations which enable us to determine the above quantities.

EMF Equation   The emf equation yields the emf induced in the armature of a dc machine. The derivation follows directly from Faraday's law of electromagnetic induction, according to which the emf induced in a moving conductor is the flux cut by the conductor per unit time.

Let us define the following symbols:

Z = number of active conductors on armature
a = number of parallel paths in armature winding
p = number of field poles
ϕ = flux per pole, Wb
n = speed of rotation of the armature, r/min

Thus, with reference to Fig. 13.14,

Flux cut by one conductor in 1 rotation = ϕp
Flux cut by one conductor in n rotations = ϕnp
Flux cut per second by one conductor = $\phi np/60$
Number of conductors in series = Z/a
Flux cut per second by Z/a conductors = $(\phi np/60)(Z/a)$

Hence, the emf induced in the armature winding is

$$E = \frac{\phi nZ}{60}\frac{p}{a} \qquad (13.7)$$

Equation (13.7) is known as the *emf equation* of a dc machine. In Eq. (13.7) we have separated p/a from the rest of the terms because p and a are related to each other for the two types of windings. For a lap winding $p = a$, and $a = 2$ in a wave winding.

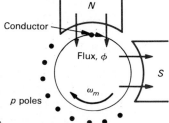

FIGURE 13.14
A conductor
rotating at a speed
n in a field of p poles.

Example 13.2   Determine the voltage induced in the armature of a dc machine running at 1750 r/min and having four poles. The flux per pole is 25 mWb, and the armature is lap-wound with 728 conductors.

Solution   Since the armature is lap-wound, $p = a$, and Eq. (13.7) becomes

$$E = \frac{\phi nZ}{60}$$

$$= \frac{25 \times 10^{-3} \times 1750 \times 728}{60}$$

$$= 530.8 \text{ V}$$

Torque Equation   The mechanism of torque production in a dc machine has been considered earlier (see Fig. 13.6), and the electromagnetic torque developed by the armature can be evaluated either from the *BlI* rule† or from Eq. (13.1). Here we choose the latter approach. Regardless of the approach, it is clear that for torque production we must have a current through the armature, as this current interacts with the flux produced by the field winding. Let I_a be the armature current and E the voltage induced in the armature. Thus, the power at the armature electrical port (see Fig. 13.1) is EI_a. Assuming that this entire electric power is transformed to mechanical form, we rewrite Eq. (13.1) as

$$EI_a = T_e \omega_m \tag{13.8}$$

where T_e is the electromagnetic torque developed by the armature and ω_m is its angular velocity in radians per second. The speed n r/min and ω_m rad/s are related by

$$\omega_m = \frac{2\pi n}{60} \tag{13.9}$$

Hence, from Eqs. (13.7) to (13.9) we obtain

$$\frac{\omega_m}{2\pi} \phi Z \frac{p}{a} I_a = T_e \omega_m$$

which simplifies to

$$T_e = \frac{Zp}{2\pi a} \phi I_a \tag{13.10}$$

which is known as the *torque equation*. An application of Eq. (13.10) is illustrated by the next example.

---

† See Eq. (12.1).

Example 13.3   A lap-wound armature has 576 conductors and carries an armature current of 123.5 A. If the flux per pole is 20 mWb, calculate the electromagnetic torque developed by the armature.

Solution   For the lap winding, we have $p = a$. Substituting this and the other given numerical values into Eq. (13.10) yields

$$T_e = \frac{576}{2\pi} \times 0.02 \times 123.5$$

$$= 226.4 \text{ N} \cdot \text{m}$$

Example 13.4   If the armature of Example 13.3 rotates at an angular velocity of 123.5 rad/s, what is the induced emf in the armature?

Solution   To solve this problem, we use Eq. (13.8) rather than Eq. (13.7). From Example 13.3 we have $I_a = 123.5$ A and $T_e = 226.4$ N·m. Hence, Eq. (13.8) gives

$$E \times 123.5 = 226.4 \times 123.5$$

or

$$E = 226.4 \text{ V}$$

Speed Equation and Back EMF   The emf and torque equations discussed above indicate that the armature of a dc machine, whether operating as a generator or as a motor, will have an emf induced while rotating in a magnetic field and will develop a torque if the armature carries a current. In a generator, the induced emf is the internal voltage available from the generator; when the generator supplies a load, the armature carries a current and develops a torque. This torque opposes the prime-mover (such as a diesel engine) torque.

In motor operation, the developed torque of the armature supplies the load connected to the shaft of the motor, and the emf induced in the armature is termed the *back emf*. This emf opposes the terminal voltage of the motor.

Referring to Fig. 13.15, which shows the equivalent circuit of a separately excited dc motor running at speed n while taking an armature current I_a at a voltage V_t, we have from this circuit

$$V_t = E + I_a R_a \tag{13.11}$$

Substituting Eq. (13.7) in Eq. (13.11), putting $k_1 = Zp/(60a)$, and solving for n, we get

$$n = \frac{V_t - I_a R_a}{k_1 \phi} \tag{13.12}$$

FIGURE 13.15
Equivalent circuit
of a separately
excited dc motor.

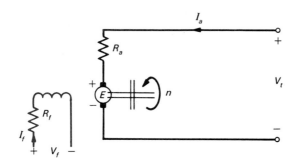

This equation is known as the *speed equation*, because it contains all the factors which affect the speed of a motor. Later we consider the influence of these factors on the speed of dc motors. For the present, let us focus our attention on k_1 and ϕ. The term k_1 replacing $Zp/(60a)$ is a design constant in the sense that Z, p, and a cannot be altered once the machine has been built. The magnetic flux ϕ is primarily controlled by the field current I_f (Fig. 13.15). If the magnetic circuit is unsaturated, then ϕ is directly proportional to I_f. Thus, we may write

$$\phi = k_f I_f \tag{13.13}$$

where k_f is a constant. We may combine Eqs. (13.12) and (13.13) to obtain

$$n = \frac{V_t - I_a R_a}{k I_f} \tag{13.14}$$

where $k = k_1 k_f$ is a constant. This form of the speed equation is more meaningful because all the quantities in Eq. (13.14) can be conveniently measured; in contrast, it is very difficult to measure ϕ in Eq. (13.12).

Example 13.5   A 250-V shunt motor has an armature resistance of 0.25 Ω and a field resistance of 125 Ω. At no load, the motor takes a line current of 5.0 A while running at 1200 r/min. If the line current at full load is 52.0 A, what is the full-load speed?

Solution   The motor equivalent circuit is shown in Fig. 13.16. The field current is

$$I_f = \tfrac{250}{125}$$
$$= 2.0 \text{ A}$$

At no load, the armature current is

$$I_a = 5.0 - 2.0$$
$$= 3.0 \text{ A}$$

FIGURE 13.16
Equivalent circuit
of a shunt motor.

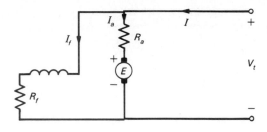

the back emf is

$$E_1 = V_t - I_a R_a$$
$$= 250 - (3 \times 0.25)$$
$$= 249.25 \text{ V}$$

and the speed is

$$n_1 = 1200 \text{ r/min} \qquad \text{given}$$

At full load, the armature current is

$$I_a = 52.0 - 2.0$$
$$= 50.0 \text{ A}$$

the back emf is

$$E_2 = 250 - (50 \times 0.25)$$
$$= 237.5 \text{ V}$$

and the speed is

$$n_2 = \text{unknown}$$

Now, since

$$\frac{n_2}{n_1} = \frac{E_2}{E_1}$$
$$= \frac{237.5}{249.25}$$

thus

$$n_2 = \frac{237.5}{249.25} \times 1200$$
$$= 1143 \text{ r/min}$$

13.6 EFFECTS OF SATURATION— BUILDUP OF VOLTAGE IN A SHUNT GENERATOR

Saturation plays a very important role in governing the behavior of dc machines. It is extremely difficult, however, to include the effects of saturation quantitatively. For the time being, then, let us consider qualitatively the consequences of saturation on the operation of a self-excited shunt generator.

A self-excited shunt machine is shown in Fig. 13.16. Writing the steady-state equations for the operation of the machine as a generator from the circuit of Fig. 13.16, we have

$$V_t = R_f I_f$$

and

$$E = V_t + I_a R_a = I_f R_f + I_a R_a$$

These equations are represented by the straight lines shown in Fig. 13.17a. Notice that the voltages V_t and E will keep building up and that no equilibrium point can

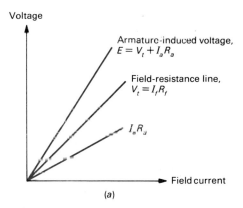

(a)

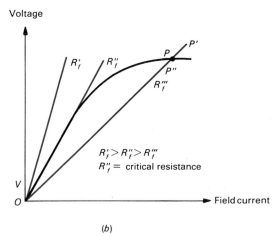

(b)

FIGURE 13.17
No-load characteristic of a shunt generator: (a) no stable operating point for the shunt generator; (b) stable no-load voltage of a shunt generator.

be reached. On the other hand, if we include the effect of saturation, as in Fig. 13.17*b*, point *P* defines the equilibrium, because at this point the field-circuit resistance line intersects the saturation curve. A deviation from *P* to *P'* or *P''* would immediately show that at *P'* the voltage drop across the field is greater than the induced voltage, which is not possible, and at *P''* the induced voltage is greater than the field-circuit voltage drop, which is not possible either.

Figure 13.17*b* shows some residual magnetism, as measured by the small voltage *OV*. Evidently, without it the shunt generator will not build up any voltage. Also shown in Fig. 13.17*b* is the critical resistance. A field-circuit resistance greater than the critical resistance (for a given speed) would not let the shunt generator build up any appreciable voltage. Finally, we should ascertain that the polarity of the field winding is such that a current through it produces a flux which aids the residual flux. If it does not, the two fluxes tend to neutralize and the machine voltage will not build up. To summarize, the conditions for the buildup of a voltage in a shunt generator are: the presence of residual flux, a field-circuit resistance less than the critical resistance, and an appropriate polarity of the field winding.

13.7 GENERATOR CHARACTERISTICS

The no-load and load characteristics of dc generators are usually of interest in determining their potential applications. Of the two, load characteristics are of greater importance. As the names imply, no-load and load characteristics, respectively, correspond to the behavior of the machine when it is supplying no power (open-circuited, in case of a generator) and when it is supplying power to an external circuit.

The only no-load (or open-circuit) characteristics which are meaningful are those of the shunt and separately excited generators. In the last section we discussed the no-load characteristic of a shunt generator as a voltage buildup process. For the

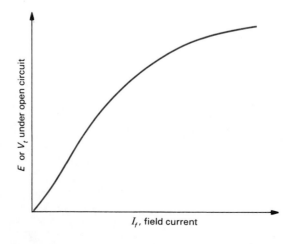

FIGURE 13.18
No-load
characteristic of a
separately excited
generator.

FIGURE 13.19
Load
characteristics of
dc generators.

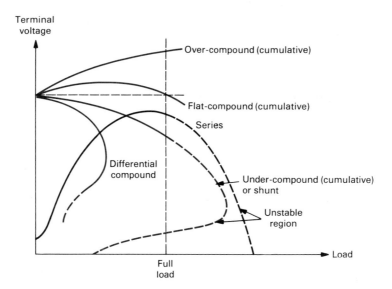

separately excited generator, the no-load characteristic corresponds to the magnetization, or saturation, characteristic—the variation of E (or V_t) under open-circuit conditions as a function of the field current I_f. This characteristic is illustrated in Fig. 13.18.

The load characteristics of self-excited generators are shown in Fig. 13.19. The shunt generator has a characteristic similar to that of a separately excited generator, except for the cumulative effect mentioned in Sec. 13.6. If the shunt generator is loaded beyond a certain point, it breaks down, in that the terminal voltage collapses. In a series generator, the load current flows through the field winding; this implies that the field flux, and hence the induced emf, increases with the load until the core begins to saturate magnetically. Thus, a load beyond a certain point would result in a collapse of the terminal voltage of the series generator, too. Compound generators have the combined characteristics of shunt and series generators. In a differential compound generator, the shunt and series fields are in opposition; hence, the terminal voltage drops very rapidly with the load. On the other hand, cumulative compound generators have shunt and series fields aiding each other. The two field mmf's may be adjusted such that the terminal voltage on full load is less than the no-load voltage, as in an under-compound generator; or the full-load voltage may be equal to the no-load voltage, as in a flat-compound generator. Finally, the terminal voltage on full load may be greater than the no-load voltage, as in an over-compound generator.

Example 13.6   A 50-kW 250-V short-shunt compound generator has the following data: $R_a = 0.06 \, \Omega$, $R_{se} = 0.04 \, \Omega$, and $R_f = 125 \, \Omega$. Calculate the induced armature emf at rated load and terminal voltage. Take 2 V as the total brush-contact drop.

FIGURE 13.20
Example 13.6.

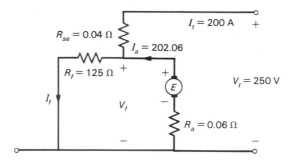

Solution   The equivalent circuit of the generator is shown in Fig. 13.20, from which:

$$I_t = \frac{50 \times 10^3}{250}$$

$$= 200 \text{ A}$$

$$I_t R_{se} = 200 \times 0.04$$

$$= 8 \text{ V}$$

$$V_f = 250 + 8$$

$$= 258 \text{ V}$$

$$I_f = \tfrac{258}{125}$$

$$= 2.06 \text{ A}$$

$$I_a = 200 + 2.06$$

$$= 202.06 \text{ A}$$

$$I_a R_a = 202.06 \times 0.06$$

$$= 12.12 \text{ V}$$

$$E = 250 + 12.12 + 8 + 2$$

$$= 272.12 \text{ V}$$

13.8   MOTOR CHARACTERISTICS

Among the various characteristics of dc motors, their torque-speed characteristics are most important from a practical standpoint. The torque and speed equations derived earlier govern the motor characteristics. From these equations (and after accounting for magnetic saturation) it follows that the shunt, series, and compound motors have the torque-speed characteristics of the forms shown in Fig. 13.21. The governing equations also yield the motor speed-current characteristics of Fig. 13.22.

In Fig. 13.21, we have also shown the developed torque versus speed for shunt and series motors. As we recall from Sec. 13.1, developed power is simply the product of developed torque (N·m) and angular speed (rad/s).

FIGURE 13.21
Torque-speed
characteristics of
dc motors.

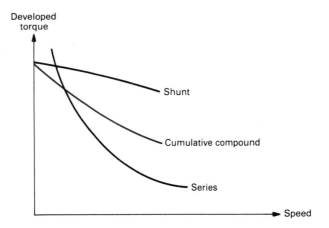

FIGURE 13.21
Torque-speed
characteristics of
dc motors.

FIGURE 13.22
Speed-current
characteristics.

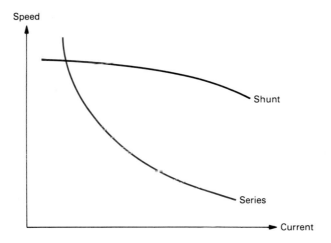

13.9 STARTING OF DC MOTORS

In addition to certain operational conveniences, the basic requirements for satis-factory starting of a dc motor are (1) sufficient starting torque and (2) armature current, within safe limits, for successful commutation and for preventing the armature from overheating. The second requirement is obvious from the speed equation, according to which the armature current I_a is given by $I_a = V_t/R_a$ when the motor is at rest (n or $\omega_m = 0$). A typical 50-hp 230-V motor having an armature resistance of 0.05 Ω, if connected across 230 V, draws a 4600-A current. This current is evidently too large for the motor, which might be rated to take 180 A on full load. Commonly, double the full-load current is allowed to flow through the armature at the time of starting. For the motor under consideration, therefore, an external resistance $R_x = [230/(2 \times 180)] - 0.05 = 0.59\ \Omega$ must be inserted in series with the armature to limit I_a to within double the rated value.

In practice, the necessary starting resistance is provided by means of a pushbutton starter.

Example 13.7   A 230-V shunt motor has an armature resistance of 0.05 Ω and a field resistance of 75 Ω. The motor draws a 7-A line current while running light at 1120 r/min. The line current at a certain load is 46 A. (*a*) What is the motor speed at this load? (*b*) At this load, if the field-circuit resistance is increased to 100 Ω, what is the new speed of the motor? Assume that the line current remains unchanged.

Solution   (*a*) On no load

$$n_o = 1120 \text{ r/min} \qquad \text{given}$$

$$I_f = \frac{230}{75}$$
$$= 3.07 \text{ A}$$

$$I_a = 7 - 3.07$$
$$= 3.93 \text{ A}$$

The speed equation gives

$$1120 = \frac{230 - (3.93 \times 0.05)}{3.07k}$$

or

$$k = 0.0668$$

On load (with $R_f = 75\ \Omega$)

$$I_f = 3.07 \text{ A}$$

$$I_a = 46 - 3.07$$
$$= 42.93 \text{ A}$$

$$n = \frac{230 - (42.93 \times 0.05)}{3.07 \times 0.0668}$$

$$= 1111 \text{ r/min}$$

(*b*) On load (with $R_f = 100\ \Omega$)

$$I_f = \frac{230}{100} = 2.3 \text{ A}$$

$$I_a = 46 - 2.3$$
$$= 43.7 \text{ A}$$

$$n = \frac{230 - (43.7 \times 0.05)}{2.3 \times 0.0668}$$

$$= 1483 \text{ r/min}$$

Example 13.8
Refer to part (*a*) of the solution to Example 13.7. The no-load conditions remain unchanged. On load, the line current remains at 46 A, but a 0.1-Ω resistor is inserted in the armature. Determine the speed of the motor. Also determine the loss in the 0.1-Ω resistor.

Solution In this case we have (from Example 13.7)

$$I_f = 3.07 \text{ A}$$

$$I_a = 42.93 \text{ A}$$

$$k = 0.0668$$

Thus (with $R_a = 0.05 + 0.1 = 0.15 \ \Omega$), $I_a^2 R_a = (42.93)^2(0.1) = 184.3$ W, and

$$n = \frac{230 - (42.93 \times 0.15)}{3.07 \times 0.0668}$$

$$= 1098 \text{ r/min}$$

13.10   SPEED CONTROL OF DC MOTORS

From the governing equations, Eqs. (13.10) and (13.12), it is clear that the motor torque and speed can be controlled by controlling ϕ (i.e., the field current), V_t, and R_a, and changes in these quantities can be accomplished as follows. In essence, the method of control involves field control, armature control, or a combination of the two.

Chopper Control   Resistance control, in either the field or armature of a dc motor, results in poor efficiency. High-power solid-state controllers, on the other hand, offer the most practical, reliable, and efficient method of motor control. The most commonly used solid-state devices in motor control are power transistors and thyristors (or SCRs). The main differences between the two are that the transistor requires a continuous driving signal during conduction, whereas the thyristor requires only a pulse to initiate conduction; and the transistor switches off when the driving signal is removed, whereas the thyristor turns off when the load current is reduced to zero or when a reverse-polarity voltage is applied.

Utilizing power semiconductors, the dc chopper is the most common electronic controller used in electric drives. In principle, a chopper is an on-off switch connecting the load to and disconnecting it from the battery (or dc source), thus producing a chopped voltage across the load. Symbolically, a chopper as a switch is represented in Fig. 13.23*a*, and a basic chopper circuit is shown in Fig. 13.23*b*; when the thyristor does not conduct, the load current flows through the freewheeling diode *D*. From Fig. 13.23*b* it is clear that the average voltage across the load V_0 is given by

$$V_0 = \frac{t_{\text{on}}}{t_{\text{on}} + t_{\text{off}}} V = \frac{t_{\text{on}}}{T} V = \alpha V \tag{13.15}$$

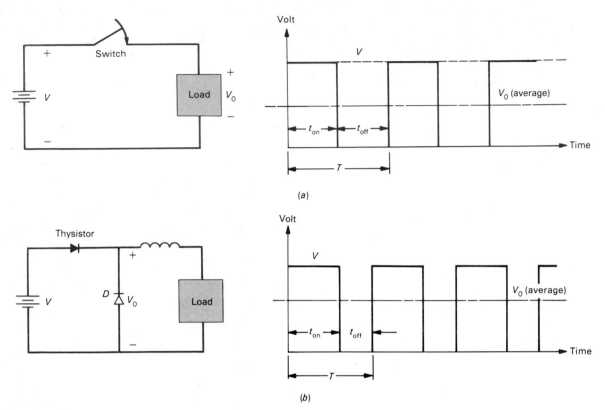

FIGURE 13.23 (*a*) Symbolic representation of a chopper switch, and output waveforms. (*b*) Basic circuits of a thyristor, and output waveforms.

where the various times are shown in Fig. 13.23, T is known as the chopping period, and $\alpha = t_{on}/T$ is called the *duty cycle*. Thus, the voltage across the load varies with the duty cycle.

There are three ways in which the chopper output voltage can be varied, and these are illustrated in Fig. 13.24. In the first method, the chopping frequency is kept constant, and the pulse width (or on-time t_{on}) is varied; this method is known as *pulse-width modulation*. The second method, called *frequency modulation*, has either t_{on} or t_{off} fixed and a variable chopping period, as indicated in Fig. 13.24*b*. The third method combines the preceding two methods to obtain pulse-width and frequency modulation, shown in Fig. 13.24*c*, which is used in current limit control. In a method involving frequency modulation, the frequency must not be decreased to a value that may cause a pulsating effect or a discontinuous armature current, and the frequency should not be increased to such a high value as to result in excessive switching losses. The switching frequency of most choppers for electric drives ranges from 100 to 1000 pulses per second. The drawback of a high-frequency chopper is that the current interruption generates a high-frequency noise.

A chopper circuit in a simplified form is shown in Fig. 13.25*a*, where the chopper is shown to supply a dc series motor. The circuit is a pulse-width modulation circuit,

FIGURE 13.24
(*a*) Constant-
frequency variable
pulse width.
(*b*) Variable-
frequency constant
pulse width.
(*c*) Variable-
frequency variable
pulse width.

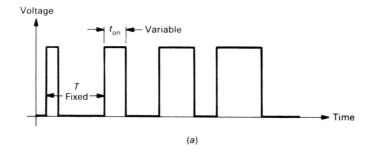

(*a*)

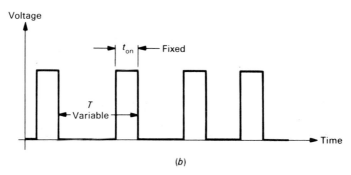

(*b*)

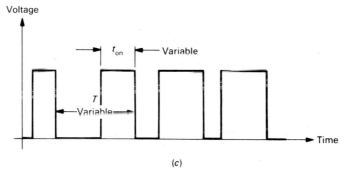

(*c*)

where t_{on} and t_{off} determine the average voltage across the motor, as given by Eq. (13.15). As mentioned earlier, the SCR cannot turn itself off, once it begins to conduct. So, to turn off the SCR, we require a commutating circuit which impresses a negative voltage on the SCR for a very short time (on the order of microseconds). The circuitry for commutation is often quite involved. For our purposes, we denote the commutating circuit by a switch in Fig. 13.25*a*.

The motor current and voltage waveforms are shown in Fig. 13.25*b*. The SCR is turned on by a gating signal at $t = 0$. The armature current i_a builds up, and the motor starts and picks up speed. After the time t_{on}, the SCR is turned off, and it remains off for a period t_{off}. During this period, the armature current continues to drop through the freewheeling diode circuit. Again, at the end of t_{off}, the SCR is turned on, and the on-off cycle continues. The chopper thus acts as a controllable voltage source.

FIGURE 13.25
(a) A dc motor
driven by a
chopper.
(b) Motor voltage
and current
waveforms.

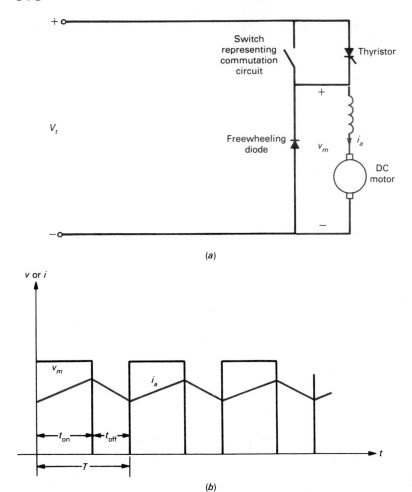

Example 13.9

For a dc series motor driven by a chopper, the following data are given: supply voltage = 440 V, duty cycle = 30 percent, motor circuit inductance = 0.04 H, maximum allowable change in the armature current = 8 A. Determine the chopper frequency.

Solution   From Eq. (13.15) the average voltage across the motor is

$$V_0 = 0.3 \times 440 = 132 \text{ V}$$

The voltage across the motor circuit inductance is

$$V_L = 440 - 132$$
$$= 308 \text{ V}$$

This voltage is also given by

$$V_L = L\frac{\Delta i}{\Delta t}$$

where $V_L = 308$ V

$\qquad L = 0.04$ H

$\qquad \Delta i = 8$ A

$\qquad \Delta t = t_{on}$

Hence,

$$t_{on} = \frac{0.04 \times 8}{308}$$

$$= 1.04 \text{ ms}$$

But

$$\alpha = 0.3$$

$$= \frac{t_{on}}{t_{on} + t_{off}}$$

$$= \frac{t_{on}}{T}$$

or

$$T = \frac{1.04}{0.3}$$

$$= 3.46 \text{ ms}$$

Hence, the chopper frequency is

$$\frac{1}{T} = 289 \text{ pulses per second}$$

Converters   A converter changes an ac input voltage to a controllable dc voltage. Converters have an advantage over choppers in that *natural*, or *line*, *commutation* is possible in converters, and so no complex commutation circuitry is required. Controlled converters use SCRs and operate on either single-phase or three-phase alternating current. The four types of phase-controlled converters commonly used for dc motor control are

1   Half-wave converters

2   Semiconverters

3   Full converters

4   Dual converters

Each of the above could be either a single-phase or a three-phase converter. Semiconverters are one-quadrant converters in that the polarities of the voltage and current at the dc terminals do not reverse. In a full converter, the polarity of the voltage reverses, but the current is unidirectional. In this sense, a full converter is a two-quadrant converter. Dual converters are four-quadrant converters.

A half-wave converter and its quadrant operation are shown in Fig. 13.26. This type of a converter is used for motors with ratings up to $\frac{1}{2}$ hp. Other types of single-phase converters have been used in drives rated up to 100 hp.

The performance of converters is measured in terms of the following parameters:

$$\text{Input power factor} = \frac{\text{mean input power}}{\text{RMS input voltamperage}} \tag{13.16}$$

$$\text{Input displacement factor} = \cos \phi_1 \tag{13.17}$$

where ϕ_1 is the angle between the supply voltage and the fundamental component of the supply current.

$$\text{Harmonic factor} = \frac{\text{RMS value of } n\text{th harmonic current}}{\text{fundamental component of supply current}} \tag{13.18}$$

To understand the operation of a converter-fed dc motor, we consider the simplest converter—the single-phase half-wave converter. Figure 13.27a shows such a system. The waveforms of motor current and voltage are given in Fig. 13.27b. By controlling the firing angle α of the SCR, we control the voltage across the motor armature and hence the motor speed. Notice, however, that the armature current does not begin to flow immediately after the SCR is turned on. Only when the line voltage v_t becomes greater than the motor voltage v_m does the armature current begin to flow. The current continues to flow for the period γ, as shown in Fig. 13.27b. This period is also known as the *conduction angle;* it is determined by the equality of the shaded areas A_1 and A_2 in Fig. 13.27b. Area A_1 corresponds to the energy stored in the inductance L of the motor circuit while the armature current is building up. This stored energy is returned to the source during the period that the armature current decreases. When the armature current becomes zero, the SCR turns off until it is turned on again. The motor speed is controlled by the firing angle α.

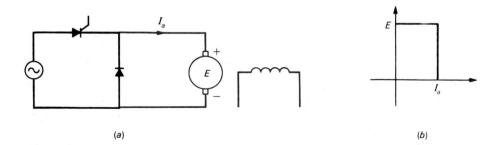

FIGURE 13.26
(a) A half-wave converter and (b) its quadrant operation.

(a)

(b)

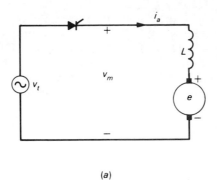

(a)

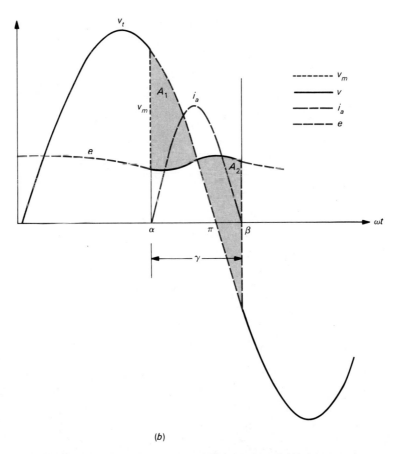

(b)

FIGURE 13.27   (a) A half-wave SCR drive and (b) its v, e, i_a, and v_m waveforms.

Example 13.10   A 1-hp 240-V dc motor is designed to run at 500 r/min when supplied from a dc source. The motor armature resistance is 7.56 Ω. The torque constant k is 4.23 N · m/A, and the back-emf constant is 4.23 V/(rad·s). This motor is driven by a half-wave converter at 200-V 50-Hz alternating current and draws a 2.0-A average current at a certain load. For the period during which the SCR conducts, the average motor voltage V_m is 120 V. Determine the torque developed by the motor, the motor speed, and the supply power factor.

Solution   The torque is

$$kI_a = 4.23 \times 2$$
$$= 8.46 \text{ N} \cdot \text{m}$$

The back emf is

$$E = 120 - 2.0 \times 7.56$$
$$= 104.88$$

But

$$E = k\Omega_m$$

or

$$104.88 = 4.23\Omega_m$$
$$= \frac{4.23 \times 2\pi n}{60}$$

Hence,

$$n = \frac{60 \times 104.88}{2\pi \times 4.23}$$
$$= 237 \text{ r/min}$$

The supply voltamperage is

$$200 \times 2 = 400 \text{ VA}$$

and the power taken by the motor is

$$V_m I_a = 120 \times 2$$
$$= 240 \text{ W}$$

The power factor is thus

$$\tfrac{240}{400} = 0.6$$

Armature control and field control are combined in the *Ward-Leonard system*, which gives a wide variation of speed. This method is presented in a simplified form in Example 13.11.

Example 13.11

The system shown in Fig. 13.28 is called the Ward-Leonard system for controlling the speed of a dc motor. Discuss the effects of varying R_{fg} and R_{fm} on the motor speed. Subscripts g and m correspond to generator and motor, respectively.

FIGURE 13.28
The Ward-Leonard system.

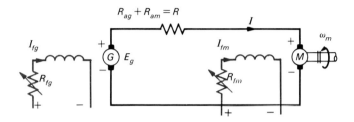

Solution Increasing R_{fg} decreases I_{fg} and hence E_g. Thus, the motor speed will decrease. The opposite will be true if R_{fg} is decreased.

Increasing R_{fm} will increase the speed of the motor, as shown in Example 13.7. Decreasing R_{fm} will result in a decrease of the speed.

13.11 LOSSES AND EFFICIENCY

Besides the voltamperage and speed-torque characteristics, the performance of a dc machine is measured by its efficiency:

$$\text{Efficiency} \equiv \frac{\text{power output}}{\text{power input}} = \frac{\text{power output}}{\text{power output} + \text{losses}} \tag{13.19}$$

Efficiency may therefore be determined from either load tests or losses. The various losses are classified as follows:

1 *Electrical.* These are the copper losses in various windings, such as the armature winding and different field windings, and the loss due to the contact resistance of the brush with the commutator.

2 *Magnetic.* These are the iron losses; they include the hysteresis and eddy-current losses in the various magnetic circuits, primarily in the armature core and pole faces.

3 *Mechanical.* These include the bearing-friction, windage, and brush-friction losses.

4 *Stray-load.* These are other load losses not covered above. They are taken as 1 percent of the output (as a rule of thumb). Another recommendation often followed is to take the stray-load loss as 28 percent of the core loss at the rated output.

FIGURE 13.29
Power flow in a
separately excited
(*a*) dc generator
and (*b*) dc motor.

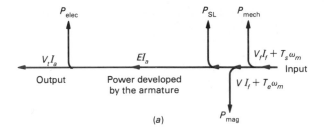

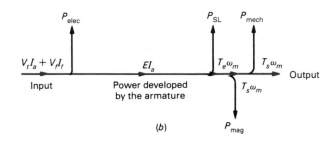

The power flow in a dc generator or motor is represented in Fig. 13.29 in which the symbols are as follows:

$$E = \text{back or induced emf in armature, V}$$
$$I_a = \text{armature current, A}$$
$$I_f = \text{current through field winding, A}$$
$$P_{\text{elec}} = \text{electric power, W}$$
$$P_{\text{mag}} = \text{magnetic loss, W}$$
$$P_{\text{mech}} = \text{rotational mechanical loss, W}$$
$$P_{\text{SL}} = \text{stray-load loss, W}$$
$$T_e = \text{electromagnetic torque developed by armature, N} \cdot \text{m}$$
$$T_s = \text{torque available at shaft, N} \cdot \text{m}$$
$$V_f = \text{voltage across field winding, V}$$
$$V_t = \text{terminal voltage, V}$$
$$\omega_m = \text{mechanical angular velocity, rad/s}$$

Example 13.12   A 230-V 10-hp shunt motor takes a full-load line current of 40 A. The armature and field resistances are 0.25 and 230 Ω, respectively. The total brush-contact drop is 2 V, and the core and friction losses are 380 W. Calculate the efficiency of the motor. Assume that the stray-load loss is 1 percent of the rated output.

Solution

Input: 40 × 230   = 9200 W

Field-resistance loss:	$\dfrac{(230)^2}{230}$	$= 230$ W
Armature-resistance loss:	$(40-1)^2(0.25) =$	380 W
Core loss and friction loss:		$= 380$ W
Brush-contact loss:	2×39	$= 78$ W
Stray-load loss:	$\frac{10}{100} \times 746$	$= \underline{75\text{ W}}$
Total losses:		$= 1143$ W
Power output:	$9200 - 1143$	$= 8057$ W
Efficiency:	$\dfrac{8057}{9200}$	$= 87.6\%$

PROBLEMS

13.1 The flux per pole of a generator is 75 mWb. The generator runs at 900 r/min. Determine the induced voltage (a) if the armature has 32 conductors connected as lap winding and (b) if the armature has 30 conductors connected as wave winding.

13.2 Determine the flux per pole of a six-pole generator required to generate 240 V at 500 r/min. The armature has 120 slots, with 8 conductors per slot, and is lap-connected.

13.3 If the armature current in the generator of Prob. 13.2 is 25 A, what is the developed electromagnetic torque?

13.4 The armature and field resistances of a 240-V shunt generator are 0.2 and 200 Ω, respectively. The generator supplies a 9600-W load. Determine the induced voltage, taking 2 V as the total brush-contact voltage drop.

13.5 The armature, shunt-field, and series-field resistances of a compound generator are 0.02, 80, and 0.03 Ω, respectively. The generator-induced voltage is 510 V, and the terminal voltage is 500 V. Calculate the power supplied to the load (at 500 V) if the generator has (a) a long-shunt connection and (b) a short-shunt connection.

13.6 A four-pole shunt-connected generator has a lap-connected armature with 728 conductors. The flux per pole is 25 mWb. If the generator supplies two

hundred 110-V 75-W bulbs, determine the speed of the generator. The field and armature resistances are 110 and 0.075 Ω, respectively.

13.7 A shunt generator delivers a 50-A current to a load at 110 V, at an efficiency of 85 percent. The total constant losses are 480 W, and the shunt-field resistance is 675 Ω. Calculate the armature resistance.

13.8 For the generator of Prob. 13.7, plot a curve for efficiency versus armature current. At what armature current is the efficiency maximum?

13.9 A separately excited six-pole generator has a 30-mWb flux per pole. The armature is lap-wound and has 534 conductors. This supplies a certain load at 250 V while running at 1000 r/min. At this load, the armature copper loss is 640 W. Calculate the load supplied by the generator. Take 2 V as the total brush-contact drop.

13.10 In a dc machine, the hysteresis and eddy-current losses at 1000 r/min are 10,000 W at a field current of 7.8 A. At 750 r/min speed and 7.8-A field current, the total iron losses become 6000 W. Assuming that the hysteresis loss is directly proportional to the speed and that the eddy-current loss is proportional to the square of the speed, determine the hysteresis and eddy-current losses at 500 r/min.

13.11 The saturation characteristic of a dc shunt generator is as follows:

Field current, A	1	2	3	4	5	6	7
Open-circuit voltage, V	53	106	150	192	227	252	270

The generator speed is 900 r/min. At this speed, what is the maximum field-circuit resistance such that the self-excited shunt generator would not fail to build up?

13.12 To what value will the no-load voltage of the generator of Prob. 13.11 build up, at 900 r/min, for a field-circuit resistance of 42 Ω?

13.13 A 240-V separately excited dc machine has an armature resistance of 0.25 Ω. The armature current is 56 A. Calculate the induced voltage for (a) generator operation and (b) motor operation.

13.14 The field- and armature-winding resistances of a 400-V dc shunt machine are 120 and 0.12 Ω, respectively. Calculate the power developed by the armature (a) if the machine takes 50 kW while running as a motor and (b) if the machine delivers 50 kW while running as a generator.

13.15 The field and armature resistances of a 220-V series motor are 0.2 and 0.1 Ω, respectively. The motor takes a 30-A current while running at 700 r/min. If the total iron and friction losses are 350 W, determine the motor efficiency.

13.16 A 400-V shunt motor delivers 15 kW of power at the shaft at 1200 r/min while drawing a line current of 62 A. The field and armature resistances are 200 and 0.05 Ω, respectively. Assuming a 1-V contact drop per brush, calculate (a) the torque developed by the motor and (b) the motor efficiency.

13.17 A 400-V series motor, having an armature circuit resistance of 0.5 Ω, takes a 44-A current while running at 650 r/min. What is the motor speed for a line current of 36 A?

13.18 A 220-V shunt motor, having an armature resistance of 0.2 Ω and a field resistance of 110 Ω, takes a 4-A line current while running on no load. When loaded, the motor runs at 100 r/min while taking a 42-A current. Calculate the no-load speed.

13.19 The machine of Prob. 13.18 is driven as a shunt generator to deliver a 44-kW load at 220 V. If the machine takes 44 kW while running as a motor, what is its speed?

13.20 A 220-V shunt motor, having an armature resistance of 0.2 Ω and a field resistance of 110 Ω, takes a 4-A line current while running at 1200 r/min on no load. On load, the input to the motor is 15 kW. Calculate (a) the speed, (b) developed torque, and (c) efficiency at this load.

13.21 A 400-V series motor has a field resistance of 0.2 Ω and an armature resistance of 0.1 Ω. The motor takes a 30-A current at 1000 r/min while developing a torque T. Determine the motor speed if the developed torque is $0.6T$.

13.22 A shunt machine, while running as a generator, has an induced voltage of 260 V at 1200 r/min. Its armature and field resistances are 0.2 and 110 Ω, respectively. If the machine is run as a shunt motor, it takes 4 A at 220 V. At a certain load the motor takes 30 A at 220 V. On load, however, armature reaction weakens the field by 3 percent. Calculate the motor speed and efficiency at the specified load.

13.23 The machine of Prob. 13.22 is run as a motor. It takes a 25-A current at 800 r/min. What resistance must be inserted in the field circuit to increase the motor speed to 100 r/min? The torque on the motor for the two speeds remains unchanged.

13.24 The motor of Prob. 13.22 runs at 600 r/min while taking 40 A at a certain load. If a 0.8-Ω resistance is inserted in the armature circuit, determine the motor speed, provided that the torque on the motor remains constant.

13.25 A 220-V shunt motor delivers 400 hp on full load at 950 r/min and has an efficiency of 88 percent. The armature and field resistances are 0.2 and 110 Ω, respectively. Determine (a) the starting resistance, such that the starting line current does not exceed 1.6 times the full-load current, and (b) the starting torque.

13.26 A 220-V series motor runs at 750 r/min while taking a 15-A current. The total resistance in the armature circuit, including the field resistance, is 0.4 Ω. What is the motor speed if it takes a 10-A current? The torque on the motor is such that it increases as the square of the speed.

13.27 A dc motor is connected to a 96-V battery through a chopper. If the duty cycle is 45 percent, what is the average voltage applied to the motor?

13.28 A chopper is on for 20 ms and off for the next 50 s. Determine the duty cycle.

13.29 A pulse-width-modulated output waveform from a chopper is shown in Fig. P13.29. This chopper is connected to a dc motor having an armature resistance of 0.25 Ω and running at 350 r/min. If the motor back-emf constant is 0.12 V/(r·min), calculate the average armature current.

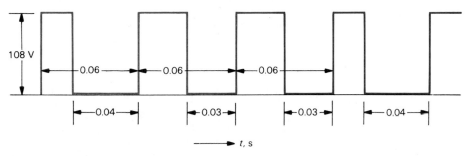

FIGURE P13.29

13.30 Repeat Prob. 13.29 if the motor is fed from the frequency-modulated voltage waveform shown in Fig. P13.30.

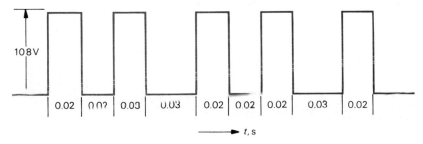

FIGURE P13.30

13.31 A chopper having the following data drives a dc motor: supply voltage = 400 V, duty cycle = 50 percent, and armature circuit inductance = 20 mH. If the chopper frequency is 270 pulses per second, what is the change in the armature current?

13.32 For the motor of Prob. 13.31, if the allowable change in the armature current is 5 A, determine the chopper frequency.

REFERENCES

1 M. G. Say and E. O. Taylor, *Direct Current Machines*, Halsted Press, New York, 1980.

2 S. A. Nasar, *Schaum's Outline of Theory and Problems in Electric Machines and Electromechanics*, McGraw-Hill, New York, 1981.

3 S. A. Nasar, *Electric Machines and Transformers*, Macmillan, New York, 1983.

4 A. E. Fitzgerald, C. Kinsley, Jr., and S. D. Umans, *Electric Machinery*, 4th ed., McGraw-Hill, New York, 1983.

Synchronous Machines

The bulk of electric power for everyday use is produced by polyphase synchronous generators, which are the largest single-unit electric machines in production. For instance, synchronous generators with power ratings of several hundred megavoltamperes (MVA) are fairly common, and it is expected that machines of several thousand megavoltamperes will be in use in the near future. These are called *synchronous machines* because they operate at constant speeds and constant frequencies under steady-state conditions.

Like most rotating machines, synchronous machines are capable of operating both as motors and as generators. They are used as motors in constant-speed drives; where a variable-speed drive is required, a synchronous motor is used with an appropriate frequency changer. As generators, several synchronous machines often operate in parallel, as in a power station. While operating in parallel, the generators share the load with each other. At a given time, one of the generators may not carry any load; in such a case, instead of shutting down the generator, it is allowed to "float" on the line as a synchronous motor on no load.

The operation of a synchronous generator is based on Faraday's law of electromagnetic induction. Thus, an ac synchronous generator works very much as a dc generator, in which emf is generated by the relative motion of conductors and magnetic flux (Fig. 13.4 illustrated a synchronous generator in its elementary form). Unlike a dc generator, however, a synchronous generator does not have a commutator. The two basic parts of a synchronous machine are the magnetic field structure, carrying a dc-excited winding, and the armature; the armature often has a three-phase winding in which the ac emf is generated. Almost all modern synchronous machines have stationary armatures and rotating field structures. The dc winding on the rotating field structure is connected to an external source through slip rings and brushes. Some field structures do not have brushes but instead have brushless excitation by rotating diodes. In some respects the stator carrying the armature windings is similar to the stator of a polyphase induction motor (discussed in the next chapter).

14.1   SOME CONSTRUCTION DETAILS

Some of the factors that dictate the form of construction of a synchronous machine are as follows:

1 *Form of excitation.* Note from the preceding remarks that the field structure is usually the rotating member of a synchronous machine and is supplied with a direct-current-excited winding to produce the magnetic flux. This dc excitation may be provided by a self-excited dc generator mounted on the rotor shaft of the synchronous machine; such a generator is known as the *exciter.* The direct current thus generated is fed to the synchronous machine field winding. In slow-speed machines with large ratings, such as hydroelectric generators, the exciter may not be self-excited; instead, a pilot exciter, which may be self-excited or may have a permanent magnet, activates the exciter. A hydroelectric generator and its rotor and exciters are shown in Fig. 14.1. The maintenance problems of direct-coupled dc generators impose a limit on this form of excitation at a rating of about 100 megawatts (MW). An alternative form of excitation is provided by silicon diodes and thyristors, which do not present excitation problems for large synchronous machines; the two types of solid-state excitation systems are

   a Static systems that have stationary diodes or thyristors, in which the current is fed to the rotor through slip rings.
   b Brushless systems that have shaft-mounted rectifiers which rotate with the rotor, thus avoiding the need for brushes and slip rings. Figure 14.2 shows a brushless excitation system.

2 *Field structure and speed of machine.* We have already mentioned that the synchronous machine is a constant-speed machine. This speed is known as the *synchronous speed.* For instance, a 2-pole 60-Hz synchronous machine must run at 3600 r/min, whereas the synchronous speed of a 12-pole 60-Hz machine is only 600 r/min. The rotor field structure consequently depends on the speed rating of the machine. Turbogenerators, which are high-speed machines, have smooth, *round cylindrical rotors*, as shown in Fig. 14.2. Hydroelectric and diesel-electric generators are low-speed machines and have protruding *salient-pole rotors*, as shown in Figs. 14.1 and 14.3. Such rotors are less expensive to fabricate than round rotors; they are not suitable for large, high-speed machines, however, because of the excessive centrifugal forces and mechanical stresses that develop at speeds around 3600 r/min.

3 *Stator.* The stator of a synchronous machine carries the armature, or load, winding. We recall from Chap. 13 that the armature of a dc machine has a winding distributed around its periphery and that this armature winding consists of slot-embedded conductors which cover the entire surface of the armature and are interconnected in a predetermined manner. Likewise, in a synchronous machine

FIGURE 14.1
A hydroelectric
generator.

the armature winding is formed by interconnecting the various conductors in the slots spread over the periphery of the stator of the machine. Often more than one independent winding is on the stator. When constructing a three-phase machine, e.g., the engineers displace the three phases from each other in space as they distribute the windings in the slots over the entire periphery of the stator; generally, each slot contains two coil sides. Such a winding is known as a *double-layer* winding.

4 *Cooling.* Because synchronous machines are often built in extremely large sizes, they arc designed to carry very large currents. A typical armature current density may be of the order of 10 A/mm^2 in a well-designed machine. Also the magnetic loading of the core is such that it reaches saturation in many regions. These severe

FIGURE 14.2
Field winding on a
round or
cylindrical rotor.

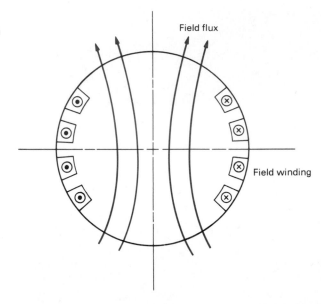

electric and magnetic loadings in a synchronous machine produce heat that must be appropriately dissipated. Thus, the manner in which the active parts of a machine are cooled determines its overall physical structures. In addition to air, some coolants used in synchronous machines include water, hydrogen, and helium.

5 *Damper bars.* So far we have mentioned only two electrical windings of a synchronous machine: the three-phase armature winding and the field winding. We have also pointed out that under steady-state conditions the machine runs at a constant speed, i.e., at the synchronous speed. However, like other electric

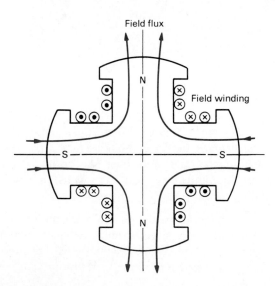

FIGURE 14.3
Field winding on a
salient rotor.

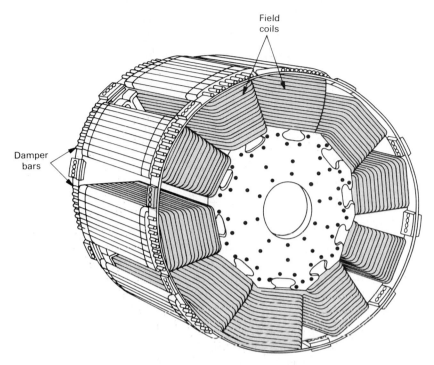

FIGURE 14.4   A salient rotor, showing the field windings and damper bars (shaft not shown).

machines, a synchronous machine undergoes transients during starting and abnormal conditions. During transients, the rotor may undergo mechanical oscillations and its speed may deviate from the synchronous speed, which is an undesirable phenomenon. To overcome this, a winding (resembling the cage of an induction motor) is mounted on the rotor; this set constitutes the damper windings. When the rotor speed is different from the synchronous speed, currents are induced in the damper winding. The damper winding acts as the cage rotor of an induction motor, producing a torque to restore the synchronous speed. Also the damper bars provide a means of starting the machine, which is otherwise not self-starting. Figure 14.4 shows the damper bars on a salient rotor.

14.2   MAGNETOMOTIVE FORCES (MMF's) AND FLUXES DUE TO ARMATURE AND FIELD WINDINGS

In general, we may say that the behavior of an electric machine depends on the interaction between the magnetic fields (or fluxes) produced by various mmf's acting on the magnetic circuit of the machines. For instance, in a dc motor the torque is produced by the interaction of the flux produced by the field winding and the flux produced by the current-carrying armature conductors. In a dc generator, too, we must consider the effect of the interaction between the field and armature mmf's. As mentioned in Sec. 14.1, the main sources of fluxes in a synchronous machine are the armature and field mmf's. In contrast to a dc machine, in which the flux due to the armature mmf is stationary in space, the fluxes due to each phase of the armature mmf pulsate in time and the resultant flux rotates in space, as will be demonstrated in Eq. (14.1) and in Sec. 15.10. For the present, we consider the mmf's produced by a single full-pitch coil† having N turns where the slot opening is negligible and the machine has two poles, as shown in Fig. 14.5a. The mmf has a constant value of Ni between the coil sides, as shown in Fig. 14.5b. Traditionally, the magnetic effects of a winding in an electric machine are considered on a per-pole basis. Thus, if i is the current in the coil, the mmf per pole is $Ni/2$; this is plotted in Fig. 14.5c. The reason for such a representation is that Fig. 14.5c also represents a flux density distribution, but to a different scale. Obviously, the flux density over one pole (say, the north pole) must be opposite to that over the other (south) pole, thus keeping the flux entering the rotor equal to that leaving the rotor surface. Comparing Fig. 14.5b and c, we notice that the representation of the mmf curve with positive and negative areas (Fig. 14.5c) has the advantage that it gives the flux density distribution, which must contain positive and negative areas. The mmf distribution shown in Fig. 14.5c may be resolved into its harmonic components. In practice, we may assume harmonics to be absent in properly designed armature windings, and the resultant mmf of each phase may be ideally taken as sinusoidal.

---

† Pole pitch was defined in Sec. 13.3.

FIGURE 14.5   Flux and mmf produced by a concentrated winding; (a) flux lines produced by an N-turn coil; (b) mmf produced by the N-turn coil; (c) mmf per pole.

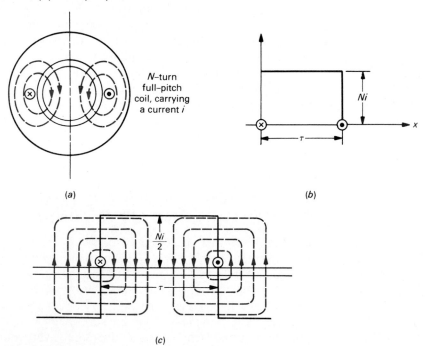

Let the three-phase stator (or armature) winding of a synchronous machine be excited by three phase currents. As a result, the mmf's produced by the three phases are displaced from each other by 120° in time and space. If we assume the mmf distribution in space to be sinusoidal, we may write for the three mmf's

$$\mathscr{F}_a = F_m \sin \omega t \cos p\theta$$
$$\mathscr{F}_b = F_m \sin (\omega t - 120°) \cos (p\theta - 120°)$$
$$\mathscr{F}_c = F_m \sin (\omega t + 120°) \cos (p\theta + 120°)$$

where F_m is the amplitude of the mmf and ω and p are defined in Sec. 14.3. The space and time variations of the resultant mmf is then the sum of the above three mmf's. Observing that $\sin A \cos B = \frac{1}{2} \sin (A - B) + \frac{1}{2} \sin (A + B)$ and adding $\mathscr{F}_a$, $\mathscr{F}_b$, and $\mathscr{F}_c$, we obtain for the resultant mmf

$$\mathscr{F}(\theta, t) = 1.5 F_m \sin (\omega t - p\theta) \tag{14.1}$$

The magnetic field resulting from the mmf of Eq. (14.1) is a *rotating magnetic field*. Graphically, the production of the rotating field is illustrated in Fig. 14.6. The

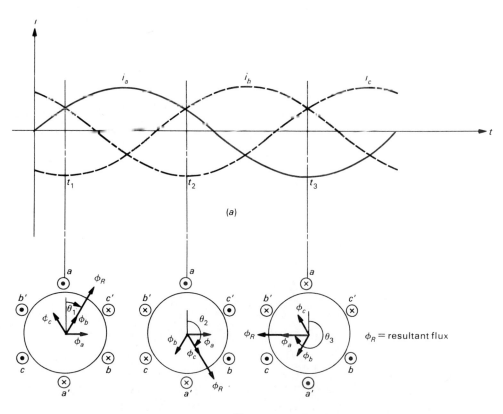

FIGURE 14.6
Production of a rotating magnetic field by a three-phase excitation; (*a*) time diagram; (*b*) space diagram.

existence of the rotating magnetic field is essential to the operation of a synchronous motor.

14.3 THE SYNCHRONOUS SPEED

To determine the velocity of the rotating field given by Eq. (14.1), imagine an observer traveling with the mmf wave from a certain point. To this observer, the magnitude of the mmf wave will remain constant (independent of time), implying that the right side of Eq. (14.1) would appear constant. Expressed mathematically, this would mean that

$$\sin(\omega t - p\theta) = \text{constant}$$

or

$$\omega t - p\theta = \text{constant}$$

Differentiating both sides with respect to t, we obtain

$$\omega - p\dot{\theta} = 0$$

or

$$\omega_m \equiv \dot{\theta} = \frac{\omega}{p}$$

(14.2)

This speed is known as the *synchronous speed*.

In Eqs. (14.1) and (14.2), ω is the frequency of the stator mmf's. What is the significance of p? Notice that the given mmf's vary in space as $\cos p\theta$, indicating that for one complete travel around the stator periphery, the mmf undergoes p cyclic changes. Thus, p may be considered as the order of harmonics, or the number of *pole pairs* in the mmf wave. If P is the number of *poles*, then obviously $P = 2p$. Writing ω_m in terms of speed n_s in revolutions per minute and ω in terms of the frequency f, we have

$$\omega_m = \frac{2\pi n_s}{60}$$

and

$$\omega = 2\pi f$$

Substituting for P, ω_m, and ω in Eq. (14.2) yields

$$n_s = \frac{120f}{P}$$

(14.3)

which is the synchronous speed in revolutions per minute.

An alternative form of Eq. (14.3) is

$$f = \frac{Pn_s}{120} \qquad (14.4)$$

which implies that the frequency of the voltage induced in a synchronous generator having P poles and running at n_s r/min is f Hz. We could reach the same conclusion from Figs. 13.4 and 13.14; namely, in a 2-pole machine, one cycle is generated in one rotation. Thus, in a P-pole machine, $P/2$ cycles are generated in one rotation; and in n_s rotations, $Pn_s/2$ cycles are generated. Since n_s rotations take 60 s, in 1 s $Pn_s/2 \times 60 = Pn_s/120$ cycles are generated, which is the frequency f.

Example 14.1   For a 60-Hz generator, list four possible combinations of the number of poles and the speed.

Solution   From Eq. (14.4) we must have $Pn_s = 120 \times 60 = 7200$. Hence, we obtain the following table:

Number of Poles	2	4	6	8
Speed, r/min	3600	1800	1200	900

14.4 SYNCHRONOUS GENERATOR OPERATION

Like the dc generator, a synchronous generator functions on the basis of Faraday's law. If the flux linking the coil changes in time, a voltage is induced in the coil. Stated in alternative form, a voltage is induced in a conductor if it cuts magnetic flux lines (recall Fig. 13.4). Consider the machine shown in Fig. 14.7. Assuming that the flux

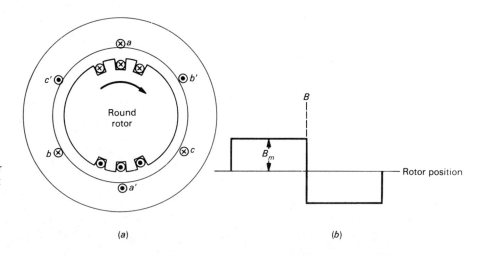

FIGURE 14.7
(*a*) A three-phase round-rotor machine. (*b*) Flux density distribution produced by the rotor excitation.

(a)   (b)

FIGURE 14.8
A three-phase
voltage produced
by a three-phase
synchronous
generator.

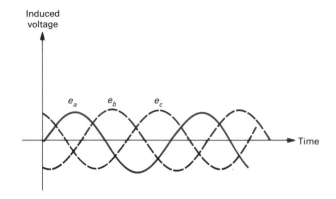

density in the air gap is uniform, we see that sinusoidally varying voltages will be induced in the three coils aa', bb', and cc' if the rotor, carrying direct current rotates at a constant speed n_s. Recall from Eq. (13.6) that if ϕ is the flux per pole, ω is the angular frequency, and N is the number of turns in phase a (coil aa'), then the voltage induced in phase a is given by

$$e_a = \omega N \phi \sin \omega t$$
$$\quad = E_m \sin \omega t \qquad\qquad\qquad (14.5)$$

where $E_m = 2\pi f N \phi$ and $f = \omega/(2\pi)$ is the frequency of the induced voltage. Because phases b and c are displaced from phase a by $\pm 120°$ (Fig. 14.7), the corresponding voltages may be written as

$$e_b = E_m \sin(\omega t - 120°)$$
$$e_c = E_m \sin(\omega t + 120°)$$

These voltages are sketched in Fig. 14.8 and correspond to the voltages from a three-phase generator.

14.5 PERFORMANCE OF A ROUND-ROTOR SYNCHRONOUS GENERATOR

At the outset we wish to point out that we study the machine on a per-phase basis, implying a balanced operation. Thus, let us consider a round-rotor machine operating as a generator on no load. Variation of V_o with I_f is shown in Fig. 14.9 and is known as the *open-circuit characteristic* of a synchronous generator. Let the open-circuit phase voltage be V_o for a certain field current I_f. Here V_o is the internal voltage of the generator. We assume that I_f is such that the machine is operating under unsaturated conditions. Now let us short-circuit the armature at the terminals, keeping the field current unchanged (at I_f), and measure the armature phase current

FIGURE 14.9
Open-circuit
characteristics of a
synchronous
generator.

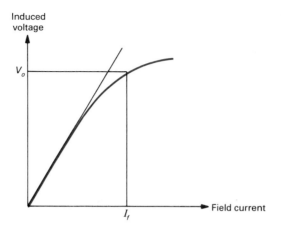

I_a. In this case, the entire internal voltage V_o is dropped across the internal impedance of the machine. In mathematical terms,

$$\mathbf{V}_o = \mathbf{I}_a \mathbf{Z}_s$$

where $\mathbf{Z}_s$ is known as the *synchronous impedance*. One portion of $\mathbf{Z}_s$ is R_a, the armature resistance per phase, and the other is a reactance X_s, which is known as *synchronous reactance*; i.e.,

$$\mathbf{Z}_s = R_a + jX_s \tag{14.6}$$

If the generator operates at a terminal voltage V_t while supplying a load corresponding to an armature current I_a, then

$$\mathbf{V}_o = \mathbf{V}_t + \mathbf{I}_a(R_a + jX_s) \tag{14.7}$$

where X_s is the synchronous reactance.

In an actual synchronous machine (except in very small ones) we almost always have $X_s \gg R_a$, in which case $Z_s \cong X_s$. We use this restriction in most of our analyses. Among the steady-state characteristics of a synchronous generator, its voltage regulation and power-angle characteristics are the most important. As for a transformer and a dc generator, we define the *voltage regulation* of a synchronous generator at a given load as

$$\text{Percentage voltage regulation} = \frac{V_o - V_t}{V_t} \times 100 \tag{14.8}$$

where V_t is the terminal voltage on load and V_o is the no-load terminal voltage. Clearly, for a given V_t, we can find V_o from Eq. (14.7) and hence the voltage regulation, as illustrated by the following examples.

Example 14.2   Calculate the percentage voltage regulation for a three-phase wye-connected 2500-kVA 6600-V turboalternator operating at full load and 0.8 lagging power factor. The per-phase synchronous reactance and the armature resistance are 10.4 and 0.071 Ω, respectively.

FIGURE 14.10
Phasor diagrams:
(a) lagging power
factor; (b) leading
power factor.

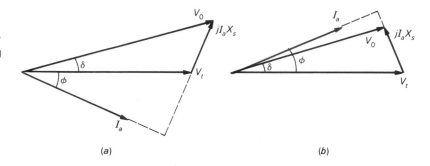

(a)   (b)

Solution   Here we have $X_s \gg R_a$. The phasor diagram for the lagging power factor, neglecting the effect of R_a, is shown in Fig. 14.10a. The numerical variables are as follows:

$$V_t = \frac{6600}{\sqrt{3}}$$

$$= 3810 \text{ V}$$

$$\mathbf{I}_a = \frac{2500 \times 1000}{\sqrt{3} \times 6600} \underline{/-36.87^\circ}$$

$$= 218.7 \underline{/-36.87^\circ} \text{ A}$$

From Eq. (14.7) we have

$$\mathbf{V}_o = 3810 + 218.7(0.8 - j0.6)j10.4$$
$$= 5485\underline{/19.3^\circ}$$

and from Eq. (14.8) the percentage regulation is

$$\frac{5485 - 3810}{3810} \times 100 = 44\%$$

Example 14.3   Repeat the preceding calculations with 0.8 leading power factor.

Solution   In this case we have the phasor diagram shown in Fig. 14.10b, from which we get

$$\mathbf{V}_0 = 3810 + 218.7(0.8 + j0.6)j10.4$$
$$= 3048\underline{/36.6^\circ}$$

and the percentage voltage regulation is

$$\frac{3048 - 3810}{3810} \times 100 = -20\%$$

We observe from these examples that the voltage regulation is dependent on the power factor of the load. Unlike what happens in a dc generator, the voltage regulation for a synchronous generator may even become negative. The angle between V_0 and V_t is defined as the *power angle* δ. To justify this definition, we reconsider Fig. 14.10a, from which we obtain

$$I_a X_s \cos \phi = V_0 \sin \delta \qquad (14.9)$$

Now, from the approximate equivalent circuit (assuming $X_s \gg R_a$) the power delivered by the generator = power developed: $P_d = V_t I_a \cos \phi$ (which also follows from Fig. 14.10). Hence, in conjunction with Eq. (14.9), we get

$$P_d = \frac{V_0 V_t}{X_s} \sin \delta \qquad (14.10)$$

which shows that the internal power of the machine is proportional to $\sin \delta$. Equation (14.10) is often said to represent the *power-angle characteristic* of a synchronous machine. A plot of Eq. (14.10) is shown in Fig. 14.11b, which shows that for a negative δ the machine will operate as a motor, discussed next.

FIGURE 14.11
(*a*) An approximate equivalent circuit. (*b*) Power-angle characteristic of a round-rotor synchronous machine.

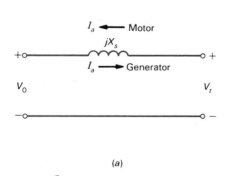

(a)

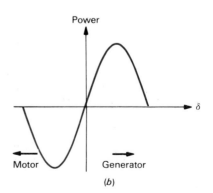

(b)

14.6   SYNCHRONOUS MOTOR OPERATION

We know from Sec. 14.2 that the stator of a three-phase synchronous machine carrying a three-phase excitation produces a rotating magnetic field in the air gap of the machine. Referring to Fig. 14.12, we have a rotating magnetic field in the air gap of the salient-pole machine when its stator (or armature) windings are fed from a three-phase source.

Now let the rotor (or field) winding of this machine be unexcited. The rotor has a tendency to align with the rotating field at all times in order to present the path of least reluctance. Thus, if the field is rotating, the rotor tends to rotate with the field. A round rotor, on the other hand, does not tend to follow the rotating magnetic field; we see from Fig. 14.7a that this is because the uniform air gap presents the same reluctance all around the air gap, and thus the rotor does not have any preferred direction of alignment with the magnetic field. The torque which we have in the salient-rotor machine of Fig. 14.12 but not in the round-rotor machine of Fig. 14.7a is called the *reluctance torque*. It is present by virtue of the variation of the reluctance around the periphery of the machine.

Now let the field winding of either machine (Fig. 14.7a or 14.12) be fed by a dc source that produces a rotor magnetic field of definite polarities, and the rotor will tend to align with the stator field and to rotate with the rotating magnetic field. We observe that both a round rotor and a salient rotor, when excited, tend to rotate with the rotating magnetic field, although the salient rotor has an additional reluctance torque because of the saliency, whether excited or unexcited.

So far, we have indicated the mechanism of torque production in a round-rotor and in a salient-rotor machine. To recapitulate, we might say that the stator rotating magnetic field has a tendency to "drag" the rotor along, as if a north pole on the stator "locks in" with a south pole of the rotor. However, if the rotor is at a standstill, the stator poles tend to make the rotor rotate in one direction and then in the other as they rotate and sweep across the rotor poles. Therefore, a synchronous motor is not self-starting. In practice, as mentioned in Sec. 14.1, the rotor carries damper bars

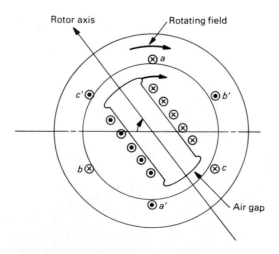

FIGURE 14.12
A salient-rotor
machine.

that act as the cage of an induction motor and thereby provide a starting torque. The mechanism of torque production by the damper bars is similar to the torque production in an induction motor (discussed in the next chapter). Once the rotor starts running and almost reaches the synchronous speed, it locks into position with the stator poles. The rotor pulls into step with the rotating magnetic field and runs at the synchronous speed; the damper bars go out of action. Any departure from the synchronous speed results in induced currents in the damper bars, which tend to restore the synchronous speed. Machines without damper bars—or very large machines with damper bars—may be started by an auxiliary motor. We discuss the operating characteristics of synchronous motors next.

14.7 PERFORMANCE OF A ROUND-ROTOR SYNCHRONOUS MOTOR

Except for some precise calculations, we may neglect the armature resistance as compared to the synchronous reactance. Therefore, the steady-state per-phase equivalent circuit of a synchronous machine simplifies to the one shown in Fig. 14.11a. Notice that this circuit is similar to that of a dc machine, where the dc armature resistance has been replaced by the synchronous reactance. In Fig. 14.11a we have shown the terminal voltage V_t, the internal excitation voltage V_0, and the armature current I_a going into the machine or out of it, depending on the mode of operation—"into" for motor and "out of" for generator. With the help of this circuit and Eq. (14.10), we will study some of the steady-state operating characteristics of a synchronous motor. In Fig. 14.11b we show the power-angle characteristics as given by Eq. (14.10). Here positive power and positive δ imply generator operation, while a negative δ corresponds to motor operation. Because δ is the angle between $\mathbf{V}_0$ and $\mathbf{V}_t$, $\mathbf{V}_0$ is ahead of $\mathbf{V}_t$ in a generator, whereas in a motor $\mathbf{V}_t$ is ahead of $\mathbf{V}_0$. The voltage-balance equation for a motor is, from Fig. 14.11a,

$$\mathbf{V}_t = \mathbf{V}_0 + j\mathbf{I}_a X_s$$

If the motor operates at a constant power, then Eqs. (14.9) and (14.10) require that

$$V_0 \sin \delta = I_a X_s \cos \phi = \text{constant} \qquad (14.11)$$

We recall from Fig. 14.9 that V_0 depends on the field current I_f. Consider two cases: (1) when I_f is adjusted so that $V_0 < V_t$ and the machine is underexcited and (2) when I_f is increased to a point that $V_0 > V_t$ and the machine becomes overexcited. The voltage-current relationships for the two cases are shown in Fig. 14.13a. For $V_0 > V_t$ at constant power, δ is greater than the δ for $V_0 < V_t$, as governed by Eq. (14.11). Notice that an underexcited motor operates at a lagging power factor ($\mathbf{I}_a$ lagging $\mathbf{V}_t$), whereas an overexcited motor operates at a leading power factor. In both cases the terminal voltage and the load on the motor are the same. Thus, we observe that the operating power factor of the motor is controlled by varying the field

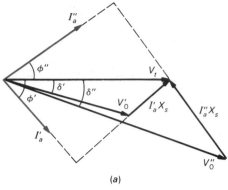

(a)

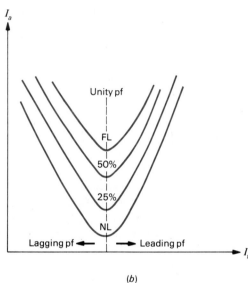

(b)

FIGURE 14.13   (a) Phasor diagram for motor operation: V'_0, I'_a, ϕ', and δ' correspond to underexcited operation; V''_0, I''_a, ϕ'', and δ'' correspond to overexcited operation. (b) V curves of a synchronous motor.

excitation, hence altering V_0. This is a very important property of synchronous motors. The armature current at a constant load, as given by Eq. (14.11), for varying field current is also shown in Fig. 14.13a. From this we can obtain the variations of the armature current I_a with the field current I_f (corresponding to V_0), and this can be done for different loads, as shown in Fig. 14.13b These curves are known as the *V curves* of the synchronous motor. One application of a synchronous motor is in power-factor correction, as demonstrated by the following examples.

Example 14.4   A three-phase wye-connected load takes a 50-A current at 0.707 lagging power factor at 220 V between the lines. A three-phase wye-connected round-rotor synchronous motor, having a synchronous reactance of 1.27 Ω per phase, is connected in parallel with the load. The power developed by the motor is 33 kW at a power angle of 30°. Neglecting the armature resistance, calculate the reactive kilovoltamperes of the motor and the overall power factor of the motor and the load.

FIGURE 14.14
(a) Circuit
diagram.
(b) Phasor
diagram.

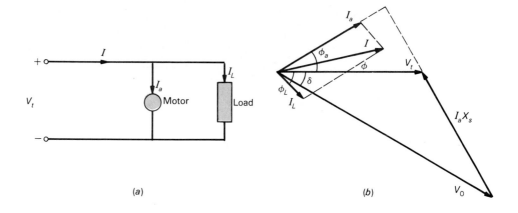

(a) (b)

Solution The circuit and the phasor diagram on a per-phase basis are shown in Fig. 14.14. From Eq. (14.10) we have

$$P_d = \frac{1}{3} \times 33,000$$

$$= \frac{220}{\sqrt{3}} \frac{V_0}{1.27} \sin 30°$$

which yields $V_0 = 220$ V. From the phasor diagram, $I_a X_s = 127$ or $I_a = 127/1.27 = 100$ A and $\phi_a = 30°$. The reactive kilovoltamperes (kVAr) of the motor $= \sqrt{3} V_t I_a \sin \phi_a = \sqrt{3} \times (220/1000) \times 100 \times \sin 30° = 19$ kVAr.

The overall power-factor angle is given by

$$\tan \phi = \frac{I_a \sin \phi_a - I_L \sin \phi_L}{I_a \cos \phi_a + I_L \cos \phi_L} = 0.122$$

or $\phi = 7°$ and $\cos \phi = 0.992$ leading.

Example 14.5

For the generator of Example 14.4, calculate the power factor for zero voltage regulation on full load.

Solution Let ϕ be the power-factor angle. Then

$$\mathbf{I}_a \mathbf{Z}_s = 218.7 \times 10.4 \underline{/\phi + 89.6}$$
$$= 2274.48 \underline{/\phi + 89.6} \quad \text{V}$$

For voltage regulation to be zero, $|V_0| = |V_t|$. Hence

$$|3810| = |3810 + 2274.48[\cos(\phi + 89.6) + j \sin(\phi + 89.6)]$$

$$3810^2 = [3810 + 2274.48 \cos(\phi + 89.6)]^2 + [2274.48 \sin(\phi + 89.6)]^2$$

from which

$$\phi = 17.76° \quad \text{and} \quad \cos \phi = 0.95 \text{ leading}$$

PROBLEMS

14.1   At what speed must a 6-pole synchronous generator run to generate a 50-Hz voltage?

14.2   The open-circuit voltage of a 60-Hz generator is 11,000 V at a field current of 5 A. Calculate the open-circuit voltage at 50 Hz and a 2.5-A field current. Neglect saturation.

14.3   A 60-kVA three-phase wye-connected 440-V 60-Hz synchronous generator has a resistance of 0.15 Ω and a synchronous reactance of 3.5 Ω per phase. At rated load and unity power factor, calculate the percent voltage regulation.

14.4   A 1000-kVA 11-kV three-phase wye-connected synchronous generator supplies a 600-kW 0.8-leading-power-factor load. The synchronous reactance is 24 Ω per phase, and the armature resistance is negligible. Calculate (*a*) the power angle and (*b*) the voltage regulation.

14.5   A 1000-kVA 11-kV three-phase wye-connected synchronous generator has an armature resistance of 0.5 Ω and a synchronous reactance of 5 Ω. At a certain field current the generator delivers rated load at 0.9 lagging power factor at 11 kV. For the same excitation, what is the terminal voltage at 0.9-leading-power-factor full load?

14.6   An 11-kV three-phase wye-connected generator has a synchronous impedance of 6 Ω per phase and negligible armature resistance. For a given field current, the open-circuit voltage is 12 kV. Calculate the maximum power developed by the generator. Determine the armature current and power factor for the maximum power condition.

14.7   A 400-V three-phase wye-connected synchronous motor delivers 12 hp at the shaft and operates at 0.866 lagging power factor. The total iron, friction, and field copper losses are 1200 W. If the armature resistance is 0.75 Ω per phase, determine the efficiency of the motor.

14.8   The motor of Prob. 14.7 has a synchronous reactance of 6 Ω per phase and operates at 0.9 leading power factor while taking an armature current of 20 A.

Calculate the induced voltage. Neglect armature resistance.

14.9   A 12.6-kV three-phase synchronous motor, having a synchronous reactance of 0.9 Ω per phase and negligible armature resistance, operates at 0.866 leading power factor while taking a line current of 1575 A. Calculate the excitation (or induced) voltage and the power angle.

14.10   The load on the motor of Prob. 14.9 is such that the power angle is 30°. Determine the armature current and the motor power factor.

14.11   A 400-V three-phase wye-connected synchronous motor has a synchronous impedance $Z_s = (0.15 + j3)$ Ω per phase. At a certain load and excitation, the motor takes 22.4-A armature current and operates at 0.8 leading factor. Calculate the excitation voltage and the power angle.

14.12   The field excitation of the motor of Prob. 14.11 is adjusted so that the motor takes a minimum armature current of 17.94 A. Determine the excitation voltage. Neglect the effect of armature resistance.

14.13   The excitation voltage of the motor of Prob. 14.11 is 280 V per phase, while the armature current is 25 A. Neglecting the effect of armature resistance, obtain the operating power factor.

14.14   If the field current of the motor of Prob. 14.11 is adjusted such that $V_0 = V_t$ and the armature current for this condition is 20 A, determine the power developed by the motor. Neglect R_a.

14.15   A 1000-kVA 11-kV three-phase wye-connected synchronous motor has a 10-Ω synchronous reactance and a negligible armature resistance. Calculate the induced voltage for (a) 0.8 lagging power factor, (b) unity power factor, and (c) 0.8 leading power factor when, in each case, the motor takes 1000 kVA

14.16   The per-phase induced voltage of a synchronous motor is 2500 V. It lags behind the terminal voltage by 30°. If the terminal voltage is 2200 V per phase, determine the operating power factor. The per-phase synchronous reactance is 6 Ω. Neglect the armature resistance.

14.17   The per-phase synchronous reactance of a synchronous motor is 8 Ω, and its armature resistance is negligible. The per-phase input is 400 kW, and the induced voltage is 5200 V per phase. If the terminal voltage is 3800 V per phase, determine (*a*) the power factor and (*b*) the armature current.

14.18   A 2200-V three-phase 60-Hz four-pole wye-connected synchronous motor has a synchronous reactance of 4 Ω and a negligible armature resistance. The excitation is adjusted so that the induced voltage is 2200 V (line to line). If the line current is 220 A at a certain load, calculate (*a*) the input power, (*b*) the developed torque, and (*c*) the power angle.

14.19   An overexcited synchronous motor is connected across a 150-kVA inductive load of 0.7 lagging power factor. The motor takes 12 kW while running on no load. Calculate the kilovoltampere rating of the motor if it is desired to bring the overall power factor of the motor–inductive load combination to unity.

14.20   Repeat Prob. 14.18 if the synchronous motor is used to supply a 100-hp load at an efficiency of 90 percent.

REFERENCES

1   S. A. Nasar and L. E. Unnewehr, *Electromechanics and Electric Machines*, 2d ed., Wiley, New York, 1983.

2   A. E. Fitzgerald, C. Kinsley, Jr., and S. D. Umans, *Electric Machinery*, 4th ed., McGraw-Hill, New York, 1983.

3   S. A. Nasar, *Electric Machines and Transformers*, Macmillan, New York, 1983.

Induction Machines

The induction motor is the most commonly used electric motor. It is the workhorse of the industry. Like the dc machine and the synchronous machine, an induction machine consists of a stator and a rotor. The rotor is mounted on bearings and separated from the stator by an air gap. Electromagnetically, the stator consists of a core made up of punchings (or laminations) carrying slot-embedded conductors. These conductors are interconnected in a predetermined fashion and constitute the armature windings, which are similar to the windings of synchronous machines (Chap. 14).

Alternating current is supplied to the stator windings, and the currents in the rotor windings are induced by the stator currents. The rotor of the induction machine is cylindrical and carries either (1) conducting bars short-circuited at both ends, as in a *cage-type* machine, or (2) a polyphase winding with terminals brought out to slip rings for external connections, as in a *wound-rotor* machine. A wound-rotor winding is similar to that of the stator. Sometimes the cage-type machine is also called a *brushless* machine and the wound-rotor machine a *slip-ring* machine. The stator and the rotor, in its three different stages of production, are shown in Fig. 15.1. The motor is rated at 2500 kW, 3 kV, 575 A, two-pole, and 400 Hz. Figure 15.2 shows the wound rotor of a three-phase slip-ring induction motor.

An induction machine operates on the basis of the interaction of the induced rotor currents and the air-gap fields. If the rotor is allowed to run under the torque developed by this interaction, the machine will operate as a motor. On the other hand, when the rotor is driven by an external source beyond a certain speed, the machine begins to deliver electric power and operates as an induction generator (instead of as an induction motor, which absorbs electric power). Thus, we see that the induction machine is capable of functioning either as a motor or as a generator. In practice, its application as a generator is less common than its application as a motor. We will first study the motor operation, then develop the equivalent circuit

548

FIGURE 15.1
Rotor of a
2500-kW 3-kV
two-pole 400-Hz
induction motor in
different stages of
production.

FIGURE 15.2
Complete rotor of a
3400-kW 6-kV
990-r/min
induction motor.

of an induction motor, and subsequently show that the complete characteristics of an induction machine, operating either as a motor or as a generator, are obtainable from the equivalent circuit.

15.1   OPERATION OF A THREE-PHASE INDUCTION MOTOR

The key to the operation of an induction motor is the production of the rotating magnetic field. We established in the last chapter (Sec. 14.3) that a three-phase stator excitation produces a rotating magnetic field in the air gap of the machine and that the field rotates at a synchronous speed given by Eq. (14.3). As the magnetic field rotates, it cuts the rotor conductors. By this process, voltages are induced in the conductors. The induced voltages give rise to rotor currents, which interact with the air-gap field to produce a torque. The torque is maintained as long as the rotating magnetic field and the induced rotor current exist. Consequently, the rotor starts rotating in the direction of the rotating field. The rotor will achieve a steady-state speed n such that $n < n_s$. When $n = n_s$, there will be no induced currents and hence no torque. The condition $n > n_s$ corresponds to the generator mode.

An alternative approach to explaining the operation of the polyphase induction motor is to consider the interaction of the (excited) stator magnetic field with the (induced) rotor magnetic field. The stator excitation produces a rotating magnetic field, which rotates in the air gap at a synchronous speed. The field induces polyphase currents in the rotor, thereby giving rise to another rotating magnetic field, which also rotates at the same synchronous speed as the stator and with respect to the stator. Thus, we have two rotating magnetic fields, both rotating at a synchronous speed with respect to the stator but stationary with respect to each other. Consequently, according to the principle of alignment of magnetic fields, the rotor experiences a torque. The rotor rotates in the direction of the rotating field of the stator.

15.2   SLIP

The actual mechanical speed n of the rotor is often expressed as a fraction of the synchronous speed n_s as related by *slip* s, defined as

$$s = \frac{n_s - n}{n_s} \tag{15.1}$$

where n_s is given by Eq. (14.3), which is repeated here for convenience;

$$n_s = \frac{120f}{P} \tag{14.3}$$

The slip may also be expressed as percent slip as follows:

$$\text{Percent slip} = \frac{n_s - n}{n_s} \times 100 \tag{15.2}$$

At standstill, the rotating magnetic field produced by the stator has the same relative speed with respect to the rotor windings as with respect to the stator windings. Thus, the frequency of rotor currents f_r is the same as the frequency of stator currents f. At synchronous speed, there is no relative motion between the rotating field and the rotor, and the frequency of rotor current is zero. At other speeds, the rotor frequency is proportional to the slip s; that is,

$$f_r = sf \tag{15.3}$$

where f_r is the frequency of rotor currents, and is known as slip frequency, and f is the frequency of stator input current (or voltage).

Example 15.1   A six-pole three-phase 60-Hz induction motor runs at 4 percent slip at a certain load. Determine the synchronous speed, rotor speed, frequency of rotor currents, speed of the rotor rotating field with respect to the stator, and speed of the rotor rotating field with respect to the stator rotating field.

Solution   From Eq. (14.3), the synchronous speed is

$$n_s = \frac{120 \times 60}{6}$$
$$= 1200 \text{ r/min}$$

From Eq. (15.1), the rotor speed is

$$n = (1 - s)n_s$$
$$= (1 - 0.04)(1200)$$
$$= 1152 \text{ r/min}$$

From Eq. (15.3), the frequency of rotor currents is

$$f_r = 0.04 \times 60$$
$$= 2.4 \text{ Hz}$$

The six poles on the stator induce six poles on the rotor. The rotating field produced by the rotor rotates at a corresponding synchronous speed n_r relative to the rotor such that

$$n_r = \frac{120 f_r}{P} = \frac{120 f}{P} s = sn_s$$

But the speed of the rotor with respect to the stator is

$$n = (1 - s)n_s$$

Hence, the speed of the rotor field with respect to the stator is

$$n_s' = n_r + n$$
$$= sn_s + (1 - s)n_s$$
$$= 1200 \text{ r/min}$$

The speed of the rotor field with respect to the stator field is

$$n_s' - n_s = n_s - n_s = 0$$

15.3 DEVELOPMENT OF EQUIVALENT CIRCUITS

To develop an equivalent circuit of an induction motor, we consider the similarities between a transformer and an induction motor (on a per-phase basis). If we consider the primary of the transformer to be similar to the stator of the induction motor, then its rotor corresponds to the secondary of the transformer. From this analogy, it follows that the stator and the rotor have their own respective resistances and leakage reactances. Because the stator and the rotor are magnetically coupled, we must have a magnetizing reactance, just as in a transformer. The air gap in an induction motor makes its magnetic circuit relatively poor, and thus the corresponding magnetizing reactance will be relatively smaller than that of a transformer. The hysteresis and eddy-current losses in an induction motor can be represented by a shunt resistance, as was done for the transformer. Up to this point, we have mentioned the similarities between a transformer and an induction motor. A major difference between the two, however, is introduced because of the rotation of the rotor. Consequently, the frequency of rotor currents is different from the frequency of stator current; see Eq. (15.3). Keeping these facts in mind, we now represent a three-phase induction motor by a stationary equivalent circuit.

Considering the rotor first and recognizing that the frequency of rotor currents is the slip frequency, we may express the per-phase rotor leakage reactance x_2 at a slip s in terms of the standstill per phase reactance X_2:

$$x_2 = sX_2 \tag{15.4}$$

Next we observe that the magnitude of the voltage induced in the rotor circuit is also proportional to the slip.

A justification of this statement follows from transformer theory because we may view the induction motor at standstill as a transformer with an air gap. For the transformer, we know that the induced voltage, say E_2, is given by

$$E_2 = 4.44fN\phi_m \tag{15.5}$$

But at a slip s, the frequency becomes sf. Substituting this value of frequency into Eq. (15.5) yields the voltage e_2 at a slip s as

$$e_2 = 4.44 sf N\phi_m = sE_2$$

We conclude, therefore, that if E_2 is the per-phase voltage induced in the rotor at standstill, then the voltage e_2 at a slip s is given by

$$e_2 = sE_2 \qquad (15.6)$$

Using Eqs. (15.4) and (15.6), we obtain the rotor equivalent circuit shown in Fig. 15.3a. The rotor current I_2 is given by

$$I_2 = \frac{sE_2}{\sqrt{R_2^2 + (sX_2)^2}}$$

which may be rewritten as

$$I_2 = \frac{E_2}{\sqrt{(R_2/s)^2 + X_2^2}} \qquad (15.7)$$

resulting in the alternative form of the equivalent circuit shown in Fig. 15.3b. Notice that these circuits are drawn on a per-phase basis. To this circuit we may now add the per-phase stator equivalent circuit to obtain the complete equivalent circuit of the induction motor.

In an induction motor, only the stator is connected to the ac source. The rotor is not generally connected to an external source, and rotor voltage and current are produced by induction. In this regard, as mentioned earlier, the induction motor may be viewed as a transformer with an air gap, having a variable resistance in the secondary. Thus, we may consider that the primary of the transformer corresponds to the stator of the induction motor, whereas the secondary corresponds to the rotor on a per-phase basis. Because of the air gap, however, the value of the magnetizing reactance X_m tends to be relatively low, compared to that of a transformer. As in a transformer, we have a mutual flux linking both the stator and the rotor, represented by the magnetizing reactance and various leakage fluxes. For instance, the total rotor leakage flux is denoted by X_2 in Fig. 15.3. Now considering that the rotor is coupled

FIGURE 15.3
Two forms of rotor equivalent circuit.

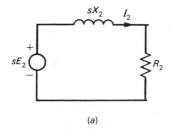

(a)

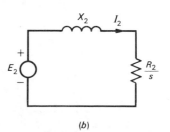

(b)

FIGURE 15.4
Stator and rotor as
coupled circuits.

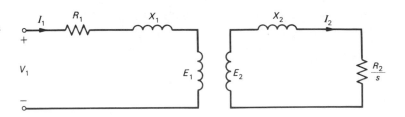

to the stator as the secondary of a transformer is coupled to its primary, we may draw the circuit shown in Fig. 15.4. To develop this circuit further, we need to express the rotor quantities as referred to the stator. The pertinent details are cumbersome and are not considered here. However, having referred the rotor quantities to the stator, we obtain from the circuit given in Fig. 15.4 the exact equivalent circuit (per phase) shown in Fig. 15.5. For reasons that will become immediately clear, we split R_2'/s as

$$\frac{R_2'}{s} = R_2' + \frac{R_2'}{s}(1 - s)$$

to obtain the circuit shown in Fig. 15.5. Here R_2' is simply the per-phase standstill rotor resistance referred to the stator, and $R_2'(1 - s)/s$ is a dynamic resistance that depends on the rotor speed and corresponds to the load on the motor. Notice that all the parameters shown in Fig. 15.5 are standstill values and that the circuit is the per-phase exact equivalent circuit referred to the stator.

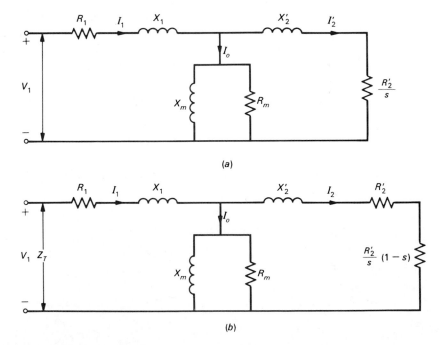

FIGURE 15.5
Two forms of
equivalent circuits
of an induction
motor.

15.4 PERFORMANCE CALCULATIONS

We now show the usefulness of the equivalent circuit in determining motor performance. To illustrate the procedure, we refer to Fig. 15.5. We redraw this circuit in Fig. 15.6, where we also show approximately the power flow and various power losses in one phase of the machine. From Fig. 15.6 we obtain the following relationships on a per-phase basis:

$$\text{Input power } P_i = V_1 I_1 \cos \phi \tag{15.8a}$$

$$\text{Stator } I^2 R_1 \text{ loss} = I_1^2 R_1 \tag{15.8b}$$

$$\text{Power crossing the air gap } P_g = P_i - I_1^2 R_1 \tag{15.8c}$$

Since this power P_g is dissipated in R_2'/s (of Fig. 15.5a), we also have

$$P_g = \frac{I_2^2 R_2'}{s} \tag{15.9}$$

Subtracting the $I_2^2 R_2'$ loss from P_g yields the developed electromagnetic power P_d. Thus

$$P_d = P_g - I_2^2 R_2' \tag{15.10}$$

From Eqs. (15.9) and (15.10) we get

$$P_d = (1 - s)P_g \tag{15.11}$$

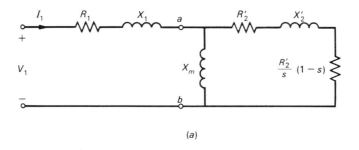

(a)

FIGURE 15.6 (a) An approximate equivalent circuit of an induction motor. (b) Power flow in an induction motor.

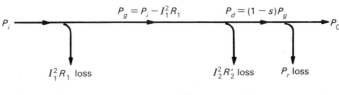

(b)

This power appears across the resistance $R'_2(1 - s)/s$ (of Fig. 15.5b), which corresponds to the load. Subtracting the mechanical rotational power P_r from P_d gives the output power P_o. Hence,

$$P_o = P_d - P_r \tag{15.12}$$

and

$$\text{Efficiency} = \frac{\text{output power}}{\text{input power}} = \frac{P_o}{P_i} \tag{15.13}$$

Torque calculations can be made from the power calculations. Thus, to determine the electromagnetic torque T_e developed by the motor at a speed ω_m rad/s, we write

$$T_e \omega_m = P_d \tag{15.14}$$

But

$$\omega_m = (1 - s)\omega_s \tag{15.15}$$

where ω_s is the synchronous speed in radians per second. From Eqs. (15.11), (15.14), and (15.15) we obtain

$$T_e = \frac{P_g}{\omega_s} \tag{15.16}$$

which gives the torque at a slip s. At standstill $s = 1$; hence, the standstill torque developed by the motor is given by

$$T_{e,\text{standstill}} = \frac{P_{gs}}{\omega_s} \tag{15.17}$$

Notice that we have neglected the core losses, most of which are in the stator. We will include core losses only in efficiency calculations. The reason for this simplification is to reduce the amount of the complex arithmetic required in numerical computations. The following example illustrates the calculation details.

Example 15.2

The parameters of the equivalent circuit in Fig. 15.6 for a 220-V three-phase four-pole wye-connected 60-Hz induction motor are

$R_1 = 0.2\ \Omega$ $R'_2 = 0.1\ \Omega$

$X_1 = 0.5\ \Omega$ $X'_2 = 0.2\ \Omega$

$X_m = 20.0\ \Omega$

The total iron and mechanical losses are 350 W. For a slip of 2.5 percent, calculate the input current, output power, output torque, and efficiency.

Solution From Fig. 15.6 the total impedance is

$$\mathbf{Z}_1 = R_1 + jX_1 + \frac{jX_m(R'_2/s + jX'_2)}{R'_2/s + j(X_m + X'_2)}$$

$$= 0.2 + j0.5 + \frac{j20(4 + j0.2)}{4 + j(20 + 0.2)}$$

$$= (0.2 + j0.5) + (3.77 + j0.95)$$

$$= 4.23\underline{/20°}\ \Omega$$

the phase voltage is

$$V_1 = \frac{220}{\sqrt{3}}$$

$$= 127\ \text{V}$$

the input current is

$$I_1 = \frac{127}{4.23}$$

$$= 30\ \text{A}$$

the power factor is

$$\cos\phi = \cos 20°$$

$$= 0.94$$

and the total input power is

$$3V_1I_1\cos\phi = \sqrt{3} \times 220 \times 30 \times 0.94$$

$$= 10.75\ \text{kW}$$

From the equivalence of Figs. 15.6a and 15.7, we obtain the total power across the air gap:

$$P_g = 3 \times 30^2 \times 3.77$$

$$= 10.18\ \text{kW}$$

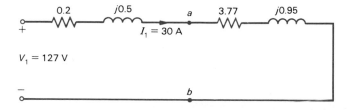

FIGURE 15.7
Example 15.2.

Notice that $3.37\ \Omega$ is the resistance between terminals a-b in the two circuits. The total power developed is

$$P_d = (1 - s)P_g$$
$$= 0.975(10.18)$$
$$= 9.93\ \text{kW}$$

the total output power is

$$P_d - P_{\text{loss}} = 9.93 - 0.35$$
$$= 9.58\ \text{kW}$$

and the total output torque is

$$\frac{\text{Output power}}{\omega_m} = \frac{9.58}{184} \times 1000$$
$$= 52\ \text{N} \cdot \text{m}$$

where $\omega_m = 0.975 \times 60 \times \pi = 184$ rad/s. Thus

$$\text{Efficiency} = \frac{\text{output power}}{\text{input power}}$$
$$= \frac{9.58}{10.75}$$
$$= 89.1\%$$

Using this procedure, we can calculate the performance of the motor at other values of the slip ranging from 0 to 1. The characteristics thus calculated are shown in Fig. 15.8. It may be verified that the approximate equivalent circuit shown in Fig. 15.9 gives almost identical results.

FIGURE 15.8.   Characteristics of an induction motor. T_m = maximum torque; T_s = starting torque.

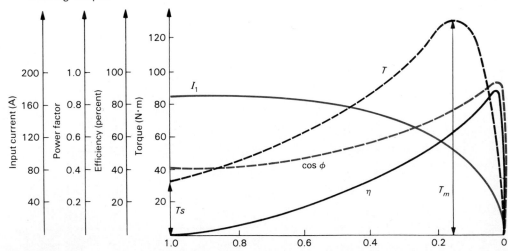

FIGURE 15.9
An approximate
equivalent circuit
(per phase) of an
induction motor.

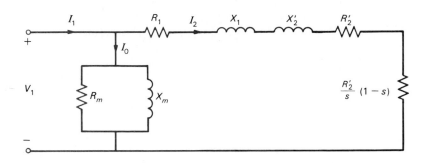

Example 15.3   A two-pole three-phase 60-Hz induction motor develops 25 kW of electromagnetic power at a certain speed. The rotational mechanical loss at this speed is 400 W. If the power crossing the air gap is 27 kW, calculate the slip and the output torque.

Solution   From Eq. (15.10) we have

$$1 - s = \frac{P_d}{P_g}$$

or

$$1 - s = \frac{25}{27}$$

Thus,

$$s - 0.074 \text{ or } 7.4\%$$

The developed torque is given by Eq. (15.16). Substituting $\omega_s = 2\pi n_s/60 = 2\pi \times 3600/60$ and $P_g = 27$ kW (given) in Eq. (15.16), we have

$$T_e = \frac{27,000}{2\pi \times 3600/60}$$

$$= 71.62 \text{ N} \cdot \text{m}$$

The torque lost due to mechanical rotation is found from

$$T_{\text{loss}} = \frac{P_{\text{loss}}}{\omega_m}$$

$$= \frac{400}{(1 - 0.074)2\pi \times 3600/60}$$

$$= 1.15 \text{ N} \cdot \text{m}$$

Hence, the output torque is

$$T_e - T_{\text{loss}} = 71.62 - 1.15$$
$$= 70.47 \text{ N} \cdot \text{m}$$

15.5 PERFORMANCE CRITERIA OF INDUCTION MOTORS

Example 15.2 shows the usefulness of the equivalent circuit in calculating the performance of the motor. The performance of an induction motor may be characterized by the following major factors:

1 Efficiency

2 Power factor

3 Starting torque

4 Starting current

5 Pullout (or maximum) torque

Notice that these characteristics are shown in Fig. 15.8. In design considerations, the heating because of I^2R losses and core losses and the means of heat dissipation must be included. It is not within the scope of this book to discuss in detail the effects of design changes and consequently of parameter variations on each performance characteristic. Here we summarize the results qualitatively. For example, the efficiency is approximately proportional to $1 - s$; thus, the motor would be most compatible with a load running at the highest possible speed. Because the efficiency is clearly dependent on I^2R losses, R'_2 and R_1 must be small for a given load. To reduce

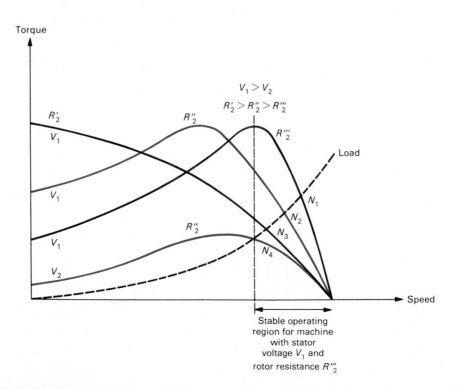

FIGURE 15.10
Effect of rotor resistance on torque-speed characteristics.

core losses, the working flux density B must be small. But this imposes a conflicting requirement on the load current I_2' because the torque is dependent on the product of B and I_2'. In other words, an attempt to decrease the core losses beyond a limit would result in an increase in the I^2R losses for a given load.

We see from the equivalent circuits (developed in Sec. 15.4) that the power factor can be improved by decreasing the leakage reactances and increasing the magnetizing reactance. However, it is not wise to reduce the leakage reactances to a minimum, since the starting current of the motor is essentially limited by these reactances. Again, we notice the conflicting conditions for a high power factor and a low starting current. Also the pullout torque would be higher for lower leakage reactances.

A high starting torque is produced by a high R_2'; that is, the higher the rotor resistance, the higher would be the starting torque. A high R_2' is in conflict with a high efficiency requirement. The effect of varying rotor resistance on the motor torque-speed characteristics is shown in Fig. 15.10, which also shows three different steady-state operating speeds for three values of the rotor resistance and two stator voltages V_1 and V_2.

15.6 SPEED AND TORQUE CONTROL OF INDUCTION MOTORS

Because of its simplicity and ruggedness, the induction motor finds numerous applications. However, it suffers from the drawback that, in contrast to dc motors, its speed cannot be easily and efficiently varied continuously over a wide range of operating conditions. We briefly review the various possible methods by which the speed of the induction motor can be varied either continuously or in discrete steps.

The speed of the induction motor can be varied by varying (1) the synchronous speed of the rotating field or (2) the slip. Because the efficiency of the induction motor is approximately proportional to $1 - s$, any method of speed control that depends on the variation of slip is inherently inefficient. On the other hand, if the supply frequency is constant, varying the speed by changing the synchronous speed results only in discrete changes in the speed of the motor. We now consider these methods of speed control in some detail.

Recall from Eq. (14.3) that the synchronous speed n_s of the traveling field in a rotating induction machine is given by

$$n_s = \frac{120f}{P}$$

where $P =$ number of poles and $f =$ supply frequency, which indicates that n_s can be varied by changing the number of poles P or by changing the frequency f. Both methods have found applications, and we consider here the pertinent qualitative details. This method applies to synchronous motors also.

A list of the different types of ac motors and of the various techniques for speed and torque control of these drives is given in Fig. 15.11. A control scheme (in the

FIGURE 15.11
AC motors and
controls.

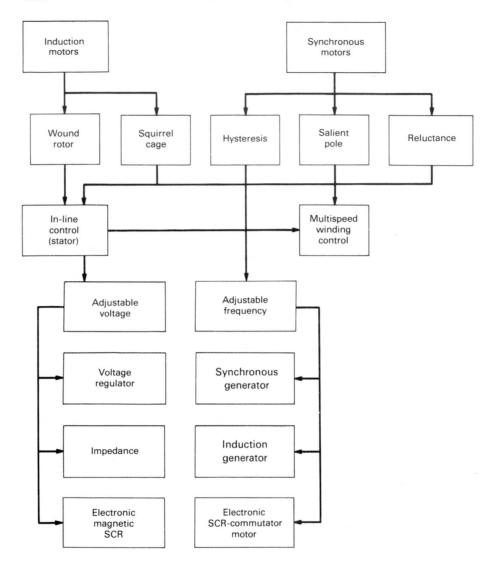

form of a block diagram) for an induction motor is illustrated in Fig. 15.12, in which
the power-conditioning unit (PCU) includes the energy source (which may even be
a dc source), a means of producing an ac variable-frequency source (the inverter)
from the available dc source, and some means of controlling the output voltage [the
adjustable voltage inverter (AVI) or the pulse-width-modulated (PWM) inverter]. In
fact, it would be desirable to keep the voltage-to-frequency (V/f) ratio fixed, as we
shall see.

Inverters   Basically, an inverter has a dc input and an ac output. We see from the
preceding paragraph that the inverter is the backbone of an ac drive system. A wide

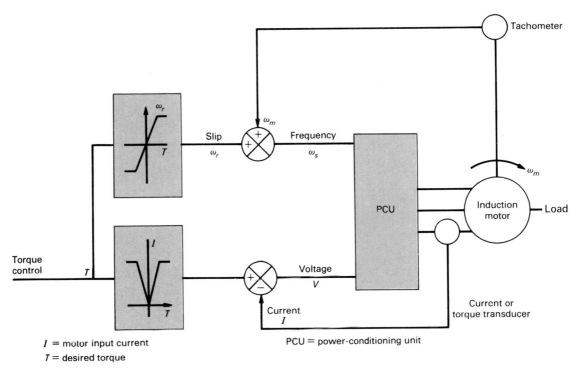

FIGURE 15.12   Block diagram for induction motor control.

variety of inverter circuits may be used for a drive motor. Four common inverter types are

1   AC transistor inverter

2   AC SCR McMurray inverter

3   AC SCR load-commutated inverter

4   AC SCR current inverter (ASCI)

For the present, however, we consider the full-bridge inverter circuit shown in Fig. 15.13, which also shows the voltage and current waveforms. When T_1 and T_3 are conducting, the battery voltage appears across the load with the polarities shown in Fig. 15.13b. But when T_2 and T_4 are conducting, the polarities across the load are reversed. Thus, we get a square-wave voltage across the load, and the frequency of this wave can be varied by varying the frequency of the gating signals. If the load is not purely resistive, the load current will not reverse instantaneously with the voltage. The antiparallel-connected diodes, shown in Fig. 15.13a, allow for the load current to flow after the voltage reversal. The principle of the bridge inverter can be extended to form the three-phase bridge inverter.

FIGURE 15.13
(a) Single-phase full-bridge inverter and (b) its load voltage waveform.

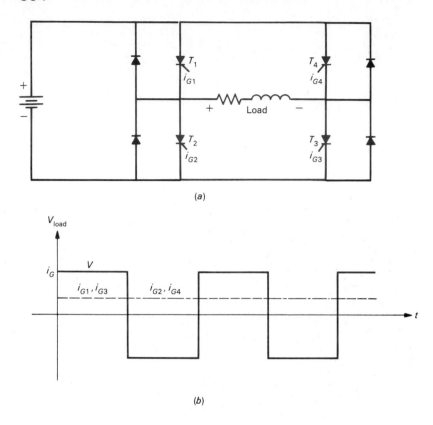

Adjustable Voltage Inverter   In an AVI, the output voltage and frequency can both be varied. The voltage is controlled by including a chopper between the battery and the inverter, whereas the frequency is varied by the frequency of operation of the gating signals. In an AVI, the amplitude of the output decreases with the output frequency, and the V/f ratio essentially remains constant over the entire operating range. The AVI output waveform does not contain second, third, fourth, sixth, eighth, and tenth harmonics. Other harmonic contents as a fraction of the total RMS output voltage are as follows:

Fundamental: 0.965

Fifth harmonic: 0.1944

Seventh harmonic: 0.138

Eleventh harmonic: 0.087

The losses produced by these harmonics tend to heat the motor and thereby decrease the motor efficiency by about 5 percent compared to a motor driven by a purely sinusoidal voltage.

Pulse-Width-Modulated (PWM) and Pulse-Frequency-Modulated (PFM) Inverters   The voltage control in PWM and PFM inverters is obtained in a manner similar to that for a chopper, as shown in Fig. 13.24. In a PWM inverter, the output-voltage amplitude is fixed and equal to the battery voltage. The voltage is varied by varying the width of the pulse "on time" relative to the fundamental half-cycle period, as illustrated in Fig. 15.14a and b. As the voltage is reduced (Fig. 15.14b), the harmonics vary rapidly in magnitude.

Low-frequency harmonics can be reduced by PFM, in which the number and width of pulses within the half-cycle period are varied. PFM waveforms for three different cases are given in Fig. 15.15. Harmonics from the output of a PFM inverter can be reduced by increasing the number of pulses per half cycle. But this requires a reduction in the pulse width, which is essentially limited by the thyristor turn-off time and the switching losses of the thyristor. Furthermore, an increase in the number of pulses increases the complexity of the logic system and thereby increases the overall cost of the PFM inverter system.

In comparing an AVI system with a PWM (or PFM) inverter system, we observe that whereas the AVI system is efficient but expensive, the PWM inverter is relatively inexpensive but inefficient. Both types (AVI and PFM) of inverter system are suitable for induction motors. However, for the control of a synchronous motor, a thyristor inverter requires a motor voltage greater than the dc link voltage in order to turn off the thyristors. Also no diodes are required in the ac portion of the circuit, thus avoiding uncontrolled current. If a synchronous motor is controlled by a transistor inverter, then the motor internal voltage must be less than the dc source voltage; otherwise, the diodes in the inverter circuit will conduct, resulting in uncontrollable currents and high losses.

Variable-Frequency Operation of Induction Motors   The envelope illustrated in Fig. 15.16a shows the torque-speed characteristics of an induction motor for several

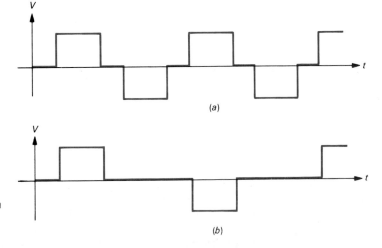

FIGURE 15.14
(a) Full-voltage output and (b) half-full-voltage output of a single-pulse PWM inverter.

FIGURE 15.15
Outputs from a
PFM inverter.

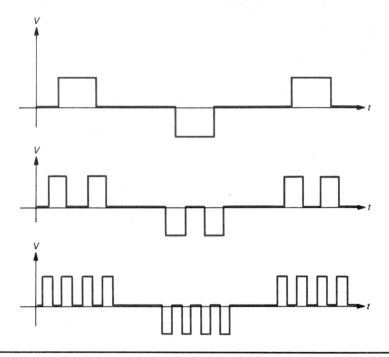

(a)

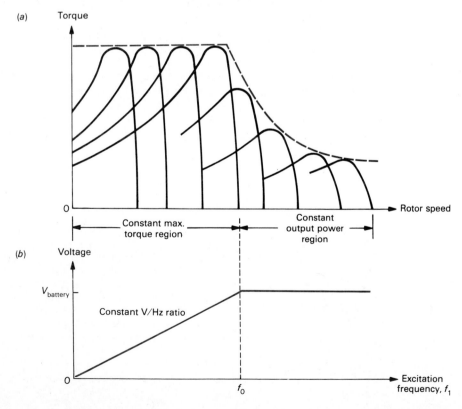

FIGURE 15.16
Typical torque,
speed, voltage
characteristic for
variable-frequency
operation of an
induction motor.

frequencies. The dashed-line envelope defines two distinct operating regions of constant maximum torque and constant output power. In the constant-maximum-torque region, the ratio of applied motor voltage to supply frequency (in volts per hertz) is held constant by increasing the motor voltage directly with frequency, as shown in Fig. 15.16b.

The volts/hertz ratio defines the air-gap magnetic flux in the motor and can be held constant for constant torque within the excitation frequency range and corresponding rotor speeds, extending to frequency f_0. Beyond frequency f_0, the motor voltage cannot be increased to maintain a constant volts/hertz ratio due to the limitation of a finite dc voltage. For motor speeds in the range f_1 is greater than f_0, the supply voltage is held constant at the battery voltage, and the supply frequency is increased to provide the motor speed demand. This is the constant-maximum-output power region of operation, since the maximum developed torque decreases nonlinearly with speed.

The inverter output voltage and current waveforms are rich in harmonics. These harmonics have detrimental effects on motor performance. Among the most important effects are the production of additional losses and harmonic torques. The additional losses that may occur in a cage induction motor owing to the harmonics in the input current are summarized below:

1 *Primary I^2R losses.* The harmonic currents contribute to the total RMS input current. The skin effect in the primary conductors may be neglected in small wire-wound machines, but it should be taken into account in motor analysis when the primary conductor depth is appreciable.

2 *Secondary I^2R losses.* When calculating the additional secondary I^2R losses, we must take the skin effect into account for all sizes of motor.

3 *Core losses due to harmonic main fluxes* These core losses occur at high frequencies, but the fluxes are highly damped by induced secondary currents.

4 *Losses due to skew-leakage fluxes.* These losses occur if there is a relative skew between the rotor and stator conductors. At 60 Hz the loss is usually small, but it may be appreciable at harmonic frequencies. Since the time-harmonic mmf's rotate relative to both the primary and the secondary, skew-leakage losses are produced in both members.

5 *Losses due to end-leakage fluxes.* As in the case of skew-leakage losses, these losses occur in the end regions of both the primary and the secondary and are a function of harmonic frequency.

6 *Space-harmonic mmf losses excited by time-harmonic currents.* These correspond to the losses that, in the case of the fundamental current component, are termed *high-frequency stray-load losses.*

In addition to these losses and harmonic torques, the harmonics act as sources of magnetic noise in the motor.

15.7 STARTING OF INDUCTION MOTORS

Most induction motors—large and small—are rugged enough that they can be started across the line without incurring any damage to the motor windings, even though about 5 to 7 times the rated current flows through the stator at rated voltage at standstill. In large induction motors, however, large starting currents are objectionable in two respects. First, the mains supplying the induction motor may not be of a sufficiently large capacity. Second, a large starting current may cause excessive voltage drops in the lines, resulting in a reduced voltage across the motor. Because the torque varies approximately as the square of the voltage, the starting torque may become so small at the reduced line voltage that the motor might not even start on load. Thus, we formulate the basic requirement for starting: The line current should be limited by the capacity of the mains, but only to the extent that the motor can develop sufficient torque to start (on load, if necessary).

Example 15.4

An induction motor is designed to run at 5 percent slip on full load. If the motor draws 6 times the full-load current at starting at the rated voltage, estimate the ratio of the starting torque to the full-load torque.

Solution The torque at a slip s is given by Eq. (15.16), which in conjunction with Eq. (15.9) becomes

$$T_e = \frac{I_2^2 R_2'}{s\omega_s}$$

At full load, with $I_2 = I_{2f}$, the torque is

$$T_{ef} = \frac{I_{2f}^2 R_2'}{0.05\omega_s}$$

At starting, $I_{2s} = 6I_{2f}$ and $s = 1$, so that

$$T_{es} = \frac{(6I_{2f})^2 R_2'}{\omega_s}$$

Hence,

$$\frac{T_{es}}{T_{ef}} = \frac{(6I_{2f})^2 R_2'}{\omega_s} \frac{0.05\omega_s}{I_{2f}^2 R_2'} = 1.8$$

Example 15.5

If the motor of Example 15.4 is started at a reduced voltage to limit the line current to 3 times the full-load current, what is the ratio of the starting torque to the full-load torque?

Solution In this case we have

$$\frac{T_{es}}{T_{ef}} = 3^2 \times 0.05 = 0.45$$

Notice that the starting torque has been reduced by a factor of 4, relative to the case of full-voltage starting. In many practical cases, the line current is limited to 6 times the full-load current, and the starting torque is desired to be about 1.5 times the full-load torque.

There are numerous types of push-button starters for induction motors now commercially available. In the following, however, we briefly consider only the principles of the two commonly used methods. We consider the current limitation first. Some of the common methods of limiting the stator current while starting are:

Reduced-Voltage Starting A reduced voltage is applied to the stator at the time of starting, and the voltage is increased to the rated value when the motor is within 25 percent of its final speed. This method has the obvious limitation that a variable-voltage source is needed and the starting torque drops substantially. The so-called wye-delta method of starting is a reduced-voltage starting method. If the stator is normally connected in delta, reconnection to wye reduces the phase voltage, resulting in less current at starting; e.g., if the line current at starting is about 5 times the full-load current in a delta-connected stator, the current in the wye connection will be less than 2 times the full-load value. But, at the same time, the starting torque for a wye connection would be about one-third its value for a delta connection. One advantage of wye-delta starting is that it is inexpensive and requires only a three-pole (or three single-pole) double-throw switch or switches, as shown in Fig. 15.17.

Current Limiting by Series Resistance Series resistances inserted in the three lines sometimes are used to limit the starting current. These resistances are short-circuited, once the motor has gained speed. Because of the extra losses in the external resistances, this method has the obvious disadvantage of being inefficient.

Turning now to the starting torque, we recall from the last section that the starting torque is dependent on the rotor resistance. Thus, a high rotor resistance results in a high starting torque. Therefore, in a wound-rotor machine external resistance in the rotor circuit may be conveniently used (see Fig. 15.18). In a cage rotor, deep slots are used, where the slot depth is 2 or 3 times greater than the slot width (see Fig. 15.19). Rotor bars embedded in deep slots provide a high effective resistance and a large torque at starting. Under normal running conditions with low

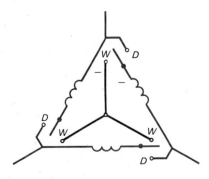

FIGURE 15.17 Wye-delta starting. Switches on *W* correspond to the wye connection, and switches on *D* correspond to the delta connection.

FIGURE 15.18
Effect of changing
rotor resistance on
the starting of a
wound-rotor
motor.

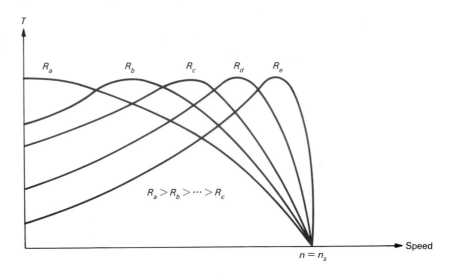

FIGURE 15.19
Deep-bar rotor
slots: (*a*) open;
(*b*) partially
closed.

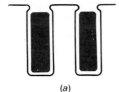 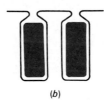

slips, however, the rotor resistance becomes lower and the efficiency high. This characteristic of rotor bar resistance is a consequence of the *skin effect*. Because of the skin effect, the current will have a tendency to concentrate at the top of the bars at starting, when the frequency of rotor currents is high; at this point, the frequency of rotor currents will be the same as the stator input frequency (for example, 60 Hz). While running, the frequency of rotor currents (= slip frequency = 3 Hz at 5 percent slip and 60 Hz) is much lower; at this level of operation, the skin effect is negligible and the current is almost uniformly distributed throughout the entire bar cross section.

The skin effect is used in an alternative form in a *double-cage* rotor (Fig. 15.20), where the inner cage is deeply embedded in iron and has low-resistance bars. The outer cage has relatively high-resistance bars close to the stator. At starting, because of the skin effect, the influence of the outer cage dominates, thus producing a high starting torque. While the motor is running, the current penetrates to full depth into

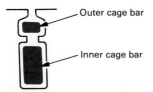

FIGURE 15.20
Form of a slot for a
double-cage rotor.

the lower cage—because of the insignificant skin effect—which results in an efficient steady-state operation. Notice that under normal running conditions both cages carry current, thus somewhat increasing the rating of the motor.

Example 15.6   A motor employs a wye-delta starter which connects the motor phases in wye at the time of starting and in delta when the motor is running. The full-load slip is 4 percent, and the motor draws 9 times the full-load current if started directly from the mains. Determine the ratio of starting torque T_s to full-load torque T_{FL}.

Solution   When the phases are switched to delta, the phase voltage, and hence the full-load current, is increased by a factor of $\sqrt{3}$ over the value it would have had in a wye connection. Then it follows from the last equation in Example 15.5 that

$$\frac{T_s}{T_{FL}} = \left(\frac{9}{\sqrt{3}}\right)^2 (0.04)$$

$$= 1.08$$

PROBLEMS

15.1   A six-pole 60-Hz induction motor runs at 1152 r/min. Determine the synchronous speed and the percent slip.

15.2   A six-pole induction motor is supplied by a synchronous generator having four poles and running at 1500 r/min. If the speed of the induction motor is 750 r/min, what is the frequency of rotor current?

15.3   A four-pole 60-Hz induction motor runs at 1710 r/min. Calculate (a) the slip in percent, (b) the frequency of rotor currents, and (c) the speed of the rotating magnetic field produced by (i) the stator, and (ii) the rotor, with respect to the stator, in revolutions per minute and in radians per second.

15.4   A two-pole 60-Hz wound-rotor induction motor has 127 V per phase across its stator. The voltage induced in the rotor is 3.81 V per phase. Assuming that the stator and the rotor have equal effective numbers of turns per phase, calculate (a) the motor speed, and (b) the slip.

15.5   The power crossing the air gap of an induction motor is 24.3 kW. If the developed electromagnetic power is 21.9 kW, what is the slip? The rotational loss at this slip is 350 W. Calculate the output torque, if the synchronous speed is 3600 r/min.

15.6   A three-phase four-pole 60-Hz induction motor develops a maximum torque of 180 N·m at

800 r/min. If the rotor resistance is 0.2 Ω per phase, determine the developed torque at 1000 r/min.

15.7   A 400-V four pole three-phase 60-Hz induction motor has a wye connected rotor having an impedance of $(0.1 + j0.5)$ Ω per phase. How much additional resistance must be inserted in the rotor circuit for the motor to develop the maximum starting torque? The effective stator-to-rotor turns ratio is 1.

15.8   For the motor of Prob. 15.7, what is the motor speed corresponding to the maximum developed torque without any external resistance in the rotor circuit?

15.9   A two-pole 60-Hz induction motor develops a maximum torque of twice the full-load torque. The starting torque is equal to the full-load torque. Determine the full-load speed.

15.10   The input to the rotor circuit of a four-pole 60-Hz induction motor, running at 1000 r/min, is 3 kW. What is the rotor copper loss?

15.11   The stator current of a 400-V three-phase wye-connected four-pole 60-Hz induction motor running at a 6 percent slip is 60 A at 0.866 power factor. The stator copper loss is 2700 W, and the total iron and rotational losses are 3600 W. Calculate the motor efficiency.

15.12 An induction motor has an output of 30 kW at 86 percent efficiency. For this operating condition, stator copper loss = rotor copper loss = core losses = mechanical rotational losses. Determine the slip.

15.13 A four-pole 60-Hz three-phase wye-connected induction motor has a mechanical rotational loss of 500 W. At 5 percent slip, the motor delivers 30 hp at the shaft. Calculate (a) the rotor input, (b) the output torque, and (c) the developed torque.

15.14 A three-phase 230-V 60-Hz wye-connected two-pole induction motor operates at 3 percent slip while taking a line current of 22 A. The stator resistance and leakage reactance per phase are 0.1 and 0.2 Ω, respectively. The rotor leakage reactance is 0.15 Ω per phase. Calculate (a) the rotor resistance, (b) power crossing the air gap, and (c) developed power. Neglect X_m and R_c.

15.15 A four-pole 60-Hz three-phase induction motor has a rotor resistance of 1.0 Ω and a leakage reactance of 2.5 Ω per phase. Using the rotor circuit only, find the slip at which the motor develops a maximum torque. Also determine the ratio of the maximum developed torque to the torque developed at 5 percent slip.

15.16 A three-phase 60-Hz four-pole induction motor runs at 1710 r/min. If the power crossing the air gap is 120 kW, what is the rotor copper loss? The motor has core and mechanical losses (at 1710 r/min) of 1.7 and 2.7 kW, respectively, and the stator copper loss is 3 kW. Calculate the motor efficiency and output torque.

15.17 A wound-rotor six-pole 60-Hz induction motor has a rotor resistance of 0.8 Ω and runs at 1150 r/min at a given load. The load on the motor is such that the torque remains constant at all speeds. How much resistance must be inserted in the rotor circuit to bring the motor speed down to 950 r/min? Neglect rotor leakage reactance.

15.18 A 400-V three-phase wye-connected induction motor has a stator impedance of $(0.6 + j1.2)\ \Omega$ per phase. The rotor impedance referred to the stator is $(0.5 + j1.3)\ \Omega$ per phase. Using the approximate equivalent circuit, determine the maximum electromagnetic power developed by the motor.

15.19 The motor of Prob. 15.18 has a magnetizing reactance of 35 Ω. Neglecting the iron losses, at 3 percent slip calculate (a) the input current and (b) the power factor. Use the approximate equivalent circuit.

15.20 On no load, a three-phase delta-connected induction motor takes 6.8 A and 390 W at 220 V. The stator resistance is 0.1 Ω per phase. The friction and windage loss is 120 W. Determine the values of parameters X_m and R_c of the equivalent circuit of the motor.

15.21 A four-pole 60-Hz three-phase wound-rotor induction motor has a rotor resistance of 0.2 Ω and a rotor leakage reactance of 2.0 Ω. Using the rotor circuit only, determine the value of the external resistance that must be inserted in the rotor circuit so that the motor develops the same torque at a 20 percent slip as at a 5 percent slip.

15.22 A 400-V 60-Hz three-phase wye-connected four-pole induction motor has the following per-phase equivalent circuit parameters (see Fig. 15.6a):

$$R_1 = 2R_2' = 0.2\ \Omega \qquad X_1 = 2.5X_2' = 0.5\ \Omega$$

$$X_m = 20.0\ \Omega$$

The total mechanical and core losses at 1755 r/min are 800 W. At this speed, calculate (a) the output torque and (b) the motor efficiency.

15.23 A four-pole 400-V three-phase 60-Hz induction motor takes a 150-A current at starting and 25 A while running at full load. The starting torque is 1.8 times the torque at full load at 400 V. If it is desired that the starting torque be the same as the full-load torque, determine (a) the applied voltage and (b) the corresponding line current.

15.24 An induction motor is started by a wye-delta switch. Determine the ratio of the starting torque to the full-load torque if the starting current is 5 times the full-load current and the full-load slip is 5 percent.

15.25 An induction motor is started at a reduced voltage. The starting current is not to exceed 4 times the full-load current, and the full-load torque is 4 times the starting torque. What is the full-load slip? Calculate the factor by which the motor terminal voltage must be reduced at starting.

Small AC Motors

By small motors we imply that such motors generally have outputs of less than 1 hp (or 746 W), although in a special case the output may be greater than 1 hp. Small motors find applications in electric tools, appliances, and office equipment. Whereas permanent magnet small dc commutator motors have numerous applications, such as in toys, control devices, and so on, in the following we consider only small ac motors.

Almost invariably, small ac motors are designed for single-phase operation. The three common types of small ac motors are

1   Single-phase induction motors

2   Synchronous motors

3   AC commutator motors

Special types of small ac motors include

1   AC servomotors

2   Tachometers

3   Stepper motors

16.1   SINGLE-PHASE INDUCTION MOTORS

Let us consider a three-phase cage-type induction motor (discussed in Chap. 15) running light. It may be experimentally verified that if one of the supply lines is disconnected, the motor will continue to run, although at a speed lower than the speed of the three-phase motor. Such an operation of a three-phase induction motor (with one line open) may be considered to be similar to that of a single-phase induction motor. Next, let the three-phase motor be at rest and supplied by a single-phase source. Under this condition, the motor will not start because we have a pulsating magnetic field in the air gap rather than a rotating magnetic field which is required for torque production. Thus we conclude that a single-phase induction motor is not self-starting, but will continue to run if started by some means. This implies that to make it self-starting, the motor must be provided with an auxiliary means of starting. Later we examine the various methods of starting the single-phase induction motor.

Not considering the starting mechanism for the present, we see that the essential difference between the three-phase and single-phase induction motor is that the latter has a single stator winding which produces an air-gap field that is stationary in space, but alternating (or pulsating) in time. On the other hand, the stator of a three-phase induction motor has a three-phase winding that produces a time-invariant rotating magnetic field in the air gap. The rotor of the single-phase induction motor is generally a cage-type rotor and is similar to that of a three-phase induction motor. The rating of a single-phase motor of the same size as a three-phase motor would be smaller, as expected, and single-phase induction motors are most often rated as fractional-horsepower motors.

Performance Analysis   The operating performance of the single-phase induction motor is studied on the basis of the double-revolving field theory. This approach is based on the concept that an alternating magnetic field is equivalent to two rotating magnetic fields rotating in opposite directions. When this concept is expressed mathematically, the alternating field is of the form

$$B(\theta, t) = B_m \cos \theta \sin \omega t \qquad (16.1)$$

Then (16.1) may be rewritten as

$$B_m \cos \theta \sin \omega t = \tfrac{1}{2} B_m \sin (\omega t - \theta) + \tfrac{1}{2} B_m \sin (\omega t + \theta) \qquad (16.2)$$

In (16.2) the first term on the right-hand side denotes a forward rotating field, whereas the second term corresponds to a backward rotating field. The theory based on such a resolution of an alternating field into two counterrotating fields is known as the *double-revolving field theory*. The direction of rotation of the forward rotating field is assumed to be the same as the direction of rotation of the rotor. Thus if the rotor runs at n r/min and n_s is the synchronous speed in revolutions per minute, the slip s_f of the rotor with respect to the forward rotating field is the same as s, defined by (15.1), or

$$s_f = s = \frac{n_s - n}{n_s} = 1 - \frac{n}{n_s} \qquad (16.3)$$

FIGURE 16.1 (a) Equivalent circuit and (b) torque-speed characteristics, based on double-revolving field theory of the single-phase induction motor.

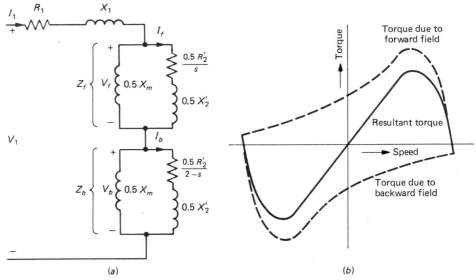

(a)

(b)

But the slip s_b of the rotor with respect to the backward rotating flux is given by

$$s_b = \frac{n_s - (-n)}{n_s} = 1 + \frac{n}{n_s} = 2 - s \tag{16.4}$$

We know from the operation of polyphase motors that, for $n < n_s$, (16.3) corresponds to a motor operation and (16.4) denotes the braking region. Thus the two resulting torques have an opposite influence on the rotor.

The torque relationship for the polyphase induction motor is applicable to each of the two rotating fields of the single-phase motor. We notice from (16.2) that the amplitude of the rotating fields is one-half the alternating flux. Thus the total magnetizing and leakage reactances of the motor can be divided equally so as to correspond to the forward and backward rotating fields. The approximate equivalent circuit of a single-phase induction motor, based on the double-revolving field theory, becomes as shown in Fig. 16.1a. The torque-speed characteristics are qualitatively shown in Fig. 16.1b. The following example illustrates the usefulness of the circuit.

Example 16.1   For a 230-V single-phase induction motor, the parameters of the equivalent circuit, in Fig. 16.1a, are $R_1 = R'_2 = 8\ \Omega$, $X_1 = X'_2 = 12\ \Omega$, and $X_m = 200\ \Omega$. At a slip of 4 percent, calculate (a) the input current, (b) the input power, (c) the developed power, and (d) the developed torque (at rated voltage). The motor speed is 1728 r/min.

Solution   From Fig. 16.1a,

$$Z_f = \frac{(j100)(4/0.04 + j6)}{j100 + 4/0.04 + j6} = 47 + j50 \ \Omega$$

$$Z_b = \frac{(j100)(4/1.96 + j6)}{j100 + 4/1.96 + j6} = 1.8 + j5.7 \ \Omega$$

$$Z_1 = R_1 + jX_1 = 8 + j12 \ \Omega$$

$$Z_{total} = 56.8 + j67.7 = 88.4 \underline{/50°} \ \Omega$$

(a)   Input current $\equiv I_1 = \dfrac{230}{88.4} = 2.6$ A

(b)   Power factor $= \cos 50° = 0.64$ lagging

   Input power $= (230)(2.6)(0.64) = 382.7$ W

(c)   Proceeding as in Example 15.2, we have

$$P_d = (I_1^2 \ \text{Re} \ Z_f)(1 - s) + (I_1^2 \ \text{Re} \ Z_b) \ [1 - (2 - s)]$$
$$= I_1^2 (\text{Re} \ Z_f - \text{Re} \ Z_b)(1 - s) = (2.6)^2 (47 - 1.8)(1 - 0.04)$$
$$= 293.3 \ \text{W}$$

(d)   Torque $= \dfrac{P_d}{\omega_m} = \dfrac{293.3}{2\pi(1728)/60} = 1.62$ N · m

---

Example 16.2   To reduce the numerical computation, Fig. 16.1a is modified by neglecting $0.5X_m$ in Z_b and taking the backward circuit rotor resistance at low slips as $0.25R_2'$. With these approximations, repeat the calculations of Example 16.1 and compare the results.

Solution

$$Z_f = 47 + j50 \ \Omega$$
$$Z_b = 2 + j6 \ \Omega$$
$$Z_1 = 8 + j12 \ \Omega$$
$$Z_{total} = 57 + j68 = 88.7 \underline{/50°} \ \Omega$$

(a) $I_1 = \dfrac{230}{88.7} = 2.6$ A

(b) $\cos \phi = 0.64$ lagging

   Input power $= (230)(2.6)(0.64) = 382.7$ W

(c) $P_d = (2.6)^2 (47 - 2)(1 - 0.04) = 292.0$ W

(d) Torque $= \dfrac{292.0}{2\pi(1728)/60} = 1.61$ N·m

Following is a comparison of the results of the preceding two examples:

Example	Input Current, A	Input Power, W	Power Factor	Developed Torque, N·m
16.1	2.6	382.7	0.64	1.62
16.2	2.6	382.7	0.64	1.61

This comparison indicates that the approximation suggested in Example 16.2 is adequate for most cases.

In the next example we show the procedure for efficiency calculations for a single-phase induction motor.

Example 16.3

A single-phase 110-V 60-Hz four-pole induction motor has the following constants in the equivalent circuit, Fig. 16.1a; $R_1 = R'_2 = 2\,\Omega$, $X_1 = X'_2 = 2\,\Omega$, and $X_m = 50\,\Omega$. There is a core loss of 25 W and a friction and windage loss of 10 W. For a 10 percent slip, calculate (a) the motor input current and (b) the efficiency.

Solution

$$\mathbf{Z}_f = \frac{(j25)(1/0.1 + j1)}{j25 + 1/0.1 + j1} = 8 + j4\ \Omega$$

$$\mathbf{Z}_b = \frac{(j25)(1/1.9 + j1)}{j25 + 1/1.9 + j1} = 0.48 + j0.96\ \Omega$$

$$\mathbf{Z}_1 = 2 + j2\ \Omega$$

$$\mathbf{Z}_{\text{total}} = 10.48 + j6.96 = 12.6\underline{/33.6^\circ}\ \Omega$$

(a) $I_1 = \dfrac{110}{12.6} = 8.73$ A

(b) Developed power $= (8.73)^2(8 - 0.48)(1 - 0.10) = 516$ W

Output power $= 516 - 25 - 10 = 481$ W

Input power $= (110)(8.73)(\cos 33.6^\circ) = 800$ W

Efficiency $= \dfrac{481}{800} = 60\%$

Starting of Single-Phase Induction Motors   We already know that because of the absence of a rotating magnetic field, when the rotor of a single-phase induction motor is at standstill, it is not self-starting. The two methods of starting a single-phase motor are to introduce commutator and brushes, such as in a repulsion motor, and to produce a rotating field by means of an auxiliary winding, such as by split phasing. We consider the latter method next.

From the theory of the polyphase induction motor, we know that to have a rotating magnetic field, we must have at least two mmf's which are displaced from each other in space and carry currents having different time phases. Thus, in a single-phase motor, a starting winding on the stator is provided as a source of the second mmf. The first mmf arises from the main stator winding. The various methods to achieve the time and space phase shifts between the main winding and starting winding mmf's are summarized below.

Split-Phase Motors   This type of motor is represented schematically in Fig. 16.2*a*, where the main winding has a relatively low resistance and a high reactance. The starting winding, however, has a high resistance and a low reactance and has a centrifugal switch, as shown. The phase angle α between the two currents I_m and I_s is about 30 to 45°, and the starting torque T_s is given by

$$T_s = KI_mI_s \sin \alpha \qquad (16.5)$$

where K is a constant. When the rotor reaches a certain speed (about 75 percent of its final speed), the centrifugal switch comes into action and disconnects the starting winding from the circuit. The torque-speed characteristic of the split-phase motor is of the form shown in Fig. 16.2*b*. Such motors find applications in fans, blowers, and so forth and are rated up to $\frac{1}{2}$ hp.

A higher starting torque can be developed by a split-phase motor by inserting a series resistance in the starting winding. A somewhat similar effect may be obtained by inserting a series inductive reactance in the main winding. This reactance is short-circuited when the motor builds up speed.

FIGURE 16.2   (*a*) Connections for a split-phase motor; (*b*) a torque-speed characteristic.

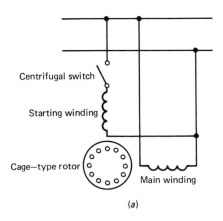

(a)

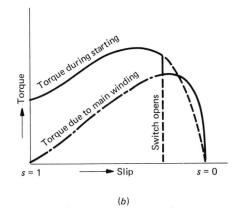

(b)

Capacitor-Start Motors   By connecting a capacitance in series with the starting winding, as shown in Fig. 16.3, the angle α in (16.5) can be increased. The motor will develop a higher starting torque by doing this. Such motors are not restricted merely to fractional-horsepower ratings and may be rated up to 10 hp. At 110 V, a 1-hp motor requires a capacitance of about 400 μF, whereas 70 μF is sufficient for a $\frac{1}{8}$-hp motor. The capacitors generally used are inexpensive electrolytic types and can provide a starting torque that is almost 4 times the rated torque.

As shown in Fig. 16.3, the capacitor is merely an aid to starting and is disconnected by the centrifugal switch when the motor reaches a predetermined speed. However, some motors do not have the centrifugal switch. In such a motor, the starting winding and the capacitor are meant for permanent operation, and the capacitors are much smaller. For example, a 110-V $\frac{1}{2}$-hp motor requires a 15-μF capacitance.

A third kind of capacitor motor uses two capacitors: one that is left permanently in the circuit together with the starting winding and one that gets disconnected by a centrifugal switch. Such motors are, in effect, unbalanced two-phase induction motors.

Shaded-Pole Motors   Another method of starting very small single-phase induction motors is to use a shading band on the poles, as shown in Fig. 16.4, where the main single-phase winding is also wound on the salient poles. The shading band is simply a short-circuited copper strap wound on a portion of the pole. Such a motor is known as the *shaded-pole motor*. The purpose of the shading band is to retard (in time) the portion of flux passing through it in relation to the flux coming out of the rest of the pole face. Thus, the flux in the unshaded portion reaches its maximum before that located in the shaded portion. And we have a progressive shift of flux from the direction of the unshaded portion to the shaded portion of the pole, as shown in Fig. 16.4. The effect of the progressive shift of flux is similar to that of a rotating flux, and because of it, the shading band provides a starting torque. Shaded-pole motors are the least expensive of the fractional-horsepower motors and are generally rated up to $\frac{1}{20}$ hp.

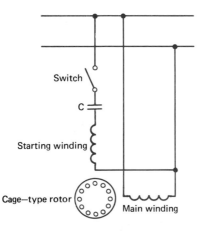

FIGURE 16.3
A capacitor-start motor.

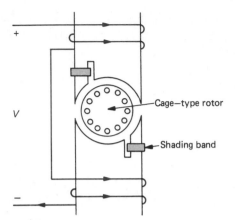

FIGURE 16.4
A shaded-pole motor.

16.2 SMALL SYNCHRONOUS MOTORS

The two common types of small synchronous motors are the *reluctance motor* and the *hysteresis motor*. These motors are constant-speed motors and are used in clocks, timers, turntables, and so forth.

Reluctance Motor   We are somewhat familiar with the reluctance motor from Chap. 14. The torque in a reluctance motor is similar to the torque arising from saliency in a salient-pole synchronous motor. Schematically, a single-phase reluctance motor is shown in Fig. 16.5a. A reluctance motor starts as an induction motor, but normally operates as a synchronous motor. The stator of a reluctance motor is similar to that of an induction motor (single-phase or polyphase). Thus, to start a single-phase motor, almost any of the methods discussed in Sec. 16.1 may be used. A three-phase reluctance motor is self-starting when started as an induction motor. After starting, to pull it into step and then to run it as a synchronous motor, a three-phase motor should have low rotor resistance. In addition, the combined inertia of the rotor and the load should be small. A typical construction of a four-pole rotor is shown in Fig. 16.6. Here the aluminum in the slots and in spaces where teeth have been removed serves as the rotor of an induction motor for starting.

FIGURE 16.5
(*a*) A single-phase reluctance motor;
(*b*) inductance of stator winding, as a function of rotor position.

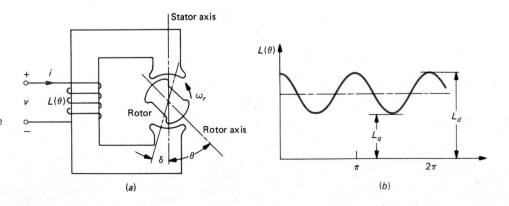

(a)

(b)

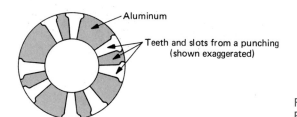

FIGURE 16.6
Rotor of a reluctance motor.

With the maximum inductance L_d and the minimum inductance L_q as defined in Fig. 16.5b, the average value of the torque developed by the motor, for a single-phase current $i = I_m \sin \omega t$, is found to be

$$T_e = \tfrac{1}{8} I_m^2 (L_d - L_q) \sin 2\delta \qquad (16.6)$$

where δ is the power angle (defined in Chap. 14).

Example 16.4

A two-pole one-phase 60-Hz reluctance motor carries 5.66 A of current. The direct- and quadrature-axis inductances, L_d and L_q, respectively, are given by $L_d = 2L_q = 200$ mH. Determine (a) the rotor speed, (b) the power angle for maximum torque, and (c) the maximum value of the developed torque.

Solution

(a) Synchronous speed:

$$n_s = \frac{120f}{P} = \frac{120 \times 60}{2} = 3600 \text{ r/min}$$

(b) For maximum torque

$$\sin 2\delta = 1$$
$$2\delta = 90°$$
$$\delta = 45°$$

(c) The maximum torque for $\delta = 45°$ is obtained (16.6), which gives

$$T_e = \tfrac{1}{8}(\sqrt{2} \times 5.66)^2(200 - 100) \times 10^{-3} = 0.8 \text{ N} \cdot \text{m}$$

Hysteresis Motor   Like the reluctance motor, the hysteresis motor does not have a dc excitation on its rotor. Unlike the reluctance motor, however, the hysteresis motor does not have a salient rotor. Instead, the rotor of a hysteresis motor has a ring of special magnetic material, such as chrome, steel, or cobalt, mounted on a cylinder of aluminum or some other nonmagnetic material, as shown in Fig. 16.7a.

FIGURE 16.7
(*a*) Rotor of a
hysteresis motor;
(*b*) motor torque
characteristic.

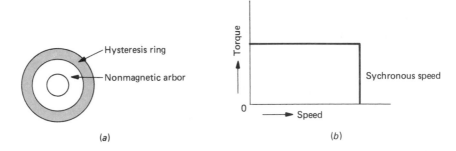

(*a*)   (*b*)

The stator of the motor is similar to that of an induction motor, and the hysteresis motor is started as an induction motor. The torque-speed characteristic of the motor is shown in Fig. 16.7*b*.

16.3   AC COMMUTATOR MOTORS

In preceding chapters we distinguished a dc motor from an ac motor by the presence of a commutator in the dc motor. However, there exist a number of types of motors that have commutators but operate on alternating current. In the following discussion we consider only the universal motor.

Universal Motors   A universal motor is a series motor that may be operated either on direct current or on single-phase alternating current at approximately the same speed and power output, while supplied at the same voltage (on direct current and on alternating current), and the frequency of the ac voltage does not exceed 60 Hz. Much of the application of universal motors is in domestic appliances and tools such as food mixers, sewing machines, vacuum cleaners, portable drills, and saws. There are two types of universal motors: *uncompensated* and *compensated*. The former is less expensive and simpler in construction. It is generally used for lower power outputs and higher speeds than the compensated motor. Typical torque-speed characteristics of the two types of motors are shown in Fig. 16.8. Notice that the compensated motor has better universal characteristics in that the operation on direct current and on alternating current at 60 Hz is not substantially different in the high-speed range. The two types of motors differ in construction also. The uncompensated motor has salient poles and a concentrated field winding, whereas the field winding of a compensated type of motor is distributed on a nonsalient-pole magnetic structure. In addition, some compensated-type universal motors have a compensating winding, and some compensated motors have only one field winding, which also serves as a compensating winding.

The principle of operation of the universal motor is similar to that of the dc series motor, the construction of the two being essentially similar. As seen from Chap. 13, the torque equation of a dc motor may be written as

$$T_e = k_m \varphi I_a$$

FIGURE 16.8
Characteristics of
(a) uncompensated
and
(b) compensated
universal motors.

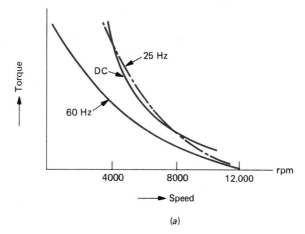

(a)

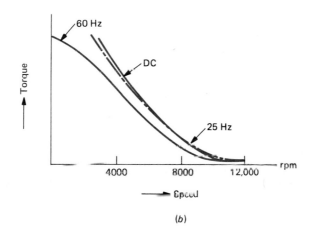

(b)

where $k_m = ZP/(2\pi a) = $ a constant. If saturation is neglected, $\varphi = k_f I_f$, where k_f is a constant and I_f is the field current. Since $I_f = I_a = I$ in a series motor, by denoting $k_f k_m = k$, the developed torque of a dc series motor is given as

$$T_e = kI^2 \tag{16.7}$$

where k is a constant and I is the current through the field and the armature, the two being in series. Under ac operation, with a current $I_m \sin \omega t$, Eq. (16.7) yields

$$T_e = I_m^2 \sin^2 \omega t \tag{16.8}$$

Since $\sin^2 \omega t = \frac{1}{2}(1 - \cos 2\omega t)$, Eq. (16.8) becomes

$$T_e = kI^2(1 - \cos 2\omega t) \tag{16.9}$$

which is an expression for the instantaneous torque. The time-average value of T_e is therefore

$$(T_e)_{\text{av}} = kI^2 \tag{16.10}$$

Hence the average torque is unidirectional and has the same magnitude as that with dc excitation, but the torque pulsates at twice the supply frequency.

In contrast to the solid poles of small dc series motors, the field structure of a universal motor is laminated to reduce hysteresis and eddy-current losses on ac operation. For a given voltage, the universal motor will draw less current (because of the reactance of the field and armature windings), develop less torque, and hence operate at a lower speed on alternating current compared to the operation of the motor on direct current. Another difference between the alternating current and direct current operations of a universal motor is that on alternating current the commutation is poorer because of the voltage induced, by transformer action, in the coils undergoing commutation.

A compensating winding is used to neutralize the reactance voltage of the armature winding. The compensating winding is connected in series with the armature, but is arranged so that the mmf of this winding opposes and neutralizes the armature mmf. Thus, the compensating winding is displaced 90° (electrical) from the field winding. The compensating winding not only neutralizes the armature reactance but also aids in commutation.

16.4 TWO-PHASE MOTORS

Two-phase ac motors are usually used in instrumentation and control systems. Two such motors, discussed next, are alternating current tachometers and two-phase servomotors.

AC Tachometers In many control applications it is desirable to express the speed of a motor as an alternating current voltage. A small two-phase induction motor, connected as shown in Fig. 16.9, is suitable for this purpose. The reference winding

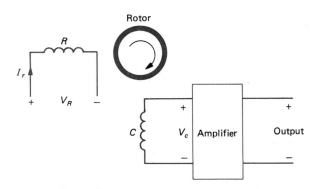

FIGURE 16.9
A two-phase
tachometer.

R is connected to an alternating current source of constant voltage and frequency. While the rotor is rotating, a voltage of the same frequency (as that of the reference voltage) is available at the control winding. The magnitude of this voltage is (ideally) linearly proportional to the speed of the rotor, and the phase of the voltage is fixed with respect to the reference voltage. In practice, the control winding is connected to a high-impedance amplifier and thus may be considered open-circuited for practical purposes. For a given current in the reference winding $\mathbf{I}_r$, the control-winding voltage $\mathbf{V}_c$ is given by

$$\mathbf{V}_c = k\mathbf{I}_r(\mathbf{Z}_f - \mathbf{Z}_b) \tag{16.11}$$

where k is a constant and $\mathbf{Z}_f$ and $\mathbf{Z}_b$ are the forward and backward impedances of Fig. 16.1a. Because $\mathbf{Z}_f$ and $\mathbf{Z}_b$ are functions of speed, $\mathbf{V}_c$ will measure the motor speed. AC tachometers are commonly used in 400-Hz systems.

Two-Phase Servomotors A two-phase servomotor, used in control applications, is very much similar to the two-phase alternating current tachometer just described. Servomotors are used to transform a time-varying signal to a time-varying motion, e.g., in an antenna-positioning system. An arrangement of a two-phase servomotor is shown in Fig. 16.10. The motor has two stator field windings displaced from each other by 90 electrical degrees. In a conventional two-phase induction motor, the voltages V_R and V_c (Fig. 16.10) are equal and 90° displaced from each other in time. The speed of the induction motor depends on the load (or slip). However, the speed of the servomotor must be proportional to an input signal voltage. The control voltage V_c results from the amplitude of the voltage of phase 2, but modulated by the signal. The motor speed, in turn, will depend on the signal or control voltage V_c. In this sense, the alternating current tachometer is a generator, whereas the servomotor functions as a motor.

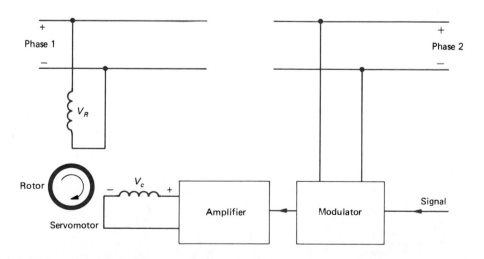

FIGURE 16.10
A two-phase
servomotor.

16.5   STEPPER MOTORS

The stepper motor is not operated to run continuously. Rather, it is designed to rotate in steps in response to electrical pulses received at its input from a control unit. The motor indexes in precise angular increments. The average shaft speed of the motor n_{av} is given by

$$n_{av} = \frac{60\,(\text{pulses per second})}{\text{number of phases in winding}} \quad \text{r/min}$$

Two basic types of stepper motors are variable-reluctance (VR) and permanent magnet (PM) stepper motors. In principle, a stepper motor is a synchronous motor and typically has three-phase or four-phase windings on the stator. The number of poles on the rotor depends on the step size (or angular displacement) per input pulse. The rotor either is reluctance type or may be made of permanent magnet. When a pulse is given to one of the phases on the stator, the rotor tends to align with the mmf axis of the stator coil. These coils are sequentially switched, and the rotor follows the stator mmf in sequence.

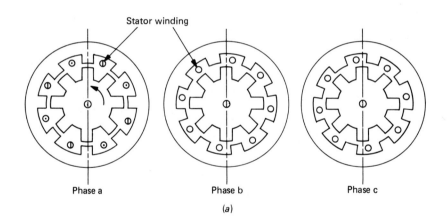

Phase a   Phase b   Phase c

(a)

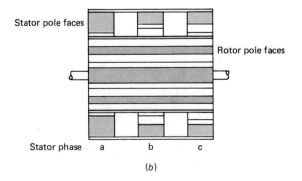

FIGURE 16.11   A variable-reluctance stepper motor: (a) phase *a* energized and relative displacements of stator phases *b* and *c*; (b) cross section of the assembled motor.

A variable-reluctance motor is shown in Fig. 16.11. The machine has a rotor with eight poles and three independent eight-pole stators arranged coaxially with the rotor as shown in Fig. 16.11*b*. When phase *a* is energized, the motor poles align with the stator poles of phase *a*. Notice from Fig. 16.11*a* that the phase *b* stator is displaced from the phase *a* stator by 15° in the counterclockwise direction, and the phase *c* stator is further displaced from the phase *b* stator by another 15° in the counterclockwise direction. Now, when the current in phase *a* is turned off and phase *b* is energized, the rotor will rotate counterclockwise through 15°. Next, when phase *b* current is turned off and phase *c* is energized, the rotor will turn by another 15° in the counterclockwise direction. Finally, turning off the current in phase *c* and exciting phase *a* will complete one step (of 45°) in the counterclockwise direction. Additional current pulses in the sequence *abc* will produce further steps in the counterclockwise direction. Reversal of rotation is obtained by reversing the phase sequence to *acb*.

The stepping motion of the rotor of a stepper motor is typical of an undamped system, as illustrated in Fig. 16.12. The initial displacement of the rotor overshoots the final position and then gradually settles down to the final position. Some means of damping these oscillations is necessary in stepper motors. As the frequency of pulsing is increased, the period τ decreases. When τ is close to the oscillatory period t_1, the motor reaches its operating limit.

As seen from the preceding discussion, the step angle of a motor is determined by the number of poles. Typical step angles are 15°, 5°, 2°, and 0.72°. The choice of step angle depends on the angular resolution required for the application. The speed at which a stepper motor can operate is limited by the degree of damping existing in the system. Speeds up to 200 steps per second are typically attainable. A steady and continuous speed of rotation (*slewing*) greater than this value can be achieved, but the motor is then unable to stop the system in a single step.

The parameters that are important in the selection of a stepper motor are speed, torque, single step response, holding torque, settling time, and slewing. In the slewing mode, the rotor does not come to a complete stop before the next pulse is switched on. The typical torque steps per second characteristic of a step motor is shown in Fig. 16.13. Stepper motors have a wide range of applications, including process control, machine tools, computer peripherals, and certain medical equipment.

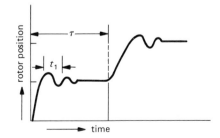

FIGURE 16.12
Undamped
response of
the rotor to
step inputs.

FIGURE 16.13
Torque-pulse
rate characteristics

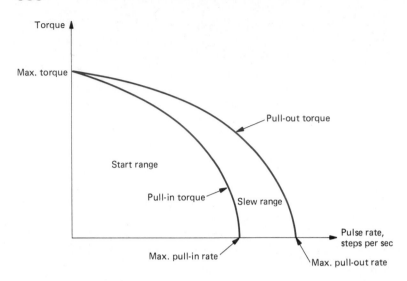

PROBLEMS

16.1 A 60-Hz single-phase induction motor has six poles and runs at 1000 r/min. Determine the slip with respect to (a) the forward rotating field and (b) the backward rotating field.

16.2 A 110-V single-phase two-pole 60-Hz induction motor is designed to run at 3420 r/min. The parameters of the motor equivalent circuit of Fig. 16.1a are $R_1 = R'_2 = 6\ \Omega$, $X_1 = X'_2 = 10\ \Omega$, and $X_m = 80\ \Omega$. Determine (a) the input current and (b) the developed torque at 3420 r/min.

16.3 The circuit of Fig. 16.1a is modified by neglecting $0.5X_m$ from the backward circuit of the rotor. In this circuit the rotor resistance is replaced by $0.3R'_2$. With these approximations, repeat the calculations of Prob. 16.2, and compare the results.

16.4 If the motor of Prob. 16.2 has a core loss of 8 W and a friction and windage loss of 10 W, determine the motor efficiency at 5 percent slip, using (a) the equivalent of Fig. 16.1a and (b) the modified equivalent circuit of Prob. 16.3.

16.5 What is the relative amplitude of the resultant forward-rotating flux density to the resultant

backward-rotating flux density for the motor of Prob. 16.2 at 5 percent slip?

16.6 The direct- and quadrature-axis inductances of a reluctance motor are 60 and 25 mH. The motor operates at a power angle of 15° while taking 2.2 A of current. Calculate the torque developed by the motor.

16.7 Determine the torque developed by the motor of Prob. 16.6 if the motor is supplied by a 110-V 60-Hz source and the power angle is 15°. Neglect the motor winding resistance.

16.8 What is the operating speed of a reluctance motor if it has four poles and is operated at 400 Hz?

16.9 A 110-V universal motor has an input impedance of $(30 + j40)\ \Omega$. The armature has 36 conductors and is lap-wound. The field has two poles, and the flux per pole is 30 mWb. Determine the torque developed by the motor.

16.10 The motor of Prob. 16.9 has a total no-load and rotational loss of 30 W at 4000 r/min. Calculate (a) the power factor and (b) the efficiency of the motor.

Control
and
Instrumentation

17

Feedback Control Systems

The title's three words—"feedback," "control," and "systems"—describe the key themes of this chapter. In reverse order, these words lead us from generalized to fairly specific ideas and concepts having to do with the control of physical systems, which is the subject of this chapter.

Systems, one of the catchwords of our age, is a term used in many fields of study—ecology, economics, and the social sciences as well as the physical sciences. We use the term to describe an assemblage of components, subsystems, and interfaces arranged or existing in such a manner as to perform a function or functions or to achieve a goal. *Control* refers to the function or purpose of the system we wish to discuss; it does not refer to such functions as power, motion, sensing, or communicating, which are the functions of most of the components and circuits previously discussed. A chief distinction is that control is almost always realized not by a single component (such as a transistor, resistor, or motor) but by an entire system of components and interfaces. Finally, we use the term *feedback* in this chapter to mean a specific system configuration—as distinct from the more generalized meaning in common use today that associates the term with human communications and interrelationships.

17.1 BASIC CONCEPTS OF CONTROL

Figure 17.1 illustrates a basic feedback control system in block-diagram format and introduces some notation which is used throughout this chapter. Each block of the diagram in Fig. 17.1 has a specific meaning. For example, relationships between an input function and an output function are defined by the equation

$$A_n(s) = \frac{V_{no}(s)}{V_{ni}(s)} \tag{17.1}$$

FIGURE 17.1
Generalized
feedback control
system.

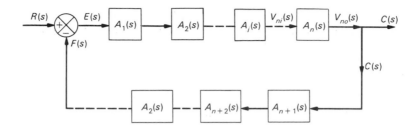

where $A_n(s)$ is defined as the transfer function, $V_{no}(s)$ is the Laplace transform of an output function, and $V_{ni}(s)$ is the Laplace transform of an input function.

The use of the argument s in the above equation indicates Laplace-transformed variables. Such a formulation results in the simplification of the "algebra of block diagrams," plus many useful theorems related to stability and response which are discussed subsequently. For a discussion of Laplace transforms, see App. C.

Other definitions associated with Fig. 17.1 are related to topography. The upper branch of the diagram is known as the *forward loop*. The output of this branch is called the *controlled variable C(s)*, or sometimes the *output variable*. The input is the *error variable E(s)*. The lower branch is termed the *feedback loop*, since it is a path for feeding back the output variable to the input. The circle at the left represents an *error detector*, or *comparator*. The incoming signal on the left is called the *reference variable R(s)*, or sometimes the *input variable*. The symbols V, C, E, and R are used in Fig. 17.1 and throughout this chapter to represent generalized variables; these may be state variables—current, voltage, velocity, position, force, etc.—or variables related to the state variable. The algebraic symbols in the error-detector circle are such as to indicate *negative feedback*; i.e., the *feedback signal F(s)* is algebraically subtracted from the reference signal $R(s)$. If the sign within the lower quadrant of the circle were positive, *positive feedback* would be indicated. We now give mathematical and physical significance to these definitions.

17.1.1 Block Diagrams for Linear Control Systems

The linearity of a control system influences not only its characteristics and performance but also the means of mathematical representation and analysis. A linear system is one that can be described by linear differential equations. In a physical sense, it is also useful to consider linearity in terms of the physical coefficients or circuit constants that compose a given system. For example, a linear resistance is defined by the familiar Ohm's law relationship $r(t) = v(t)/i(t) = R$ (a constant). If in a certain control system the voltage across this resistance is considered an output variable and the current through this resistance the input variable, then the resistance function could be described by the Laplace-transformed equation

$$V(s) = RI(s) \tag{17.2}$$

and the transfer function

$$A(s) = \frac{V(s)}{I(s)} = R \tag{17.3}$$

A nonlinear resistance might be described by the equation $r(t) = R_o[i(t)]^2$, indicative of the variation of resistance with current magnitude (or temperature), or by the equation $r(t) = f[v(t)]^a$, typical of the varistor and other metal-oxide resistors. Note that the resistor defined by the relationship $r(t) = tR_o$ would be considered a linear resistor with time-varying coefficients and could be treated by means of linear time-variant differential equations. In conclusion, a linear coefficient or circuit constant is one that is independent of the dependent variables. In general, most circuit constants in electrical systems contain some degree of nonlinearity: Amplifiers saturate at some level of the dependent variables, most circuits containing the common magnetic steels and alloys show many nonlinear characteristics (saturation, hysteresis, etc.), and electric motor torque is a function of the *product* of state variables. Many mechanical systems also exhibit nonlinearities, such as breakaway-type friction, deadband or "windup" in gearing, and fluid heating in hydraulic systems. Although there are some very elegant and interesting mathematical methods for treating nonlinear control systems analytically and/or graphically, such systems are better and more efficiently treated by computer simulation techniques. However, many systems containing nonlinearities may be treated as piecewise-linear systems by considering variations over a relatively small variation of the dependent variables or within a region in which the nonlinearity is insignificant (such as occurs when the degree of magnetization of a transformer or motor is limited to magnetic intensities below the knee of the saturation characteristic). In the remaining sections, linear systems are assumed.

The generalized block diagram of Fig. 17.1 may be conveniently represented by the simplified diagram shown in Fig. 17.2. In this diagram, the symbol $G(s)$ represents the product of all forward-loop transfer functions

$$G(s) = \prod_{j=1}^{n} A_j(s) \tag{17.4}$$

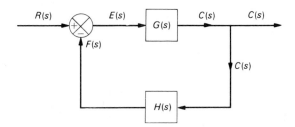

FIGURE 17.2
Classical linear control theory block diagram.

where the symbol $\prod$ indicates the multiplication, or product, of the n terms. Likewise, the feedback branch is simplified by a similar product term $H(s)$

$$H(s) = \prod_{j=n+1}^{z} A_j(s) \tag{17.5}$$

Let us look at the algebra associated with Fig. 17.2. Note that it represents negative feedback. The relationship between the controlled variable and the error variable is

$$C(s) = G(s)E(s) \tag{17.6}$$

Other relationships include

$$F(s) = H(s)C(s) \tag{17.7}$$

$$E(s) = R(s) - F(s) \tag{17.8}$$

$$F(s) = G(s)H(s)E(s) \tag{17.9}$$

By combining Eqs. (17.6) to (17.9), we obtain

$$\frac{C(s)}{R(s)} = \frac{G(s)}{1 + G(s)H(s)} \tag{17.10}$$

and

$$\frac{E(s)}{R(s)} = \frac{1}{1 + G(s)H(s)} \tag{17.11}$$

Equation (17.9) is known as the *open-loop* or return function, Eq. (17.10) is called the *closed-loop* transfer function, and Eq. (17.11) is the error response transfer function.

The individual transfer function terms, or $A_j(s)$, consist of numerator and denominator functions of polynomials in s:

$$A_j(s) = \frac{A_{jn}(s)}{A_{jd}(s)} \tag{17.12}$$

In linear systems, both the numerator and denominator will consist of products of one or more of four types of polynomials: constants, first-degree polynomials

$$s \pm a$$

nth-degree polynomials

$$(s \pm a)^n$$

and complex pairs

$$(s^2 \pm bs + \omega^2)^n$$

The forward-loop transfer function $G(s)$ and the feedback-loop transfer function $H(s)$ consist of numerators and denominators of products of similar terms. When multiplied, the numerators and denominators are said to be expressed as polynomials of the nth degree in s (or $j\omega$ or p):

$$A_{jn}(s), A_{jd}(s) = s^n \pm c_1 s^{n-1} \pm c_2 s^{n-2} \pm \cdots \pm c_i s^{n-i} \pm \cdots \pm c_{n-1} s \pm c_n \qquad (17.13)$$

These various forms of the transfer function all have mathematical significance, as illustrated in subsequent sections of this chapter.

We have now talked about the concept of feedback and have illustrated this concept with Figs. 17.1 and 17.2 and Eqs. (17.6) to (17.11). There are many useful and significant methods of controlling physical processes that do not involve feedback, but this discussion is limited to systems in which feedback *is* involved. Notice that feedback is involved *naturally* in many physical systems—particularly in biological and ecological systems—and is commonplace in many manufactured systems, such as the dc motor. Whereas natural or inherent feedback is most significant in many aspects of our lives and our society and is illustrated in some examples below, we are particularly concerned with systems in which feedback has been purposefully added. The fact that feedback is purposefully added to systems raises the question of why. There are a number of reasons for the use of feedback, and usually several of these reasons are involved in choosing this mode of control:

1 Precise control around a given set point or reference variable

2 Control of impedances in electronic systems

3 Improvement of the stability of a system

4 Reduction of the sensitivity of the system characteristics to changes in one or more parameters

5 Improvement of system response time and reducing overshoot

Related to the concept of impedance changes in electrical systems is the concept of *return difference*. To describe this concept, it is necessary to return to the feedback circuit of Fig. 17.2. This circuit is redrawn as Fig. 17.3, which shows several breaks, or "cuts," in the circuit. The reader should verify that if an input signal of 1.0 is applied in a clockwise direction at any cut, the difference between the input signal and the returned signal will be

Return difference $= 1 - G(s)H(s)$ (17.14)

This value, as expressed in Eq. (17.14), is known as the *return difference*. Note that this same result occurs regardless of where the break in the closed-loop circuit is made in Fig. 17.3.

FIGURE 17.3
Classical block
diagram with X's
marking "cuts" for
evaluating return
difference.

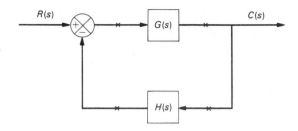

The form of the transfer function mathematical expression depends on the nature of the physical system being analyzed. There is a broad dichotomy in control systems based on the nature of the signals that occur in the system; these signals may be *continuous* or *discontinuous*, and these two classes of systems are often rather loosely called *analog* and *digital*, respectively. Although these terms are not in any sense rigorous definitions, they are very helpful in describing the types of components or hardware used in constructing a system and, surprisingly, the type of software or mathematics used to analyze a system. Much of classical control theory has been developed for continuous systems, but it can also be applied to many digital or discontinuous systems by using average or mean signals or by applying classical theory to individual discrete time intervals of a discontinuous system.

Transfer functions are obtained by means of the mathematical and physical theory associated with the particular device or system. To obtain the frequency-response form of a transfer function, the first step is to write the differential equation describing the device or system; the second is to obtain the Laplace transform of the differential equations, setting all initial conditions to zero; and the third is to obtain the output/input ratio of the device or system. A few examples will illustrate this process.

Example 17.1

Figure 17.4 illustrates a simple RC network. Determine the transfer function $E_o(s)/E_i(s)$ for this circuit.

Solution   The differential equations along the current paths shown are

$$e_i = i_1 R_1 + \frac{1}{C_1} \int i_1 \, dt - \frac{1}{C_1} \int i_2 \, dt$$

$$0 = -\frac{1}{C_1} \int i_1 \, dt + R_2 i_2 + \frac{1}{C_1} \int i_2 \, dt + \frac{1}{C_2} \int i_2 \, dt$$

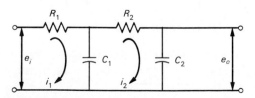

FIGURE 17.4
Example 17.1: a
simple RC network.

The output voltage is (assuming infinite load impedance)

$$e_o = \frac{1}{C_2} \int i_2 \, dt$$

Transforming these equations and eliminating i_1 yield

$$E_i(s) = I_2(s) \left[C_1 s \left(R_2 + \frac{1}{C_1 s} + \frac{1}{C_2 s} \right) \left(R_1 + \frac{1}{C_1 s} \right) - \frac{1}{C_1 s} \right]$$

$$E_o(s) = \frac{1}{C_2 s} I_2(s)$$

The ratio of $E_o(s)$ to $E_i(s)$ is

$$A_1(s) = \frac{E_o(s)}{E_i(s)} = \frac{1}{R_1 R_2 C_1 C_2 s^2 + (R_1 C_1 + R_2 C_2 + R_1 C_2)s + 1}$$

Example 17.2
A common analog control element is the dc tachometer. This is basically a permanent magnet generator and is shown schematically in Fig. 17.5. For a tachometer, the input is speed and the output is voltage. Determine the transfer function of this device.

Solution The tachometer input is mechanical speed ω_m. The electromagnetically induced armature voltage is

$$e_a = K_a \omega_m = i_a(R_a + R_L) + I_u \frac{di_a}{dt} \qquad (17.15)$$

The output of this transfer function is the voltage across an instrument, usually a dc millivoltmeter, and is

$$v_o = i_a R_L \qquad (17.16)$$

Taking the Laplace transform of Eqs. (17.15) and (17.16) and solving for the transfer function give

$$A_2(s) = \frac{V_o(s)}{\Omega_m(s)} = \frac{R_L K_a}{R_a + R_L + sL_a}$$

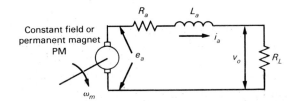

FIGURE 17.5
Example 17.2: the dc tachometer.

Note that if an electronic instrument with very high impedance ($R_L \to \infty$) were used to sense the generator output, the transfer function would be

$$A_2(s) \to K_a$$

$$R_L \to \infty$$

Example 17.3

The dc servomotor is another common analog element in control systems; it is shown schematically in Fig. 17.6. Again the constant-field configuration is assumed, and this is frequently realized by means of permanent magnet excitation. Linear elements are assumed.

FIGURE 17.6
Example 17.3: the dc servomotor.

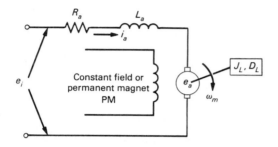

Solution   The differential equations describing system performance are

$$e_i = i_a R_a + L_a \frac{di_a}{dt} + e_a$$

$$e_a = K_a \omega_n \tag{17.17}$$

$$T_m = K_a i_a = D_L \omega_m + J_L \frac{d\omega_m}{dt}$$

in which T_m is used to represent the motor electromagnetic torque, ω_m is the motor output speed, and J_L and D_L are the inertia and viscous friction coefficients of the load, respectively. Transforming the equation set of Eq. (17.17) and solving for the transfer function give

$$A_3(s) = \frac{\Omega_m(s)}{E_i(s)} = \frac{K_a}{(D_L + sJ_L)(R_a + sL_a) + K_a^2}$$

Example 17.4

This example illustrates a more complex device and also introduces the concept of *on-off control*, or digital control. Figure 17.7 illustrates a basic semiconductor power controller known as a *chopper*. A chopper-controlled motor is frequently used in control systems, and the differential equations and transfer functions of such systems must be known. A chopper has two distinct states—the on state, when the switch or semiconductor is fully conductive, and the off state, when the switch or semiconductor is essentially open. For control systems analysis, it is usually adequate to assume that these two states can be represented by two idealized equivalent circuits: a short circuit for the on state and an open circuit for the off state. This assumption is made in the following analysis.

FIGURE 17.7
(*a*) A chopper
controller, (*b*) its
current waveforms,
and (*c*) its control
characteristic.

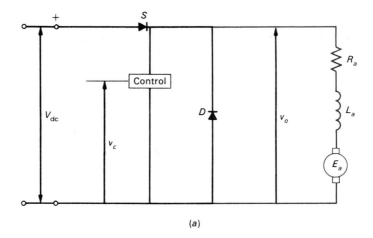

(*a*)

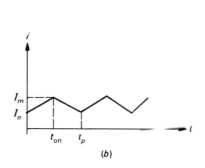

(*b*)

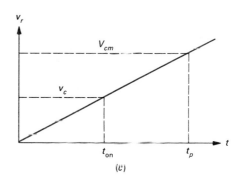

(*c*)

Figure 17.7*a* illustrates schematically the equivalent resistance R_a, leakage inductance L_a, and back emf or speed voltage V_{dc} of the armature of a separately excited motor or of the armature plus field of a series motor; it also shows a freewheeling diode D, the chopper S (neglecting commutation circuitry), and the dc source voltage E_a. Figure 17.7*b* illustrates the steady-state waveform of the motor current and an assumed chopper control characteristic, which is the familiar sawtooth capacitor charge characteristic. Note that *continuous current* in the motor is assumed; the case for discontinuous current is left as a problem at the end of this chapter. The chopper period is the time t_p; for time t_{on}, the SCR S is the on state; for time $t_p - t_{on}$, the SCR is off and current circulates through the motor and freewheeling diode circuit. When the reference voltage v_r equals the capacitor voltage v_c, S is commutated off. Determine the chopper transfer function $V_o(s)/V_r(s)$.

Solution   From the geometry of Fig. 17.7*b*, note that the average motor current is

$$I_{av} = \tfrac{1}{2}(I_m + I_o)$$

$$v_r = v_c = V_{cm}\frac{t_{on}}{t_p}$$

from Fig. 17.7c. The average voltage across the motor during the on time (which is also the voltage across D) is

$$V_m = I_{av}R_m + E_m = V_{dc}$$

since the *average* voltage across the inductance over a full period is zero. The average voltage across the motor (or D) during the off time is zero, neglecting the diode voltage drop. Therefore, the average motor or diode voltage (which is also v_o in Fig. 17.7) for a full period is

$$v_o = V_{dc}\frac{t_{on}}{t_p}$$

Therefore, the chopper transfer function is

$$A_4(s) = \frac{V_o(s)}{V_r(s)} = \frac{V_{dc}}{V_{cm}}$$

This is often called the *chopper gain;* the dynamic terms or time constants associated with the chopper itself are almost always very small compared to the time constants of other portions of the motor control system and can be ignored.

17.1.2 Multiple-Loop and Multiple-Input Systems

The block-diagram algebra introduced in Eqs. (17.6) to (17.11) can be extended as a means for simplifying block-diagram topography and for treating systems with more than one reference or input function. First note that for a single-loop system of the configuration of Fig. 17.2 but with *positive feedback*, Eq. (17.10) becomes

$$\frac{C(s)}{R(s)} = \frac{G(s)}{1 - G(s)H(s)} \tag{17.18}$$

Positive feedback generally results in instability, as we will soon show. However, it is frequently employed at low magnitudes or in portions of a large control system, such as in an inner loop.

Figure 17.8 illustrates a multiple-loop feedback control system. In the transfer function notation used in this figure, the argument s is omitted for simplicity. Three basic loops are shown in Fig. 17.8a. The positive feedback loop 2 can be eliminated by use of Eq. (17.18) and redrawn as a single transfer function (as shown in the subsequent figures) as

$$G_4' = \frac{G_2}{1 - G_2 H_2}$$

FIGURE 17.8
A multiple-loop
feedback control
system.

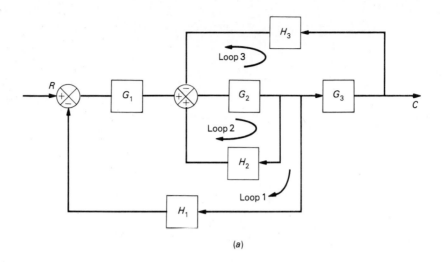

(a)

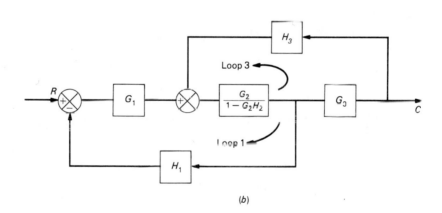

(b)

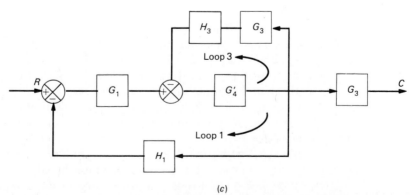

(c)

(*Continued*)

FIGURE 17.8
(*Continued*)

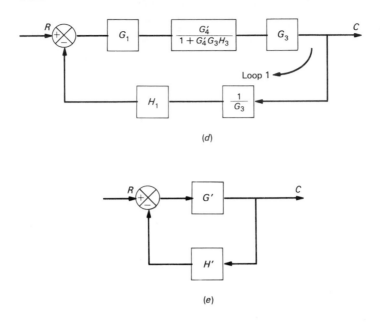

(d)

(e)

The takeoff point of a feedback loop can be relocated by including in the relocated feedback path the forward-loop transfer functions that have been bypassed. This process is illustrated by the relocation of the takeoff point for loop 3 in Fig. 17.8c. An inverse process is also illustrated in the relocation of loop 1 in Fig. 17.8d. Loop 3 in Fig. 17.8c is eliminated and expressed as a single transfer function in Fig. 17.8d by use of Eq. (17.10). Finally, a single forward and feedback path representation can be made as shown in Fig. 17.8e, where

$$G' = G_1 G_3 \frac{G'_4}{1 + G'_4 G_3 H_3}$$

$$H' = \frac{H_1}{G_3}$$

Some additional loop topography that is relatively common in control systems is illustrated in Fig. 17.9a. The upper loop is a feed-forward loop (in contrast to a feedback loop), quite common in digital control systems. The lower loop with parallel branches represents the combination of velocity and acceleration feedback that is used in many position control systems. The algebra developed in Sec. 17.1.1 is generally adequate for simplifying the system topography of Fig. 17.9a, as well as of many other configurations. The single-loop representation for Fig. 17.9a is given in Fig. 17.9b. The details of obtaining the simplified form are left for a problem at the end of this chapter.

In many types of feedback control systems there is more than one input or reference function. In some systems, multiple inputs may be designed into the systems as part of the control philosophy; in others, one or more inputs may be reactions

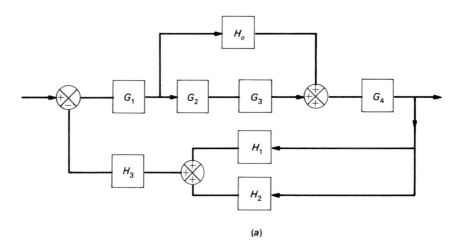

(*a*)

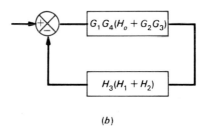

(*b*)

produced by the physical elements which comprise the control system. A common example of the latter type of input is the load torque of a motor used as a control element. This type of input is often called a *load disturbance*. Examples of the former type of multiple-input systems (i.e., where all input functions are designed or planned) generally result from systems in which the reference function, $R(s)$ in Fig. 17.1, is varying in response to other references. For example, a positioning system may have a reference that is moving as a function of another reference position. A classic example of such a system is the air-fuel ratio control of an internal combustion engine in which the reference for the magnitude of the air-fuel ratio is a function of engine speed, vehicle speed, engine torque, and engine temperature.

A generalized 2-input system is shown in Fig. 17.10. Again, we have omitted the argument s in all functions for simplicity. In *linear control systems*, multiple inputs are treated by means of the *principle of superposition*, which states that the response of a system with multiple inputs is the sum of the responses of the system to each input individually. To evaluate the system of Fig. 17.10, we first simplify the inner loop system and replace it with a single transfer function:

$$G_2' = \frac{G_2}{1 + G_2 H_2}$$

FIGURE 17.10
Multiple-loop
system with two
input functions, R_1
and R_2.

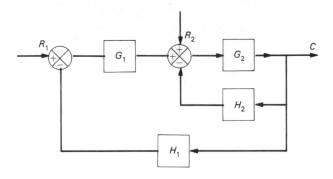

The response to reference function R_1 is

$$C_1 = \frac{G_1 G_2'}{1 + G_1 G_2' H_1} R_1$$

The response to R_2 is

$$C_2 = \frac{G_2'}{1 + G_1 G_2' H_1}$$

The system response is

$$C = C_1 + C_2 = \frac{G_1 G_2' R_1 + G_2' R_2}{1 + G_1 G_2' H_1}$$

The following example illustrates how a load torque on a dc motor may be treated as an input function.

Example 17.5

Add load torque to the load on the servomotor of Fig. 17.6, and from the resulting equations develop a block diagram with motor speed as reference and load torque as a second input or load disturbance.

Solution A voltage signal will serve as a speed reference; also we illustrate the inherent feedback due to the motor back emf by rearranging the equations developed in Example 17.3. The only change in these equations describing motor performance is the addition of the load torque T_L to the right side of the third equation of Eq. (17.17), giving

$$T_m = K_a i_a = D_L \omega_m + J_L \frac{d\omega_m}{dt} + T_L \tag{17.19}$$

To show the inner feedback path, we maintain the second equation of Eq. (17.17); also to introduce the load disturbance input, we maintain the summation shown by Eq. (17.19). Transforming Eq. (17.19) and the first two equations of Eq. (17.17) and adding the reference signal $E_R(s)$ gives the configuration of Fig. 17.11.

FIGURE 17.11
Example 17.5:
block diagram of
dc servomotor
with load torque.

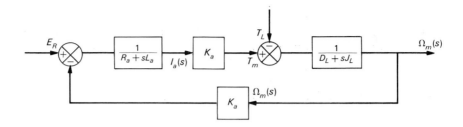

Remember that the multiple-loop and multiple-input techniques discussed above are valid only in linear systems. In general, there are no similar reduction or superposition methods for nonlinear systems, and computer modeling is generally required.

17.2   SOME ELEMENTARY FEEDBACK SYSTEMS

A few of the basic electrical or electronic feedback systems are described in this section by means of examples. In developing the differential equations and transformed equations for these systems, we use lowercase letters for time functions and uppercase letters for frequency-response or Laplace-transformed functions. Also circuit constants (resistances, amplifier gains, etc.) are represented by uppercase letters.

Example 17.6:
Feedback
Amplifier

The feedback, or operational, amplifier has been described in detail. The purpose of this example is to relate control theory concepts and nomenclature to the feedback amplifier. The operational amplifier is one of the most widely used components in control systems and is the cornerstone of many analog control systems. It is a means of simulating many transfer functions in control systems as well as in analog computers. Recall from Chap. 9 that the basic operational amplifier has a tremendously large gain and that the input current to the amplifier can be considered zero. Figure 17.12a illustrates an amplifier with feedback and input impedances Z_{fb} and Z_i, respectively. It was shown in Chap. 9 that the transfer function for this system is

$$G(s) = \frac{E_o(s)}{E_i(s)} = -\frac{Z_{fb}}{Z_i}$$

With the two impedances taking on the RC network configurations as shown in Fig. 17.12b, the transfer function becomes

$$G(s) = -\frac{R_2}{R_1} \frac{1 + R_1 C_1 S}{1 + R_2 C_2 S}$$

This function is frequently used in control systems to modify phase characteristics.

FIGURE 17.12
(*a*) Feedback
amplifier with
input impedance
(s.j. = summing
junction). (*b*)
Feedback
amplifier with *RC*
networks.

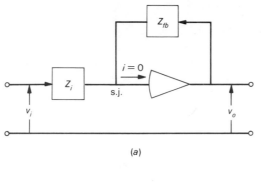

(a)

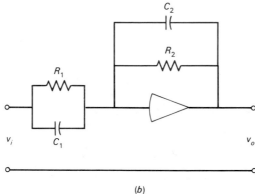

(b)

**Example 17.7:
Wien-Bridge
Oscillator**

This device is primarily used as an oscillator or variable-frequency source rather than as a control system, but it is a good example of the use of feedback in an electronic circuit—in this case, positive feedback, a characteristic of electronic oscillators. The basic Wien-bridge oscillator circuit is shown in Fig. 17.13*a*. This figure illustrates the source of the name of this circuit by showing the circuit elements in the form of the familiar Wien bridge, a circuit long in use as a means of accurate measurement of a capacitance. A bridge balance (when used as a capacitance-measuring circuit) is achieved when the voltage $v_o = 0$. Note that this condition corresponds to the assumption of infinite gain in the operational amplifier, as noted in Chap. 9 and the previous example, and is a condition for oscillator operation in the Wien-bridge oscillator. The bridge form of the circuit may be redrawn to illustrate the feedback characteristics, as shown in Fig. 17.13*b*. Note the existence of positive feedback. The forward-loop transfer function is

$$G(s) = \frac{R_2 + R_1}{R_2}$$

The feedback transfer function is

$$H(s) = \frac{1}{3 + RCS + 1/(RCS)}$$

FIGURE 17.13
(a) Wien-bridge
oscillator. (b)
Wien-bridge
oscillator,
illustrating positive
feedback.

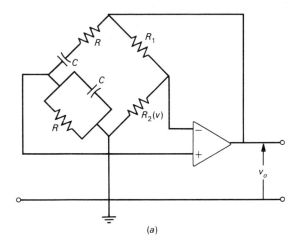

(a)

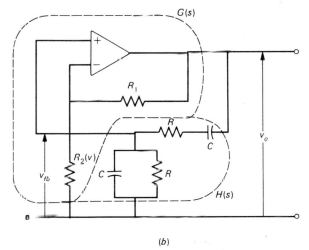

(b)

Oscillation occurs when the following conditions are satisfied:

$$f_o = \frac{1}{2\pi RC}$$

$$R_2 = 2R_1 \tag{17.20}$$

Resistances R_1 and R_2 are nonlinear, the former consisting of an incandescent lamp or other impedance whose resistance increases with temperature and the latter consisting of a voltage-sensitive resistance whose resistance decreases with voltage, such as a field-effect transistor. Stable operation exists when the voltages across the two resistive arms are equal as a result of variations in the nonlinear resistances in these arms, which result in satisfying Eq. (17.20).

**Example 17.8:
Ward-Leonard
Control**

This means of very precise control of huge electromagnetic machines has been in use for many years and is one of the early examples of feedback control applications. The Ward-Leonard system is used in the control of large dc motors used in rolling mills and similar applications. With feedback, this system is capable of such astonishing feats as reversing a 10,000-hp motor in less than 1 s while carrying full load. A simplified schematic circuit is shown in Fig. 17.14. For a given load and base speed condition, it is frequently possible to assume that the field current (and hence machine magnetic excitation) is held constant. Therefore, the simplified constant-field motor and generator (tachometer) transfer functions developed in Examples 17.3 and 17.2, respectively, are used here. We also ignore load torque in the development, but it may be included in the analysis by the methods discussed in Sec. 17.1.2. Likewise, it can be assumed that the main generator speed is held constant under all conditions of operation. In Fig. 17.14a, the tachometer, error detector, and amplifier are represented symbolically rather than by electric circuit connections. Parameters R_e and L_e represent the equivalent resistance and inductance, respectively, of the generator and motor armature circuits, which effectively are in series. Let us determine the transfer functions and block diagram for this system.

First the power generator will be developed. The differential equations describing this element are (referring to Fig. 17.14a)

$$e_f = i_f R_f + L_f \frac{di_f}{dt}$$

$$e_g = K_e i_f \qquad \text{for } \omega_g = \text{constant}$$

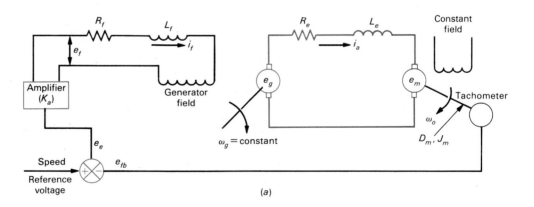

(a)

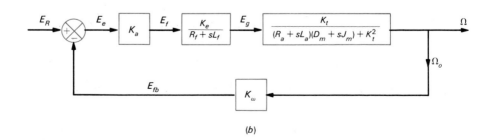

(b)

FIGURE 17.14
(a) Schematic representation of Ward-Leonard dc motor control. (b) Block diagram of the Ward-Leonard system.

Transforming and rearranging give the transfer function for the generator operated at constant speed:

$$\frac{E_g}{E_f} = \frac{K_e}{R_f + sL_f}$$

The differential equation in the armature circuits is

$$e_g = i_a R_e + L_e \frac{di_a}{dt} + e_m \tag{17.21}$$

The motor differential equations are

$$e_m = K_t \omega_o \qquad \text{for constant-field current} \tag{17.22}$$

$$T_d = K_i i_a = D_m \omega_o + J_m \frac{d\omega_o}{dt} \qquad \text{neglecting load torque} \tag{17.23}$$

Transforming Eqs. (17.21) to (17.23) and rearranging give the armature circuit and motor transfer function:

$$\frac{\Omega_o}{E_g} = \frac{K_t}{(R_a + sL_a)(D_m + sJ_m) + K_t^2}$$

The remaining component in the forward loop is the amplifier. The transfer function of electronic amplifiers (such as the operational amplifiers described in Chap. 9) can generally be approximated by a gain term with no dynamic or time constant terms, since the time constants of these devices are extremely small compared to electric power or electromechanical time constants. With this assumption, the forward-loop transfer function of the Ward-Leonard system is

$$G(s) = \frac{\Omega_o}{E_e} = \frac{K_a K_e K_t}{(R_f + sL_f)[(R_a + sL_a)(D_m + sJ_m) + K_t^2]} \tag{17.24}$$

All symbols are defined in Fig. 17.14a. The feedback loop consists of the tachometer. This could be described in terms of the dc generator with constant-field excitation transfer function developed in Example 17.2. However, due to the small physical size of the typical dc or ac tachometer, the dynamic terms shown in Example 17.2 can be ignored. The typical tachometer is generally treated as a speed-to-volts transducer with a transfer function K_ω which is a gain term. The feedback-loop transfer function is

$$H(s) = K_\omega$$

The block diagram of the Ward-Leonard system is shown in Fig. 17.14b.

**Example 17.9:
Voltage
Regulator**

Figure 17.15a illustrates the simplified circuit diagram of the buck regulator circuit, one of many types of circuits used to control the output voltage of electronic power supplies. The control mechanism used to control output voltage in this—and many other—circuits is

FIGURE 17.15
(*a*) Schematic
circuit for
electronic voltage
regulator. (*b*)
Logic relationships
for PWM
control. (*c*) Block
diagram showing
control signal flow
for electronic
regulator.

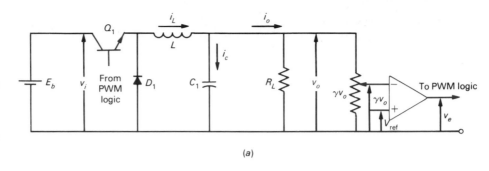

(a)

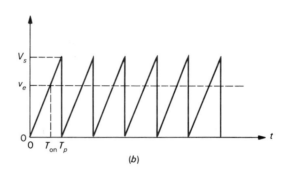

(b)

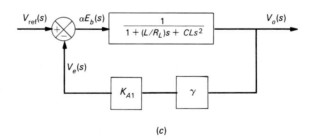

(c)

pulse-width modulation (PWM), a technique for controlling *average* voltage by varying the amount of time the input voltage v_i is connected to the circuit by means of transistor Q_1. The output voltage is controlled by the width of the "on" pulse of v_i, as shown in Fig. 17.15*b*. Also shown in Fig. 17.15*b* is one of the common means of controlling the pulse period—which is to compare the output of the error-detector amplifier A_1 with the voltage of a standard capacitor charging voltage (sawtooth) wave and to turn on Q_1 when these two signals are equal. Note that this gives a limited range of control determined by the sawtooth peak voltage V_s, which is one example of the nonlinearity known as *saturation* that occurs in most electronic circuits. We will determine the transfer functions and block diagram for this circuit.

Neglecting the internal impedances of the voltage source E_b, the average voltage applied to the regulator during the period T_p is

$$V_i = \frac{T_p - T_{on}}{T_p} E_b = \alpha E_b$$

Since continuous current is maintained by the freewheeling diode D_1, it can be assumed that this average voltage is applied to the smoothing inductor and output capacitor each period, resulting in the differential equations

$$v_i = L \frac{di_L}{dt} + v_o$$

$$i_c = C \frac{dv_o}{dt} \tag{17.25}$$

$$i_L = i_o + i_c = \frac{v_o}{R_L} + i_c$$

Transforming Eq. (17.25) and solving for the ratio of output to input voltage, we have

$$G(s) = \frac{V_o(s)}{V_i(s)} = \frac{1}{1 + (L/R_L)s + CLs^2}$$

The block diagram is shown in Fig. 17.15c.

17.3 LINEAR CONTROL SYSTEMS ANALYSIS

Classical linear control theory—which has fostered the rapid growth of control systems in industrial societies—includes a large body of theorems and analytical techniques. Portions of this theory encompass and define some of the fundamental concepts that underlie the nature of systems operations, including the basic concepts of electrical systems theory. Much of linear control theory is based on the frequency-response formulation of the systems equations, and a large body of quasi-graphical and algebraic techniques have been developed to analyze and design linear control systems based on frequency-response methods. An introduction to several of these techniques is given in this section; for more complete discussion of frequency-response methods, a great many articles and texts are available, including Refs. 1 to 5 at the end of this chapter.

With the development and widespread use of discrete (digital) control systems and the advent of relatively low-cost digital computers, time-response methods have become both more necessary and more available. Time-response systems analysis may be divided into two broad methodologies: (1) the actual simulation or modeling of the system differential equations by either analog or digital computers and (2) the state-variable formulation of the system state equations and their solution by a digital computer. A good basic reference for the state-variable formulation (among very many) is Ref. 6. State-variable methods are probably the most general approach to system analysis and are useful in the solution of both linear and nonlinear systems equations. The brief introduction to multiple-loop and multiple-input systems given in Sec. 17.1.2 may have already a'erted the reader to the conclusion that this approach—using signal-flow block diagrams and frequency-response formulation—is very limited, generally to not more than three inputs and probably the same number of inner loops. The matrix formulations associated with state-variable techniques have largely supplanted the block-diagram formulations as illustrated in Sec. 17.1.2, and computer software for solving a great variety of state equation formulations is available on most computer systems today. However, much of the physical reality of any system is lost in the state-variable formulation, including the relationships between system response and system parameters. Furthermore, a great many of the concepts and design parameters used in many practical control systems of industry and aerospace today are based on both the nomenclature and the theory of the frequency-response methods. Therefore, although frequency-response techniques are limited to relatively simple systems—and, in the rigorous mathematical sense, to linear systems—they are still most useful in system design and stability analysis of practical systems and can give a great deal of information about the relationships between system parameters (time constants, gains, etc.) and system response. Under certain conditions, frequency-response methods can be adapted to nonlinear systems (see Refs. 1 and 7). We now examine the basic concepts of frequency response.

17.3.1 Frequency-Response Formulation

Most of the equations presented so far in this chapter have been in the *frequency-response formulation*. This term refers to the Laplace-transformed form of the differential equations describing systems performance. In electric circuits and machines, the frequency-response formulation can also be developed (with care) by expressing the steady-state ac equations in terms of the frequency variable $j\omega$. The differences between these two variables, the Laplacian s and the Fourier $j\omega$, are very meaningful and should be understood by the reader. However, *as a methodology for developing frequency-response transfer functions and in total disregard of mathematical rigor*, either formulation is acceptable and should be used when appropriate for a given problem, and the interchange of variables

$$s \leftrightarrow j\omega \tag{17.26}$$

is generally acceptable. We illustrate this approach by means of a return look at Example 17.1.

Example 17.10

Develop the transfer function of the *RC* network shown in Fig. 17.4 by means of steady-state ac network analysis

Solution Assuming that the input function e_i can be represented by an RMS voltage function E_i, the network equations are

$$E_i = (R_1 - jX_{C1})I_1 + jX_{C1}I_2$$
$$0 = jX_{C1}I_1 + [R_2 - j(X_{C1} + X_{C2})]I_2 \tag{17.27}$$
$$E_o = jX_{C2}I_2$$

w^1 $_\cdot X$ has the meaning $1/(\omega C)$. Solving for the ratio of output voltage over input voltage yields

$$A_1(j\omega) = \frac{E_o(j\omega)}{E_i(j\omega)} = \frac{1}{\dfrac{R_1 R_2}{j^2 X_{C1} X_{C2}} + 1 + j\left(\dfrac{R_1}{X_{C2}} + \dfrac{R_2}{X_{C2}} + \dfrac{R_1}{X_{C1}}\right)} \tag{17.28}$$

By interchanging the frequency variables, as in Eq. (17.26), the use of steady-state ac network theory as presented earlier can be most helpful in deriving transfer functions for electrical and electromechanical networks and systems.

However, frequency-response expressions are generally formulated in terms of the Laplace transform variable s. Generalized forms of frequency-response expressions have already been presented below Eq. (17.12). An additional standard form for the complex second-order expression is given by

$$s_2 \pm 2\zeta\omega_o s + \omega_o^2 \tag{17.29a}$$

where ω_o is called the natural frequency, s^{-1}, and ζ is called the *damping ratio*, dimensionless. The solutions for s, obtained by setting the polynomial equal to zero, are known as *roots* of the polynomial. The complex roots of Eq. (17.29a) are

$$s = \pm\zeta\omega_o \pm j\omega_o\sqrt{1 - \zeta^2} \tag{17.29b}$$

Roots of polynomials are frequently illustrated graphically by being plotted on the complex plane of complex variable theory known as the S plane, a concept useful in frequency-response analysis.

Two polynomial expressions of significance in the analysis of linear feedback control systems are the *open-loop response*, given by Eq. (17.9), and the *closed-loop response*, given by Eq. (17.10). A number of examples given previously have illustrated the methods for obtaining the forward-loop transfer function $G(s)$ and the feedback-loop transfer function $H(s)$ for various types of electrical and electromechanical systems. The next step in the analysis is to rearrange these equations into the standard polynomial forms given in Sec. 17.1.1. Let us perform this step for the Ward-Leonard system of Example 17.8, with $G(s)$ given by Eq. (17.24) and $H(s)$ given in Example 17.8. Taking the product of G and H and factoring through by the coefficient of the term with highest order in s give, for the open-loop function,

$$G(s)H(s) = \frac{K_a K_t K_\omega K_e / L_a L_f J_m}{s^3 + \left(\dfrac{R_a}{L_a} + \dfrac{R_f}{L_f} + \dfrac{D_m}{J_m}\right)s^2 + \left[\dfrac{R_f}{L_f}\left(\dfrac{D_m}{J_m} + \dfrac{R_a}{L_a}\right) + \dfrac{R_a D_m + K_t^2}{L_a J_m}\right]s + \dfrac{R_f(R_a D_m + K_t^2)}{L_a L_f J_m}} \tag{17.30}$$

Note that this is the unfactored form of the polynomial Eq. (17.13), which is the form usually obtained from determination of the transfer functions, except in the simplest configurations. This type of expression is often written in the form

$$G(s)H(s) = \frac{K_{GH}}{s^3 + C_1 s^2 + C_2 s + C_0} \tag{17.31}$$

From basic Laplace transform theory, the reader should note that the *steady-state gain* of the open-loop system when excited by a unit step function is

$$\lim_{s \to 0} [sG(s)H(s)] = \lim_{s \to 0} \frac{sK_{GH}}{s(s^3 + C_1s^2 + C_2s + C_0)} = \frac{K_{GH}}{C_0} \tag{17.32}$$

The closed-loop response is found by inserting Eqs. (17.30) into the right-hand side of Eq. (17.10). This is a most tedious process to perform by hand for even the relatively simple third-order function of the Ward-Leonard example; using the simplified notation of Eq. (17.31), we obtain

$$\frac{C(s)}{R(s)} = \frac{G(s)}{1 + G(s)H(s)} = \frac{K_{GH}/K_\omega}{s^3 + C_1s^2 + C_2s + C_0 + K_{GH}} \tag{17.33}$$

Again we have ended up with an expression in the unfactored form of Eq. (17.13).

Now let us look once again at Eqs. (17.9) and (17.10). The left-hand side of both equations is a ratio of a transformed output function to a transformed input function. Cross multiplying by the input function gives the transformed differential equation, from which the output function or output response can be determined. We will repeat these equations in this form:

$$F(s) = G(s)H(s)E(s) \qquad \text{open loop} \tag{17.34}$$

$$C(s) = \frac{G(s)R(s)}{1 + G(s)H(s)} \qquad \text{closed loop} \tag{17.35}$$

To help understand these equations, refer to the system diagrams (Figs. 17.1 to 17.3). Also note that the denominators of the open-loop and closed-loop transfer functions, examples of which are given in Eqs. (17.30) to (17.33), represent the characteristic equations of Eqs. (17.34) and (17.35) from which the form of the system response or output function is determined. The transfer functions result from the transformed homogeneous differential equations describing the system, which, it is recalled, determines the particular solution (often termed the *transient* or *impulsive response*) of the differential equations. The concept of impulsive response (i.e., the response to the unit impulse input) is particularly useful in the evaluation of linear control systems since the Laplace transform of the unit impulse is unity.

17.3.2   The S Plane; Poles and Zeros

Much of this material should be familiar to the reader and is inserted here primarily for completeness and as a reference for subsequent developments.

Briefly, the S plane is merely the two-dimensional graph on which the roots of the transformed differential equation are plotted. The roots of differential equations, such as Eqs. (17.34) and (17.35), are classified as poles or zeros depending on their location in the denominator or numerator, respectively, of the equation. The name *zero* is obvious, because when the transform variable s is set equal to the root of the numerator polynomial, the output function becomes zero. The name *pole* is less obvious; when the transform variable is set equal to the root of the polynomial in a denominator term, the transformed equation "blows up"; i.e., it mathematically takes on the value of infinity. From the methods of solving linear differential equations by the Laplace transform, note that the nature of the denominator roots, or poles, determines the form of the time response. Simple poles of the form $s + a$ result in the time response e^{-at}; multiple poles $(s + a)^n$ result in the time response $t^{n-1}e^{-at}/n!$; the time response resulting from the complex pairs given in Eqs. (17.29a) and (17.29b) result in damped sinusoidal functions. Note the inverse relationship between the sign of the pole and the sign of the damping coefficient in the exponential terms. Negative real roots or negative real parts of complex roots result in exponential terms with negative damping coefficients ($s = -a$, or $s = -\zeta \pm j\omega_o$). From the final value theorem noted in the previous section and by observation of the form of the time-response exponential terms, the final value (steady-state value) of the time-response term resulting from negative real poles or from complex pair poles with negative real parts is zero. Time-response terms resulting from poles with positive real parts become infinite mathematically.

Figure 17.16 illustrates a typical S-plane graph of the poles and zeros of a transformed equation. Poles are designated by the symbol $\times$ and zeros by the symbol $\bigcirc$. Complex-pair roots always occur symmetrically about the horizontal axis on the S plane. Real roots lie on the horizontal axis. Purely imaginary roots (complex pairs with zero real part) lie on the vertical axis. The origin represents the location of the real root $s = 0$, which in electrical systems represent a response equivalent to the steady-state response of a circuit excited by direct current. In regard to control system response, the vertical axis represents a very significant dividing line: Roots with negative real parts lie to the *left* of this line, and this portion of the plane is, not surprisingly, called the *left-hand S plane*.

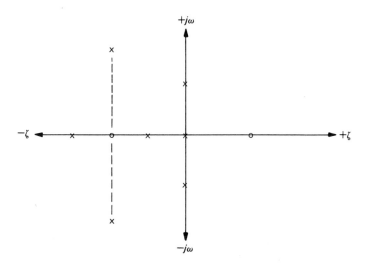

FIGURE 17.16
Typical pole-zero plot on the S plane, using the notation $s = \zeta \pm j\omega$.

17.3.3 Control System Stability

The systems concept of stability refers to the ability of a system to arrive at a stable condition of operation—what we have been referring to as the *steady state*—in which input and output functions are related by numerical constants derived from system coefficients in the complementary solution of the system differential equation. Mathematically, the concept of stability in a linear system can be precisely defined: The system characteristic homogeneous equation must contain no poles in the right-half S plane, i.e., no poles with positive real parts. Note that this mathematical criterion is an either-or criterion; the system is either stable or it blows up (goes to an infinite output). In a practical system this condition of *infinite response* seldom, if ever, occurs, for it is usually prevented by the onset of nonlinearities in system parameters, such as saturation in electronic amplifiers, stress in a structural member beyond the yield strength, or actual fracturing of a structural member. However, the tendency toward infinite response until something saturates or breaks is very real in linear systems and must be eliminated for a system to operate properly.

The linear stability of negative feedback systems constructed from basically stable components and subsystems is primarily a function of system or "loop" gain. This is the parameter that has already been defined as steady-state gain in Eq. (17.32) and is the product of gain terms in the forward and feedback loops. The word *gain* is used here somewhat loosely and actually refers to the magnitude terms in the open-loop transfer function; the system gain may, in actuality, be a number less than 1.0, depending on the choice of system units. Because of the importance of the gain or magnitude function upon system stability—as well as upon other important aspects of system response—the typical open-loop transfer function, Eq. (17.31), is often written in the form

$$G(s)H(s) = KG'(s) = \frac{K_{GH}/C_0}{s^3/C_0 + s^2 C_1/C_0 + s C_2/C_0 + 1} \qquad (17.36)$$

where K is the open-loop or steady-state gain and $G'(s)$ contains all terms involving s and has a final value of 1.0.

A negative feedback system that is found to be unstable at a given value of K can usually be made stable by reducing the value of K. In electrical systems, this is normally accomplished by adjustment of the gain of an electronic amplifier in the forward loop, such as K_a in the Ward-Leonard system of Fig. 17.14. By contrast, in a system exhibiting nonlinear instability, stability can be achieved in many cases by *increasing* the forward-loop gain. In linear systems (and most nonlinear systems) the reduction of system gain always results in some compromises or tradeoffs with other aspects of system response. In linear systems, the speed of response or response time generally deteriorates as the system gain is reduced, i.e., the system responds more slowly. Also there may be some deterioration in the accuracy of the steady-state response as the system gain is reduced. Therefore, the principal problem of control systems analysis is to achieve stable operation and the required or specified response time and steady-state accuracy.

The either-or stability criterion for linear systems can be most easily evaluated by an analysis of the coefficients (including algebraic sign) of the transformed characteristic equation, i.e., the denominators of such expressions, as illustrated in Eqs. (17.30) to (17.36). We have been calling the form of these expressions the *unfactored form;* i.e., the roots are not known. A general form for the unfactored polynomial in the transformed variable s is given by the right-hand side of Eq. (17.13). Observe that the polynomial must be arranged in decreasing powers of s. In the following discussions, the algebraic sign in front of coefficients c_i is assumed to be included as part of the coefficient. Linear instability is indicated by the presence of negative real roots in the *characteristic*, or homogeneous, equation of the form

$$s^n \pm c_1 s^{n-1} \pm c_2 s^{n-2} \pm c_i s_{n-i} \pm \cdots \pm c_{n-1} s \pm c_n = 0 \tag{17.37}$$

Visual examination of the form of Eq. (17.37) can give considerable information on the presence of roots with negative real parts, which is indicated by

1 A negative c_i

2 The absence of a c_i (other than c_n itself), which means that the polynomial can be factored to the $n - 1$ form

Further analysis when all c_i's are positive and nonzero is made by means of a coefficient manipulation technique known as *Routh's criterion* (see Refs. 1 and 2).

17.4 TYPES OF FEEDBACK CONTROL SYSTEMS

There are several means of classifying feedback control systems. We have already noted the use of the terms *multiloop, multiple-input, feed-forward,* and *positive feedback* in relation to the block diagram analysis in Sec. 17.1. Two broader classifications are widely used and are based on the nature of the *open-loop transfer function* $G(s)H(s)$.

17.4.1 System Type

Let the open-loop transfer function be expressed as

$$G(s)H(s) = \frac{KN(s)}{s^q D(s)} \tag{17.38}$$

where $N(s)$ and $D(s)$ are polynomials in s of the form of Eq. (17.13) (with all coefficients positive), K is a numerical constant, and s corresponds to a Laplace-transformed variable. The exponent of s, or q, in the denominator represents the number of integrations in the open loop. With $q = 0$, there are no integrators; $q = 1$, there is one integrator; etc. *The exponent q defines the system type.* This classification is generally applied only in systems with $q = 0, 1$, or 2, and these are often called *position*, *velocity*, and *acceleration systems*, respectively, because of the system error response, to be discussed.

The error response of a feedback control system is defined in Eq. (17.11). It is another performance characteristic based on frequency-response analysis, and it is a measure of the control accuracy that the system can exhibit. Both dynamic (transient) and steady-state error response may be of concern in the design of a control system. Dynamic response is best obtained from computer simulation of the system time response. However, steady-state error response (i.e., after the transients have disappeared) can be readily obtained from the frequency-response form, Eq. (17.11), and the application of the final-value theorem, according to which

$$\lim_{t \to \infty} f(t) = \lim_{s \to 0} sF(s) \tag{17.39}$$

Error and output response in the frequency domain are frequently evaluated by assuming standard unit function inputs initiated at $t = 0$:

Unit step: $f(t) = 1$ $F(s) = \dfrac{1}{s}$

Unit ramp: $f(t) = t$ $F(s) = \dfrac{1}{s^2}$ (17.40)

Unit acceleration: $f(t) = t_2$ $F(s) = \dfrac{1}{s^3}$

Example 17.11

Determine the steady-state error of a type 0 system with a unit step reference input function for the Ward-Leonard system of Fig. 17.14.

Solution   The open-loop transfer function is

$$G(s)H(s) = \frac{K_a K_e K_t K_w}{(R_f + sL_f)[(R_a + sL_a)(D_m + sJ_m) + K_t^2]} \tag{17.41}$$

which is a type 0 system. A unit step function reference input has the transform

$$R(s) = \frac{1}{s} \tag{17.42}$$

Substituting (17.42) and (17.41) into (17.11) gives

$$\begin{aligned}
E(s) &= \frac{R(s)}{1 + G(s)H(s)} \\[2mm]
&= \frac{(R_f + sL_f)[(R_a + sL_a)(D_m + sJ_m)] + K_t^2}{s\{K_a K_e K_t K_w + (R_f + sL_f)[(R_a + sL_a)(D_m + sJ_m)] + K_t^2\}}
\end{aligned} \tag{17.43}$$

Applying the final-value theorem, Eq. (17.39), results in the steady-state system error

$$\text{Steady-state error} = \lim_{t \to \infty} e(t) = \lim_{s \to 0} sE(s)$$

$$= \frac{R_f(R_a D_m + K_t^2)}{K_a K_e K_t K_w + R_f(R_a D_m + K_t^2)} \tag{17.44}$$

The last expression in (17.44) represents the steady-state error. i.e., the difference between the reference or "set" voltage E_R and the voltage proportional to motor speed E_{fb} in Fig. 17.14b. The term *steady state* refers to the condition in time when all transients due to the application of the step reference function have disappeared, and it is mathematically represented by a very large value of t or letting $t \to \infty$.

Although Eq. (17.44) appears to be rather complex, note that most of the terms in it are fixed by the choice of motor generator and load: the load friction coefficient D_m, the machine resistances R_a and R_f, and the machine volt and torque constants K_e and K_t. For a given motor, generator, and load, the system performance is controlled by the amplifier constant K_a and, to a lesser degree, the feedback tachometer constant K_w. From Eq. (17.44), we see that the steady-state error can be decreased by increasing the amplifier gain K_a. However, frequency-response stability analysis and the more complex time-response simulation will show that increasing the forward-loop gain increases the oscillatory nature of the output (greater overshoot, more cycles of transient oscillation, etc.) and may lead to system instability.

Table 17.1 summarizes the error response for other system types and for other unit input functions. The development of these results is left as problems at the end of this chapter.

TABLE 17.1			
	Error Response for Unit Inputs		
System Type	Unit Step	Unit Ramp	Unit Acceleration
0	Finite	Infinite	Infinite
1	0	Finite	Infinite
2	0	0	Finite

17.4.2 Classification by Control Action

A more common means of describing industrial and process controllers is by the manner in which the error signal $E(s)$ is utilized in the forward loop. The basic control elements are a proportional device, such as an amplifier, a differentiating device, such as an inductor, and an integrating device, such as an operational amplifier with feedback. These are summarized in Table 17.2, using the nomenclature of Fig. 17.2.

TABLE 17.2

Action	$G(s)$
Proportional	K_p
Differential	sK_d
Integral	K_i/s

Many industrial controllers utilize two or more of the basic elements, and these are summarized in Table 17.3.

TABLE 17.3

Action	$G(s)$
Proportional and differential (PD)	$K_p + sK_d$
Proportional and integral (PI)	$K_p + K_i/s$
Proportional and differential and integral (PID)	$K_p + sK_d + K_i/s$

Proportional, integral, and differential controllers were developed many years ago using pneumatic, hydraulic, and electromechanical hardware and are still widely used in industrial and process control applications. In modern control applications, this more rugged hardware is frequently tied in with various electronic compensation circuits, microprocessor supervisory control, electronic data gathering systems, etc.

Figures 17.17 and 17.18 illustrate simplified gas-turbine speed governors exhibiting proportional and integral action, respectively. The transfer functions of these governors as the ratio of fuel input to the turbine to the actuating signal or speed

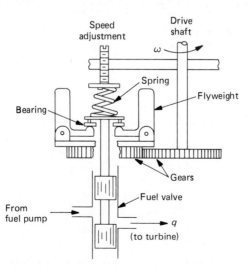

FIGURE 17.17 Speed governor exhibiting proportional control.

FIGURE 17.18
Speed governor
exhibiting integral
control.

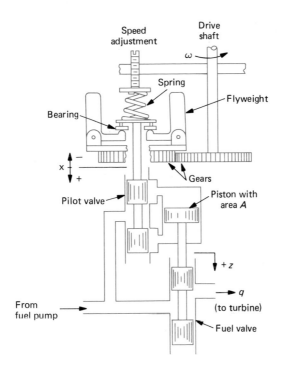

error can be shown to be K_p and K_i/s, respectively. A simplified block diagram for a turbine-speed control system is shown in Fig. 17.19 for the proportional governor, with load transfer function.

FIGURE 17.19
An elementary
speed control
system using
proportional
control.

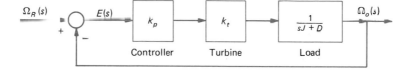

Example 17.12 Determine the system type, steady-state error, and closed-loop time constant for the system of Fig. 17.19.

Solution The forward-loop transfer function is $G(s) = \Omega_0(s)/E(s) = K_p K_t/(D + sJ)$. It is a type 0 system.

The error response is $E(s)/\Omega_R(s) = (D + sJ)/(K_p K_t D + sJ)$. For a unit step reference function, $\Omega_R(s) = 1/s$. And by the final-value theorem, the steady-state error is $D/(D + K_p K_t)$. The output response is $\Omega_0(s)/\Omega_R(s) = K_p K_t/(K_p K_t + D + sJ)$. The time constant of the characteristic equation (denominator) is $J/(K_p K_t + D)$.

FIGURE 17.20
A speed control
system using
integral control.

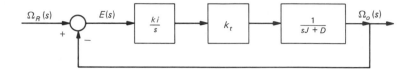

This very simple example illustrates clearly that by increasing the forward loop gain (K_pK_t), steady-state error is reduced and response time is improved (since the time constant is decreased). Let us now compare this type of response with that when an integral governor (Fig. 17.18) is used. The block diagram for a simple turbine speed control system using the integral governor is shown in Fig. 17.20.

Example 17.13

Repeat Example 17.12 for the speed control system shown in Fig. 17.20.

Solution The forward-loop transfer function is $G(s) = K_iK_t/s(D + sJ)$; this is a type 1 system.

The error response is $s(D + sJ)/(K_iK_t + sD + s^2J)$. Using the final-value theorem, we see that the steady-state error for a unit step reference function is zero, as expected from Table 17.1. The characteristic equation for the output response is the same as for the error response, as seen in Example 17.12. This is now a second-order equation, and there are two time constants. However, it is more convenient to observe system performance by using the form of Eq. (17.29a). Manipulating the characteristic equation to the form of Eq. (17.29a) gives $s^2 + sD/J + K_iK_t/J$, from which the damping ratio is found as

$$\zeta = \frac{D}{2\sqrt{K_iK_tJ}} \tag{17.45}$$

The form of the characteristic equation in Example 17.13 can be used to typify second-order control systems. The response of a system may be underdamped or overdamped, depending on the value of the damping ratio ζ. If $\zeta = 0$, the system will oscillate continuously at frequency ω_0 although this is a hypothetical situation since there is some damping in all physical systems. Control systems are generally underdamped to achieve fast response time, and the standard definitions of the response parameters are illustrated in Fig. 17.21. We have already referred to the concepts defined in this figure in our discussions of the tradeoff between steady-state error and response time. These definitions, such as the rise time, overshoot, and settling time, can also be applied to systems with characteristic equations of order higher than 2, although the shape of the response curve will be different.

If a system is highly oscillatory (long settling time), the response can obviously be improved by increasing the physical damping, such as the friction coefficient D in the above examples, electrical resistance in electrical systems, fluid resistance in fluid systems, etc. But there is always an energy loss associated with physical damping, and the rise time is also generally increased. Let us see what happens when we add proportional control to integral control (PI control), which is one of the most common types of industrial controllers.

FIGURE 17.21
Typical response
of an
underdamped
second-order
system.

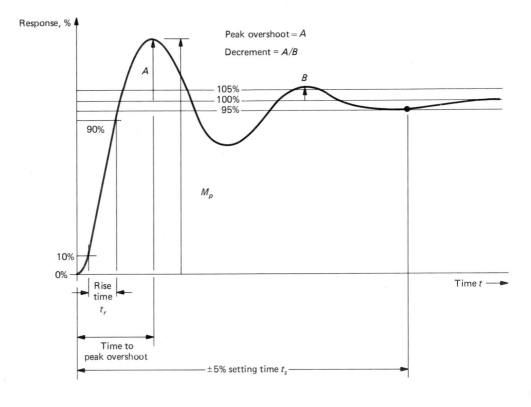

Example 17.14   Add proportional control to the system of Example 17.13, and determine the new damping ratio.

Solution   The new block diagram is shown in Fig. 17.22. Solving for the characteristic equation of the output response $G(s)/[1 + G(s)]$ gives $Js^2 + (D + K_t K_p)s + K_t K_i$. Factoring this expression into standard form gives a damping ratio

$$\zeta = \frac{D + K_t K_p}{2\sqrt{K_t K_i J}}$$ (17.46)

FIGURE 17.22
A speed control
system with PI
control.

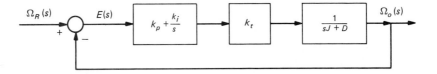

Proportional plus integral control gives the best of both the error and time-response worlds: zero steady-state error and a damping characteristic that can be controlled in a relatively loss-free manner by means of the constant K_p.

PROBLEMS

17.1   Determine the transfer functions
$G(s) = V_0(s)/V_i(s)$ of the networks shown in Fig. P17.1.

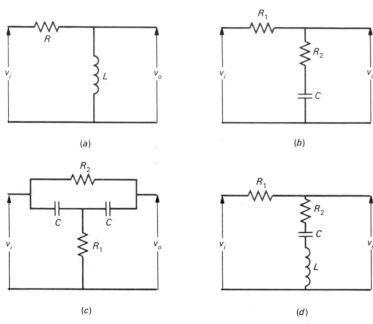

FIGURE P17.1

17.2   Determine the transfer function $G(s)$ for the platform position $Y(s)$ for the spring-mounted platform shown in Fig. P17.2. Consider only the translation motion in a vertical plane for both table and platform.

17.3   The network shown in Fig. P17.3 is an example of a nonminimum phase transfer function. Derive the transfer function $G(s) = V_0(s)/V_i(s)$, and show that zeros exist in the right half of the S plane.

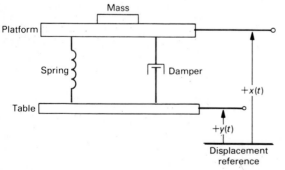

FIGURE P17.2

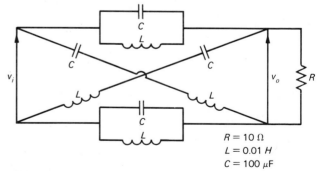

FIGURE P17.3

17.4 Figure P17.4 illustrates a practical op-amp circuit used for differentiation in analog computers. Assuming an ideal op amp, determine the transfer function $G(s) = V_o(s)/V_i(s)$.

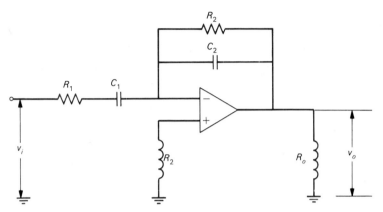

FIGURE P17.4

17.5 Repeat the analysis of Example 17.4 (Fig. 17.7) for the on-off controller, or chopper, and derive the transfer function for the case of discontinuous load current flow.

17.6 Figure P17.6 shows a common-emitter three-stage transistor amplifier. The small-signal transistors Q are similar to the 2N718 described in Chap. 9, and the dc bias voltages are such that the transistors are operated in the linear range of operation with $h_{FE} = 90$. Other circuit values are as follows: $C_1 = 10\ \mu F$, $C_2 = 50\ \mu F$, $R_1 = 20\ k\Omega$, $R_2 = 6400\ \Omega$, $R_3 = 2000\ \Omega$, $R_4 = 1000\ \Omega$, $R_L = 600\ \Omega$. Determine the transfer function for ac operation of the amplifier $V_o(s)/V_i(s)$, including the effects of the coupling capacitors. The small-signal equivalent common-emitter circuit developed in Chap. 9 should be used; neglect h_{ie} and all bias currents and voltages

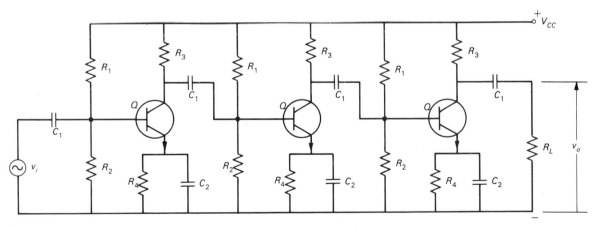

FIGURE P17.6

17.7   Figure P17.7a represents the block diagram of a motor speed control system with a load disturbance T. Show that the block diagram of Fig. P17.7a can be reduced to the form shown in Fig. P17.7b, and determine the transfer functions for each block in Fig. P17.7b.

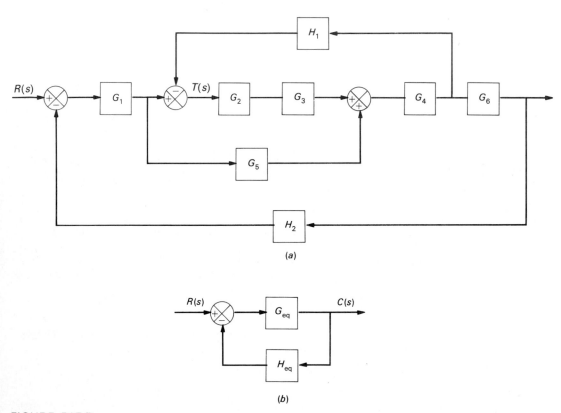

(a)

(b)

FIGURE P17.7

17.8   Determine the characteristic equation for the *closed-loop* response $C(s)/R(s)$ for the system shown in Fig. P17.8.

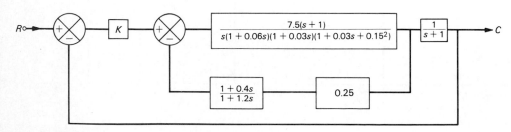

FIGURE P17.8

17.9 Determine the open-loop transfer function $G(s)$ for the system shown in Prob. 17.8.

17.10 Derive the simplified block diagram of Fig. 17.9*b* from the original multiple-loop diagram shown in Fig. 17.9*a*.

17.11 Many simple liquid level controllers use proportional control action, as illustrated in Fig. P17.11. The inlet valve has a stem travel of 3.0 cm and a capacity of 0 to 40 liters per minute (L/min) depending on the position of the valve stem. Determine the numerical value for the proportional gain constant K_p, and state its units.

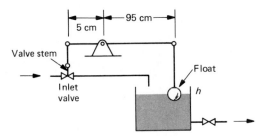

FIGURE P17.11

17.12 Figure P17.12 illustrates a speed control system with proportional control. (*a*) Obtain the expression for the time response $\omega_c(t)$ if a unit step function $u(t)$ is applied to the reference input: assume that the disturbance torque $Td(s)$ is zero. (*b*) Repeat with a unit step function applied to the disturbance input and with the reference input assumed to be zero.

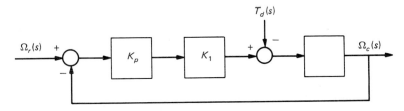

FIGURE P17.12

17.13   Show that the block-diagram representation for the tank shown in Fig. P17.13 is correct where q_m is the input variable, q_u is a disturbance flow rate, and the liquid level h_c is the controlled variable. The liquid level is to be controlled by proportional action as shown in the lower block diagram of Fig. P17.13. Determine the minimum proportional gain K_p such that, with a sustained disturbance q_u of 28.5 L/s, the deviation of liquid level will not exceed 2.5 cm. Let $C = 0.85$ m^2 and $R = 108$ s/m^2.

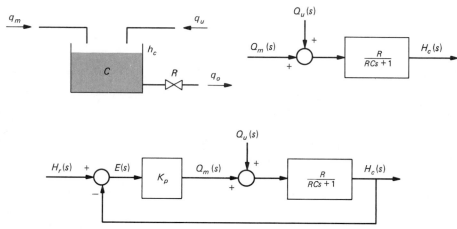

FIGURE P17.13

17.14   The block diagram for a speed control system with proportional plus integral control action is shown in Fig. P17.14. Prove that this combination of control actions will result in the steady-state elimination of system error in response to a unit step disturbance torque input $T_d u(t)$.

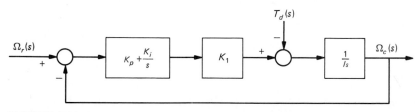

FIGURE P17.14

17.15   Verify the error-response conclusions listed in Table 17.1.

17.16 Figure P17.16 illustrates two type 1 systems with the integrator located in two different positions. Will both systems eliminate the steady-state error for unit step function disturbance input $U(s)$?

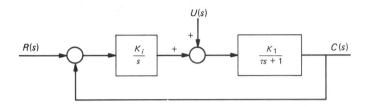

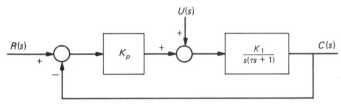

FIGURE P17.16

REFERENCES

1 H. Chestnut and R. W. Mayer, *Servomechanisms and Regulating System Design*, 2d ed., vol. 1, Wiley, New York, 1959.

2 R. C. Dorf, *Modern Control Systems*, 3d ed., Addison-Wesley, Reading, Mass., 1980.

3 J. J. D'Azzo and C. H. Houpis, *Linear Control System Analysis and Design*, 2d ed., McGraw-Hill, New York, 1981.

4 M. E. El-Hawary, *Control Systems Engineering*, Reston Publishing Co., 1984.

5 H. L. Harrison and J. G. Bollinger, *Introduction to Automatic Controls*, 2d ed., International Textbook Co., Scranton, Pa., 1969.

6 P. M. DeRusso, R. J. Roy, and C. M. Close, *State Variables for Engineers*, Wiley, New York, 1965.

7 G. J. Thaler and M. P. Pastel, *Analysis and Design of Nonlinear Feedback Control Systems*, McGraw-Hill, New York, 1962.

8 B. C. Kuo, *Digital Control Systems*, Holt, Rinehart & Winston, New York, 1980.

9 G. H. Hostetter, C. J. Savant, Jr., and R. T. Stefani, *Design of Feedback Control Systems*, Holt, Rinehart & Winston, New York, 1982.

Electrical Instrumentation

The measurement of signals, parameters, and dimensions is one of the most significant branches of science and engineering. The activity of measuring tells each of us much about our universe, our environment, our economy, and ourselves. Much of the world's industry, commerce, transportation, and even human relationships is based upon accurate measurements. Moreover, the basis for all scientific principles and engineering concepts rests, ultimately, on their verification by test and measurement. The theory and principles presented so far in this text have little meaning except as verified by measured test data—and since this is a relatively basic text, the reader can be assured that almost all the theories and concepts presented here do rest on sound experimental evidence. Therefore, our aim in this chapter is not to explore the methods of verification of the fundamental theorems of science or philosophy, but rather to describe the usefulness of accurate measurement techniques in "everyday" engineering, to explore some of the limitations of measurements, especially those of electrical parameters, and to describe some of the more common electrical measuring devices.

18.1 SOME BASIC
CONCEPTS OF ELECTRICAL MEASUREMENTS

The measurement of physical parameters can be broadly divided into two categories—analog and digital. This classification is perhaps more appropriately defined in terms of the nature of the signals used in the measurement process—continuous and discrete. Analog instruments and analog computers generally share common hardware (or circuit components), common performance characteristics, and common methods of mathematical analysis. The same can also be said of digital instruments and digital computers. In fact, instrumentation of both types is often a link in a computing system or a control system.

630

Digital meters are generally more accurate and can be equipped with more scales and broader ranges than analog meters, and they frequently have faster time response and greater frequency response. On the other hand, analog meters are generally less expensive and less susceptible to electromagnetic noise. A more subtle difference is the information gleaned by the person reading the meter. In digital meters, one sees only one specific reading or number; in analog meters, this reading or number is seen in light of an entire range or scale of reading, which is often very informative. When the reading is changing, one has no idea where a digital meter is going, whereas an analog meter shows an increasing or decreasing reading toward some other numerical value; this is especially helpful when one is trying to set a parameter, such as motor speed, at a specific value. Digital meters can also be confusing or at least annoying when the reading is oscillating around a certain value for one reason or another. However, in general, more skill and experience are required to read an analog meter, especially if precision is required, since interpolation between scale markings is generally required and since the angle at which the reader looks at the instrument may cause parallax effects. Therefore, if an unequivocal, accurate reading is required—and especially one by an unskilled reader, as is the case in many assembly-line measurements—digital meters are usually preferred.

18.1.1 Accuracy, Precision, and Errors

Accuracy is perhaps the first characteristic that one thinks of when selecting or applying an electrical meter. Accuracy is defined as "freedom from mistake or error" or "conformity to truth or a standard." It is the second of these definitions that most closely defines the meaning of the term *accuracy* as applied to meters and instrumentation. Conformity to truth is the fundamental concept of accuracy, and a meter of perfect accuracy is perfectly truthful in indicating the value of the quantity being measured. In the strict sense of this meaning, however, there is no such thing as a "perfect" meter, even though certain measurements of time and distance do approach perfection. Therefore, the concept of conformity to a *standard of measure* is closer to defining the function of a meter. The degree to which a meter or instrumentation system conforms to a standard of measure is known as the *accuracy* of the meter. Standards of the basic MKS or SI units are maintained in various national standards laboratories. The standard for the secondary unit upon which most electrical units rest, the ampere, is maintained in the French National Laboratory.

Accuracy is often confused with or interchanged with the concept of precision, which may be simply defined as the "quality of being sharply or concisely defined." Related to meters and instruments, precision describes the degree to which a scale, dial, or gauge can be read. For example, on a simple ruler marked with $\frac{1}{8}$-in markings, a distance can be read to a precision of only $\frac{1}{8}$ in. Or we often say, "To the nearest $\frac{1}{8}$ in." This means not that the reading of distance using this scale is *accurate* to $\frac{1}{8}$ in, only that the meter (in this case, a ruler) can be read with a *precision* of $\frac{1}{8}$ in. This author once had a summer job between college terms as a surveyor working the "nonreading" end of a surveyor's tape, the finest scale markings of which were 0.1 in. On the opposite "reading" end of the tape was an old-time employee who would consistently call out such readings as "43 ft 6.578 in" and would defend them all most passionately to the young doubting engineering student. Most of us made similar faux pas at one time or another, but in the recording or reporting of technical data this is a trap to be avoided.

Related to the concept of precision in measurement and meter reading is the concept of the number of significant figures in a numerical value of a meter or instrument reading, a concept probably quite familiar to the reader. When measured data are used in calculations or data processing, it is important to maintain only the significant figures that can be derived from the basic precision with which the measurements were made. In this era of performing most data manipulation on calculators or computers, it is often too easy to present an answer using the 8- or 10-digit accuracy of the calculating device based upon meter readings that have only three significant figures. This is another trap to be avoided in the presentation of data based upon measured values.

Precision in digital meters is much more straightforward and is equal to the number of digits in the readout. Many digital meters have the number of readout digits equal to an integer plus one-half, such as a "$3\frac{1}{2}$-digit readout." In this example, the three digits can register a count of 0 to 9; the $\frac{1}{2}$ refers to the fourth digit, which is always the leftmost digit and which can register 0 or 1, as well as the sign of the reading ($+$ or $-$). Thus, a $3\frac{1}{2}$-digit readout can register a maximum count of 1999.

All practical meters have a certain degree of inaccuracy, owing to the presence of several types of errors that may occur while an operator is reading and recording an indicated value from a meter. A very general summary of measurement errors is given in Table 18.1.

Most of these errors are obvious from their definition. The first class of error is called *instrumental* rather than *instrument*, since it may result from an improper use of an instrument or an improper planning of a test procedure as well as from the inherent error of the instrument itself. Environmental errors result from the effects of temperature, pressure, humidity, etc., and corrections for such errors can usually be made if the environmental conditions are known. Random or erratic errors frequently occur in the best-designed schemes of measurement and are often unexplainable; allowances for such errors should be made in all measurement planning. Instrument errors as specified by the instrument manufacturer usually take into

TABLE 18.1 Measurement Errors

1 Instrumental errors

2 Errors due to environmental conditions

3 Human errors in reading and recording

4 Random or unexplainable errors

consideration this possibility of random or unexplainable errors. In other words, manufacturer-specified instrument errors are generally conservative. Instrument errors are generally specified (1) as a percent of full-scale reading, (2) as a percentage of the actual reading, or in some cases (3) as a fixed error. Also errors due to temperature deviations are often specified.

18.1.2 Errors in Digital and Analog Meters

Analog meter error is generally stated as a percent of full scale. Digital meter error is generally specified as a combination of errors, including a percent of full scale. However, since this latter is generally very small, the comparison of errors in the two types of instruments may be somewhat misleading. As an example, a good-quality analog voltmeter may be specified as having an instrument error of 0.5 percent of full scale and 0.002 percent of reading per degree Celsius. A digital meter of equivalent rating may have the following error specifications: 0.05 percent of reading, 0.05 percent of scale, 0.005 percent of reading per degree Celsius, and 0.005 percent of scale per degree Celsius. Let us assume that both meters have a full scale of 200 V and that a reading of voltage is being made at 35.0 V in an environment consisting of a temperature of 40°C above ambient. Note that the above error specifications are maximum or guaranteed maximum errors; actual errors of instruments manufactured by reliable dealers are generally far below these values. Continuing with the above example, the analog voltmeter could have an error of $0.005 \times 200 + 0.00002 \times 35 \times 40 = 1.0 + 0.028 = 1.028$ V. The error as a percent reading is $1.028 \times \frac{100}{35} = 2.94$ percent. The digital meter will have an error of $0.0005 \times 200 + 0.0005 \times 35 + 0.00005 \times 200 \times 40 + 0.00005 \times 35 \times 40 = 0.5875$ V. As a percent reading, this is an error of 1.68 percent. Figure 18.1 gives a picture of the relative

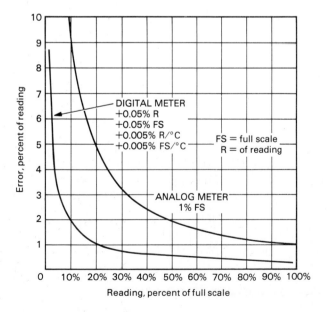

FIGURE 18.1
Comparison of
digital and analog
meter accuracies.

differences between analog and digital meter accuracy. Readings on the low end of a meter scale will contain a large percent error.

Most meter manufacturers use the word "accuracy" to describe what has been properly termed "error" in the above paragraph—probably to avoid the use of the word "error" in describing their product. Also it is assumed that the algebraic sign of the error may be either positive or negative. Thus, the analog meter discussed in the previous paragraph might be specified by the manufacturer as having an "accuracy" of ± 0.5 percent of its full-scale value. This specification is often termed a *known* instrument error in contrast to a *probable* error, which has a considerably different meaning. By this specification, the manufacturer guarantees that the meter reading will be within an increment around the true value, the increment being equal to ± 0.5 percent of full scale. Note that the increment (when specified as a function of full scale) is constant over the entire scale. The fact that there may be an error in every measured value (with maximum possible value equal to known error) must be considered when one is performing mathematical manipulation on measured values. For example, if a measured voltage is to be multiplied by a measured current to obtain power or apparent power, the actual multiplication should be

$$P = (V + e_v)(I + e_i) = VI(1 + e_v + e_i + e_v e_i) \tag{18.1}$$

Since the latter term in Eq. (18.1) is very small, it may be dropped, and the difference between the true power (VI) and the measured power (i.e., the error in the measured power) is

$$e_p = e_v + e_i \tag{18.2}$$

Similar relationships can be derived for the other mathematical manipulations.

Instruments should be checked periodically to ensure that indicated readings are within this error (or accuracy) increment. This process is known as *calibration*. A meter is calibrated by first comparing its indicated value with that of a standard meter and then making adjustments, if necessary, to bring the meter's reading within the error increment at major divisions of each scale. A *standard meter* is one of high accuracy which is used *only* for calibration purposes (i.e., not for general measurements), which itself is frequently calibrated against primary standard meters maintained by national standards laboratories and which is usually kept within a controlled environment. As a rule of thumb, the standard meter should have an error increment, or accuracy, one-tenth that of the meter being calibrated. Certain classes of instruments, such as voltmeters and frequency counters, are often calibrated against standard *sources* of the parameter being measured.

18.2 MEASUREMENT OF VOLTAGE, CURRENT, AND POWER

A wide variety of instruments are available for measuring voltage and current over a broad spectrum of frequencies, from direct current to the gigahertz range. Digital meters have gradually become more popular for this function, although analog meters are still widely used, especially in the dc and power frequency range.

In selecting a *voltmeter*, the following parameters must be considered: ranges of voltage to be measured, accuracy, scale precision, effect of current drain (which is a function of the meter's impedance), frequency of the signals to be measured, waveform of the signals to be measured, power required to operate the meter, location of measurements (panel or portable), available accessories, and cost. Accuracy and precision have been discussed in Sec. 18.1, and the ranges required for the signals to be measured are usually obvious, keeping in mind that most instruments give greater accuracy if the readings are near full scale. The other parameters deserve some further discussion:

1 *Current drain:* Electronic meters generally have very high internal resistance, usually in the megohm range, and current drain is therefore negligible in most applications; Table 18.2 lists the specifications of a typical digital multimeter (DMM), probably the most common type of laboratory meter in use today, and the impedance of the voltmeter portions of this meter are seen to be very high on both ac and dc signals. Analog meters, on the other hand, may have relatively low internal impedance, often as low as 100 Ω/V. Therefore, current drain by the meter may be significant, and the voltmeter may have a significant effect on the circuit being measured. Meter capacitance must also be considered when signals of high frequency or short pulses are measured.

2 *Frequency range:* The typical multimeter maintains good accuracy well into the upper audio ranges. *Dynamometer* and *iron-vane* analog meters are generally restricted to measurements at power frequencies, i.e., below 1000 Hz. *Thermocouple* analog meters are suitable for frequency ranges well into the upper audio ranges and are probably the most versatile type of meter available for measuring signals with pulses and highly irregular waveforms. *Rectifier* analog and special electronic instruments are available for measuring high-frequency signals above the audio-frequency range. Most electronic instruments have digital circuitry and digital readouts. Analog meters use continuous circuits and have analog scale readouts. Some of these meters are *true analogs* of the signals being measured. For example, the dynamometer meter is a miniature "motor" which models the square of the root-mean-square (RMS) expression

$$v^2 = \frac{1}{\tau} \int_0^\tau v^2 \, dt \tag{18.3}$$

The natural form of the scale on a dynamometer meter is therefore a "square-law" scale, which on a typical meter is made into a quasi-linear scale by the choice of scale markings; because of this square-law action, dynamometer meters measure both direct and alternating current. Obviously, a dynamometer meter has very poor accuracy at low-scale readings. The thermocouple is also a true analog of RMS voltage (and current), since the equivalent heating effect is the very basis for the definition of an RMS parameter. Thermocouple meters are

TABLE 18.2 Specifications of a Typical DMM (*Courtesy of Keithley Instrument Co.*)

		130	169	178	179 (179-20A)	177	191
					Model		
	Counts	1999	1999	19,999	19,999	19,999	199,999
	Digits	$3\frac{1}{2}$	$3\frac{1}{2}$	$4\frac{1}{2}$	$4\frac{1}{2}$	$4\frac{1}{2}$	$5\frac{1}{2}$
DC Volts							
	Maximum sensitivity	100 μV	100 μV	100 μV	10 μV	1 μV	1 μV
	Maximum reading	1000 V	1000 V	120 V	1200 V	1200 V	1200 V
	Basic long-term accuracy	0.5%	0.25%	0.04%	0.04%	0.03%	0.007%
AC Volts							
	Method	Average	Average	Average	TRMS	TRMS	Average
	Maximum sensitivity	100 μV	100 μV	100 μV	10 μV	10 μV	10 μV
	Maximum reading	750 V	1000 V	1000 V	1000 V	1000 V	1000 V
	Basic long-term accuracy	1%	0.75%	0.3%	0.5%	0.5%	0.1%
Ohms							
	Special features	Diode test on 20-kΩ range	Two low-voltage ranges	—	Hi-lo switch selectable	Front panel lead compensation	Auto 2/4 terminal
	Maximum sensitivity	100 mΩ	100 mΩ	100 mΩ	100 mΩ	1 mΩ	1 mΩ
	Maximum reading	20 MΩ	20 MΩ	20 MΩ	20 MΩ	20 MΩ	20 MΩ
	Basic long-term accuracy	0.5%	0.02%	0.04%	0.04%	0.04%	0.012%

DC Amperes

Maximum sensitivity	1 µA	100 nA	—	10 nA	1 nA	—
Maximum reading	10 A	2 A	—	2 A (20 A)	2 A	—
Basic long-term accuracy	1%	0.75%	—	0.2%	0.2%	—

AC Amperes

Method	Average	Average	—	TRMS	TRMS	—
Maximum sensitivity	1 µA	100 nA	—	10 nA	10 nA	—
Maximum reading	10 A	2 A	—	2A (20 A)	2 A	—
Basic long-term accuracy	2%	1.5%	—	1.0%	0.8%	—
Ranging	Manual	Manual	Manual	Manual	Manual	Manual
Outputs	—	—	—	BCD opt IEEE opt	Analog BCD opt IEEE opt	—
Battery operation	Standard	Standard	Optional	Optional	Optional	—
Capsule comment	Pocket size; 0.6-in digits; rugged case	Function and range annunciators; 0.6-in digits; 1-year battery; float up to 1.4 kV	Float up to 1.4 kV	1-kV protection on Ω; float up to 1.4 kV; 20-A alternating and direct current as 20 A	1-µV, 1-nA, 1-mΩ sensitivity	µP control allows up to four conversions per second; digital filter; push-button null

actuated by the signal from a thermocouple as a function of the heat generated by a current (or current proportional to a voltage) and hence measure the *true RMS value* of the signal; this is discussed further below. The principal analog meter used to measure dc signals is known as the *d'Arsenvol meter* and like the dynamometer meter, it is based on the motor action of a current-carrying conductor in a magnetic field produced by a permanent magnet. The force on the conductor is basically a function of the instantaneous current in the coil, although, due to the relatively high inertia of the moving-coil system in this instrument, the meter reading is normally the average of the instantaneous forces on the coil except for low-frequency variations; hence, this meter indicates the average or dc values of the signal. Table 18.3 summarizes some of the characteristics of the common analog instruments.

3   *Waveform:* The waveform of the signal being measured may have a profound influence on the numerical value of the meter reading. Many electronic instruments use sampling techniques for obtaining the numerical value of the signal. The sampled values—values measured at each 10 μs, e.g.,—are combined mathematically by means of digital circuitry to give a meter reading of the average or RMS value of the signal. To perform this mathematical manipulation, a fixed waveform must be assumed, and for ac measurements this is almost always a sinusoidal waveform; therefore, such instruments read *true* RMS values only for sinusoidal signals. Other electronic voltmeters actually perform the root-mean-squaring process by analog or digital multiplication and square-rooting and can therefore indicate true RMS values regardless of the waveform. One feature of electronic voltmeters that is not available on analog meters is the ability to measure and hold *peak* values of a signal, such as occurs during motor starting. Most analog meters are *inherently true RMS meters;* this has already been noted for the thermocouple meter, and it is also true of the dynamometer analog meters. However, the frequency response of any circuit, instrument, or machine is primarily a function of its inductance and capacitance. Dynamometer meters have relatively high inductance, as compared to thermocouple and rectifier instruments, and hence their frequency response is limited to the low-audio frequencies, generally to less than 1000 Hz; their ability to respond accurately to signals

TABLE 18.3   Analog Voltmeter Characteristics

Type	Frequency Range, Hz	Resistance, Ω/V	Accuracy, %
D'Arsenvol	Direct current	100–5000	0.01–0.5
Iron vane	25–125	75–200	0.75–2.0
Dynamometer	0–3200	10–100	0.1–1.5
Thermocouple	0–50,000	100–500	0.1–1.0
Rectifier	$20–50 \times 10^6$	1000–5000	1.0–5.0
Electrostatic	0–10,000	>10,000	0.5–2.0

Note: Analog ammeters generally have frequency and accuracy capabilities similar to their voltmeter counterparts.

containing sharply rising wavefronts is thus very limited. The iron-vane meter is likewise limited to near sinusoidal signals at power frequencies.

4 *Required power:* Analog meters generally require no external power for operation, but all electronic meters do. Portable electronic meters are battery-operated, which can be a nuisance in some applications. It is important that electronic instruments have some means of indicating when the battery voltage level is below that required for proper meter operation.

5 *Meter location:* All types of meters may be generally classified as either portable or panel types. Panel instruments are used in a great variety of industrial applications, and in terms of numbers of meters sold per year, they probably far outsell the portable type, although the latter is probably much more familiar to the student reader. In general, the differences between these two classes are structural, and most of the comments concerning other parameters, such as accuracy or frequency range, apply equally to panel or portable instruments. Vibration may be of concern in many applications of analog panel instruments and must be considered in such applications.

6 *Available accessories:* Electronic meters generally have many alternative functions which may or may not be included in the price of the "standard package" meter. For example, the typical electronic multimeter (or DMM, as it is commonly called) contains a milliammeter and ohmmeter as well as ac and dc voltmeter functions. Many DMMs, such as the one shown in Fig. 18.2, and Table 18.4, also can be used to measure temperature with the purchase of the required thermocouple. Most electronic meters have facilities for providing an output signal proportional to the signal being measured for use in data acquisition and/or manipulation in a digital computer.

7 *Cost:* For pure voltmeter applications, electronic and DMM instruments offer many advantages over analog instruments, the main advantages being their inherent high accuracy, multirange, and ac/dc capability. Analog meters of comparable accuracy are generally very expensive and have limited voltage and frequency range. However, where accuracy is not important and the range of voltages to be measured is rather limited, such as on many panel meter applications, an analog voltmeter will usually be much less costly than an equivalent electronic or digital meter. The differences in how a meter is read must also be considered in panel meter applications, as discussed earlier.

The above comments on voltmeters *generally apply to ammeters*, too. There is one major difference related to measuring current, however: Electronic or digital ammeters are generally limited in range to milliamperes or a few amperes. Also the resistance that an electronic ammeter places in series with the current being measured is generally much higher than that of an equivalent analog meter. Therefore, in contrast to the situation in voltage measurement, an electronic ammeter may have more effect on the circuit being measured than an analog ammeter does. The lower-frequency meters, such as the dc d'Arsenvol, iron vane, and dynamometer, generally have negligible effect (i.e., negligible series resistance) on the circuit being

TABLE 18.4* Specifications for the Voltmeter and Ohmmeter Sections of a Typical Digital Multimeter (*Courtesy of John Fluke Mfg. Co., Inc.*)

DC Volts†

Range	Resolution	Accuracy for 1 year
±200 mV	100 μV	
±2 V	1 mV	
±20 V	10 mV	±(0.1% of reading + 1 digit)
±200 V	100 mV	
±1000 V	1 V	

AC Volts (True RMS Responding)‡

Range	Resolution	Accuracy for 1 Year			
		45 Hz to 1 kHz	to 10 kHz	10 kHz to 20 kHz	20 kHz to 50 kHz
200 mV	100 μV				±(5% of reading +3 digits)
2 V	1 mV			±(1.0% of readding +2 digits)	
20 V	10 mV	±(0.5% of reading + 2 digits)			
200 V	0.1 V				
750 V	1 V	±(0.5% of reading +2 digits)			

Resistance§

Range	Resolution	Accuracy for 1 Year	Full-Scale Voltage	Maximum Test Current
200 Ω	0.1 Ω		0.25 V	1.3 mA
2 kΩ	1 Ω		1.0 V	1.3 mA
20 kΩ	10 Ω	±(0.2% of reading +1 digit)	<0.25 V	10 μA
200 kΩ	100 Ω		1.0 V	35 μA
2000 kΩ	1 kΩ		<0.25 V	0.10 μA
20 MΩ	10 kΩ	±(0.5% of reading +1 digit)	1.5 V	0.35 μA

FIGURE 18.2
Clip-on ammeter
with associated
voltage- and
resistance-
measuring circuits.
(Courtesy
of the Triplett
Corp., Blufton,
Ohio.)

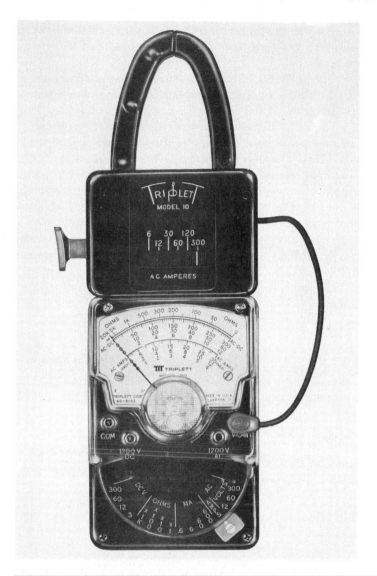

* The electrical specifications given assume an operating temperature of 18 to 28°C, humidity up to 90 percent, and a 1-year calibration cycle. The functions are dc volts, ac volts, direct current, resistance, and conductance.

† Input impedance: 10 MΩ, all ranges. Normal mode rejection ratio: >60 dB at 60 Hz (at 50 Hz on 50-Hz option).

‡ Volt-hertz product: 10^7 max (200 V max @ 50 kHz); extended frequency response: typically ±3 dB at 200 kHz; common-mode noise rejection ratio (1-kΩ unbalance); >60 dB at 50 and 60 Hz; crest factor range: 1.0 to 3.0; input impedance; 10 MΩ in parallel with <100 pF.

§ Overload protection: 300 V dc/ac RMS on all ranges. Open-circuit voltage: less than 3.5 V on all ranges. Response time: 1 s all ranges except 2000-kΩ and 20-MΩ ranges; 4 s these two ranges. Diode test: These three ranges have enough voltage to turn on silicon junctions to check for proper forward-to-back resistance. The 2-kΩ range is preferred and is marked with a larger diode symbol on the front panel of the instrument. The three nondiode test ranges will not turn on silicon junctions, so in-circuit resistance measurements can be made with these three ranges.

measured. The thermocouple and rectifier instruments have somewhat higher series resistance, yet nevertheless a much lower one than that of an equivalent electronic ammeter.

Analog meters can be constructed to measure current magnitudes of very large values, into the thousands of amperes. At some point, however, it becomes desirable from the technical and economic standpoints to limit the range of an ammeter and use a *current transformer* to achieve the higher-magnitude capabilities. Current transformers also permit the use of the low-scale capabilities of electronic ammeters and DMMs. Current transformers, briefly discussed in Chap. 12, are basically two-winding transformers connected in series with the current being measured. There is an error associated with a current transformer (an error in both the magnitude and the phase angle of the current being measured) which must be considered in determining the overall error of a measuring system.

One of the most useful instruments for measuring currents in the ampere range is the *clip-on ammeter* which combines the current transformer (with a 1-turn primary) and the measurement functions into one device. Figure 18.2 illustrates a typical clip-on ammeter.

18.2.1 Power Measurements

Power is perhaps the most difficult electrical parameter to measure with any degree of accuracy. *Instantaneous power* is defined as the product of the voltage across a given element and the current through this element, or

$$p = ei \tag{18.4}$$

Average power is defined as

$$P = \frac{1}{\tau} \int_0^\tau p \, dt \tag{18.5}$$

where τ is a specific time period, such as the period of a sinusoidal signal.

The dynamometer analog instrument is an excellent power-measuring device since, as has been noted in Eq. (18.3), the motor action of the dynamometer performs the product function required by Eq. (18.4) and the inertia of the movement and the proper design of the meter's scale perform the averaging function indicated in Eq. (18.5). However, given the high inductance of the moving and stationary coils of this instrument, its frequency response is limited to power and low-audio frequencies. The maximum frequency for which dynamometer wattmeters have been adapted is approximately 3200 Hz.

Hall device wattmeters have been developed to provide a means of measuring power in the frequency range above 1000 Hz. The Hall phenomenon is utilized in obtaining the product described in Eq. (18.4). A number of commercial Hall wattmeters is available today. However, the frequency response of Hall wattmeters is limited to the midaudio frequency range. For power measurement at frequencies (or waveforms having frequency components) above this range, other circuits are required.

Electronic multiplier wattmeters extend the frequency response of power measurements to approximately 50 kHz. Figure 18.3 shows the block diagram for an electronic multiplier wattmeter designed to measure both instantaneous and average power. This type of meter has been applied successfully in the measurement of the nonsinusoidal low-power-factor power associated with power electronic circuits, such as the power loss in thyristor commutating circuits (Ref. 1) and core losses in motors excited from power semiconductor choppers.

The accurate measurement of power at low power factors is particularly difficult even with low-frequency sinusoidal signals. For sinusoidal signals, Eq. (18.5) becomes

$$P = \frac{1}{\pi} \int_0^\pi (E_m \sin \omega t)[I_m \sin(\omega t - \theta)] \, d(\omega t)$$

$$= \tfrac{1}{2} E_m I_m \cos \theta$$

$$= EI \cos \theta \tag{18.6}$$

Average power is thus the product of three parameters—RMS voltage, RMS current, and $\cos \theta$ or power factor (pf) (in sinusoidal circuits)—and, according to the relationships derived in Sec. 18.1.2, the total wattmeter error is the *sum* of the errors in the measurement of all three parameters. At low power factors, θ approaches $90°$ and $\cos \theta$ approaches zero. At a power-factor angle of $80°$, which is not untypical of many reactive circuits and motor and transformer circuits at light-load conditions, an error in θ of $0.5°$ results in an error in $\cos \theta$ of almost 5 percent. Therefore, care must be taken in the design of both the current and the voltage circuits in any type of wattmeter to ensure minimum magnitude and angle errors. In electronic multiplier and Hall wattmeters, a large number of scales can be made available by means of variable-gain circuitry, and power even at very low power factors can be read accurately. However, on standard dynamometer wattmeters, scales are designed for unity power-factor measurements, which means that full scale in watts occurs at rated

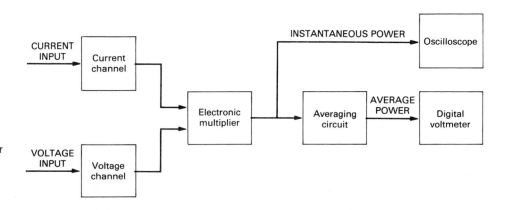

FIGURE 18.3
Block diagram for an electronic multiplier wattmeter.

or full scale in the voltage and current circuits. When power is measured at a low pf, this will require reading the meter in the lower—and hence the less accurate—portion of the watts scale. Therefore, wattmeters with *low-power-factor scales* (full-scale watts occurs at rated current, rated voltage, and 20 percent power factor) are available and should be used to measure power in reactive circuits. In many power electronic circuits, in which the signals are usually nonsinusoidal, the simple concept of the power factor as the cosine of the angle between the voltage and current signals, as shown in Eq. (18.6), has little meaning. The more general definition of *power factor* is

$$pf = \frac{\text{average power}}{(\text{RMS volts})(\text{RMS amperes})} \tag{18.7}$$

Example 18.1   A certain low-power-factor wattmeter is rated at 200 V and 10 A, and the full scale on the watts scale is 400 W. The known errors are ± 0.5 percent of full-scale volts and ± 0.5 percent of full-scale amperes, and there is an angle error of 0.5°. The wattmeter is used to measure power in a sinusoidal system in which the voltage is 175 V, the current is 5 A, and the power factor is 0.15. Assuming errors at the maximum value, what is the maximum possible error in the power reading?

Solution   The power factor of 0.15 implies a power-factor angle of 81.37°. The maximum possible power reading is therefore

$$(175 - 1)(5 - 0.05) \cos 81.87° = 122.56 \text{ W}$$

The power is truly equal to

$$175 \times 5 \times 0.15 = 131.25$$

The error is therefore

$$131.25 - 122.56 = 8.68 \text{ W}$$

or

$$8.69 \times \frac{100}{131.25} = 6.6\%$$

Proper connection of the two circuits on a wattmeter is essential for preventing damage to the meter and for ensuring accurate readings. Most commercial wattmeters have the terminals distinguished by polarity markings, usually the $(\pm)$ marking. Obviously, it is most essential to connect the voltage and current circuits in the proper orientation since the current circuit on most wattmeters is a very low-resistance delicate circuit and is easily damaged by excessive current. But it is also

FIGURE 18.4
Correct connection
for a single-phase
wattmeter.

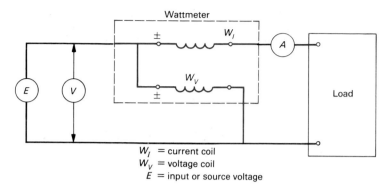

W_I = current coil
W_V = voltage coil
E = input or source voltage

important to connect the two circuits with the proper polarity connections to ensure an upscale reading on the meter and to minimize stray capacitive coupling between the coils or electronic circuits that make up the voltage and current channels of the meter. Figure 18.4 illustrates the correct connection of a wattmeter. Frequently, it is necessary to measure voltage and current as well as power, and so the correct locations of the voltmeter and ammeter are also shown in Fig. 18.4. Note that either the voltmeter current drain or the ammeter voltage drop will result in a slight error in the wattmeter reading, no matter where these instruments are located in the circuit. It is usually desirable to eliminate the voltmeter error and accept the ammeter error, as shown in Fig. 18.4, since the latter is usually much smaller than the former.

The connection of wattmeters to measure power in three-phase circuits can become rather cumbersome, since so many circuits are involved. In four-wire three-phase circuits, three wattmeters must be used, each connected as a single-phase meter between open line and neutral, using the polarity markings as shown in Fig. 18.4. In three-wire three-phase circuits, only two wattmeters are required. The two-wattmeter method of measuring three-phase power is shown in Fig. 18.5.

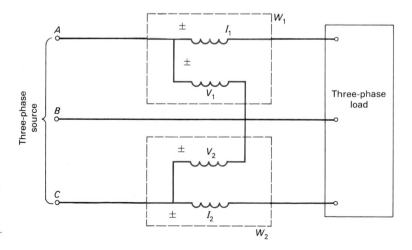

FIGURE 18.5
Connections for
the two-wattmeter
method of
measuring
three-phase power.

18.3   MEASUREMENT OF IMPEDANCE

Electrical impedance is composed of real and reactive components, as described in earlier chapters of this text. The concept of impedance generally implies sinusoidal excitation of the electric circuit. With the advent and widespread use of semiconductors and other switching and nonlinear elements in a great many types of circuits, single-frequency, sine-wave impedance or reactance may not be adequate in some applications. It is generally more important to know the value of the circuit elements—i.e., resistance, inductance, and capacitance (or their reciprocals)—rather than the reactance value (which is a product of the circuit element value and the frequency). Impedance over a range of frequency is usually required in electronic and pulse-circuit applications.

18.3.1   Resistance Measurements

Resistance, of course, is theoretically unaffected by the frequency of the applied signal, and therefore many of the above comments would be expected to be inappropriate in relation to resistance measurements. In actuality, however, resistance is probably the most elusive of the three electric circuit parameters. The resistances of almost all practical resistors are frequency-dependent and often include major components of the other circuit elements—inductance and capacitance. For example, the typical wire-wound resistor which will appear purely resistive at power frequencies (below $\simeq 1 \text{ kHz}$) may appear as primarily inductive at audio frequencies, and it will almost certainly be primarily capacitive at frequencies in the radio-frequency (RF) range and above. Most resistances are measured at zero (dc) or power frequencies. In general, this type of measurement is not even adequate for evaluating resistances in power circuits, due to the phenomenon known as the skin effect. Therefore, if at all possible, resistance should be measured under the excitation that will be used in the ultimate application of this resistance.

However, notwithstanding the above comments, a number of so-called resistance-measuring devices are among the most useful instruments available for many types of experimental work. Among these is the *ohmmeter*, which is often one element in digital multimeters. One function of an ohmmeter is to show circuit continuity (i.e., lack of an open circuit), and in this application it is invaluable. For measurement of the high-resistance values found in many electronic circuits, it gives good accuracy. For low-resistance measurements, its accuracy may be poor, and ohmmeter scales are often difficult to read with precision. Table 18.2 gives the specifications of a typical ohmmeter circuit in a DMM. In terms of measurement, the ohmmeter should never be used below resistance values of about $1.0 \ \Omega$. The ohmmeter is basically a dc device measuring the quotient of voltage (at a set value) and current.

More accurate resistance measurements are performed by means of resistance bridges. By far the most common is the *Wheatstone bridge*, the circuit diagram of which is shown in Fig. 18.6. In the Wheatstone bridge, as in all other bridge circuits, the measurement is made by means of a comparison with known values. In Fig. 18.6, the measurement of the unknown resistance R_x is performed by balancing the

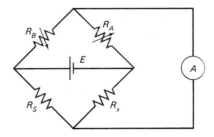

FIGURE 18.6
Elementary
Wheatstone-bridge
circuit for
measurement of
resistance.

variable resistances R_A and R_B until zero current flows through meter A. At this condition,

$$R_x = \frac{R_A}{R_B} R_S \tag{18.8}$$

The Wheatstone bridge is generally accurate for resistance measurements of $1.0\,\Omega$ and above; in most bridges, the accuracy of the measurement is equal to the accuracy to which the standard resistance R_S is known.

To measure resistances below $1.0\,\Omega$ (which incudes the resistance of a great many rotating machine and transformer circuits used for power and control), the *Kelvin*, or *double, bridge* is required. When resistances of very low magnitude are measured, the voltage drop across the contacts connecting the resistance to the external circuit may be appreciable and cause an error in the resistance measurement. To eliminate contact voltage drop, a four-terminal resistance is used: two outer terminals for connection to the external circuit (i.e., the terminals carrying high current), and two inner terminals connecting to the instrumentation circuit. Thus, the contact potential of the current-carrying terminals is eliminated from the measuring circuit. This principle is used in the Kelvin bridge and in instrument shunts.

18.3.2   Impedance Measurements

Bridge circuits are also used to measure the two reactive circuit elements, inductance and capacitance. The two most common bridge circuits—which are used to measure inductance in the common impedance, or *universal*, bridge—are the Maxwell and Hay bridge circuit, shown in Fig. 18.7.

Both of these circuits measure inductance in terms of a standard capacitance C_1. At balance—the condition of zero current in the sensing meter A—the unknown resistance and inductance are found in the Maxwell bridge to be

$$L_x = R_2 R_3 C_1 \qquad R_x = \frac{R_2 R_3}{R_1} \tag{18.9}$$

The Maxwell bridge is suited for the measurement of coils of relatively low-quality factor $Q = \omega L_x / R_x$. When it is used to measure coils with higher Q, a very large

FIGURE 18.7
Inductance-
measuring circuits.

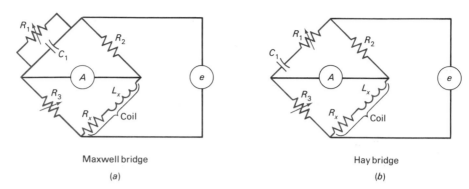

Maxwell bridge

(a)

Hay bridge

(b)

value of R_1 is required and the resistive balance (to obtain R_x) will not be well defined. For high-Q coils, the Hay circuit is preferred, although its balance equations [similar to Eq. (18.9)] are more complicated and are a function of the frequency of the signal source e.

In general, both the inductance and the resistance of a coil may vary significantly with frequency, and therefore the frequency at which the measurement is performed should be stated. The magnitude of the signal source in all commercial bridges, i.e., the value of e in Fig. 18.7, is very low. This introduces another problem when the inductance of an iron-core coil is measured, which is that the core will be magnetized at its level of *initial* magnetization near the origin of its magnetic characteristic, or *B-H*, curve (see Chap. 12). The slope of the *B-H* curve, which determines inductance, is usually small at the origin and much different from the slope in the linear operating region of the curve; therefore, the inductance of an iron-core coil as measured on an inductance bridge will almost always be too low. The inductance of iron-core coils should be measured by a voltmeter, ammeter, and wattmeter with the coil exciting at the frequency and current level at which it is to be used.

Capacitance is measured by bridge circuits similar to the inductance bridge circuits shown in Fig. 18.7. The two most common circuits are the Wien-bridge circuit and the capacitance comparison bridge. Capacitance is generally the most constant and stable of the three circuit elements and can be measured with a high degree of accuracy. It is also probably the "purest" circuit element in that the typical capacitor contains very little resistance and no inductance. For this reason, it is used as a standard of measurement in most bridge circuits, including those of Fig. 18.7.

18.4 OSCILLOSCOPES

The electronic oscilloscope is probably the most versatile instrument in the electronics laboratory, and it also serves many useful functions on the production line, in the automotive repair shop, in the TV and radio repair shop, and in the households of many engineers and "tinkerers." The oscilloscope is a means of not only *measuring* signals—and with generally excellent accuracy—but also giving a picture of signals, which is perhaps its greatest asset.

The basic element of an oscilloscope is the cathode-ray tube (CRT), although in recent years liquid-crystal displays (LCDs) have been used; LCDs result in a much smaller and more compact scope package. The CRT consists of three major elements: the electron source, often called an electron *gun;* a means of focusing the ray of electrons emitted by the gun, using both electric and magnetic fields; and a screen on which the traverse of the electron beam is exhibited. The CRT screen is constructed of a phosphor layer which emits visible light with a very short time constant upon being excited by the electron beam. The yellow-green portion of the visible light spectrum is the most commonly used phosphor on CRTs since the human eye is most sensitive to this frequency range; however, phosphors emitting many colors can be used, and of course are used in the sophisticated relatives of the CRT screen: the color TV screen, CAD/CAM computer terminal screens, etc. The electron beam is controlled at any given instant by two inputs; the *horizontal control,* which even in complex oscilloscopes is basically the field resulting from the simple triangular waveform produced by the charging of a resistive-capacitive (*RC*) series circuit, and the *vertical control,* which is determined by the instantaneous magnitude of the signal being observed through the activation of magnetic and/or electric fields as a function of this signal. Secondary control is also exerted by various fields that focus and improve the quality of the beam and by other signals that modify the horizontal control to synchronize it with various external processes or signals, a process generally referred to as *triggering* the oscilloscope. Many oscilloscopes have provision for a wide variety of modifications of the incoming signal being measured; these include a wide range of attenuators for adjusting the input-voltage level, current probes (actually current transformers) for converting current signals to voltage signals of the proper magnitude, circuits for resolving the incoming signal into its Fourier components, and filtering circuits. Thus, a total oscilloscope system is a versatile system capable of reproducing almost any type of signal, providing that the oscilloscope has been designed to handle the frequency, voltage, and rise-time range of the signal.

The two most important parameters in the process of choosing an oscilloscope are generally *bandwidth* and *rise time*—factors which determine the frequency response of the scope. Rise time is shown in Fig. 17.21; bandwidth in a scope is the reciprocal of the frequency of repetitive signals. However, bandwidth is often stated as a range of frequencies, or the maximum frequency that can be displayed on the scope. Various constraints in scope construction and design make these two parameters interrelated. As a rule of thumb, the product of bandwidth (MHz) and rise time (ns) should be approximately 0.35. Also the rise-time rating or capability of an oscilloscope should be approximately one-fifth that of the fastest signal to be measured. Using these two approximate relationships gives a desired relationship between the two important oscilloscope parameters of (see Ref. 2)

$$\text{Minimum bandwidth (MHz)} \simeq \frac{1.7}{\text{fastest rise time (ns)}} \qquad (18.10)$$

An oscilloscope should have a flat response over its rated bandwidth and a smooth roll-off at its upper frequency.

Other important scope parameters are *sensitivity* and *sweep rate*. In an oscilloscope, sensitivity refers to the input voltage required to produce a certain deflection of the spot on the CRT. Most oscilloscopes have a very wide range of voltage sensitivities, but for special applications other ranges must be considered. Sensitivity is the vertical calibration on an oscilloscope. Most oscilloscopes have a built-in sawtooth sweep generator that causes the spot on the CRT to move in the horizontal direction at a certain rate. In modern oscilloscopes, the sweep rate is calibrated in the unit of time, usually s/div on the horizontal scale. The sweep rate should be adequate to properly represent the highest frequency signal (or fastest rise-time signal) to be displayed on the oscilloscope. As a rule of thumb, a sweep is considered adequate if it is capable of displaying one cycle across the full horizontal scale (see Ref. 2). Most modern oscilloscopes are equipped with means for calibrating the sweep rate in terms of a direct unit of time; hence, they are generally very accurate in measuring the frequency and rise time.

Many oscilloscopes are capable of displaying several signals concurrently. One technique used for this purpose is to electronically switch the deflection system among two or more input signals; since the switching is done at a very high rate, two or more input signals can be displayed virtually simultaneously. This electronic switching between two input signals is known as a *dual-trace* process. *Dual-beam* oscilloscopes contain two independent beams and deflection systems and, in general, can be operated as two independent oscilloscopes, although some triggering options may be shared by the two systems. A dual-beam oscilloscope with dual-trace switching can display four signals simultaneously.

The oscilloscope trace seen on the screen of an analog-type oscilloscope is transitory unless it depicts a repetitive steady-state signal. Traces of transient signals disappear with the end of the transient and are often too fast to be seen by the naked eye. There are several means of obtaining permanent pictures of transient signals:

1 Photographs. Cameras that can use Polaroid-type film are available for all common types of analog oscilloscopes.

2 Long persistent screens. Sometimes called *storage oscilloscopes*, these instruments use a type of phosphor that retains the image on the screen for a relatively long time—from 10 s to several minutes.

3 Digital oscilloscopes.

18.4.1 Digital Oscilloscopes

The digital oscilloscope represents the combination of the best of analog oscilloscope technology and the vast technology of modern digital computers. By digital sampling techniques, the oscilloscope trace is digitized and stored in the digital memory included with the digital oscilloscope. This storage is as permanent as the memory. The digitized trace can be reviewed by putting it back on the screen at any time, but it can also be printed on an ink or photographic recorder, processed mathematically (such as for determining RMS, average, or peak values of the signal), and inputted into a computer for further processing or combining with other signals. Digital

FIGURE 18.8
A digital
oscilloscope
(Courtesy of
Hewlett
Packard
Corporation.)

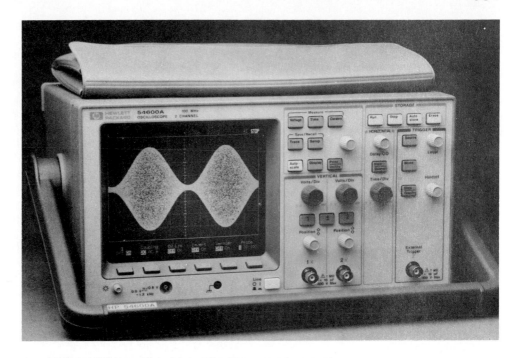

oscilloscopes can operate on repetitive signals just as an analog oscilloscope, or trigger, and record in response to transient signals with pulse widths as small as several nanoseconds. Many digital oscilloscopes include some mathematical processing hardware within the oscilloscope package. The common mathematic functions included are addition, subtraction, multiplication, and averaging of signals.

Figure 18.8 illustrates a typical digital oscilloscope and some of the capabilities of this particular model. For the capture of short transient signals, or "glitches" as they are often called, the *bandwidth* and *rise time* of the digital scope must be chosen carefully according to the guidelines given above and in Eq. (18.10). These two parameters are a function of the sampling rate of the oscilloscope, its ability in processing digital bits per second, and its memory size. Digital oscilloscopes are generally more costly than analog ones, but their capability in the analysis and processing of signals is vastly superior.

PROBLEMS

18.1 A voltmeter has a full scale of 500 V and a probable error of 0.1 percent of full scale. When this voltmeter is used to measure a signal of 75 V, what percent of probable error can exist?

18.2 Compare the percent error in two 10-A ammeters, one digital and one analog, when measuring (*a*) a current of 5 A. The error specifications on the digital meter are 0.07 percent of the reading, 0.05 percent of full scale, 0.005 percent of reading per degree Celsius, and 0.002 percent of full scale per degree Celsius; the analog meter error specifications are 0.5 percent of full scale and 0.001 percent of the reading per degree Celsius. Assume that the temperature at the time of the measurement is 20°C above the ambient. Repeat this comparison when measuring (*b*) a current of 2 A at the same temperature.

18.3 The ohmic (I^2R) loss in a power resistor is to be measured. The current is measured with an analog ammeter having a probable error of ± 0.5 percent of full scale of 100 A. The resistance is measured with an ohmmeter having a probable error of ± 1 percent of a 10-Ω scale. The current is measured to be 63 A, and the resistance is 1.2 Ω. What is the maximum probable error (in watts and in percentage of measured value) in the loss measurement?

18.4 Figure 18.6 shows the two-wattmeter connection for making power measurements on balanced three-phase loads. Derive the expression for the reading of each wattmeter in terms of the line voltage V_1, the line current I_1, and the functions of the load power-factor angle θ. Show that the sum of the two wattmeter readings equals $\sqrt{3}V_1I_1 \cos \theta$. At what power factor does one wattmeter indicate a zero scale reading?

18.5 A standard wattmeter has a current coil rated at 20 A, a voltage coil rated at 200 V, and a full scale of 4000 W. This meter is to be used to measure the no-load losses on an induction motor; the losses are expected to be about 1500 W at a 25 percent power factor. If the measurement is made at 200 V_1, will the current coil be operated over its rating? If the meter's probable error is 0.75 percent of the scale, what is the percentage of probable error in the reading?

18.6 A certain DMM reads true RMS values of current. If the *peak* value of each of the following periodic current waves is 1.0 A, what will the meter reading be for (*a*) a sine wave, (*b*) a square wave, (*c*) a triangular wave, (*d*) a sawtooth wave?

18.7 In the electronic multiplier-type wattmeter, Fig. 18.4, the output of the multiplier is $p = e \cdot i$, instantaneous power. Derive an RC op-amp network for the block labeled "Averaging circuit" in Fig. 18.4 to convert this signal to a signal proportional to average power, as given in Eq. (18.5).

REFERENCES

1 D. R. Hamburg and L. E. Unnewehr, "An Electronic Wattmeter for Nonsinusoidal Low Power Factor Measurements," *IEEE Transactions on Magnetics*, vol. MAG-7, no. 3, September 1971.

2 *Tektronix Reference Manual*, Tektronics, Inc., Beaverton, Ore., 1981.

3 T. G. Beckwith, N. L. Buck, and R. D. Marangoni, *Mechanical Measurements*, 3d ed., Addison-Wesley, Reading, Mass., 1982.

4 A. Desa, *Principles of Electronic Instrumentation*, Halsted Press, New York, 1981.

Digital Logic Circuits

The concepts broadly categorized by the words "digital" and "analog" have been introduced and discussed in several previous chapters. A brief insight into analog computation was given in Chap. 9, and analog and digital instrumentation were discussed and compared in Chap. 18.

Digital circuits of any type (electric, mechanical, hydraulic, acoustical, etc.) can operate properly in only *two* discrete states, or values. These are generally called one (1) and zero (0). A digital circuit either is in one of these states or is malfunctioning. The state of 1, called *logic 1*, means that the system is in the higher of two signal levels. For example, in many electrical logic systems, this state may have a voltage level in the range of 2 to 10 V. It is the "on" state, or high state. Conversely, logic 0 is the "off" state, or low state. Thus, the concept of logic 1 and logic 0 also means on-off. Much of the nomenclature in digital circuits has come from an earlier field of study known as *switching theory* and is related to concepts used to describe the operation of on-off devices such as relays and bang-bang controllers.

19.1 BINARY BITS

The output of a digital circuit is a single binary digit, either logic 1 or logic 0. These two output states are also frequently termed *true* (1) and *false* (0). The abbreviation for *binary digit* is *bit*.

Two advantages to this type of output are that it is not necessary to measure the output precisely—it is either on or off, 1 or 0, true or false—and that electromagnetic interference (emi) is generally less detrimental to digital circuits than to analog circuits. Obviously, the single bit is a good means of answering questions that have a yes/no answer or of operating a control function that is either on or off. However, for more complex questions and more sophisticated control schemes, the single bit is inadequate; thus, almost all practical circuits require a multitude of bits.

653

Because digital circuits are most commonly used in informational systems, such as computers, calculators, and control systems, the question naturally arises as to the relationship between the informational content and the number of bits required. A general theory (see Ref. 2) states that

> The minimum number of bits required to express n different things is k, where k is the smallest positive integer such that

$$2^k \geq n \tag{19.1}$$

The base 2, or binary, system of numbers is thus obviously well suited to the mathematics required for digital systems.

19.1.1 Decimal-to-Binary Conversion

It is assumed that the reader is familiar with both binary and decimal systems of numbers. The following rules and examples explain how to convert from one system to another. The theory behind these conversions, not included here, can be found in Refs. 3 and 4.

Let N refer to a specific whole decimal number:

1 Determine if N is odd or even.
   a If odd, write 1 and subtract 1 from N; go to step 2.
   b If even, write 0; go to step 2.

2 Divide the N obtained from step 1 by 2.
   a If $N > 1$, go back to step 1 and repeat.
   b If $N = 1$, write 1.

The number written by this process is the binary (or base 2) equivalent of the original decimal number. The number written first is the least significant bit; the number written last is the most significant bit.

Example 19.1   Find the binary equivalent of the decimal number 332.

Solution   The decimal number 332 is even, so we write a 0; dividing 332 by 2 gives 166, which is also even; write another 0; dividing 166 by 2 gives 83, which is odd; write a 1 and subtract 1 from 83, giving 82; divide 82 by 2, giving 41, which is odd; write a 1 and subtract 1 from 41, giving 40; divide 40 by 2, giving 20, which is even; write a 0; divide 20 by 2, giving 10, which is even; write a 0; divide 10 by 2, giving 5, which is odd; write a 1 and subtract 1 from 5, giving 4; divide 4 by 2, giving 2, which is even; write a 0; divide 2 by 2, giving 1; write a 1. The final result is 101001100.

A decimal fraction may be converted to binary by converting the numerator and denominator to binary by the method described above. Decimals may not have an exact binary equivalent, and the decimal-to-binary conversion often results in a repetitive sequence of binary bits (Ref. 3). The method for directly converting decimals to binary equivalent is as follows, with the decimal equal to N:

1   Double N.
   a   If this gives a value greater than 1, write a 1 and subtract 1 from the doubled value of N and go back to step 1.
   b   If the doubled value of N is less than 1, write a 0 and go back to step 1.

2   Repeat step 1 until the number of significant figures of the binary fraction is sufficient for the application at hand; note that the first number obtained from step 1 is the first number to the right of the decimal point, i.e., the most significant digit.

Example 19.2

Find the binary equivalent of decimal 0.654.

Solution   Doubling N gives 1.308, which is greater than 1; write a 1 and subtract 1 from 1.308, giving 0.308; double 0.308, giving 0.616, which is less than 1; write a 0 and double 0.616, giving 1.232, which is greater than 1; write a 1 and subtract 1, giving 0.232; double, giving 0.464; write a zero; double, giving 0.928; write a zero; double, giving 1.856; write a 1 and subtract 1, giving 0.856; double, giving 1.712; write a 1 and subtract 1, giving 0.712. As of this point, our binary equivalent is 0.1010011; the process could be continued to give the desired accuracy of the decimal equivalent. There is no assurance that the binary process will end with a firm 1, and therefore the above process might continue indefinitely.

Note that binary equivalents can be checked out in a manner similar to that used in the decimal system from the basic theory of numbers. This checkout process is, of course, applicable to numbers expressed in any base. For example, the binary number derived in Example 19.1 can be checked out as follows:

$$0 \times 2^0 = \quad 0$$
$$0 \times 2^1 = \quad 0$$
$$1 \times 2^2 = \quad 4$$
$$1 \times 2^3 = \quad 8$$
$$0 \times 2^4 = \quad 0$$
$$0 \times 2^5 = \quad 0$$
$$1 \times 2^6 = \quad 64$$
$$0 \times 2^7 = \quad 0$$
$$1 \times 2^8 = 256$$
$$\overline{332}$$

Likewise, checking out the decimal conversion of Example 19.2 gives

$$1 \times 2^{-1} = 0.5$$
$$0 \times 2^{-2} = 0.0$$
$$1 \times 2^{-3} = 0.125$$
$$0 \times 2^{-4} = 0.0$$
$$0 \times 2^{-5} = 0.0$$
$$1 \times 2^{-6} = 0.015625$$
$$1 \times 2^{-7} = \overline{0.0078125}$$
$$0.6484375$$

19.2 INTRODUCTION TO BOOLEAN ALGEBRA

Boolean algebra is a useful tool in the analysis and design of digital circuits, and we now introduce a few of this methodology's basic concepts. Boolean algebra is the mathematical expression of the logic and decision-making aspects of digital circuits.

Boolean algebra is a system of mathematics which is abstract and independent of its applications. However, it has been largely associated with the mathematics of logic circuits, and our brief presentation is aimed at this application. In boolean algebra, variables have only two possible values, 1 and 0. This must be remembered in studying the summary of basic boolean relationships listed below, for some of these relationships are in marked contrast to the common algebra associated with real and complex continuous variables.

19.2.1 Basic Laws of Boolean Algebra

Let x, y, and z be boolean variables in the set T: $\{0, 1\}$.

1 The symmetric law: If $x = y$, then $y = x$.

2 The transitive law: If $x = y$ and $y = z$, then $x = z$; and x may be substituted for y or z in any formula involving y or z.

3 Joining, the OR function: The boolean sum expression $x + y$ has the meaning of "joining x and y," the "union of x and y," or, more specifically in logic circuits, x *or* y.

4 Multiplication, the AND function: The boolean expression xy has the meaning x *and* y.

5 Commutative laws:

$$x + y = y + x \tag{19.2a}$$

$$xy = yx \tag{19.2b}$$

6 Distributive laws:

$$x(y + z) = xy + xz \tag{19.3a}$$

$$x + yz = (x + y)(x + z) \tag{19.3b}$$

Note that the second of the distributive laws has no counterpart in conventional algebra.

7 Associative laws:

$$(x + y) + z = x + (y + z) \tag{19.4a}$$

$$(xy)z = x(yz) \tag{19.4b}$$

8 Idempotent laws:

$$x + x = x \tag{19.5a}$$

$$xx = x \tag{19.5b}$$

There is no equivalent to these relationships in conventional algebra.

9 Multiplication and addition by 1 or 0:

$$0 + x = x \tag{19.6a}$$

$$1 + x = 1 \tag{19.6b}$$

$$0 \cdot x = 0 \tag{19.6c}$$

$$1 \cdot x = x \tag{19.6d}$$

10 The NOT function: This is often called the *complementary function*: it refers to the complement (or opposite) of the listed function. The complement of 1 is, of course, 0, and vice versa. This is a most important concept in the development of electronic logic circuits, the physical embodiment of NOT being the logic inverter. NOT is further described by the laws of complementarity:

$$x\bar{x} = 0 \tag{19.7a}$$

$$x + \bar{x} = 1 \tag{19.7b}$$

where $\bar{x}$ is the complement of x. The logic inverter may be represented in logic diagrams as a separate entity, as in the third and fourth diagram of Fig. 19.1, or by means of a circle on a logic symbol, often called a *bubble*.

11 De Morgan's theorem: Two important relationships between functions and their inverses, or complements, are useful in reducing logic expressions and simplifying logic hardware:

$$\overline{(x + y)} = \bar{x}\bar{y} \tag{19.8a}$$

$$\overline{(xy)} = \bar{x} + \bar{y} \tag{19.8b}$$

FIGURE 19.1
Examples of
Boolean equations
and their logic
circuit translations.

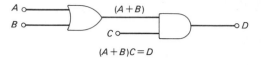

$(A+B)C = D$

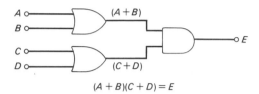

$(A + B)(C + D) = E$

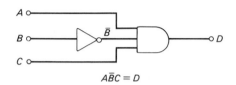

$A\bar{B}C = D$

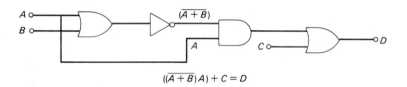

$((\overline{A+B})A) + C = D$

Note that other symbols for the OR and AND functions are frequently found in the literature: v and $+$ to represent OR and $\cdot$ to represent AND. There are several other mathematical functions not included in the above boolean formulas that are of importance in electronic logic circuits. These are:

12   Exclusive OR: This has the meaning "x or y, but not both" and is designated by the symbol $x \oplus y$.

13   NAND: This has the meaning "NOT (x and y)" and is designated by the formula $\overline{x \cdot y} = z$

19.2.2   Truth Tables

A simple graphical means of describing the logic associated with the boolean expressions given above is the *truth table*. The adjective "truth" arises from the frequently used true/false notation for the logic variables 1/0. For example, the OR function is defined by the truth table of Table 19.1. The exclusive OR (XOR) function is likewise defined by the truth table of Table 19.2. The product, or AND, function is defined by the truth table of Table 19.3. Truth tables are also sometimes called *connectives*. Note in Tables 19.1 to 19.3 that each truth table has four rows.

TABLE 19.1 Truth Table for OR Logic Function

x	y	$x + y$
0	0	0
0	1	1
1	0	1
1	1	1

TABLE 19.2 Truth Table for Exclusive OR Function

x	y	$x \oplus y$
0	0	0
0	1	1
1	0	1
1	1	0

TABLE 19.3 Truth Table for AND Function

x	y	xy
0	0	0
0	1	0
1	0	0
1	1	1

The four rows are combinations of the two states of the boolean variable, 1 and 0. From Eq. (19.1) there should be $2^4 = 16$ possible ways to fill out a column of truth tables (consult Ref. 4), and this is illustrated in Table 19.4.

In this table, the columns of the previous three tables are represented by rows. For example, the truth table of connective 6 in Table 19.4 is the exclusive OR of the right column of Table 19.2; the truth table for x (connective 3) is seen to be the left column of Tables 19.1, 19.2, and 19.3; and so forth.

TABLE 19.4

Connective	Truth Table	Meaning
0	0000	Totally false
1	0001	xy
2	0010	$x\bar{y}$
3	0011	x
4	0100	$\bar{x}y$
5	0101	y
6	0110	$(x\bar{y} + \bar{x}y) = x \oplus y$
7	0111	$x + y$
8	1000	$(\bar{x} \cdot \bar{y})$
9	1001	$(xy + \overline{xy})$
10	1010	$\bar{y}$
11	1011	$(\bar{y} + x)$
12	1100	$\bar{x}$
13	1101	$(\bar{x} + \bar{y})$
14	1110	$(\bar{x} + \bar{y}) = \overline{xy}$
15	1111	Totally true

Example 19.3

Prepare a truth table for the boolean expression $A = x + \bar{y}z$.

Solution This equation is read, "A equals x OR (complement of y AND z)." The truth table is as follows:

x	y	z	$\bar{y}$	$\bar{y}z$	A
0	0	0	1	0	0
0	0	1	1	1	1
0	1	0	0	0	0
0	1	1	0	0	0
1	0	0	1	0	1
1	0	1	1	1	1
1	1	0	0	0	1
1	1	1	0	0	1

This truth table has been prepared in more detail than is usually necessary to illustrate the individual steps in solving a boolean equation.

Example 19.4

Determine $\bar{A}$ of A in Example 19.3 in equation form from De Morgan's theorems and from the truth table of Example 19.3.

Solution In words, De Morgan's theorems [Eqs. (19.8a) and (19.8b)] state that the complement of a function A can be found by replacing each variable by its complement and by interchanging all AND and OR signs. Therefore,

$$\bar{A} = \overline{x + \bar{y}z} = \bar{x}(y + \bar{z})$$

The truth table for $\bar{A}$ can be found by replacing the last column in the table of Example 19.3 by the complement of the individual element. Shown as rows rather than columns (to conserve space), this is

A	0	1	0	0	1	1	1	1
$\bar{A}$	1	0	1	1	0	0	0	0

Example 19.5

Simplify the expression

$$B = (x + \bar{y})(\overline{y + z\alpha}) + (\beta y + \alpha\bar{x})(\overline{y + z})$$

Note that the complement sign over an entire parenthetical expression refers to the complement of the entire expression, not to the complements of the individual elements.

Solution In the analysis of this problem, we show at the right (in parentheses) the numbers of the boolean laws from Sec. 19.2.1:

$$\left.\begin{array}{l} (\overline{y + z\alpha}) = \bar{y}(\bar{z} + \bar{\alpha}) \\[2em] (\overline{y + z}) = \bar{y}\bar{z} \end{array}\right\} \quad \text{De Morgan's theorem (11)}$$

Substituting,

$$
\begin{aligned}
B &= (x + y)y(z + \alpha) + (\beta y + \alpha\bar{x})\bar{y}\bar{z} \\
&= (x\bar{y} + \bar{y})(\bar{z} + \bar{\alpha}) + (\beta y + \alpha\bar{x})\bar{y}\bar{z} && \text{(6 and 8)} \\
&= (x\bar{y} + \bar{y})(\bar{z} + \bar{\alpha}) + \alpha\bar{x}\bar{y}\bar{z} && \text{(10)} \\
&= \bar{y}(\bar{z} + \bar{\alpha}) + \alpha\bar{x}\bar{y}\bar{z} && \text{(6 and 9)} \\
&= \bar{y}\bar{z} + \bar{y}\bar{\alpha}(1 + \bar{x}\bar{z}) && \text{(6)} \\
&= \bar{y}\bar{z} + \bar{y}\bar{\alpha} && \text{(9)}
\end{aligned}
$$

19.3 IMPLEMENTATION OF BOOLEAN LOGIC

We have been discussing mainly theoretical concepts of logical decision making so far. The arithmetic of these processes is based on the base 2, or binary, number system. However, the physical realization of this system of mathematics is one of the most exciting developments in modern culture, influencing almost every aspect of industrialized society.

The three elementary logic elements, often called *gates*, are items 1, 7, and 14 in Table 19.4—the AND, OR, and NAND gates, respectively. The NAND function is read as "not $(x$ and $y)$"; it is the inverted, or complementary, AND function and is represented by the AND symbol with a circle, or bubble, on its output. Likewise, the NOR function is the inverted OR function, defined by the OR symbol with a bubble on its output.

Various symbols for the hardware implementation of logic gates and logic functions have been used through the years, just as there has been some variation in the mathematical symbols used to express boolean functions. Recently, the hardware symbols have been standardized by the American National Standards Institute (ANSI), the Institute of Electrical and Electronics Engineers (IEEE), and other standards agencies. Table 19.5 (in Sec. 19.4) lists the ANSI standard symbols for a number of logic functions, along with the meaning of these functions (Ref. 5). The functions designated FF in Table 19.5 are flip-flop functions—switching functions, each associated with various types of logic—which are discussed in Sec. 19.4. Figure 19.1 illustrates the use of these hardware symbols in developing logic circuits; it shows models of several simple boolean equations. The triangle in Fig. 19.1 represents the electronic inverter—the complementary, or NOT, function described in Sec. 19.2.1.

19.3.1 Standard Forms of Boolean Equations

Figure 19.1 introduces the concept of boolean equations or boolean functions which are useful in the analysis, design, and simplification of digital electronic circuits. A general form for a boolean equation can be given in terms of the fourth function shown in Fig. 19.1:

$$f(A, B, C) = D = (\overline{A + B})A + C \tag{19.9}$$

In Eq. (19.9), A, B, and C are variables. Symbols representing variables (A, B, $\overline{A}$, etc.) are called *literals*, a term often used in boolean algebra. We have been using x, y, and z as variables in most of the previous expressions, as in conventional algebra, but almost all letters of the Latin and Greek alphabets seem to be acceptable for boolean expressions.

There are two forms of expression which occur frequently in electronic logic circuits: sum of products (SOP) and products of sums (POS). A good example of the latter form is given in the second circuit in Fig. 19.1; an example of the former (sum of products) results by multiplying out the output of the first circuit of Fig 19.1, giving $D = AC + BC$. Product terms and sum terms are in *standard form* if they contain one literal from every variable in the domain. For example, the POS given in the second circuit of Fig. 19.1 indicates that output E is a function of A, B, C, and D. These variables define the *domain* of E, and this is indicated by writing E as $E(A, B, C, D)$. Obviously, the second equation in Fig. 19.1 is *not* in standard form by the above definition, since neither sum term contains all four literals. Likewise, the output of the first circuit of Fig. 19.1 would be written as $D(A, B, C) = AC + BC$, and this is *not* in standard SOP form since neither sum term contains A, B, and C. For many mathematical manipulations, including the important task of minimizing the number of logic functions required, it is important to describe logic outputs in standard forms. This can usually be done by means of some of the boolean identities listed in the previous section without altering the value of the output. We illustrate this procedure by several examples.

Example 19.6

Convert the form of the output of the first circuit in Fig. 19.1 to standard SOP.

Solution This output is $D(A, B, C) = AC + BC$: we see that B is missing from the first product and A from the second. The first term can be multiplied by $B + \bar{B}$, which equals 1 [from Eq. (19.7b)], giving $ABC + A\bar{B}C$. The second term is altered by multiplying it by $A + \bar{A}$, giving $ABC + \bar{A}BC$. The standard form of this equation is therefore

$$D(A, B, C) = ABC + A\bar{B}C + \bar{A}BC$$

Note that although the term ABC appears twice from this expansion, it is written only once in the standard form. This expression is standard SOP since each variable appears in each product term.

Example 19.7

Convert the form of the output of the second circuit in Fig. 19.1 to standard POS.

Solution The output of this circuit is given as $E(A, B, C, D) = (A + B)(C + D)$. Here C and D are missing from the first product term. It can be shown that the missing terms can be introduced by the following product terms:

$$(A + B + C + D)(A + B + C + \bar{D})(A + B + \bar{C} + \bar{D})(A + B + \bar{C} + D)$$

The reader should ascertain that this substitution is indeed an identity by multiplying out the terms and using the equations in Sec. 19.2.1. The second term in the original POS is missing the literals A and B. This can be introduced by means of a substitution similar to that used for the first term. The result in standard form (with all literals in each product term) is

$$E(A, B, C, D) = (A + B + C + D)(A + B + C + \bar{D})(A + B + \bar{C} + D)(A + B + \bar{C} + D)$$
$$(A + \bar{B} + C + D)(\bar{A} + \bar{B} + C + D)(\bar{A} + B + C + D)$$

Again note that the term $A + B + C + D$ appears twice in the expansion but is written only once.

Example 19.8

Convert the expression $f(a, b, c, d) = a\bar{b} + ac\bar{d}$ to standard SOP form.

Solution The literals c and d are missing from the first term; they can be introduced by multiplying first by $c + \bar{c}$ and then by $d + \bar{d}$, as explained in Example 19.6. The second term is missing the literal b and can be put in standard form by multiplying this term by $b + \bar{b}$. Performing these expansion operations gives

$$f(a, b, c, d) = a\bar{b}\bar{c}\bar{d} + a\bar{b}\bar{c}d + a\bar{b}c\bar{d} + a\bar{b}cd + abc\bar{d}$$

The term $a\bar{b}c\bar{d}$ appeared twice as a result of the above expansion operations but needs to be written only once in the standard form.

19.3.2 Numerical Representation of Standard Forms

In many operations with boolean algebra, it is useful to represent the standard forms by the sum ($\sum$) of a group of numbers, as will be illustrated in the minimization process of Sec. 19.3.4. Without getting into the theory behind this process, we now illustrate the technique for converting the standard forms to numerical representations.

Standard SOP In converting this form to numerical representation, literals are given the value of 1 and complements of literals are represented by 0. The resultant sums represent a series of binary numbers which are then converted to decimal numbers and expressed as a summation. For example, the standard SOP derived in Example 19.6 is converted to numerical form as follows:

$$
\begin{aligned}
D(A, B, C) &= ABC + A\bar{B}C + \bar{A}BC \\
&= 111 + 101 + 011 \\
&= \sum(7, 5, 3)
\end{aligned}
$$

The standard SOP of Example 19.8 is converted as

$$
\begin{aligned}
f(a, b, c, d) &= a\bar{b}\bar{c}\bar{d} + a\bar{b}\bar{c}d + a\bar{b}c\bar{d} + a\bar{b}cd + abc\bar{d} \\
&= 1000 + 1001 + 1010 + 1011 + 1110 \\
&= \sum(8, 9, 10, 11, 14)
\end{aligned}
$$

Standard POS In this conversion, the numeric value of 0 is assigned to literals and the value 1 to complements of literals—just the opposite of the conversion used for standard SOPs. The resulting product of binary numbers is expressed as a product of decimal numbers. For the standard POS of Example 19.7,

$$
\begin{aligned}
E(A, B, C, D) &= (0000)(0001)(0011)(0010)(0100)(1100)(1000) \\
&= \prod(0, 1, 3, 2, 4, 12, 8)
\end{aligned}
$$

It is possible to go to the numeric form directly from the original form of a logic equation by inserting a symbol, say X, for the missing literals and letting X take on the values of 1 and 0. For example, the original form of the POS equation of Example 19.7 is $(A + B)(C + D)$. The first term can be written as $A + B + X + X$; letting X take on both a 0 and 1 gives 0000, 0001, 0010, 0011. The second term can be written as $X + X + C + D$; letting X take on 0 and 1 gives 0000, 1000, 1100, 0100. Using only one of the 0000 terms, this gives the same set of seven terms in the product as was obtained from the standard POS form used above. The same can be done in converting standard SOP terms to numeric value. For example, the original equation in Example 19.8 is $a\bar{b} + ac\bar{d}$. Expressing the first term as $a\bar{b}XX$ gives 1000, 1001, 1010, 1011; the second term can be written as $aXc\bar{d}$, which results in 1010 and 1110. Using the term 1010 only once, the same five terms are present as were in the conversion using standard SOP above.

The numerical representation is a good shorthand method of writing a function and can easily be converted to the original equation (in standard form) by the reverse of the process described above.

Example 19.9 Write the equation in standard POS for the numerical representation $f(x, y, z) = \prod (0, 2, 4, 5, 7)$.

Solution First express the numbers in binary, giving 000, 010, 100, 101, 111. Then use the reverse of the rule for converting literals to numerals. For POS, a 0 becomes a literal and a 1 becomes a complement. Therefore, 000 becomes $x + y + z$, 010 becomes $x + \bar{y} + z$, and so on, giving

$$f(x, y, z) = (x + y + z)(x + \bar{y} + z)(\bar{x} + y + z)(\bar{x} + y + \bar{z})(\bar{x} + \bar{y} + \bar{z})$$

Example 19.10 Write the equation in standard form for $f(w, x, y, z) = \sum (1, 5, 13, 15)$.

Solution The binary forms of the numerics are 0001, 0101, 1101, and 1111. For SOP, a 1 becomes a variable and a 0 becomes a complement. Therefore,

$$f(w, x, y, z) = \bar{w}\bar{x}\bar{y}z + \bar{w}x\bar{y}z + wx\bar{y}z + wxyz$$

19.3.3 Karnaugh Maps

Logic functions which are expressed in standard form can be displayed graphically by a technique known as *Karnaugh mapping* (see Refs. 2, 4, and 8).

Logic circuits are initially developed from a verbal description of the functions or processes to be performed; from this description, logic diagrams and their associated equations are developed, simple examples of which are shown in Fig. 19.1. But there is generally no way of ascertaining during this early development stage that the resulting logic circuit is the best possible circuit for performing the desired functions. It is desirable from both cost and reliability standpoints to use the minimum number of logic functions to perform the required functions—and circuits can indeed be simplified on a somewhat hit-or-miss basis by use of some of the equations listed in Sec. 19.2 or from past experience. The Karnaugh mapping technique, on the other hand, is a systematic circuit-reduction technique that generally gives the minimum-component system. It is applicable to domains with 3, 4, and 5 variables. Other techniques are available for larger domains, but these are more cumbersome and time-consuming.

For a three-variable domain, the Karnaugh map is drawn in three stages, as shown in Fig. 19.2. The map layout is illustrated in Fig. 19.2a; it is a 2×4 matrix. The three variables are separated into two groups—the most significant variable (usually the first listed: x in this case) and the remaining two variables. Along the 4-square side are listed the four possible values of the two remaining values; on the 2-square side are listed the two possible values of the significant variable. The three binary numbers represented at each square (in the sequence of the most

FIGURE 19.2
Karnaugh
mapping: (*a*)
matrix for
3-variable domain;
(*b*) matrix with
decimal
equivalent; (*c*)
subcubes for SOP
example discussed
in Sec. 19.3.3.

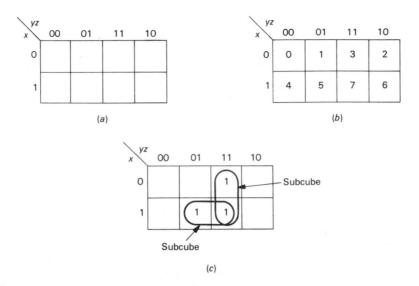

significant variable's number first and then the other two) represent an equivalent decimal number, which is then listed in each square, as shown in Fig. 19.2*b*. Note that the order of the two-binary-number sequence for *y* and *z* has been altered, with 11 coming before 10; this is done so that no coordinate differs from its adjacent coordinate by more than one bit position. The next step is to convert the logic expression to standard form and then to obtain the numerical representation of this standard form; taking the SOP used in Example 19.6, for instance, the numerical representation of this expression was found in Sec. 19.3.2 to be $\sum (7, 5, 3)$. The final step in preparing the Karnaugh map is to place a 1 in the squares represented by the decimal numbers of the numerical representation, in this case 7, 5, and 3. This is shown in Fig. 19.2*c* and represents the Karnaugh map for the domain (with *x, y, z* replacing *A, B, C*)

$$D(x, y, z) = xyz + x\bar{y}z + \bar{x}yz$$
$$= \sum (7, 5, 3)$$

Using the Karnaugh map, we reduce and simplify a logic equation by the following three steps:

1 Form subcubes that cover all 1s on the Karnaugh map. Subcubes are groups of 1s that are adjacent to each other; squares located along a diagonal are not considered to be adjacent. In Fig. 19.2*c* there are two subcubes, which are indicated by the rounded rectangles.

2 For 2-square subcubes, it can be shown that one variable changes value and that the other two have a common value: From our example in Fig. 19.2*c* the binary values of the two adjacent squares in the horizontal subcube are 101 and 111; it is the middle variable, *y*, that changes. For the vertical subcube, the two adjacent binary values are 011 and 111; it is the first variable, *x*, that changes.

3 Write the resulting logic expression. The SOP expression for a subcube consists of the variables that do not change. In our example, the two resulting variables that do not change in the horizontal subcube are x and z, and both have values of 1; therefore, the literals appear in the equation. For the vertical subcube, y and z do not change and both have values of 1. Therefore, the resulting equation is $D(x, y, z) = xz + yz$, which is identical to the equation written in the nomenclature of Example 19.6 as $D(A, B, C) = AC + BC$. This represents the minimum-component system, which we might have expected due to the simplicity of the first circuit in Fig. 19.1. When a square containing a 1 is all alone (i.e., not adjacent to any other square containing a 1), the literal for each variable appears in the resulting equation. For example, a 1 in the upper leftmost square of Fig. 19.2c would add the term $\bar{x}\bar{y}\bar{z}$ to the equation since all three variables have the value of 0 in this square.

We have gone through the above example and accompanying explanation to give the general method of preparing a Karnaugh map for logic equations. This method is applicable to most systems with a domain size of 3, 4, and 5. However, the end result was not very meaningful in this case since the system was already a minimum-component system and there was only one possible result. In the general case, there will be a choice of subcubes in step 1 above. We also still have to discuss the handling of POS forms. Therefore, let us return to step 1 above and try to gain some insight into choosing the best subcubes that result in the most economical equation— and hence in the most economical logic system.

19.3.4 Minimization of Logic Functions

It has been stated that the main purpose of Karnaugh mapping is to derive the "best" or "most economical" or "minimum" logic system to accomplish the given function. All the above terms in quotation marks are used to describe the goal of Karnaugh mapping, and the first two terms are introduced here to remind us that this is not a minimization process in the rigorous mathematical sense but one that requires some judgment and selection. Proceeding on this basis, we introduce some more interesting Karnaugh maps.

Figure 19.3 illustrates another 3-domain system with a number of possible subcubes. *Subcubes must be a power of 2, that is, 2, 4, 8, 16, etc.* Subcubes with three adjacent squares are not allowed and must be broken up into two 2-square subcubes. In Fig. 19.3 there are five possible 2-square subcubes and one 4-square subcube that

FIGURE 19.3
Karnaugh map
with many
subcubes (as
discussed in Sec.
19.3.4).

x \ yz	00	01	11	10
0	1	1	1	1
1	1			1

are obvious. There are also more subcubes that are not obvious until we point out that a Karnaugh map is a continuous surface—i.e., the 3-domain map "wraps around" on itself, the right edge of the column headed by $yz = 10$ also being the left edge of the column headed by $yz = 00$. Therefore, there is also a 4-square subcube composed of the squares in the rightmost and leftmost columns, and two more 2-square subcubes making up the big square subcube. We have listed nine subcubes in this 3-domain map. In higher-order domains, the situation is generally even more complex. What is the best choice of subcubes, and is there a best choice? The answer to the latter part of this question is, "Sometimes yes." To the first part of the question, the following steps are added to step 2 in the preceding section for determining the best SOP equation:

2a   All 1s must be covered by subcubes, including the "longer" 1s which will introduce the product of all variables into the equation.

2b   Start the process of covering 1s with subcubes by finding the *essential subcube*—the subcube that covers a certain 1 that no other subcube can cover.

2c   All subcubes should be as large as possible.

2d   There should be as few subcubes as possible.

2e   Sometimes additional factoring, using the equation and techniques of Sec. 19.2, will further reduce the system after the system equations have been written (step 3).

In Fig. 19.3 there are two 4-square subcubes that cover all the 1s; the choice of these subcubes satisfies steps 2c and 2d above, and it is these two steps that are the greatest aid in finding the system with the *minimum* number of components. The first 4-square subcube is the obvious one and contains the entire row $x = 0$. In this subcube, y and z vary and $x = 0$ throughout; therefore, its equation is $\bar{x}$. The second 4-square subcube is the square composed of the two squares at each end of the matrix of Fig. 19.3; it results when the two ends are wrapped around together. The binary values in the four subsquares of this subcube are 000, 010, 100, and 110. Note that the third digit, representing z, is 0 in all four squares and that x and y vary. Therefore, the equation from this subcube is $\bar{z}$. The resulting logic equation is

$$f(x, y, z) = \bar{x} + \bar{z} \tag{19.10}$$

The 4-variable domain is handled in the same manner as the 3-variable one, and the same rules generally apply, although there are usually many more choices of subcubes. The decimal representation of squares for a 4-variable matrix is shown in Fig. 19.4a: this has been obtained by the same means as Fig. 19.2b for the 3-variable matrix. The 1s from Example 19.8 are plotted in Fig. 19.4b with the general variables w, x, y, z replacing the variables used in the example; the numerical representation for this example was derived in the second paragraph (standard SOP) of Sec. 19.3.2. Here there are only two possible subcubes, and these are encircled in Fig. 19.4b. For the 4-square subcube, both w and x are constant, giving the product $w\bar{x}$; for the

FIGURE 19.4
Karnaugh map for
4-variable domain:
(*a*) decimal
representation; (*b*)
Karnaugh, map for
example discussed
in Sec. 19.3.4.

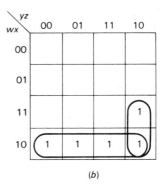

(*a*) (*b*)

2-square subcube, w, y and z are constant with the values 1, 1, and 0, respectively. Therefore, the resulting equation is composed of the SOP

$$f(w, x, y, z) = w\bar{x} + wy\bar{z}$$

which agrees with the original equation of Example 19.8.

Example 19.11   Determine the minimum equation for

$$f(A, B, C, D) = \sum (1, 4, 5, 6, 8, 12, 13, 15)$$

This is somewhat of a contrived example to illustrate steps 2*a* to 2*e* above.

Solution   The Karnaugh map is obtained by plotting 1s in the cubes of Fig. 19.4*a*, represented by the numbers in the summation above. This is shown in Fig. 19.5*a*. By step 2*c*, one might start with the 4-square subcube at the left center of Fig 19.5*a*. This gives the equation $B\bar{C}$, since $B = 1$ and $C = 0$. However, this leaves a lot of 1s dangling around the 4-square subcube, and these also must be covered by 2-square subcubes. Note that the 4-variable map wraps around on its edges to give a continuous surface (both vertical and horizontal edges), just as does the 3-variable map. Thus, the 1 in square 0110 is joined with that in square 0100

FIGURE 19.5
(*a*) Karnaugh map
for Example 19.11;
(*b*) Karnaugh map
for Example
19.12.

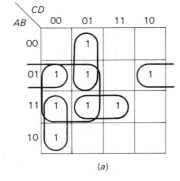   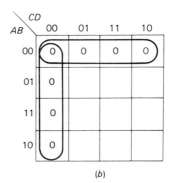

(*a*) (*b*)

to give a 2-square subcube, as shown. Adding the equations of the four 2-square subcubes gives a final equation

$$f(A, B, C, D) = B\bar{C} + \bar{A}B\bar{D} + \bar{A}\bar{C}D + A\bar{C}\bar{D} + ABD$$

which has five terms and requires five functions to implement. However, note that all four 2-square subcubes are essential—i.e., they cover a 1 that no other subcube can cover. Also the 4-square subcube is *not* essential. Therefore, starting with the essential subcubes (step 2*b*), we find that one term can be eliminated (since the 4-square subcube will not be needed), and the minimum equation is

$$f(A, B, C, D) = \bar{A}B\bar{D} + \bar{A}\bar{C}D + A\bar{C}\bar{D} + ABD$$

19.3.5   Karnaugh Map for POS Expressions

For plotting POS functions, a 0 is placed in the squares for which the function has a numerical value. Keep in mind that POS functions are represented in an inverse manner to SOP functions, i.e., a variable is represented by a 0, and a complement of a variable is represented by a 1. We illustrate this process with an example.

Example 19.12

Verify the equation for the logic circuit of Fig. 19.1, second figure; the standard form and numerical representation of this figure have been evaluated in Example 19.7 and in the third paragraph (standard POS) of Sec. 19.3.2, respectively.

Solution   The numerical representation of the second circuit of Fig. 19.1 is

$$f(A, B, C, D) = \Pi(0, 1, 2, 3, 4, 8, 12)$$

To obtain the Karnaugh map, we place a 0 on the squares of Fig. 19.4*a* that are designated in the product above. This is shown in Fig. 19.5*b*. There are two 4-square subcubes. For the horizontal subcube, *A* and *B* are constant and equal to zero. This gives the equation $A + B$. The vertical subcube supplies the equation $C + D$. The resulting equation is

$$f(A, B, C, D) = (A + B)(C + D)$$

which agrees with that of the second diagram of Fig. 19.1.

19.3.6   Don't Care Logic State

To complete our discussion of electronic logic and boolean algebra, we note a third logic state called, very simply, *don't care*. This will aid the reader in evaluating the logic circuit specifications found in most manufacturers' catalogs.

Many types of logic devices have certain combinations of input variables which give outputs that are not important or meaningful. This may be due to a variety of

FIGURE 19.6
SOP Karnaugh
map with don't
care logic states
designated by the
symbol *d*.

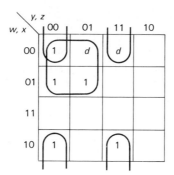

reasons: e.g., these combinations never occur in the proposed application, the outputs have no effect on the output circuitry, or the outputs "should not exist." An example of the latter is a binary logic circuit designed to identify decimal numbers 0 through 9; since 4 bits are needed for this and a 4-bit system can identify 0 through 15, the outputs associated with decimal numbers 10 through 15 are actually not wanted. Unwanted outputs are designated by the symbol *d* and called *don't care logic states*. The same term is also often used to designate input signals—i.e., input signals whose state (0 or 1) is not important for a specific output condition—and is usually designated by the symbol ×. The logic circuit truth table shown in Table 19.6 illustrates the use of don't care input states ×.

Don't care outputs are plotted on the Karnaugh map along with the 1s for SOP or 0s for POS. They can be used to form subcubes *if their use simplifies the resulting equation:* otherwise, they may be ignored. Usually, don't care outputs help in the simplifying process by creating *larger* subcubes (step 2c above). An output function containing don't cares can be expressed in numeric form as

$$F(w, x, y, z) = \sum (0, 4, 5, 8, 11) + d(1, 3)$$

The associated Karnaugh map and the three resulting subcubes are shown as Fig. 19.6. The resulting logic equation is

$$F(w, x, y, z) = \bar{w}\bar{y} + \bar{x}yz + \bar{x}\bar{y}\bar{z}$$

19.4   SEQUENTIAL LOGIC

Thus far we have been concerned with a class of logic circuits often called *combinational logic*. In this class of logic, the device output is a function of the device input only. A second broad category of devices is known as *sequential logic*. In this class of devices, device output may be a function of past history, or of the sequence in which certain signals change logic state, or, in the broadest sense, of time itself.

There are many examples of sequential logic in many devices and activities associated with our daily lives. Combination locks are a simple example in which the *sequence* of setting the numbers is as important as the value of the numbers. Most card games, chess, checkers, and many other games depend on the sequences of plays.

A few years ago, certain automobiles could not be started—at least without the sound of a bothersome buzzer—unless a certain sequence of activities by the operator occurred: sitting in or depressing the driver's seat, closing the left front door, fastening the seat belt, putting the transmission in neutral or park, and finally turning the ignition key. And the automotive assembly line is a sequential system on a grandiose scale.

In most computer and control applications, sequential functions are required along with the gating functions described in the previous section of this chapter. We also include in the sequential category many timing or clock functions and memory functions.

19.4.1 Flip-Flops

Flip-flops are basic sequential logic devices and are shown in Table 19.5. A flip-flop is a *bistable* device; i.e., it has two stable modes of operation, which again correspond to the two binary values 0 and 1, on and off, etc.

The *RS* flip-flop is perhaps the basic or elementary flip-flop. The symbols R and S signify "reset" and "set," respectively, and these two words actually describe the basic function of a flip-flop. In the truth table given in Table 19.5 for the *RS* flip-flop, the symbol h generally stands for the symbol 0 as used elsewhere in this text, i.e., the low state or low output; the symbol "n.c." stands for "no change from the previous condition of inputs." An active (or 1) signal on the input of the *RS* flip-flop ($S = 1$) causes the output to SET, or causes Q to be equal to 1. Once SET, the flip-flop remains in SET even after the signal on S is removed. It remains in SET ($Q = 1$) until an active signal of $R = 1$ is applied. Now $\bar{Q}$ is always the inverse or complement of Q. Likewise, the flip-flop remains in the reset condition ($Q = 0$ and $\bar{Q} = 1$) until the next SET signal is applied to the S terminal. The states called *indeterminate* in Table 19.5 would be included in the don't care states discussed above. The logic shown for this flip-flop is realized by two cross-coupled NAND gates, some embodiments of which are discussed in later sections. When constructed of NOR gates, the logic for the $R = S = 0$ and $R = S = 1$ states will be the inverse of that shown in Table 19.5.

The *JK* flip-flop can be constructed from two NAND devices; this eliminates some of the undefined states of the *RS* flip-flop. The basic characteristics are shown in the truth table of Table 19.5. The logic of the upper *JK* diagram in Table 19.5 can be described as follows:

1 When $J = K = 1$, the flip-flop will not change state; i.e., logic state Q_{n+1} = logic state Q_n.

2 If $J = 1$ and $K = 0$, the flip-flop will set on the next $(n + 1)$ S pulse; $Q_{n+1} = 1$.

3 If $J = 0$ and $K = 1$, the flip-flop will reset (or clear) the R pulse; $Q_{n+1} = 0$.

4 If $J = K = 0$, the flip-flop "toggles," i.e., changes state in response to each successive R or S pulse.

TABLE 19.5 Logic Symbol Formats (ANSI)*

Symbol	Function	Description
	Amplifier	Output active only when the input is active (can be used with polarity or logic indicator at input or output to signify inversion.)
or	AND	Output assumes indicated active state only when all its inputs assume their indicated active levels.
or	OR	Output assumes its indicated active state only when any of its inputs assume their indicated active levels.
or	Exclusive OR	Output assumes its indicated active level if, and only if, only one of the inputs assumes its indicated active level.
	NAND	Inverted AND
	NOR	Inverted OR
	Bilateral switch	A binary-controlled circuit that acts as an on-off switch to analog or binary signals flowing in both directions.
≥m	Logic threshold	Output will assume its active state if m or more inputs are active.
= m	M and only M	Output will be active when m and only m inputs are active (for example, exclusive OR).
>n/2	Majority function	Output will be active only if more than half the inputs are active.
mod 2	Odd function	Output is active only if an odd number of inputs are active.
x/y	Even function	Output is active only if an even number of inputs are active.
00	Signal-level converter	Input levels are different from output levels.

Symbol	Function	Description			
R FF S	RS	R	S	Q	$\overline{Q}$
		1	1	n.c.	n.c.
		1	h	h	1
		h	1	1	h
		h	h	undetermined	
T FF	T	Toggling occurs with every clock pulse			
C FF D R S	D	Data output follows data input; input is gated by C			
J FF K R S	JK	J	K	Q	$\overline{Q}$
		1	1	n.c.	n.c.
		1	h	1	h
		h	1	h	1
		h	h	toggles	
J_G FF G K_G R S	JK (gated)	J and K inputs are gated by C			
J_G FF G K_G R S	JK (master-slave)	Outputs are dependent on the negative-going edge of the clock			

ANSI polarity convention

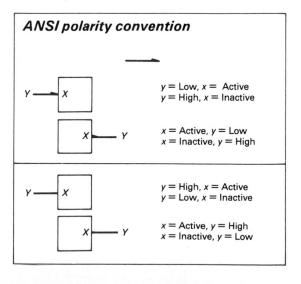

$y =$ Low, $x =$ Active
$y =$ High, $x =$ Inactive

$x =$ Active, $y =$ Low
$x =$ Inactive, $y =$ High

$y =$ High, $x =$ Active
$y =$ Low, $x =$ Inactive

$x =$ Active, $y =$ High
$x =$ Inactive, $y =$ Low

* Based on *Machine Design* [Ref. 5]. Used by permission.

The second and third versions of the *JK* flip-flop in Table 19.5 are gated by a clock signal (to be described below), which is a useful feature in many computer applications.

19.4.2   Clocks and Registers

Timing in logic circuits is performed by a wide variety of circuits and systems known, not surprisingly, as *clocks*. The simplest clock circuits operate as a function of transient rise times in *RC* networks. Examples of such timers are those used in automotive turn signals, some sweep circuits used in simple oscilloscopes, and highway flashers. Such timing circuits are, of course, susceptible to probable errors in the value of *R* and *C* and are subject to temperature variations, only portions of which can be compensated.

A much greater accuracy is required for most logic circuit applications, and this is generally supplied by temperature-compensated crystal-controlled *RC* oscillators which are constructed of an integrated circuit. Since most microprocessors and computer circuits require many clock signals, the output of a *master clock* is divided by means of a multiphase generator to provide many output timing signals. The block diagram for a clock in a typical microprocessor is shown in Fig. 19.7. The output signal from a clock used in logic timing applications appears generally as shown in Fig. 19.8 and has a square output waveform. When the value of the clock signal increases to its high (or positive) value, that rising portion of the waveform is called the *leading edge*, or *positive edge*, of the clock output; the decreasing portion of each pulse, when the clock signal returns to zero or negative or low value, is called the *lagging edge*, or *negative edge*. Logic devices which are triggered by clock pulses may be designed to operate on either the positive or negative edge of the clock pulses,

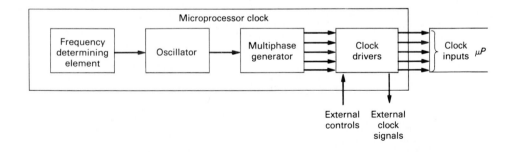

FIGURE 19.7
Block diagram showing major elements of a microprocessor clock.

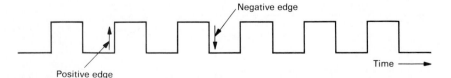

FIGURE 19.8
Output pulse train of a block.

although the latter is generally more common. These are edge-triggered logic devices. Level-sensitive logic devices are designed to operate on either the high or low level of the clock pulses; one such device is the polarity-hold SRL (shift-register latch) described in Ref. 14. Note the several versions of the JK flip-flop shown in Table 19.5. The rise and fall times of a clock pulse are of the order of 50 ns or less. Typical frequencies of the master clock are of the order of 20 MHz.

Another clock-triggered flip-flop is the *type-D* flip-flop, also shown in Table 19.5. The complete truth table for the type-D flip-flop is shown in Table 19.6, using the terminal designation as shown in Table 19.5. The upward arrow under the C (clock) input terminal column indicates the positive edge of the clock pulse; the downward arrow indicates the negative edge. Also note the use of don't care inputs designated by the $\times$. The D input channel is called the *data input*.

One of the principal uses of the type-D flip-flop is as a temporary storage or memory element in digital control and computer systems. A grouping of type-D flip-flops used for temporary storage is known as a *register* or sometimes a *shift register*. A grouping of N flip-flops is called an *N-bit register* and is used for storing N logic bits as defined in Sec. 19.1. Other flip-flops and other hardware are commonly used for this purpose, but the grouping of four type-D flip-flops shown in Fig. 19.9 illustrate the general concept of register operation.

Figure 19.9a illustrates the circuit diagram of the basic elements of a 4-bit register using type-D flip-flops. Figure 19.9b describes a typical timing sequence that might go along with this type of register. Note that this timing diagram shows the actual finite rise- and fall-time characteristics of a practical timer. Its operation is as follows: Data, in this case the value of a bit (1 or 0), come in on the D_n ($n = 1$ to 4) channels at the rate shown on the upper trace of Fig 19.9b. This particular register is triggered on the positive edge of the clock pulse train. Therefore, when the clock signal is changing from its low to its high state, D_n is transferred to the Q_n output port and held there until changed either to an opposite value of D_n or a reset signal. The reset (R) signal *asynchronously* (i.e., at any time and not in synchronism with the clock pulse train) resets Q_n to zero or low whenever $R =$ low. The output $\bar{Q}_n$ is always at the complement state of Q_n. This type of register operation is known as the *parallel mode*.

TABLE 19.6 Truth Table for Type D Flip-Flop

Inputs				Outputs		
C	D	R	S	Q	$\bar{Q}$	
↑	0	0	0	0	1	
↑	1	0	0	1	0	
↓	$\times$	0	0	Q_n	$\bar{Q}_n$	(No change)
$\times$	$\times$	1	0	0	1	
$\times$	$\times$	0	1	1	0	
$\times$	$\times$	1	1	1	1	

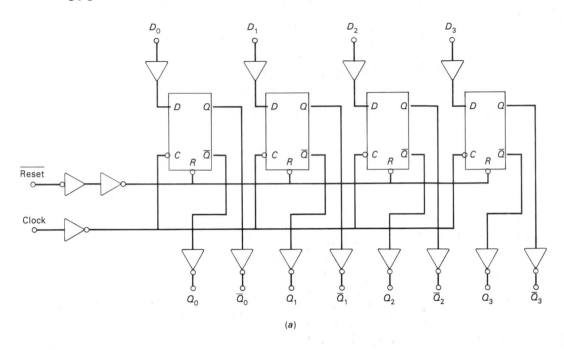

(a)

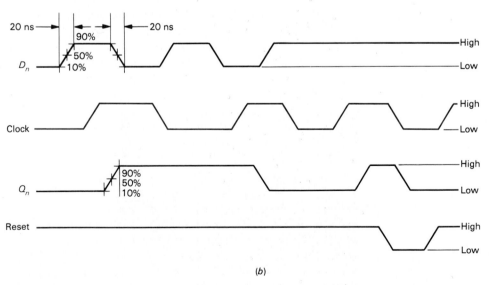

(b)

FIGURE 19.9 Four-bit register using type-D flip-flops and inverters: (a) circuit diagram; (b) timing diagram.

The same circuit shown in Fig. 19.9 can also be caused to operate in a *series mode*, in which mode it is known as a *shift register*. In this mode, the bit in one flip-flop is shifted to an adjacent flip-flop, either to the right or to the left. This serial shifting process can be continued sequentially to the last flip-flop in the sequence, or it can be halted at another flip-flop, depending on the system logic.

19.4.3 Counters

Counting is a digital (in contrast to an analog) function, and the logic circuits discussed so far are ideally suited for this application. Digital counters are used to count pulses, events (such as the operation of a particular gate), frequency, etc., and are widely used in both control and computing applications. Counters are sequential circuits, as is the process of counting itself. The size of a counter—i.e., the highest count that can be reached—is called the counter *modulus* and is designated by N. This number also equals the total number of states of the counter logic. Counting is almost universally performed in binary logic, starting with the least significant bit. The modulus N depends on the number of bits contained in the counter circuitry. A 1-bit counter can obviously count the numbers 0 and 1, and $N = 1$; a 2-bit counter can count from binary 0 up to 11, or decimal 3, and $N = 3$; a 3-bit counter has an N of binary 111 or decimal 7. In general, when a counter has counted up to the value of its modulus, the contents of the counter are reset or cleared, and the counting process starts over. Certain classes of counters, known as *up-down counters*, effectively double the modulus by counting up to N and down to 0 before the reset process is applied.

Figure 19.10 illustrates a simple 3-bit counter using the type-JK flip-flop described above and defined in Table 19.5. For this counter application, the J and K inputs are set at $J = K = 1$, or high. In Fig. 19.10a, the flip-flop on the left with output Q_1 is the least significant bit, and the bit significance increases in each successive flip-flop to the right. Some waveforms of the counting process are shown in Fig. 19.10b. These waveforms assume an initial reset or cleared condition of the counter in which all three flip-flops are set at 0, giving the bit count of 000. The input is applied to the flip-flop representing the least significant bit, the flip-flop on the left of Fig. 19.10a. Each flip-flop changes state on the negative edge of the input pulse train—which is the pulses to be counted. The first input pulse is counted when Q_1 changes from 0 to 1 on the negative edge of the first input pulse. Now Q_1 returns to 0 on the negative edge of the second pulse; however, since Q_1 is applied to the input of the second flip-flop and this change from 1 to 0 represents a negative edge, Q_2 becomes 1. This gives a binary count (reading from right to left, that is, $Q_3 Q_2 Q_1$) of 010, which is decimal 2, which correctly indicates a count of two pulses. The next input pulse causes Q_1 to change to 1, giving binary 011, or decimal 3, which is correct. The next, or fourth, input pulse causes Q_1 to return to 0, which inputs a negative edge to the second flip-flop, causing Q_2 to change to 0—which, in turn, inputs a negative edge to the third flip-flop, causing it to change state to a 1, giving a binary count of 100, or 4. This process is continued until the binary count is 111, or decimal

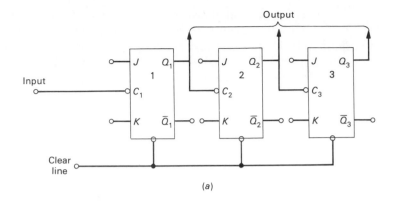

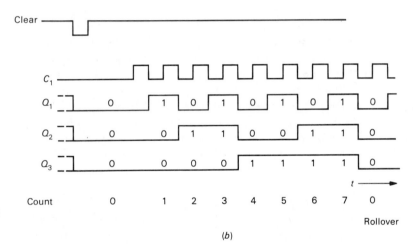

FIGURE 19.10
Three-bit up counter: (*a*) circuit diagram: (*b*) timing diagram.

7. The next pulse causes the counter to revert to 000, and the process of counting up to 7 (actually a count of 8) is repeated.

Counters can be constructed from many other logic devices and circuit configurations and may have logic considerably different from that illustrated above. The *down* counter has already been mentioned and can be simply derived from the up counter of Fig. 19.10 by using the $\bar{Q}$ outputs instead of the Q outputs. *Synchronous* counters are controlled by a clock such that all flip-flops change with the clock pulse, regardless of the number of stages. *Divide-by-N counters* are useful in relating the count to the time period of the count and hence in determining pulses per second, or frequency. *Johnson counters* consist of D flip-flops arranged in a ring; the input pulses are applied to all clock terminals as in synchronous counters, but the complemented output of the last stage goes back into the D terminal of the first stage. The five-stage Johnson counter is called a *divide-by-10 counter* since the waveforms complete one full cycle for every 10 input pulses.

Example 19.13 Design a counter that can count up to 32.

Solution Since $N = 32 = 2^5$, a 5-bit counter is required. This can be realized by adding two more flip-flops to the circuit shown in Fig. 19.10a. With five flip-flops the maximum binary count is 11111, which equals 31, giving a count of 32 when 0 is included.

19.5 HARDWARE IMPLEMENTATION OF LOGIC CIRCUITS

Thus far we have dealt primarily with theoretical and mathematical concepts in describing the properties of digital electronic logic. Now we look at the means of realizing these properties by physical circuits and devices.

We have introduced the term *hardware* here because it has come into common use in the engineering lexicon and has led to the formulation of a relatively new antonym, *software*. Software generally describes the nonhardware components of a computer, in particular the programs needed to make computers perform their intended tasks (Ref. 10). The first software systems were libraries of mathematical routines. Software thus includes the mathematical formulation of boolean algebra, truth tables, and many other mathematical operations, as well as "housekeeping" functions for proper operations of the computer or control system and the input and output functions, discussed in the next chapter. The means of realizing all these functions and operations in a practical physical sense is commonly called hardware in today's lexicon.

Hardware refers to physical devices, circuits, and systems. Associated with this noun are a number of adjectives *developmental*, meaning still in the research or development stage; *state-of-the-art*, meaning available today but at no specified cost; and *commercial*, meaning available from conventional markets today—with many other nuances defining these basic three divisions of marketplace reality. Electronic logic hardware costs generally decrease rapidly following introduction to the marketplace. Software costs remain almost unchanged. As a result of these cost relationships, electronic logic is dominant in almost all aspects of computer and control applications, from space war games and numerical control of machine tools to all types of engineering and scientific mathematical calculations.

Much of the early interest in boolean algebra resulted from the widespread use of electromagnetic relay systems in power systems and motor protection schemes; this type of hardware is still in common use in many applications, and there is generally direct analogy between relay logic and electronic logic circuits, as discussed in Ref. 6. The early computer circuits were basically mechanical computers of various types, and some of the earliest computer languages that led to such widely used electronic computer languages as FORTRAN (formula translation) were originally developed for mechanical computers (Ref. 10). Mechanical counters are still in very common use. Fluidic computers and controllers also generally preceded electronic systems and are still used in many applications. Superconducting computers have been proposed, and a few prototypes have been constructed.

19.6 IC LOGIC FAMILIES

The first thing one encounters when trying to use or understand electronic logic devices and systems is an unholy array of acronyms. We try to alleviate this alphabetic stumbling block by listing some of the more common acronyms, although this is somewhat of a dangerous step, since IC logic systems come and go and become outdated in a matter of months. Some of the following may be ancient history by the time you read this.

IC—integrated circuit, an assemblage consisting of a number of discrete devices (transistors, diodes, etc.), circuit constants (resistors, capacitors), and interconnected leads integrated to form a complete system in a small region of semiconductor material; usually called a *chip*

DIP—dual in-line package—a common means of packaging semiconductor chips (see Fig. 19.11)

SMT—surface mount technology

SOP—small-outline package

LCC—leadless chip carriers

FET—field-effect transistor

MOS—metal-oxide semiconductor

MOSFET—metal-oxide-semiconductor field-effect transistor

PC—printed circuit

LSI—large-scale integration (a chip containing densely packed functions, or the process of developing such ICs)

The principal logic families are

CMOS—complementary MOS

HMOS—high-speed CMOS

ECL—emitter-coupled logic

I^2L—integrated injection logic

*N*MOS—*N*-channel MOS

*P*MOS—*P*-channel MOS

RTL—resistor-transistor logic

TTL(T^2L)—transistor-transistor logic

In the remaining sections of this chapter we present the external characteristics and take some very brief looks at the internal circuitry of the electronic logic used to achieve the mathematical functions described previously. The semiconductor

FIGURE 19.11
A comparison of SMT and DIP packaging of ICs. The package on the left is a small-outline package (SOP), housing the same IC as the DIP on the left. (Courtesy of North American Phillips SMD Technology, Inc.).

physics and electronic circuit theory associated with ICs have been briefly discussed in earlier chapters of this text. For further study of IC design, structure, and performance, consult Refs. 2, 7, and 11 to 17.

ICs are used in most logic applications today because of their low cost, high-volume density, and suitability for many computer and control systems. The internal structure of ICs, i.e., the electrical circuitry, more or less delineates what are called *logic families*. Many other "families" are in use today besides those listed above, and new families are continuously being introduced as the state of the art of semiconductor manufacture improves. The choice of a particular family depends on the associated components to be used in the system—the power supplies (and associated bus voltages), output devices, frequency response, type of PC board or other mounting structure, etc. The volume, or packing, density of all types of ICs is generally very high compared with mechanical or electromagnetic logic, so this is generally not an issue in the choice of logic family. ICs are available in a number of physical forms, although the DIP (dual in-line package) is probably the most popular. We use the term *package* to describe the physical and structural members used to contain, support, and protect the actual semiconductor logic member. Figures 19.11 and 19.12 illustrate several typical IC packages in common use (Ref. 5). SMT packages are gaining acceptance in semiconductor applications. The principal advantages of SMT are reduced package size, improved thermal characteristics, simplification of automatic mounting on PC boards, and elimination of hole-drilling steps required for conventional DIP. Common SMT packages are known as leadless chip carriers (LCCs), tape-automated bonding (TAB), and small-outline package (SOP). See Fig. 19.11.

Major IC packages

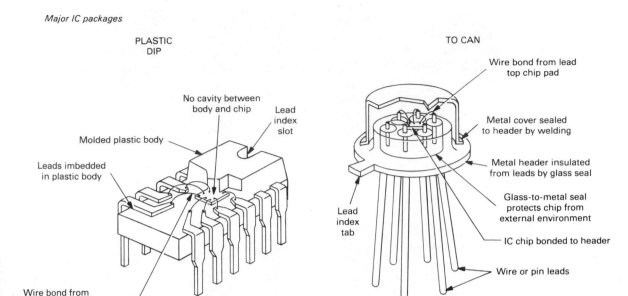

PLASTIC DIP

No cavity between body and chip

Lead index slot

Molded plastic body

Leads imbedded in plastic body

Wire bond from lead to chip

TO CAN

Wire bond from lead top chip pad

Metal cover sealed to header by welding

Metal header insulated from leads by glass seal

Glass-to-metal seal protects chip from external environment

IC chip bonded to header

Lead index tab

Wire or pin leads

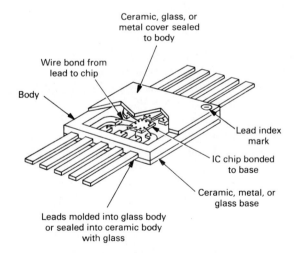

FLAT PACK

Ceramic, glass, or metal cover sealed to body

Wire bond from lead to chip

Body

Lead index mark

IC chip bonded to base

Ceramic, metal, or glass base

Leads molded into glass body or sealed into ceramic body with glass

FIGURE 19.12 Three common means of IC packaging.

19.6.1 CMOS Logic

The acronym *CMOS* refers to complementary-symmetry metal-oxide semiconductor. The letter C contrasts this semiconductor's structure and internal circuitry with the earlier *NMOS* and *PMOS* structures. Note that the symbols for logic functions, such as those shown in Fig. 19.1 and Table 19.5, are common for *all* logic families, as are the mathematical relationships developed earlier in this chapter.

The CMOS (complementary MOS) is constructed of metal-oxide semiconductors (MOS), and the basic structure of CMOS logic makes use of both N-channel and P-channel enhancement-mode transistors. Figure 19.13 illustrates the cross-sectional structure of recent CMOS devices and clearly shows their size reduction. The advantages of CMOS logic are low power dissipation and a wide range of power-supply voltages. CMOS switching circuits operate at switching rates up to 15 MHz, and CMOS memories are used up to 50 MHz. The disadvantages of CMOS are that this logic system is generally more costly than other systems, it does not offer all the circuits available in other systems, particularly TTL logic, and does not offer the frequency response of TTL.

We have devoted much of our study of logic algebra in this chapter to systems involving the direct gates, the AND and OR gates, since the concepts and operation of these gates are, in general, easily grasped. The complementary gates, NAND and NOR, have been introduced in several examples and are implied from De Morgan's theorems (Sec. 19.2.1). In practice, the complementary gates are more widely used than the direct gates. This is due to the inherent inversion property of many logic circuits as well as other electronic circuits and devices, such as operational amplifiers. The standard symbol to indicate inversion is the circle (bubble) at the output terminal of the NOR gates, as shown in the logic diagram of Fig. 19.14. In this same diagram, referring now to the numbered terminals of the dual NOR gate, the logic states that the output at terminal 1 is the *inverse* of the results of the OR process at input

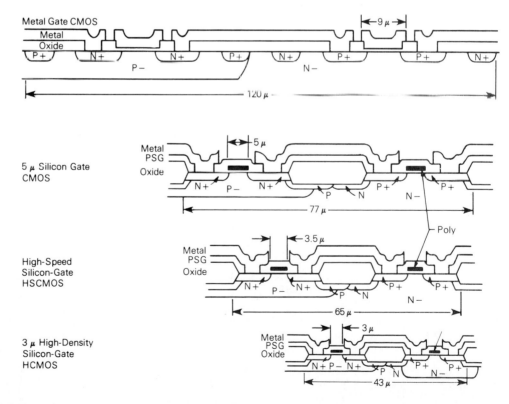

FIGURE 19.13 Evolution of CMOS chips, illustrating size reduction and structural changes. (Courtesy of Motorola Integrated Circuits Div., Austin, Texas.)

 **MOTOROLA**

MC54/74HC266

Advance Information

QUAD 2-INPUT EXCLUSIVE NOR GATE

The MC54/74HC266 is identical in pinout to the LS266, but does not have open-drain outputs. The device inputs are compatible with standard CMOS outputs; with pullup resistors, they are compatible with LSTTL outputs.

- Low Power Consumption Characteristic of CMOS Devices
- Output Drive Capability: 10 LSTTL Loads Minimum
- Operating Speeds Similar to LSTTL
- Wide Operating Voltage Range: 2 to 6 Volts
- Low Input Current: 1 μA Maximum
- Low Quiescent Current: 20 μA Maximum (74HC series)
- High Noise Immunity Characteristic of CMOS Devices
- Diode Protection on All Inputs

HIGH-PERFORMANCE
CMOS
LOW-POWER COMPLEMENTARY MOS
SILICON-GATE

**QUAD 2-INPUT
EXCLUSIVE NOR GATE**

J SUFFIX
CERAMIC PACKAGE
CASE 632

N SUFFIX
PLASTIC PACKAGE
CASE 646

ORDERING INFORMATION

54 Series −55°C to +125°C
  MC54HCXXJ (Ceramic Package Only)
74 Series −40°C to +85°C
  MC74HCXXN (Plastic Package)
  MC74HCXXJ (Ceramic Package)

FUNCTION DIAGRAM

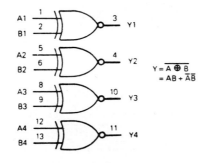

$$Y = \overline{A \oplus B}$$
$$= AB + \overline{A}\,\overline{B}$$

V_{CC} = Pin 14
GND = Pin 7

PIN ASSIGNMENT

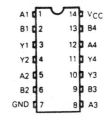

A1 □ 1	14 □ V_{CC}
B1 □ 2	13 □ B4
Y1 □ 3	12 □ A4
Y2 □ 4	11 □ Y4
A2 □ 5	10 □ Y3
B2 □ 6	9 □ B3
GND □ 7	8 □ A3

TRUTH TABLE

Inputs		Outputs
A	**B**	**Y**
L	L	H
L	H	L
H	L	L
H	H	H

FIGURE 19.14
Data sheet for
CMOS NOR gate.
(Courtesy of
Motorola
Semiconductor
Products Div.)

terminals 2, 3, 4, and 5. Calling the output Y and the inputs A, this may be stated in boolean form as

$$Y_1 = \overline{(A_2 + A_3 + A_4 + A_5)} \qquad (19.11)$$

From the first De Morgan theorem, Eq. (19.8a), we see that Eq. (19.11) can also be written as

$$Y_1 = \bar{A}_2 \bar{A}_3 \bar{A}_4 \bar{A}_5 \qquad (19.12)$$

which reads, "Y_1 equals not A_2 and not A_3 and not A_4 and not A_5." Pause for a moment or two to be sure that you agree that these two forms are equivalent—that they are stating the same logic.

There is a similar relationship for the NAND gate, which can be expressed by two equations analogous to the above (for a 4-input gate):

$$Y_1 = \overline{A_1 A_2 A_3 A_4} \qquad (19.13)$$

or

$$Y_1 = \bar{A}_1 + \bar{A}_2 + \bar{A}_3 + \bar{A}_4 \qquad (19.14)$$

Symbolically, the two forms are represented by (1) changing the bubbles from input to output, or vice versa, and (2) changing AND to OR, or vice versa.

Example 19.14

Determine the alternative form for the NOR gate of Fig. 19.15, and develop the truth table.

Solution The alternative form is an AND gate with complementary (NOT) inputs, i.e., with the bubbles on the input leads; this form corresponds to Eq. (19.12), and is shown in Fig. 19.15. The truth table follows from *either* logic form Eq. (19.11) or Eq. (19.12). In this table, 1 and 0 are used symbolically to represent high and low logic states, respectively.

FIGURE 19.15
Example 19.14:
alternative form
and truth table for
the 4-input NOR
gate of Fig. 19.14:
(*a*) alternative
form; (*b*) truth
table.

(a)

A_2	A_3	A_4	A_5	Y_1
0	0	0	0	1
1	0	0	0	0
0	1	0	0	0
0	0	1	0	0
0	0	0	1	0
1	1	0	0	0
1	1	1	0	0
1	1	1	1	0

(b)

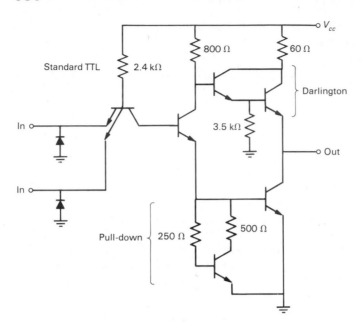

FIGURE 19.16
Conventional
circuitry for a
2-input NAND
gate in the TTL
logic family.

19.6.2 Transistor-Transistor Logic

Transistor-transistor logic (TTL, or T^2L) is a very popular type of logic. It is generally lower in cost and faster in switching speeds than CMOS logic, and it is still a widely available logic technology. Figure 19.16 illustrates the circuit for a simple 2-input NAND gate. Both inputs, A and B, come into the input transistor Q_1. The output voltage Y is derived from a Darlington circuit, often called a "totem pole" arrangement. The Darlington circuit provides a low output impedance for increased on-off switching speed and hence a higher operation frequency. In the basic TTL gates, transistors are driven into saturation, and speed limitations result primarily from storage time delays associated with transistor turnoff. TTL gates operate at frequencies up to approximately 40 MHz.

19.6.3 Emitter-Coupled Logic

Emitter-coupled logic (ECL) is faster than TTL and is used in applications where a very high switching speed is required. ECL has high switching speeds since the transistors are operated in a differential mode as emitter followers and are not driven to saturation. Switching speeds up to 100 MHz are possible with ECL. Because of these high frequencies, the ancillary leads and circuit connections associated with ECL systems must be designed with extreme care in order to minimize circuit inductance. Figure 19.17 illustrates a basic ECL circuit which can be used as either an OR or a NOR gate or both. Figure 19.18 illustrates a development by Motorola to improve the switching speed by means of series-coupled logic (Ref. 13). This

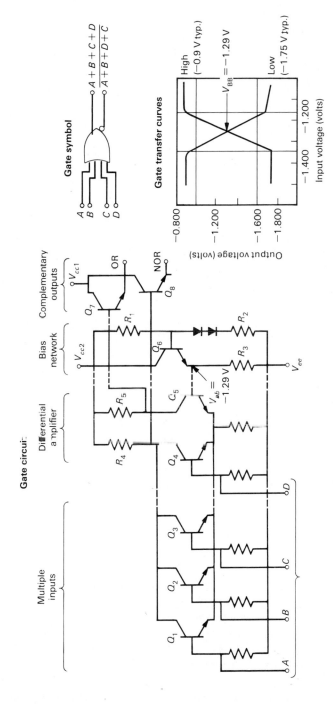

FIGURE 19.17 Emitter-coupled logic (ECL) OR and/or NOR gate.

687

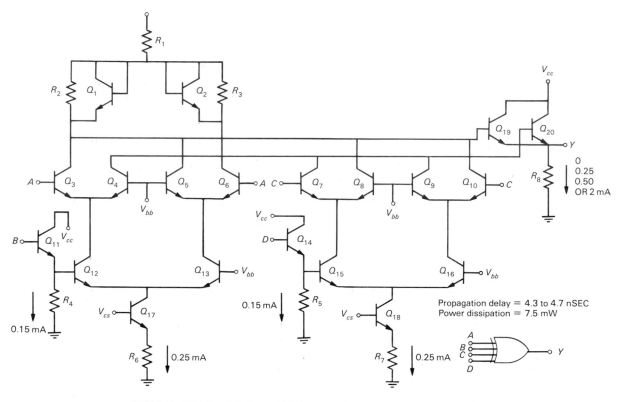

FIGURE 19.18   A 4-input XOR gate using series-coupled logic to improve response time. (Courtesy of Motorola Semiconductor Products Div.).

figure also describes the exclusive OR gate, which has some very interesting character-istics and will be our last sortie into the realm of boolean algebra.

The 2-input exclusive OR circuit has been defined in Sec. 19.2.1, Table 19.2. It differs from the regular OR circuit by eliminating from the conditions resulting in a high or true or 1 output that condition when *all* inputs are high or true or 1. The 4-input exclusive OR gate (often called the XOR gate) shown in Fig. 19.18 has the equation

$$Y = A\bar{B}CD + \bar{A}BCD + AB\bar{C}D + ABC\bar{D} + A\bar{B}\bar{C}\bar{D} + \bar{A}B\bar{C}\bar{D} + \bar{A}\bar{B}C\bar{D} + \bar{A}\bar{B}\bar{C}D \quad (19.15)$$

In the symbolic form suggested in Sec. 19.2.1, this could also be described by the equation

$$Y(A, B, C, D) = A \oplus B \oplus C \oplus D \quad (19.16)$$

where the $\oplus$ symbol means exclusive OR. The numerical representation of the 4-input XOR gate can be found by the methods of Sec. 19.3.1 [note that Eq. (19.15) is in

FIGURE 19.19
Karnaugh map for
a 4-input XOR
gate.

AB \ CD	00	01	11	10
00		1		1
01	1		1	
11		1		1
10	1		1	

standard SOP form] as

$$Y(A, B, C, D) = \sum (1, 2, 4, 7, 8, 11, 13, 14) \tag{19.17}$$

The Karnaugh map for this function is shown in Fig. 19.19. This gives an interesting checkerboard type of map, and no 1s can be combined into subcubes. From Eq. (19.12) it seems that eight NAND circuits would be required to achieve this logic equation . However, from Eq. (19.16) we see that three 2-terminal XOR circuits can also simulate this logic, which is what is done in Fig. 19.18.

PROBLEMS

19.1 Find the binary equivalent of the following decimal numbers: (*a*) 76, (*b*) 1218, (*c*) 750, (*d*) 0.061, (*e*) 0.89, (*f*) 0.175,

19.2 Verify that the binary numbers derived in Prob. 19.1 are equivalent to the decimal numbers from which they were derived.

19.3 Develop truth tables for the (*a*) 2-input NAND function, (*b*) 4-input NOR function, (*c*) fourth logic circuit of Fig. 19.1.

19.4 Write the logic equation $Y = f(A, B, C)$ and prepare a truth table for the circuit shown in Fig. P19.4.

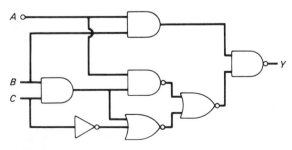

FIGURE P19.4

19.5 Using AND and OR gates, develop a logic circuit to express the logic equation

$$X = A + \bar{B}D + BD + CD$$

with A, B, C, and D being the input variables.

19.6 Does

$$X + WZ = WX + \bar{W}X\bar{Y} + W\bar{X}Z + \bar{W}Y?$$

19.7 Does $BC + ABD + A\bar{C} = BC + A\bar{C}$?

19.8 Convert the following expressions to standard SOP form:

(*a*) $f(ac, b, c) = a\bar{b} + c$
(*b*) $f(a, b, c, d) = b + \bar{a}\bar{c} + a\bar{b}cd$
(*c*) $f(W, X, Y, Z) = W\bar{X} + XYZ + \bar{W}\bar{Y}Z$

19.9 Determine the numerical representation (summation of a group of numbers) for the three equations of Prob. 19.8.

19.10 Convert the three expressions in Prob. 19.9 to standard POS form.

19.11 Determine the numerical representation (product of a group of numbers) for the three resulting equations in Prob. 19.10.

19.12 Prepare the Karnaugh maps for the three logic equations of Probs. 19.9, 19.10, and 19.11, and find the minimum expression.

19.13 Find the minimum expression for

$$f(A, B, C, D) = \sum (0, 2, 3, 4, 8, 9, 10, 11, 15)$$

19.14 Find the minimum expression for the system defined by the Karnaugh map of Fig. P19.14.

wx \ yz	00	01	11	10
00	1		1	1
01	1		1	1
11	1	1	1	
10	1	1	1	

FIGURE P19.14

19.15 Find the minimum expression for the system defined by the Karnaugh map of Fig. P19.15.

wx \ yz	00	01	11	10
00	0		0	0
01		0		
11				
10	0		0	0

FIGURE P19.15

REFERENCES

1 *Webster's Ninth New Colleagiate Dictionary*, Merriam-Webster, Springfield, Mass., 1984.

2 J. D. Greenfield, *Practical Digital Design Using ICs*, 2d ed., Wiley, New York, 1983.

3 A. Barna and D. I. Porat, *Integrated Circuits in Digital Electronics*, Wiley, New York, 1973.

4 F. E. Hohn, *Applied Boolean Algebra—Elementary Introduction*, Macmillan, New York, 1960.

5 1983 Electrical and Electronics Reference Issue, "Digital Logic," *Machine Design*, May 1983.

6 J. Prioste and T. Balph, "Relay to IC Conversion," *Machine Design*, May 28, 1970.

7 D. G. Fink and D. Christiansen, *Electronics Engineer's Handbook*, 2d ed., McGraw-Hill, New York, 1982.

8 M. Karnaugh, "The Map Method for Synthesis of Combinational Logic Circuits," *AIEE Transactions*, vol. 72, pt. 1, 1953.

9 A. Rappaport, "Automated Design and Simulation Aids Speed Semicustom IC Development," *EDN*, vol. 27, no. 15, August 4, 1982.

10 J. W. Hunt, "Programming Languages," *Computer* (IEEE), April 1982.

11 Motorola CMOS Handbook, *CMOS Integrated Circuits*, Motorola Semiconductor Products, Austin, Tex., 1980.

12 J. Kasper and S. Feller, *Digital Integrated Circuits*, Prentice-Hall, Englewood Cliffs, N.J., 1982.

13 L. Teschler, "Update of Electronic Logic," *Machine Design*, August 20, 1981.

14 E. B. Eichelberger, "A Logic Design Structure for LSI Testability," *14th Design Automation Conference*, June 1977, pp. 462–468.

15 Anthony Ralston (ed.), *Encyclopedia of Computer Science and Engineering*, 2d ed., Van Nostrand Reinhold, New York, 1982.

16 A. B. Williams, *Designer's Handbook of Integrated Circuits*, McGraw-Hill, New York, 1984.

17 D. G. Ong, *Modern MOS Technology*, McGraw-Hill, New York, 1984.

20

Digital Systems

In several previous chapters we noted a distinct dichotomy in the nature of electric circuits and devices. This dichotomy encompasses two broad circuit classifications which have gradually become known as analog and digital, based on the means of energy control—either in a continuous manner (analog) or in discrete bundles (digital). Circuits and systems in which circuit control is exclusively of an analog nature and composed exclusively of analog circuits and components are known as *analog* systems. Where control components are exclusively digital, the system is classified as *digital*. A great many circuits and systems are composed of both digital and analog circuits or components, and these are referred to as *hybrid systems*. This chapter is concerned primarily with digital and hybrid circuits and systems.

The basic building blocks of digital circuits—gates, relays, switches, etc.—are described in Chap. 19. Obviously, there are countless hosts of additional digital components, circuits, and devices, as witnessed by the multitude of very thick circuit description books provided by each electronic circuit supplier. There are also many interesting electromechanical, mechanical, and hydraulic digital devices, such as the electric stepper motor, which are mated with the digital circuits discussed here. Digital circuits are generally thought of as low-signal low-power circuits used in control and information applications, such as calculators, computers, and radar; but digital circuitry is also widely used in many high-power motor, machine tool, and energy conversion applications, usually as part of a hybrid system. An example of this type of application, in the control of an electric motor, is shown in Fig. 20.1. In the control system of Fig. 20.1, two main subsystems, the microprocessor and the digital tachometer used in the feedback branch, are digital, whereas the power amplifier and motor are analog. The link between these two types of circuits is known as a digital-to-analog (D/A) converter. The inverse, analog-to-digital (A/D) conversion, is equally important in hybrid systems. Both systems are discussed in this chapter.

Digital circuits may be assembled or manufactured by wiring together a number of discrete elements (transistors, relays, resistors, capacitors, etc.) or by integrating

FIGURE 20.1
Block diagram of a
simple hybrid
control system
containing both
digital
and analog
subsystems.

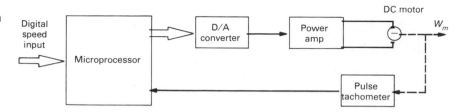

the entire assembly of devices and associated interconnections into a small semiconductor chip known as an *integrated circuit* (IC), which has been defined in the
previous chapter. It is this latter form of digital circuit assemblage that has fostered
the tremendous growth of digital circuit use in control and computational applications. One feature of IC development that has contributed to this dazzling growth
of digital circuit use is the rapid increase during recent years in the IC packing
density—i.e., the number of digital functions (OR gates, NAND gates, transistors,
etc.) that can be integrated into a given volume. Figure 20.2*a* and *b* gives further
illustration of the past and projected increase in chip density and complexity.

   A major difference between analog and digital electric circuits is the nature of
the waveforms of the electrical signals in the circuits. The signals (voltage or current)
in analog systems are continuous and have classically been analyzed on the basis of
either sinusoidal or continuous dc waveforms; much of classical electric circuit theory
is based on the assumption of one or the other of these waveforms. In contrast,
waveforms in digital systems are composed of a series of pulses separated by periods
of zero signal level, or off time. In most cases the pulses can be represented as square

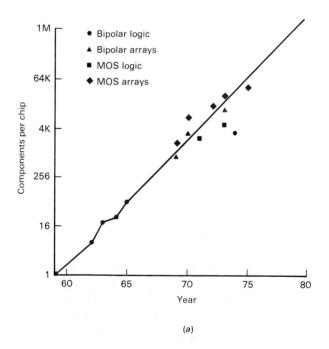

FIGURE 20.2
(*a*) Circuit
complexity in
relation to time of
introduction.

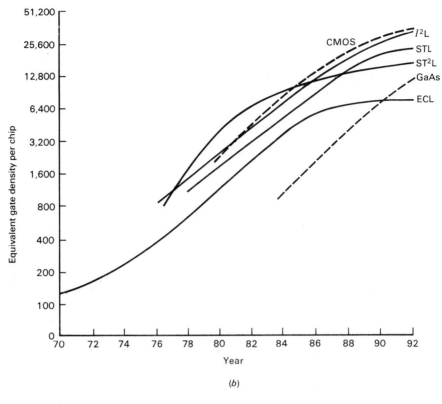

(b)

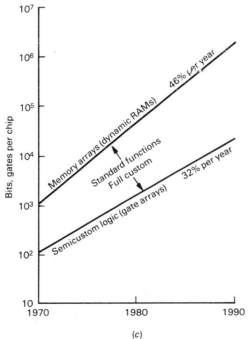

FIGURE 20.2 (b) Gate array chip complexity. (c) Very large-scale integration (VLSI) chip complexity trends.

(c)

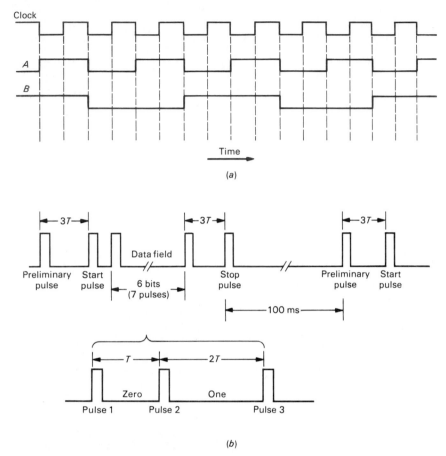

FIGURE 20.3
Examples of digital
circuit waveforms:
(*a*) clock and
derived pulses;
(*b*) an example of
pulse-position
modulation (PPM)
(see Sec. 20.1).

or rectangular. Typical digital signals are shown in Fig. 20.3. These signals are unipolar, but digital signals, in general, may be either unipolar or bipolar.

The control of analog signals is accomplished by varying the magnitude or frequency of the signal. Information transferred by analog signals is sent by superimposing magnitude or frequency variation on sinusoidal signals, a process known as *modulation*. The processes of control and information transfer in digital systems are roughly analogous to those used in analog systems, although there is a greater flexibility of control and modulation in digital systems due to the nature of digital waveforms.

Another difference is that digital systems generally have a greater immunity from extraneous signals, or *noise*, than do analog systems. This again is due primarily to the nature of the digital waveforms. Figure 20.4 is an attempt to illustrate in a very simple manner the noise immunity of digital signals. In this figure, the square pulse represents an original digital pulse that represents a binary bit of information. The second pulse shape v_2 represents the change in the original pulse due to transmission and transformer impedances, induced noise caused by switching of adjacent circuits, noise due to imperfect grounding (ground loops, etc.), and other effects along the

FIGURE 20.4
Elimination of
signal noise by
sampling
technique.

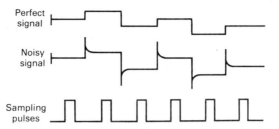

paths of a digital signal between its source and its sink. As has been noted, binary signals are detected by either their absence or their presence (1 or 0, true or false, etc.); this detection is often performed by a process called *sampling*, which is discussed in Sec. 20.1.1. We see from the third pulse of Fig. 20.4 that the distortion of the waveshape caused by system impedances and noise effects will have little effect on detecting the existence of the pulse, except for a possible shift in the pulse location (frequency).

20.1 PULSE MODULATION AND SAMPLING

Pulse modulation in the most general sense refers to the control of one or more of the parameters associated with a digital pulse. The term has most frequently been applied to the process of encoding information from digital pulse trains, but more recently, similar techniques have been applied to control the power associated with digital signals, such as the average or RMS value of pulse currents or voltages.

The three principal parameters of the pulse train used in modulation and demodulation schemes are the pulse amplitude (or magnitude), the pulse width, and the frequency of repetition of the pulse. All three parameters and several combinations of them are used in control and communications applications. Power applications have generally been based on pulse-width modulation (PWM).

Modulation refers to the process of superimposing a signal containing intelligence on a second signal known as the *carrier signal*—or sometimes known in pulse modulation as the *pulse train*. The reader is doubtless familiar with the well-known processes—used in radio and telephone communications, amplitude modulation (AM) and frequency modulation (FM)—of modulating sinusoidal carriers with a spectrum of sinusoidal signals, such as the spectrum resulting from a human voice. The intent of pulse modulation is similar to that of AM and FM, but the mechanism of the modulating techniques and the resulting waveforms are much different. Again, as we have often done on previous pages, we might use the terms *analog* and *digital* to contrast the two groups of modulating techniques; this terminology is, in fact, used in many practical applications, such as in distinguishing digital and analog recording methods. Demodulation is, of course, the reverse of the modulation process; i.e., demodulation is the extracting of the intelligent signal from the modulated signal.

Pulse modulation schemes are generally named according to the pulse parameter being controlled or modulated—pulse-amplitude modulation (PAM); pulse-time modulation (PTM), which is so named because modulation is achieved by operating on the pulse's time parameters and which includes the techniques with more familiar names, pulse-width modulation (PWM) and pulse-position modulation (PPM); pulse-frequency modulation (PFM); and pulse-code modulation (PCM), which includes such subgroupings as delta modulation (DM). Before we discuss a few of these techniques in more detail, it is necessary to introduce a process required in all pulse modulation schemes, the process of sampling.

20.1.1   Sampling

Sampling is the process of evaluating one signal at discrete time intervals for the purpose of deriving another signal. Sampling is used in many areas of technology, such as in electrical measurements and digital instrumentation, in reliability and failure-rate analysis, and even in many socioeconomic areas, such as public opinion polls. Our primary concern in this discussion, however, is the role of sampling in pulse modulation as used in digital systems of all types.

In this context, sampling generally refers to the evaluation of an analog, or continuous, signal at discrete time intervals. The ideal sampler is an *impulse sampler*, which multiplies a continuous signal $f(t)$ with a train of unit impulses $P(t)$ of period T. Such a process is illustrated in Fig. 20.5. The train of unit impulses has been converted to a train of impulses of varying magnitude, or amplitude, varying according to the analog function $f(t)$. Mathematically, the process of taking one sample of $f(t)$ is described by

$$f(t) = \int_{-\infty}^{\infty} f(\lambda)\delta(t - \lambda)\, d\lambda \qquad (20.1)$$

where $\delta(t - \lambda)$ is the unit impulse at $t = \lambda$. This integral is known as the *sampling property* of the unit impulse and is related to the *impulsive response* of a system, a concept used in circuit and control theory. The amount of information contained in the resulting pulse train of Fig. 20.5 (the sequence of vertical arrows under the analog function) depends on the *sampling rate* (or its reciprocal, the sampling period T). A large body of knowledge has developed around this concept of the relationship between the information and the technical signals representing or transmitting that information, and this body of knowledge is known as *information theory*. A key theorem of information theory, known as *Shannon's sampling theorem*, addresses the relationship between the sampling period and the informational content (Ref. 1). If the frequency spectrum of the sampled function $f(t)$ contains no frequencies above f_c, then Shannon's sampling theorem states that the continuous signal $f(t)$ can be reproduced if and only if $1/T > 2f_c$. The minimum sampling frequency $2f_c$ is known as the *Nyquist rate*. In practice, due to noise and other effects, a signal may not be completely limited to the frequency band, so the sampling frequency[2] $f_s = 1/T$ is usually made greater than twice f_c.

Also in the practical case, the ideal unit impulse (of zero pulse width) is not realizable, and narrow but finite-width pulses represent the practical sampling pulse

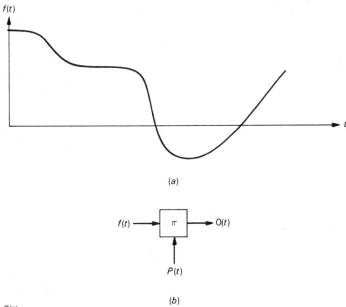

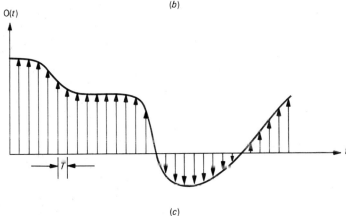

FIGURE 20.5
Ideal impulse sampler: (*a*) analog or modulating signal $f(t)$; (*b*) simplified representation of sample process where $P(t)$ represents an impulse strain of hint magnitude and of repetition period T; (*c*) modulated impulse train $0(t)$.

train. Sampling pulses may be unipolar or bipolar, or flat-topped or conformal-topped, depending on the required fidelity, cost restraints, and frequency spectrum of the analog signal. Figure 20.6 illustrates two different types of sampling pulses used in a PAM system. The pulse width of the sampling pulse τ must be small compared to the sampling period T.

A basic component of information and control systems using sampling techniques is the *sample-and-hold circuit*. The function of this type of circuit is to take a sample of an analog function at a given instant and to hold the amplitude of this function for a short interval. An obvious purpose of such a device is to keep the value of a sample available for a longer time than the sampling period T, to permit processing of this value by the circuit or system. As will be seen, various sample-and-hold circuits are useful in the demodulation process. Figure 20.7 illustrates a simple

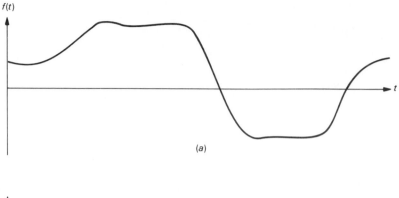

(a)

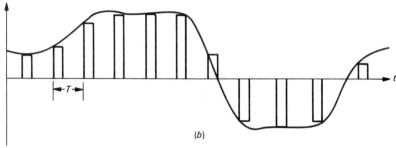

(b)

FIGURE 20.6
Pulse-amplitude
modulation: (*a*)
modulating signal;
(*b*) pulses from
bipolar
square-topped
PAM; (*c*) pulses
from unipolar
conformal PAM.

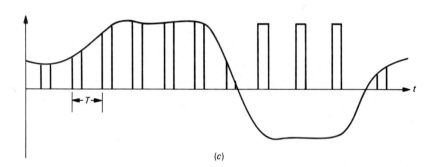

(c)

electronic sample-and-hold circuit and its associated functioning on a typical analog
signal $f(t) = v_s(t)$. This type of hold is known as *zero-order hold*, indicating that the
value of $f(t)$ held at each sampling instant is the value of $f(t)$ at that instant. As seen
from Fig. 20.7, the zero-order hold tends to convert a continuously varying analog
signal to a staircase-type signal.

20.1.2   Pulse-Amplitude Modulation

In this scheme, the modulating process consists of deriving a train of pulses, the
amplitudes of which vary with the amplitude of the analog signal $f(t)$. This process
is shown in Fig. 20.6. PAM is analogous to AM in the continuous-signal case and

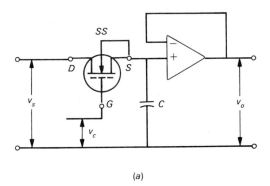

(a)

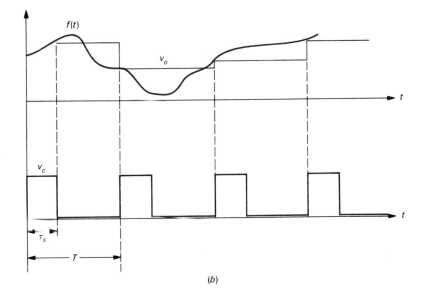

(b)

FIGURE 20.7
Sample-and-hold
circuit: (*a*) typical
circuit diagram
using MOSFET
and operational
amplifier; (*b*)
waveforms of
voltage signals.

has the same relatively low signal-to-noise ratio as does AM. Neither method is generally applicable to wide-band (f_c very large) data transmission. PAM is frequently used in multiplex systems, with the sampling of the individual channels made consistent with the sampling theorem given in the previous section. The pulse train in PAM is analogous to the carrier in AM. The hardware for performing PAM, called an *encoder*, is based on an electronic multiplication system (as shown in Fig. 20.5), with filtering, various auxiliary subsystems, and synchronizing circuits, depending on the specific type of PAM (unipolar or bipolar, square or conformal-topped, etc.) In *square-topped PAM* the pulse has a flat top, the magnitude of which represents the average value of the analog signal during the pulse period τ. In conformal or *top-modulated PAM*, the pulse conforms to the changing amplitude of the analog signal during the pulse period, as shown in Fig. 20.6.

An important class of control system based on PAM is known as *sample-data control*. In sample-data control, a sampling of the system error signal (or signal related to system error) of any waveform similar to those shown in Fig. 20.6 is fed back to the reference signal rather than a continuous function proportional to system error. This permits the processing of the error signal and the development of feedback signals to be performed by means of a digital system rather than an analog one, with the associated reduction in the effects of system noise and often with reduced system costs. A transformation calculus known as *Z transforms* has been developed for the analysis and design of sample-data systems.

20.1.3 Pulse-Width Modulation

This scheme is a specific form of the more general pulse-time modulation (PTM) methods. Information or control is transmitted by means of the time width of pulses having a constant amplitude and (usually) having a constant frequency of repetition. Pulse-width modulation is also known as *pulse-duration modulation* (PDM) or *pulse-length modulation* (PLM). There are two basic methods of PWM generation—the *uniform sampling* method in which the pulse-frequency or pulse-repetition rate is fixed, as in PAM, and the *natural sampling* method in which there is some slight modification (causing a small amount of distortion of the output signal) of the time at which the samples are actually taken.

A basic block diagram of the uniformly sampled PWM is shown in Fig. 20.8, and a typical waveform of PWM is shown in Fig. 20.9. This waveform results from what is called *leading edge synchronization* and can be described as follows: Each pulse in the pulse train (Fig. 20.9c) is initiated by the leading edge of the triangular wave (Fig. 20.9b), which is the sum of a sawtooth wave of fixed frequency and the staircase wave created from a sample-and-hold circuit operating on the analog signal $f(t)$ (Fig. 20.9a); each pulse is ended each time the sawtooth goes through its vertical wavefront; thus, the pulse width is determined by the time between the leading edge of the sawtooth (i.e., the instant at which the triangular wavefront crosses the horizontal) and the time the vertical wavefront crosses the horizontal—i.e., each pulse width is proportional to the width of the sawtooth at the horizontal axis of Fig. 20.9b. The importance of synchronizing the sample-and-hold circuit with the sawtooth wave generator is obvious from the pulse development shown in Fig. 20.9. The resulting pulse train has pulses of constant amplitude and constant frequency of repetition. Information is contained in the relative widths of the individual pulses. Note that zero on the modulating signal is well defined by the pulse of mean width. There are many variations in methods of setting pulse widths proportional to the

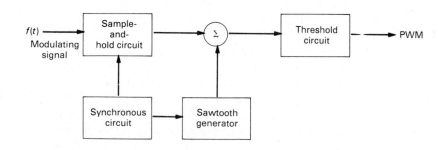

FIGURE 20.8
Block diagram of
uniform-sampling
type of
pulse-width
modulation.

FIGURE 20.9
Waveforms
associated with
uniform sampling
PWM: (*a*)
modulating
function and
sample-and-hold
output; (*b*) sum of
sample-and-hold
output and
sawtooth wave;
(*c*) generated
pulses using
leading edge
modulation.

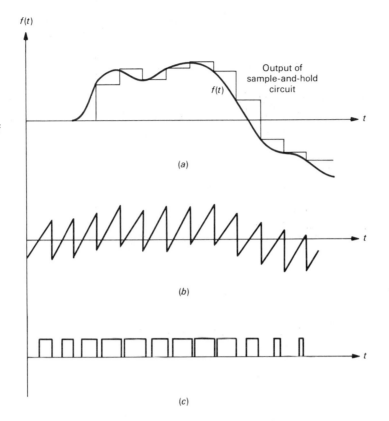

amplitude of the modulating signal $f(t)$, which results in minor differences in the frequency spectrum of the pulse train, the signal-to-noise ratio, the demodulation technique, etc., but the basic concepts of PWM are similar to those shown in Figs. 20.8 and 20.9.

PWM, through the use of constant-amplitude pulse trains, allows the circuitry creating the pulses to operate at its saturated level, which is usually the condition of maximum power efficiency. This is especially important where modulation of high-power signals is required, as has been noted. PWM is used to control motors and other power devices through modulation of power semiconductors. Power applications of PWM are described elsewhere in the text. The minimum off interval between pulses, known as the *guard interval*, must be nonzero even though the sampling theorem is satisfied with a zero interval, owing to finite pulse rise times. Overlapping of pulses will result in a distortion of the demodulated signal and in cross talk in multichannel systems.

20.1.4 Pulse-Code Modulation

PCM is markedly different from the two modulation techniques previously discussed in that it converts the modulating signal to a group of pulses, each group of which represents the numerical value of the modulating signal at the sampling instant. The translation of the quantized analog signal at the sampling instant into a "code" of pulses is the basis for the naming of this modulation technique. It has become more feasible in recent years owing to great improvements in technical characteristics and to the reduced cost of integrated circuits. PCM lends itself readily to time multiplexing (the transmitting of several signals over the same medium at distinct time intervals in a given period) and generally has a high signal-to-noise ratio. PCM signals are readily amplified by regenerative repeater stations for long-distance transmission with a minimum reduction as compared with most other modulation techniques. Error-detecting and -correcting processes are generally more easily accomplished in PCM systems.

Figure 20.10 illustrates waveforms at various stages in a PCM system. The analog, or modulating, signal is quantized by the use of sample-and-hold circuitry, giving a distinct amplitude of the analog signal at each sampling instant. This amplitude is then converted to a group of pulses by a specific instruction code. A wide variety of codes are possible, and the resulting pulses may be binary (consisting of magnitude levels 1 and 0), tertiary (1, 0, and −1), or, for that matter, n-ary. Binary pulses are the most common. The *number* of pulses in each group is related to the frequency spectrum desired for the transmission of the analog signal. In Fig. 20.10, three pulses per group have been used, which permits the encoding of eight levels of the analog signal in binary representation. Using an n-ary code with m pulses per

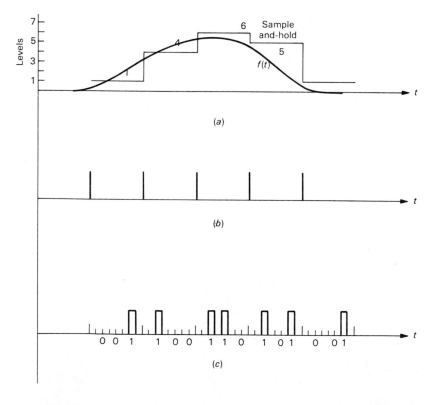

FIGURE 20.10
Pulse-code
modulation (PCM)
signals: (*a*)
modulating signal;
(*b*) synchronizing
pulse; (*c*)
output-coded
pulses.

group permits transmission of n^m values of the analog, or modulating, signal. Keeping in mind the conclusions of the sampling theorem in Sec. 20.1.1, we see that signals using PCM require a relatively wide-band frequency spectrum. The pulse capacity in bits per second for pulse-modulated systems can be shown to be an n-ary, m-pulse/group code

$$C = mf_{sN} \log_2 n^2 \qquad \text{bits/s} \tag{20.2}$$

where f_{sN} is a sampling rate based on the twice-critical-frequency criterion (the Nyquist rate, defined in Sec. 20.2.1). Shannon has shown that the maximum possible rate of binary pulses is

$$C = B \log_2\left(1 + \frac{S}{N}\right) \qquad \text{bits/s} \tag{20.3}$$

where S/N is the ratio of signal power to noise power in decibels and B is the bandwidth in hertz. The *bandwidth* of PCM is mf_{sN}, as seen from Eq. (20.2).

In PCM, there is an inherent error in the modulation process due to the quantization of the modulating signal, as shown in Fig. 20.10a. This error is often referred to as *quantization noise* and is the major noise component of N in the above equations. Quantization noise is expressed as a function of the noise at the demodulator Ku, where u is the RMS noise at the demodulator input and $N = u^2$. Using this notation, we can show that the pulse capacity of PCM is

$$C = B \log_2\left(1 + \frac{12S}{K^2N}\right) \tag{20.4}$$

In PCM there is a fairly optimum S/N ratio beyond which the quantization error decreases less rapidly. This ratio is about 9.2, or 20 dB. Using this ratio in Eq. (20.4) and typical values of K, we can show that the power S required to transmit a given bit capacity is about 7 times that of the ideal binary system described in Eq. (20.3). However, this is still a more efficient means of pulse data transmission than PAM, PWM, or PFM.

Delta modulation (DM) is a simplified version of PCM in which $n = 1$. The one digit (in contrast to binary or tertiary, etc.) is used to transmit information about the differences of successive samples of the modulating signal. At the receiver, or demodulator, the pulses are integrated to obtain the original signal. DM circuitry is very simple and inexpensive, but DM requires a sampling rate much higher than the Nyquist rate of $2f_c$ and a wider bandwidth than binary PCM.

20.1.5 Signal Processing, Demodulation

Digital systems used for the transmission and processing of intelligence signals generally include a great number of subsystems with many diverse functions. These functions come under the general classification of signal processing and handling. Some of these functions are similar to their analog system counterparts, such as data transmission from one location in space to another and amplification at various intervals during transmission, although there will obviously be differences in the required frequency ranges and in other details of the hardware implementation. Other subsystems devoted to signal shaping and filtering are generally totally different in analog and digital systems. Included in this latter category are the processes of signal modulation and demodulation.

The purpose of demodulation, or decoding a digital signal, is to recover the analog, or modulating, signal which has modulated the carrier pulse train. In PAM and PTM, demodulation generally consists of low-frequency filtering, which filters out the higher frequencies associated with the carrier pulse train and passes the low-frequency analog signal carrying the intelligence. Low-frequency or low-pass filtering is generally acceptable where the interpulse period is relatively large. In many cases, however, it is necessary to modulate on a pulse-by-pulse basis. Figure 20.11a illustrates the block diagram of a PWM demodulation system in which each pulse is processed, first with an integration stage. Signals at various stages in the demodula-

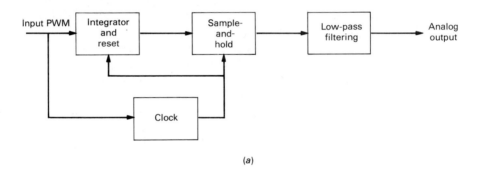

(a)

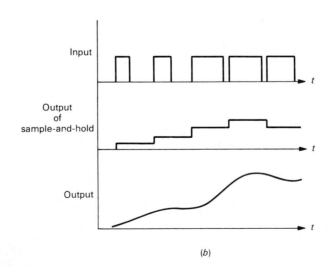

(b)

FIGURE 20.11
(a) Block diagram for PWM demodulations.
(b) Signal waveforms for PWM demodulator.

tion process are shown in Fig. 20.11*b*. The sample-and-hold circuitry sums the pulse integrals and resets the integrator in synchronization with the PWM waveform. The smoothing filters smooth the staircase-type waveform of the hold circuit to form an analog signal.

PCM demodulation, usually called *decoding*, is almost exactly the inverse of the modulation process. Each group of pulses arriving at the receiver or decoder represents an analog level. These pulses are decoded—e.g., by means of a counter—and an analog signal proportional to the count is produced. The demodulation process must be exactly synchronized with the pulse-coding process.

The modulation-demodulation process begins and ends with a signal, generally analog, containing or representing some form of information or intelligence. The use of digital transmission and signal processing is advantageous because of generally faster response times, better immunity from noise, compatibility with all-digital systems of microprocessors and computers, and generally lower hardware costs as compared to analog systems.

20.2   SIGNAL CONVERSION

We have seen in previous sections of this chapter the need for conversion of signals from analog to digital or from digital to analog forms. These processes should not be confused with the modulation-demodulation process discussed in the previous section, in which a signal of one form—usually analog—is used to vary the parameters of the other form—digital, in the cases discussed in the previous section. However, in many applications and in many subsystems, it is desirable to convert a signal in one form to the other form. This was discussed, for example, in Chap. 18 relating to electrical measurements. In many types of measurements, the measured parameter (such as motor speed in revolutions per unit time) is counted, which is a purely digital process, but it is desired to present a readout or display in analog form, or a signal proportional to motor speed is to be used in an analog-type control system. In such cases the count of revolutions per unit time would be converted to a continuous signal, a process known as digital-to-analog conversion, or DAC. However, a more common DAC application is the conversion of a binary number to an analog voltage or other form of continuous signal. The reverse process, analog-to-digital conversion, or ADC, is generally more common than is DAC, since more and more signal and data processing is being performed by digital circuits, whereas a great many sensors of data, including most human processes, are of a generally analog nature. The point at which there is a conversion between analog and digital signals is an example of a system *interface*. This term, *interface*, is, of course, equally applicable to other types of interconnections among subsystems making up a larger system in which digital or analog signals exist on both sides of the interconnection.

20.2.1   Digital-to-Analog Conversion (DAC)

DAC is generally a less complex process than its inverse, ADC, and consists of developing a specific, continuous signal level related to a specific number of incoming

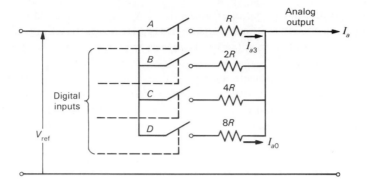

(a)

Decimal	Gate or binary column				I_a, pu
	A	B	C	D	
0	0	0	0	0	0
1	0	0	0	1	0.125
2	0	0	1	0	0.25
3	0	0	1	1	0.375
4	0	1	0	0	0.5
5	0	1	0	1	0.625
6	0	1	1	0	0.75
7	0	1	1	1	0.875
8	1	0	0	0	1.0
9	1	0	0	1	1.125
10	1	0	1	0	1.25
11	1	0	1	1	1.375
12	1	1	0	0	1.5
13	1	1	0	1	1.625
14	1	1	1	0	1.75
15	1	1	1	1	1.875

(b)

FIGURE 20.12 Relationships in a simple, weighted network DAC: (a) schematic circuit diagrams; (b) per-unit (pu) values of analog output (1 pu = V_{ref}/R); (c) analog output waveform.

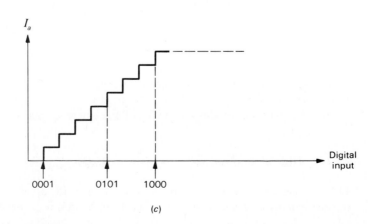

(c)

discrete digital signals. Most commonly, the digital input signals are in binary form, having the discrete values 1 and 0, or high and low, etc., as discussed in Chap 19.

The majority of DACs available today are based on the use of *weighted networks.* A schematic arrangement of a weighted network for DAC application is illustrated in Fig. 20.12*a.* This is a summing network in which the analog output (shown simply as a current here, but usually the voltage across a fixed resistor) is the sum of a number of branch currents, each of which represents a signal proportional to a column in a binary number. This relationship is illustrated in Fig. 20.12*b.* Figure 20.12*c* illustrates the staircase shape of the analog voltage output of a DAC. The operation is as follows: Each binary input channel (or binary number column) is controlled by logic gates (shown in Fig. 20.12*a* as on-off switches), which can be of any of the types discussed in Chap. 19. When a gate is in the on state, the reference voltage is applied to a *weighted resistor* associated with that gate. For example, the analog output current resulting when the gate representing the most significant digit (MSD, the leftmost digit of the binary number) is energized or is in the on state is

$$I_{a3} = \frac{V_{\text{ref}}}{R} \tag{20.5}$$

When the gate for the least significant digit is energized, the output is

$$I_{a0} = \frac{V_{\text{ref}}}{8R} \tag{20.6}$$

The general analog output is the sum of these currents (or a voltage proportional to the sum):

$$I_a = \sum_{i=0}^{n} I_{ai} \tag{20.7}$$

where it is recognized that each of the currents I_{ai} may have one of two discrete values, 0 or $V_{\text{ref}}/[(i+1)R]$. It is seen that for the case depicted in Fig. 20.12, $n = 3$ and represents a binary number of maximum length of four digits, or a decimal number of value up to 15. The decimal number is also the measure of the analog output signal. Note that this signal is a function of the reference voltage V_{ref}, which is usually determined by the voltage level for which the particular logic family is designed. Eight volts is a common logic family voltage level and is shown in Fig. 20.12*b*; and 5-, 10-, and 15-V levels are commonly used in various logic families.

Example 20.1   Prepare a weighted network diagram and determine the levels of analog voltage output for a 6-bit DAC using a 15-V logic family.

Solution   From Fig. 20.12 and Eqs. (20.5) to (20.7) we see that the multiplier for the resistance in the least-significant-digit path is

$$R_0 = 2^n R \tag{20.8}$$

FIGURE 20.13
Example 20.1:
6-bit resistor
network.

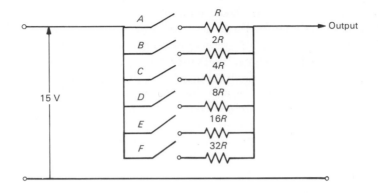

Therefore, for 6-bit conversion, the largest resistor is $2^5(R)$ or $32R$, and the others are in descending binary multiples of R. This is shown in Fig. 20.13. The voltage levels can be reasoned as follows: Starting from zero output voltage, the next step on the staircase must be one thirty-second of the maximum possible voltage and continue in increments of $V_{ref}/32$; therefore, the voltage levels of the output function are 0, 0.46875, 0.9375, ..., 15.0 V.

The disadvantage of the weight network scheme is the large number of precision resistors of many different values required which are costly even where the network is part of an IC chip. The resistance values must be relatively small and the output impedance must be very large in order to minimize the value of the current components I_{ai} and the associated voltage drops through leads and connections. Also variations due to temperature changes are relatively large in a multivalue resistance network.

A resistor system that achieves the same weighting characteristic as the networks of Figs. 20.12 and 20.13 without requiring so many different resistance values is a *ladder network*; an example is shown in Fig. 20.14. In ladder DACs, precision resistors

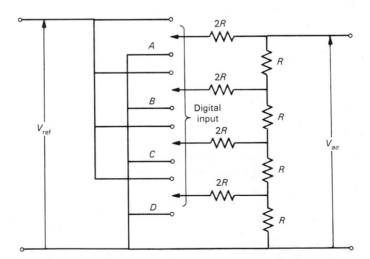

FIGURE 20.14
Ladder network
DAC. The digital
input controls the
operation of
switches
A, B, C, D; V_{a0} is
analog output.

of only two different values—values in the ratio of 2:1—are required. However, it is now necessary to use two different reference values, and a larger number of resistors are required. One of the two reference voltages is zero (in contrast to "NOT V_{ref}," as in Fig. 20.12). This zero reference value is electrically equivalent to grounding the gates associated with the specific binary channels. While this simplifies the hardware associated with achieving the DAC, it does complicate the analysis of the electric circuit. A few problems related to the DAC ladder network realization are included at the end of this chapter for those interested in the analysis of resistive networks.

Now, with a simple example based on a 2-bit ($n = 1$) DAC, we illustrate how such two-value resistor networks achieve binary conversion.

Example 20.2

Determine the relative (or per-unit) output voltage of the 2-bit ladder network shown in Fig. 20.15 as a function of the digital switch (gate) settings. Switch A represents the most significant digit (MSD), B the least significant. To produce an analog output voltage V_a for a single digital channel, the switch for that channel is in the up position of Figs. 20.14 and 20.15; all other switches are in the down position, which connects those circuits to the system ground.

Solution From the general relations given above, a 2-bit or two-channel DAC can provide three different analog output signal levels, representing the binary numbers 01, 10, and 11 and

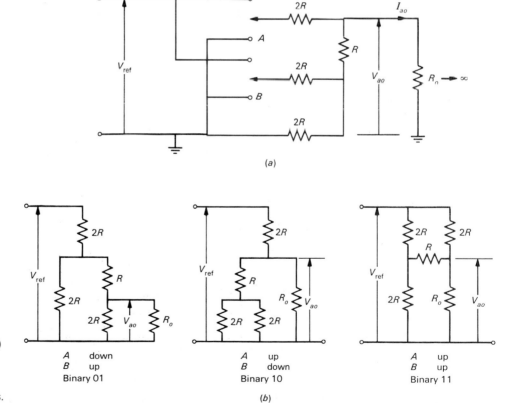

FIGURE 20.15
Example 20.2:
2-bit network: (*a*)
circuit diagram of
2-bit ladder net-
work; (*b*)
equivalent circuits.

the decimal numbers 1, 2, and 3. The first of these is with *A* down and *B* up; the second number, binary 10 or decimal 2, results from *A* up and *B* down; the third, binary 11 and decimal 3, is with both *A* and *B* up. The three networks that result from these three switch settings are shown in Fig. 20.15*b*. From simple network analysis the three resulting values of V_a are, respectively, $V_{ref}/4$, $V_{ref}/2$, and $3V_{ref}/4$. These are in the proper decimal relationship of 1:2:3 and represent the first three steps of the staircase analog output signal.

Since only two values of resistance are required in the ladder DAC, the system characteristics are more stable with temperature variations. Also the ladder configuration has a constant output resistance, the proof of which is left for a problem at the end of this chapter. The digital input circuits of a DAC are often impedance-matched to the external logic circuits by means of operational amplifiers or *RC* networks or time-matched by means of a register, a process called *buffering*.

To introduce a practical DAC and some of the parameters associated with ICs, Fig. 20.16 illustrates a specific DAC embodiment and circuit layout. Table 20.1 describes a series of DACs available in the 1990 marketplace and lists some of the nomenclature of importance in many ICs. The TDC1112 (Fig. 20.16) is a weighted type (see Fig. 20.12) of DAC capable of accepting a data rate of up to 50 megasamples per second—shown as megahertz in Table 20.1. It has 12-bit resolution which gives it high linearity (the difference between the mean analog signal and the "bumps" in the digital steps, as shown in Fig. 20.12). The 12 digital inputs are shown as terminals *D*1 to *D*12. The input registers of buffers are built into the chip and eliminate "glitches" (voltage spikes) and data skew that may result from the use of external registers. Two types of chip "packages" are illustrated in Fig. 20.16, and several others are also listed in Table 20.1 (see Sec. 19.6). The package and its heat sink and the internal chip circuitry influence the temperature rating of the package (Table 20.1). The rise time is primarily a function of the inductances of the tiny lead lengths on the chip. The TDC1112 is constructed of ECL (see Sec. 19.6).

20.2.2 Analog-to-Digital Conversion (ADC)

Analog-to-digital conversion is generally more complex than DAC, and there is a wider variety of circuit configurations and techniques, each with its own distinct merits. The most common as well as one of the simplest configurations is known as the *method of successive approximations*. Figure 20.17 depicts a block-diagram representation of this ADC configuration. This method requires a comparator, a DAC, and a clock-controlled counter. The inputs to the comparator are the analog voltage and the output of the DAC. When the analog input V_a equals the output of the DAC (i.e., when the two comparator inputs are equal), the comparator stops the clock and the counter. At this instant, the input to the DAC is the digital output of the counter, which is the digital equivalent of V_a. This is the output of the ADC system and represents the digital value of the analog input V_a. A simple up counter is acceptable, but because it has to count through the full range of counts for every negative increment of the input V_a, it is therefore slow to track decreasing analog

Functional Block Diagram

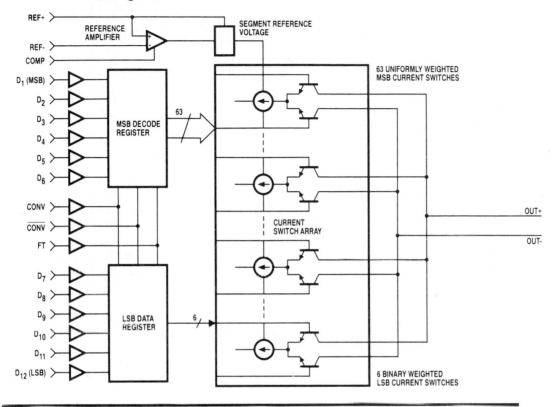

Pin Assignments

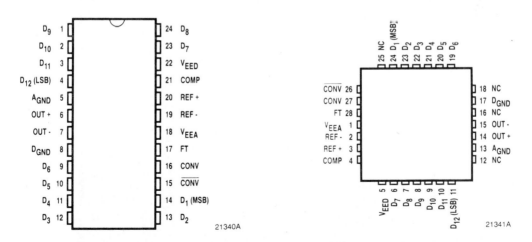

24 Pin Hermetic Ceramic DIP — J7 Package
24 Pin Plastic DIP — N7 Package

28 Contact Chip Carrier — C3 Package
28 Leaded Plastic Chip Carrier — R3 Package

FIGURE 20.16   Block diagram and packaging for the TDC 1112 DAC. (Courtesy of TRW Products.)

TABLE 20.1 Typical Commercial DACs (Courtesy TRW LSI Products, Inc)

Product	Bits	Differential Linearity Error, ±%	Data Rate, MHz	Rise Time, ns	Package	Package Designator	Available Temp/Testing*	Comments
TDC1034	4	0.80	200	2	18-pin DIP	B8	C	Low cost, ECL graphics ready.
TDC1334	Triple 4	0.80	200	2	28-pin DIP	B6	C	Low cost, ECL graphics ready.
TMC0171	Triple 6	0.78	35	8	44-lead PLCC 28-pin DIP	R2 N6	C C	RAMDAC. 256 × 18 lookup table. INMOS 171 compatible.
TDC1018	8	0.20	200	1.7	24-pin DIP 28-contact CC	B7 C3	C C	Low cost, ECL graphics ready.
TDC1318	Triple 8	0.20	200	2	40-pin DIP	B5	C	Low cost, ECL graphics ready.
TDC1016-10	10	0.05	20^5	5.5	24-pin DIP 40-pin DIP	N7 N5	C C	TTL and ECL compatible.
−9	10	0.10	20^5	5.5	24-pin DIP 40-pin DIP	B7, N7 B5, N5	C, A C, A	TTL and ECL compatible.
−8	10	0.20	20^5	5.5	24-pin DIP 40-pin DIP	B7, N7 B5, N5	C, A C, A	TTL and ECL compatible
TDC1041-1	10	0.048	20	4	28-lead PLCC	R3	C	Low cost, 10-bit video DAC.
TDC1041	10	0.096	20	4	28-lead PLCC	R3	C	Very low cost video DAC.
TDC1012-3	12	0.012	20	4	24-pin DIP	J7, N7, R3	C	Lowest glitch, TTL compatible.
−2	12	0.024	20	4	24-pin DIP	J7, N7, R3	C, V	Ideally suited for signal synthesis.
−1	12	0.048	20	4	24-pin DIP	J7, N7, R3	C, V	−70 dBc spurs. Drives 50 Ω directly.
TD1012	12	0.096	20	4	24-pin DIP	J7, N7	C, V	
TDC1112-3	12	0.012	50	4	24-pin DIP	J7, N7, R3	C	Lowest glitch, ECL compatible
−2	12	0.024	50	4	24-pin DIP	J7, N7, R3	C, V	Ideally suited for signal synthesis.
−1	12	0.048	50	4	24-pin DIP	J7, N7, R3	C, V	−70 dBc spurs Drives 50 Ω directly.
TDC112	12	0.096	50	4	24-pin DIP	J7, N7	C, V	

* C = Commercial; T_A = 0 to 70°C; A = high reliability; T_C = −55 to 125°C; V = MIL-STD-883 compliant; I_C = −55 to 125°C.

FIGURE 20.17
Example 20.3:
block diagram of
ADC based upon
method of
successive
approximations.

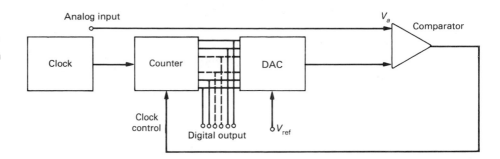

signals. An up-down counter is preferred; this type of counter can operate at clock rates of 1 MHz and above in practical ADCs. To further illustrate the operation of this type of ADC, we introduce some numerical relationships in the following example.

Example 20.3

An ADC based on the circuit shown in Fig. 20.17 has an analog input that has a value of $+6$ V at a given instant; the counter is a binary 6-bit counter (see Chap. 19) which develops a value of 0.175 V per count (often called the counter *resolution*). Determine the digital output for this 6-V input and the time required for the counter to arrive at this input value if the clock frequency is 1 MHz.

Solution The clock will stop when the output of the DAC equals the analog input V_a, which is 6 V. This will occur after $6/0.175 = 34.286$, or 35, counts. Therefore, the output (which is actually the input to the DAC) is the binary equivalent of 35, or 100011. The time required for the count up to 35 is $35 \times 1 \times 10^{-6}$, or 35 μs. Note that the accuracy in this particular example is $34.286/35$, or 98 percent. By using counters with smaller resolution and amplification of the digital output, accuracies of 99.9 percent are realizable.

20.3 DATA ACQUISITION

A major application of the vast array of digital circuitry now available is that of acquiring and processing measurements made on practically all types of technical processes. Digital circuitry and instrumentation are available for evaluation of almost any measurement conceivable, whether it be in outer space or within a vital human organ or in very mundane environments such as steam electric power plants. Both the measuring function and the data processing function can be accomplished by a wide variety of both analog and digital circuitry at present. However, digital circuitry appears to offer more advantages—particularly economic advantages—owing to the rapid development of IC chips. Also the significant developments in digital display techniques and the proliferation of mini- or microcomputers to almost every corner of industrialized society have further fostered the use of digital circuits for data acquisition and processing. However, the availability of a wide variety of low-cost ADC chips, such as discussed in the previous section, has resulted in a continuation of analog measurements in a great variety of applications.

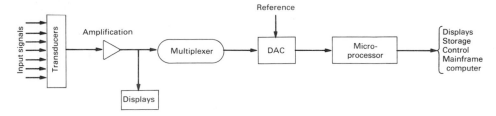

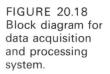

FIGURE 20.18
Block diagram for
data acquisition
and processing
system.

Figure 20.18 illustrates a very simplified block diagram of a hybrid data acquisition system, with analog sensing of the measured signals and conversion to digital form for data processing, storage, and possible use in control and further computer manipulation. The measured signals include temperatures, speeds, voltages, currents, powers, and such mechanical parameters as strain, pressure, velocity, and position, as well as most other types of measurements (chemical, optical, acoustical, etc.). It is often desirable to display the measurements in their "raw" analog form, as shown at the lower left of Fig. 20.18. For transmission to remote locations, the signals not only are amplified but also frequently require multiplexing (as shown). They can be transmitted at either the analog or digital stage, and many transmission systems can handle either type of signal, but usually at some stage the signals are converted to digital form by means of ADCs. The timesharing at the multiplexer stage permits the processing of a large quantity of information with a high degree of accuracy. Most present-day data acquisition systems are designed to be compatible with microprocessors or other digital systems for processing the measured data.

The output of the ADC stage may be used in a variety of applications. The most obvious are those that truly process the information obtained. For example, assume that a data acquisition system such as illustrated in Fig. 20.18 is used to measure the parameters of electric motors being manufactured on a major motor manufacturer's assembly line. The parameters measured, as a rule, would be those specified by the standards agencies, such as the National Electrical Manufacturers Association (NEMA). Thus, for an induction motor, the measurements might include the voltage, current, and power in all phase windings; the speed; the temperature at selected motor regions; the output torque or power; the response to high-potential insulation tests; and perhaps the mechanical vibration at the no-load and full-load conditions. Many of these measurements might be displayed at the instant of measurement on analog or digital meters. All measurement data would be sent on through the amplification, multiplexing, and ADC stages to a microprocessor. Typically the microprocessor processes these data by calculating such parameters as average voltages and currents, RMS voltages and currents, average power and torque and efficiencies and slip; comparing the readings of a given motor with previous motors, establishing trends, and noting unusual changes in any parameters; and recording and storing details about the machine rating, the various verifications of the manufacturing process made along each motor's assembly stages, and perhaps even such marketing data as cost and type of customer. As indicated in Fig. 20.18, the output of the microprocessor can be used in a variety of ways. The obvious output is some form of data storage—a printout of all the data processing calculation results, e.g., or storage on magnetic tape or magnetic disks. But in many cases the micro-processor output is used in a more dynamic mode, such as to alert manufacturing personnel about a deterioration in certain specific materials, tolerances, or other manufacturing processes—perhaps even to control modifications and changes in materials or manufacturing processes in an adaptive manner—or to transmit all the measured data to a larger computer system for subsequent analysis of long-term corporate trends in costs, sales potential, energy improvement, and so forth.

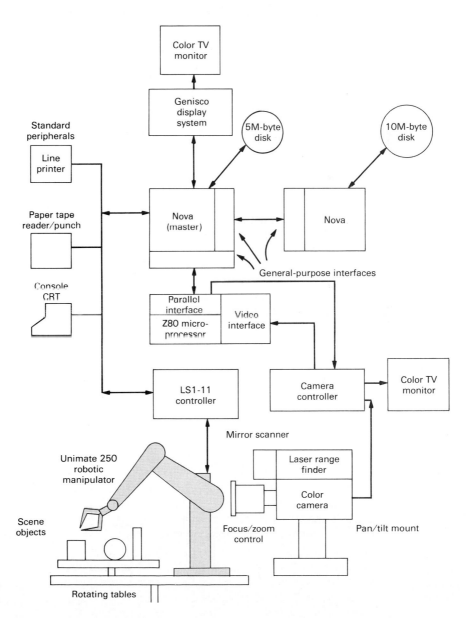

FIGURE 20.19 Visual data acquisition used in robot control. (Courtesy of Unimation Div., Westinghouse Corp.)

One of the most interesting applications of data acquisition systems is where the output of the microprocessor in Fig. 20.18 is fed into a robot or other type of automatic processor. The sensors or transducers for robot systems tend to be of the visual type—sensing position, distance, three-dimensional location, color, etc.—rather than of the more prosaic type that indicates such parameters as voltage, current, or temperature. Figure 20.19 illustrates a very generalized block diagram of a robot information acquisition system.

20.4   ARITHMETIC DIGITAL SYSTEMS

One of the earliest—and still one of the principal—functions of digital circuits is to perform arithmetic calculations. Computer arithmetic is performed in the binary system or in alternative binary codes related to the binary system, a few of which are discussed in Sec. 20.4.5. The purpose of the following discussion is to illustrate the use of digital circuits to perform arithmetic operations. The reader may wish to review binary arithmetic before proceeding with this topic.

20.4.1   Addition

Addition in any number system results in two classes of sums: one with a *carry* to the next column, such as $6 + 8$ in decimal, and one without a carry. In binary, there is only one type in each class: $1 + 1$, which has a carry of 1 and a *sum* of 0, and $0 + 1$, which has no carry and a sum of 1. Two digital circuit configurations are required to perform these two classes of summations; these configurations are known as the *full adder* and the *half-adder*, respectively. The complete binary adder for adding two *n*-bit numbers consists of one half-adder and $n - 1$ full adders. Figure 20.20 shows the truth table and basic schematic circuit diagram for a half-adder; Fig. 20.21 illustrates the same for the full adder. The half-adder accepts as inputs only the two numbers to be added, X and Y, whereas the inputs to the full adder are X, Y, and C_{in}, where C_{in} represents the carry input. To perform binary addition, the half-adder is used to add the *least significant digits* of X and Y, since there will never be a carry input with these digits. We denote the least significant digits by X_0 and Y_0. The carry output of the half-adders is fed into a full-adder stage which also has the inputs X_1 and Y_1, that is, the digit in the second column from the right in each number; the carry from this stage is fed into the third stage, a full adder, which also has the digits X_2 and Y_2 as inputs; and so on. This process is illustrated in

FIGURE 20.20
Binary half-adder:
(*a*) truth table; (*b*)
circuit diagram.

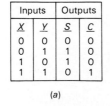

Inputs		Outputs	
X	*Y*	*S*	*C*
0	0	0	0
0	1	1	0
1	0	1	0
1	1	0	1

(*a*)

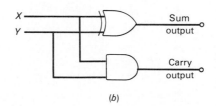

(*b*)

Inputs			Outputs	
C_{in}	X	Y	S	C_{out}
0	0	0	0	0
0	0	1	1	0
0	1	0	1	0
0	1	1	0	1
1	0	0	1	0
1	0	1	0	1
1	1	0	0	1
1	1	1	1	1

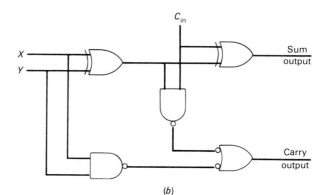

(a)
(b)

FIGURE 20.21
Binary full adder:
(*a*) truth table;
(*b*) circuit diagram.

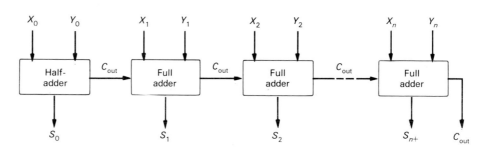

FIGURE 20.22
Block diagram for
addition of *n*-digit
binary numbers.

block-diagram form in Fig. 20.22, where each block represents the complete circuit as shown in either Fig. 20.20 or Fig. 20.21. The sum output of each stage is read out and represents the arithmetic sum for that column of digits.

Example 20.4 Develop an adder system to add the numbers 10 and 14. Show the system in block-diagram form, and show the readout of each stage.

Solution Converting 10 and 14 to binary numbers gives 1010 and 1110, respectively. These are 4-bit numbers and will require four stages of adders—one half-adder and three full adders. These are shown in Fig. 20.23 along with the inputs, sums, and carries of each stage.

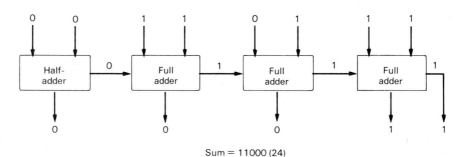

FIGURE 20.23
Example 20.4:
block diagram and
numerical values.

Sum = 11000 (24)

By developing Karnaugh maps from the truth table of the full adder (Fig. 20.21), we can find the boolean equations expressing the sum and carry outputs. These are, respectively,

$$S = X \oplus Y \oplus C_{in} \tag{20.9}$$

$$C_{out} = XY + C_{in}(X + Y) \tag{20.10}$$

20.4.2 Subtraction

Finding the difference of two binary numbers X and Y is quite similar to finding their sum. Two similar digital circuits are used, the half-subtracter for the least significant digit and the full subtracter for the remaining digits of the binary numbers. The circuit configuration for the half-subtracter is shown in Fig. 20.24; the truth table and circuit for the full subtracter are shown in Fig. 20.25. In these figures, B stands for *borrow*, and B_{out} represents a number borrowed from the column to the left when required. The truth table of Fig. 20.25 is for the difference $X - Y$. The least-significant-digit difference, and hence the half-adder, requires no B_{in}. Note that when borrowing is required in the binary system, the difference performed is $10 - 1$, giving a difference D of 1 and a borrow out B_{out} of 1.

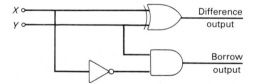

FIGURE 20.24
Binary
half-subtractor.

Inputs			Outputs	
B_{in}	X	Y	D	B_{out}
0	0	0	0	0
0	0	1	1	1
0	1	0	1	0
0	1	1	0	0
1	0	0	1	1
1	0	1	0	1
1	1	0	0	0
1	1	1	1	1

(a)

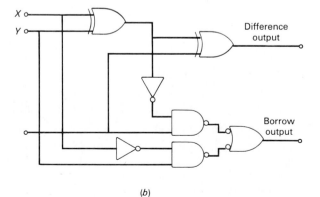

(b)

FIGURE 20.25
Binary full
subtractor: (*a*)
truth table; (*b*)
circuit diagram.

Example 20.5   As a review, subtract the number 9 from 22 in binary arithmetic.

Solution   Converting 9 and 22 to binary gives 1001 and 10110, respectively. The subtraction is performed as follows:

$$
\begin{array}{r}
10110 \\
1001 \\
\hline
1101
\end{array}
$$

In the rightmost column, a 1 is borrowed from the second-from-the-right column, giving $10 - 1 = 1$. In the second-from-the-right column, the subtraction is $0 - 0 = 0$ (since the 1 has been borrowed). The third-from-the-right column is $1 - 0 = 1$, with no borrow required. In the fourth-from-the-right column, the subtraction is again $10 - 1 = 1$, with a 1 borrowed from the fifth-from-the-right column. The binary answer is 1101, which is decimal 13, the correct answer.

The implementation of the subtraction process with electronic logic is illustrated in the next example.

Example 20.6   Show in block-diagram form a system to subtract 10 from 14, and show the resulting inputs, outputs, and borrows of each stage.

Solution   These are the same two binary numbers used in Example 20.4 and require a four-stage subtractor. This is shown in Fig. 20.26 along with the input, output, and borrow values.

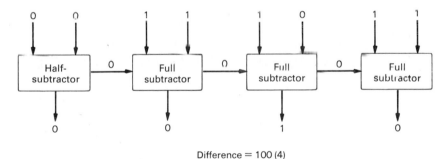

FIGURE 20.26
Example 20.6:
block diagram and
numerical values.

Difference = 100 (4)

20.4.3 Signed Integers and Word Length

The process of subtraction introduces the concept of negative numbers and how they can be handled in digital circuits. Obviously, it is important to be able to perform addition, subtraction, multiplication, and division using negative numbers or mixtures of negative and positive numbers in calculators, computers, and similar digital systems. For instance, if the number Y were subtracted from the number X in Example 20.6, we would get a negative number, in this case -4. But from examination of the block diagram of Fig. 20.26, we see that if X and Y were interchanged, there would be a borrow output from the right subtractor, which has no meaning, and the number represented by the four difference (D) signals would be binary 1100, or decimal 12, which is also meaningless. One method used on some computers to solve this dilemma is to use another digit to represent the sign of the number. For the case just discussed, where we want to find $10 - 14$, the subtraction process would be performed just as in Example 20.6 and Fig. 20.26, but another digit would be carried along with both numbers (X and Y) and with the difference to indicate their sign, usually 0 for positive and 1 for negative.

In computer-type systems, the range of numbers that can be handled in arithmetic calculations and other mathematical manipulation and processing is a function of word length. The *word length* is the number of digits or bits making up the maximum binary number that can be processed with the hardware from which the computing system is constructed. Specifically, the temporary storage devices or registers in a computing system are sized exactly according to word length. Thus, an 8-bit word length requires an 8-bit register for storing 8 binary digits. An n-digit binary number can represent up to 2^n decimal numbers; an 8-bit word length can therefore process positive decimal numbers up to $256 - 1$ (since 0 is one of the numbers), or 255. It is customary to use the most significant bit (MSB) of the word to represent the sign of a number, a process called the *signed-modulus* method for representing negative numbers. Thus, in the previous examples the number X, decimal 14 or binary 1110, would be stored in an 8-bit register as 00001110; the decimal number -14 would be stored as 10001110.

There are several attributes of the signed-modulus method of sign representation that are not desirable. First, using one digit to represent sign reduces the number of decimal numbers considerably (from 255 to 127 in the case of 8-bit word length), although the range of numbers is essentially the same (from -127 to $+127$). Second, a signed-modulus positive number is not the additive inverse of the negative number of the same value; i.e., the two do not add up to zero. Also the signed-modulus numbering system results in the rather strange situation of having two zeros—a negative zero and a positive zero. A method which overcomes some of these disadvantages of signed-modulus notation and which is more widely used in computer systems is known as 2's complement notation.

The 2's complement method of forming signed binary numbers is somewhat analogous in methodology to the 9's complement method used in decimal number subtraction. To form a decimal 9's complement, each digit in the decimal number is replaced by a number found by subtracting the digit from 9. To form a 2's complement in binary, each digit of the binary number is replaced by a number found by subtracting the digit from 1. This has the effect of replacing all 0s by 1s and all 1s by 0s. The replacement number is called the *complement* of the original number. *Negative numbers in 2's complement notation are found by adding 1 to the complement of the positive number.* Thus, to determine the binary representation of decimal -14, the normal binary representation of 14 is first found. For an 8-bit word length, this is 00001110; the complement is 11110001; adding 1 gives 11110010, which is the 2's complement notation for decimal -14. One representation for zero is used. Table

TABLE 20.2 2's Complement Notation	
Binary	**Decimal Equivalent**
00001010	10
00001001	9
00001000	8
00000111	7
00000110	6
00000101	5
00000100	4
00000011	3
00000010	2
00000001	1
00000000	0
11111111	-1
11111110	-2
11111101	-3
11111100	-4
11111011	-5
11111010	-6
11111001	-7
11111000	-8
11110111	-9
11110110	-10

20.2 shows 2's complement notation for decimal numbers between -10 and $+10$ in an 8-bit word length system. We see that the negative number is the additive inverse of the positive number (the carry from the MSB is thrown away when adding inverses). The range of number representation is about the same as for the signed-modulus method, $2^{n-1} - 1$ on the positive side and 2^{n-1} on the negative side. For an 8-bit word length, this gives a range of $+127$ to -128, which is 256 when zero is included.

Example 20.7 Find the 2's complement representation of -31, assuming an 8-bit word length.

Solution Positive 31 has a binary representation of 00011111. The complement is 11100000. Adding 1 gives 11100001.

Example 20.8 Subtract 14 from 10, using 2's complement notation for an 8-bit word length.

Solution This is equivalent to adding -14 to 10. Negative 14 was found to be 11110010 above; adding this to positive 10 gives

$$
\begin{array}{r}
11110010 \\
00001010 \\
\hline
11111100
\end{array}
$$

which is seen to be -4 in Table 20.2.

Obviously, with finite word lengths in a computing system, adding or subtracting can result in numbers larger than can be represented by the word length. This causes a condition known as *overflow* when the number is larger than the positive range of allowable numbers and *underflow* when the number is smaller than the negative range. For example, in a system with an 8-bit word length, adding the decimal numbers 100 and 50 gives decimal 150, which is beyond the allowable range of positive numbers. In 2's complement notation, this would result in $01100100 + 00110010 = 10010110$. This is equivalent to the decimal number -106, which is, of course, a ridiculous answer. To detect such situations, the computing system must contain circuitry that indicates when an overflow or underflow exists. This circuitry is based on the logic that goes along with the mathematical process. In the process of addition, e.g.,

1   If the two numbers being added are of opposite sign, there can be no overflow.

2   If two positive numbers are being added and the sum is negative, an overflow condition exists (as illustrated above).

3   If two negative numbers are being added and the sum is positive, an underflow condition exists.

Thus, overflow-underflow detection requires detection of the MSB (MSB is 1 or is 0) of the sum and the sign of the numbers to be added. The design of a logic circuit to detect these conditions is given as a problem at the end of this chapter. Overflow or underflow can occur during subtraction, too, and similar logic relations can be developed for such conditions.

20.4.4   Additional Add/Subtract Circuits

The adders shown in Figs. 20.20 to 20.23 are parallel adders since the inputs to each stage are processed nearly simultaneously. However, a finite time is required for each carry to propagate from one stage to the next. An adder circuit known as the *look-ahead carry* anticipates whether there is to be a carry across all the stages. The actual addition is performed by the same hardware as shown in Figs. 20.20 and 20.21. The carry anticipation circuitry adds considerable circuit complexity to an adder and increases its cost, but it greatly increases the response time of the adder. Separate look-ahead carry ICs are also available to add to existing adders to improve their response time. The carry anticipation circuitry is based on an implementation of Eq. (20.10); the carry in the ith stage can be written

$$C_i = G_i + P_i C_{i-1} \tag{20.11}$$

where $G_i = X_i Y_i =$ the input to the ith stage, called the *generate term*. In Fig. 20.27, $P_i = A_i B_i$ and is called the *propagate term*; $C_{i-1} = C_{in}$ for the ith stage. The carry anticipation circuitry requires the inputs from each stage and the initial carry C_o (as shown) and from this information determines almost instantaneously what the

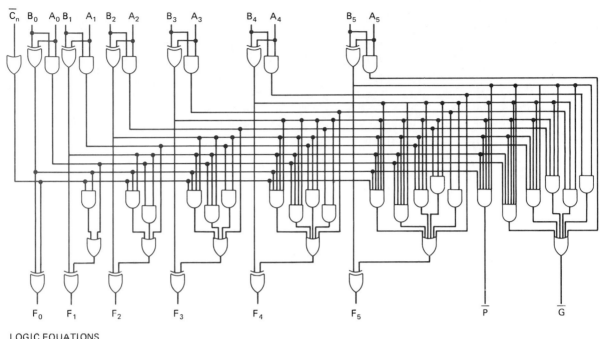

LOGIC EQUATIONS
$P_1 = A_1 \oplus B_1$
$G_1 = A_1 B_1$
$I = 0, 1, 2, 3, 4, 5$
$F_0 = P_0 \oplus C_n$
$F_1 = P_1 \oplus (G_0 + P_0 C_n)$
$F_2 = P_2 \oplus (G_1 + P_1 G_0 + P_1 P_0 C_n)$
$F_3 = P_3 \oplus (G_2 + P_2 G_1 + P_2 P_1 G_0 + P_2 P_1 P_0 C_n)$
$F_4 = P_4 \oplus (G_3 + P_3 G_2 + P_3 P_2 G_1 + P_3 P_2 P_1 G_0 + P_3 P_2 P_1 P_0 C_n)$
$F_5 = P_5 \oplus (G_4 + P_4 G_3 + P_4 P_3 G_2 + P_4 P_3 P_2 G_1 + P_4 P_3 P_2 P_1 G_0 + P_4 P_3 P_2 P_1 P_0 C_n)$
$\overline{P} = \overline{P_0 P_1 P_2 P_3 P_4 P_5}$
$\overline{G} = \overline{G_5 + P_5 G_4 + P_5 P_4 G_3 + P_5 P_4 P_3 G_2 + P_5 P_4 P_3 P_2 G_1 + P_5 P_4 P_3 P_2 P_1 G_0}$

FIGURE 20.27 Schematic diagram for a 6-bit adder using look-ahead carry.
(Courtesy of Signetics Co., Sunnyvale, CA.).

carry from each stage to the next will be. Figure 20.27 illustrates the logic diagram
for a 6-bit adder that uses look-ahead carry. This logic (also ECL) is that for the
Signetics 100180 high-speed adder. It performs full 6-bit addition of two operands
in 2 ns. The inputs are carrying (C_N) and operands A (A_n) and B (B_n); the outputs
are function (F_n), carry generate (G), and carry propagate (P).

We can see from Figs. 20.20 to 20.26 that the hardware implementation of adders
and subtractors is very similar. It is possible to use the basic full-adder circuit to
perform both addition and subtraction operations with very little addition of
hardware. Such combined circuits are called, not surprisingly, *adder/subtractors*.
Figure 20.28 illustrates the basic elements of a 4-bit adder/subtractor circuit. The
input to the C_o terminal determines the arithmetic mode of the circuit. When C_o is
zero, the circuit performs as an adder, since a zero on the left terminals of the XORs

causes the XORs to operate as normal OR gates, and the inputs on the Y terminals are unchanged. The initial carry is 0 in this case. When the input to the C_o terminal is 1, the circuit operates in the subtraction mode as follows: Placing a 1 on the left terminals of the XORs causes the complement of the digits of Y to be inputted to the adders; a 1 from C_o is added to the complement, giving the 2's complement notation negative value of Y. Thus, negative Y is added to X, and this addition is, of course, the difference $X - Y$. The digits shown in Fig. 20.28 are for the values $X = 1001$ (9) and $Y = 0101$ (5). The addition process forms the sum $1001 + 0101 = 1110$ (14). The subtraction process first forms the 2s complement negative of Y, which is $1010 + 1 = 1011$; this is added to X, giving 10100; the left digit of this number is the carry out and appears on the terminal $C4$ in Fig. 20.28 and is ignored. The remaining four digits appear on the S terminals and are 0100; this is binary 4 and is the correct answer.

20.4.5 Alternative Binary Codes

In various types of digital circuits, alternative representations of binary numbers are used. These *binary codes* are usually used to simplify hardware or software or both, or to make more efficient use of the word-length limitations. Codes are also used to improve or simplify the human-computer interface, since binary representation is not the most suitable form for writing a computer program or feeding in data to a computer.

The *Gray code*, used in many computer systems and digital circuits (DCs), is listed here primarily for references purposes. In this code, an alternative to conventional binary counting, only one digit (or bit) changes between successive numbers. This feature simplifies the logic and hence the electronic circuitry in some applications.

The *octal code* (*base* 8) is used to reduce the problems of dealing with long strings of binary 1s and 0s. It is a shorthand for replacing three binary digits with a

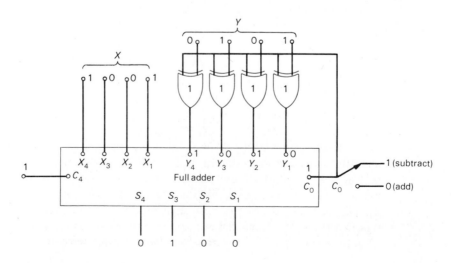

FIGURE 20.28
Block diagram for adder/subtractor.

single octal number from 0 to 7. For example, a 6-bit binary word is split into two groups of three; thus, 100010 is represented in octal as 42 and is written as 42_8. The octal number 576_8 is equal to 101111110 in binary. Octal numbering follows the rules for conventional numbering in base 8 and is not listed here.

Hexadecimal coding is widely used in computer systems with 8- and 16-bit word lengths. Hexadecimal numbering, shown in Table 20.3, uses the decimal digits plus the first six letters of the alphabet. In hexadecimal coding, binary numbers are divided into groups of four binary digits, and each group of 4 is assigned a hexadecimal digit. For example, binary 11000101 becomes (from Table 20.3) C5 and is often written $C5_{16}$ because it is essentially a base-16 numbering system with some new definitions of numbers from 10 to 15. Conversely, $A2F_{16}$ is converted to binary as 101000101111.

The *binary-coded decimal (BCD) system* codes each digit of *decimal* numbers with its *binary* equivalent. For example, the BCD equivalent of 65 is 01100101. Although this is a type of binary notation for decimal numbers, it is not the same as conventional binary counting, as seen by converting decimal 65 to the conventional binary number which is 1000001. The decimal number 472 becomes 010001110010 in BCD. Likewise, BCD 100101100101 equals 965 in decimal. A number in BCD code is much easier to produce than the pure binary equivalent of a decimal number, which requires the tedious process of continuous divide-by-2. The reverse process is also much simpler; e.g., the super-long binary string 0110001110000110110 is converted to BCD by splitting it up into groups of four binary digits (starting on the right) and assigning the decimal equivalent number to each group, which gives 53836. On the other hand, BCD "wastes" a lot of digits since four binary digits can count up to decimal 15, whereas BCD codes only the decimal digits from 0 to 9. Thus, as we have seen, conventional binary 8-bit words cover a range of decimal numbers

TABLE 20.3 The Gray Code and Hexadecimal

Decimal Number	Binary Number	Gray Code Number	Hexadecimal
0	0000	0000	0
1	0001	0001	1
2	0010	0011	2
3	0011	0010	3
4	0100	0110	4
5	0101	0111	5
6	0110	0101	6
7	0111	0100	7
8	1000	1100	8
9	1001	1101	9
10	1010	1111	A
11	1011	1110	B
12	1100	1010	C
13	1101	1011	D
14	1110	1001	E
15	1111	1000	F

from 0 to 255, whereas an 8-bit word in BCD can cover a range of only 0 to 99. However, BCD is widely used in many computing systems, particularly calculators. Note that this waste of digits does not occur in octal and hexadecimal coding.

BCD adders and subtractors must differ in circuitry from conventional binary systems (such as illustrated in Figs. 20.20 to 20.28) owing to the differences in the meaning of digit magnitude in the two systems. For example, 110111 in conventional binary represents decimal 55; 00110111 in BCD represents decimal 37. An adder/subtractor must be able to detect this difference in the meaning of the digits. The basic BCD adder can be constructed from the same circuit elements as used in the basic binary adder, but logic must be added to the BCD adder to cause the sum to be presented in BCD format rather than conventional binary. This is illustrated by the 4-bit BCD adder in Fig. 20.29. The full-adder blocks in this diagram are the conventional binary full adders of Fig. 20.21. The function of the carry-detection circuit in Fig. 20.29 is to detect when a sum of the upper full adder is greater than decimal 9 or binary 1001. Thus, the operation of a BCD basic adder is as follows: One full adder accepts the sum of two BCD numbers and a carry from the addition of BCD digits of lower significance; the output from this adder of the three most significant digits is supplied to the carry-detection circuitry; if the values of these three digits indicate a sum greater than decimal 9 (note that the least significant digit doesn't matter in this decision), the lower full adder is modified to give the correct BCD summation; note that the lower-adder carry C_4 represents the left digit of summation (1 or 0001 in Fig. 20.29) and is applied to the next stage of BCD digit addition, if such exists. If the summation from the upper full adder is decimal 9 or less, there is no modification of the lower full adder and the summing process is equivalent to conventional binary summation. In Fig. 20.29, the BCD representations

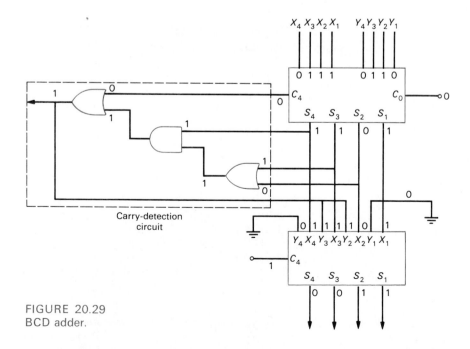

FIGURE 20.29
BCD adder.

of decimal 7 (0111) and decimal 6 (0110) are being added. A conventional binary adder gives the sum 1101 (13), which is correct, of course. But the BCD representation of decimal 13 is 00010011. The carry-detection circuitry of a BCD adder accomplishes this change in output, as shown in Fig. 20.29.

20.4.6 Multiplication

There are many different circuit configurations used to perform binary multiplication. These are generally based on bit-by-bit multiplication and storing with the proper

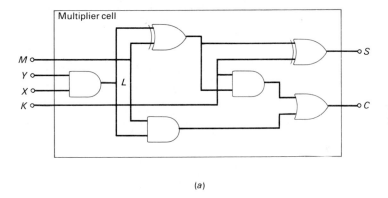

(a)

LOGIC DIAGRAM

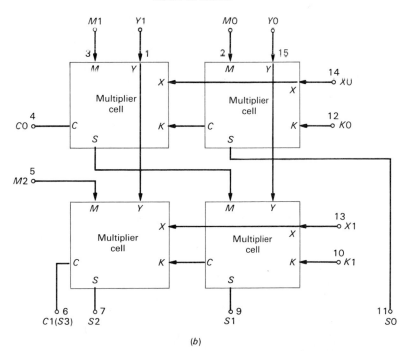

FIGURE 20.30 CMOS multiplier circuit and specifications: (a) circuit diagram of basic 1-bit multiplier; (b) 2-bit multiplier using basic circuit of a.

(b)

EQUATIONS

$S = (X \times Y) + K + M$

Where:

$\times$ Means Arithmetic Times.

$+$ Means Arithmetic Plus.

$S = S3\ S2\ S1\ S0,\ X = X1\ X0,\ Y = Y1\ Y0.$

$K = K1\ K0,\ M = M1\ M0$ (binary numbers).

Example:

Given: $X = 2(10),\ Y = 3(11)$

$K = 1(01),\ M = 2(10)$

Then: $S = (2 \times 3) + 1 + 2 = 9$

$S = (10 \times 11) + 01 + 10 = 1001$

Note: $C0$ connected to $M2$ for this size multiplier.

See general expansion diagram for other size multipliers.

(c)

2-BIT BY 2-BIT PARALLEL BINARY MULTIPLIER

The MC14554B 2 x 2-bit parallel binary multiplier is constructed with complementary MOS (CMOS) enhancement-mode devices. The multiplier can perform the multiplication of two binary numbers and simultaneously add two other binary numbers to the product. The MC14554B has two multiplicand inputs ($X0$ and $X1$), two multiplier inputs ($Y0$ and $Y1$), five cascading or adding inputs ($K0$, $K1$, $M0$, $M1$, and $M2$), and five sum and carry outputs ($S0$, $S1$, $S2$, $C1$ [$S3$], and $C0$). The basic multiplier can be expanded into a straightforward m-bit–by–n-bit parallel multiplier without additional logic elements.

Application areas include arithmetic processing (multiplying/adding, obtaining square roots, polynomial evaluation, obtaining reciprocals, and dividing), fast Fourier transform processing, digital filtering, communications (convolution and correlation), and process and machine controls.

- Diode protection on all inputs
- All outputs buffered
- Quiescent current = 5.0 nA typical @ 5 V dc
- Straightforward m-bit–by–n-bit expansion
- No additional logic elements needed for expansion
- Multiplies and adds simultaneously
- Positive logic design
- Supply voltage range = 3.0 to 18 V dc
- Capable of driving two low-power TTL loads, one low-power Schottky TTL load, or two HTL loads over the rated temperature range

FIGURE 20.30 CMOS multiplier circuit and specifications: (c) mathematical processes performed by multiplier; (d) data specifications. (Courtesy of Motorola Semiconductor Products Div.)

MAXIMUM RATINGS (voltages referenced to V_{SS})

Rating	Symbol	Value	Unit
DC supply voltage	V_{DD}	−0.5 to +18	V dc
Input voltage, all inputs	V_{in}	−0.5 to V_{DD} +0.5	V dc
DC current drain per pin	I	10	mA dc
Operating temperature range—AL device CL/LP device	T_A	−55 to +125 −40 to +85	°C
Storage temperature range	T_{stg}	−65 to +150	°C

(d)

column alignment in a shift register. Some computing systems perform multiplication by means of software rather than hardware, and multiplication can, of course, be performed by repetitive summations, using the adder circuits described earlier in this chapter. However, with the decreasing cost profile of ICs, the use of hardware multiplication has become very inexpensive and has resulted in the replacement of some analog multipliers by binary multipliers in certain applications.

Figure 20.30a illustrates the circuit configuration for a 1-bit, or "bit-by-bit," multiplier. This circuit, which is composed of CMOS AND, OR, and XOR gates, which should be quite familiar to the reader by now, is the core of the IC chip forming the 2-bit multiplier and adder shown in Fig. 20.30b. The mathematical equations performed by the IC are listed in Fig. 20.30c, and a partial data sheet listing some characteristics and input-output parameters are shown in Fig. 20.30d. This chip and many similar chips by other manufacturers perform the multiplying function in many microprocessors, calculators, and computers. The reader should work through the circuit of Fig. 20.30a and b to further understand the mechanism of binary multiplication and to observe how the most elemental binary multiplication—1×0 and 1×1—is performed. The outputs of the 2-bit multiplier—$S0$, $S1$, and $S3$—are the digits of the product and are generally stored in a register of some sort.

The basic 2-bit multiplier is commonly used in tandem with many other similar circuits to multiply large binary numbers. Note that the basic 2-bit multiplier of Fig 20.30 can be used for a variety of mathematical processes—division, obtaining square roots, etc. For further study of these more complex binary mathematical embodiments, consult Refs. 7 to 9.

20.5 TYPES OF INTEGRATED CIRCUITS

The integrated circuit has already been introduced in several previous discussions in this text but without much of a definition. This lack of definition is due in part to the fact that the name largely defines what it is. Also because ICs are so much a part of life in modern industrial societies a definition seems superfluous. Certainly, every reader is aware of the countless ICs in the car, office, kitchen, timepiece, etc. IC logic families and a brief description of IC internal construction are presented in Sec. 19.6. IC external packages are also discussed in that section and have been illustrated in several device discussions, such as Fig. 20.16. However, between the choice of the logic application (such as an adder, buffer, microprocessor, flip-flop, etc.) and that final package that we buy at the local electronics store, there is a very important stage involved with how the IC, or chip, as it is commonly called (i.e., the actual piece of silicon or gallium arsenide) is created or manufactured. The method of IC manufacture is largely an economic consideration and is a function of the quantity of ICs required and, to a certain extent, the application in which the IC will be used.

An IC is basically an electronic circuit in miniature, and all the circuit laws and analytical techniques presented in previous sections of this text apply without exception to this tiny circuit on a semiconductor chip, usually too small to be observed by the naked eye. The techniques of IC manufacture are one of the great achievements of twentieth-century technology, far too rich and diverse to be discussed in this text. CMOS has become the major IC architecture and is distinguished among the many manufacturers by the width of the conductors etched on the chip and other details of circuit construction, such as crossover techniques, turn radii, etc. In general, the thinner the conductor, the greater the "packing density," i.e., number of logic functions such as gates per unit area of the chip. In multilayer chips, this architecture also affects the chip depth and hence its volume. The conductor widths and circuit architecture also determine the electrical parameters of the electronic circuit on the chip—resistance, inductance, and capacitance—which, in turn, determine the chip parameters such as rise time, delay time, required drive power, etc., just as in large circuits. The various levels of electronic circuit manufacture are as follows:

1 *PC board*, using discrete components. This is obviously the first step in developing an electronic circuit, computer, or system, and it is often called *preprototype*. The circuit parameters (R, L, and C) of a PC board layout are vastly different from those of an IC.

2 *Programmable read-only memories (PROMs)*. The PROM is one of the most common types of chips manufactured and consists of a large number of AND gates. The programmable version has fused links in the interconnecting leads which can be broken by ultraviolet light or electrical signals to form the desired logic circuit. PROMs are suitable for prototype or small production applications with annual production volumes of 20,000 units or less.

3 *Programmable logic devices (PLDs)*. This technique is viable for prototypes or small production quantities of about 100,000 or less per year. It is further described below.

4 *Standard cells*. These are chip layouts or plans from which the user can design a wide variety of digital systems at relatively low cost. The plans consist of a "library" of available logic functions, such as gates, adders, flip-flops, registers, etc., which the manufacturer can supply; the wiring architecture available; input-output blocks, including buffers; and pin count for connection to external circuits. Figure 20.31 shows a typical standard cell layout offered by one manufacturer. The customer can design a digital system from these lists of available components and wiring instructions, and it can be manufactured at much less cost than a custom chip since the manufacturing process has been set up by the manufacturer. Chips manufactured in this manner are also called *semicustom* and are widely used in the development of *application-specific ICs* (ASICs)—ICs designed for one specific application. Generally, standard cell technology is viable for annual production quantities of 50,000 to 1,000,000.

5 *Custom ICs*. This is the technology of manufacture for very high-volume applications, such as the assembly of ICs used in commercial appliances, automotive applications, radio, TV, videos, cameras, etc. The cost of setting up the manufacturing process is high, but the actual assembly cost is extremely low. Custom assembly is therefore suited for production quantities of 1,000,000 or more.

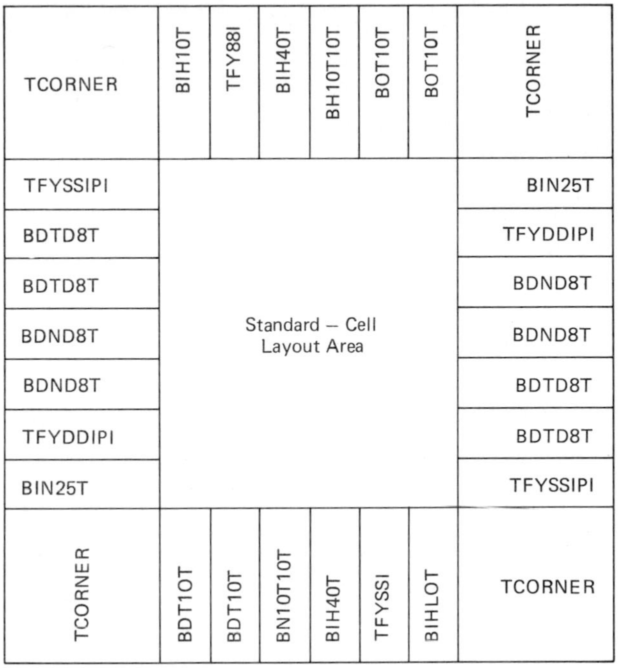

FIGURE 20.31
Layout of plan for
a semicustom
chip. (Courtesy
AT&T.)

TCORNER	BIH10T	TFY88I	BIH40T	BH10T10T	BOT10T	BOT10T	TCORNER
TFYSSIPI							BIN25T
BDTD8T							TFYDDIPI
BDTD8T		Standard — Cell Layout Area					BDND8T
BDND8T							BDND8T
BDND8T							BDTD8T
TFYDDIPI							BDTD8T
BIN25T							TFYSSIPI
TCORNER	BDT10T	BDT10T	BN10T10T	BIH40T	TFYSSI	BIHLOT	TCORNER

20.5.1 Semiconductor Memories

By far the most common digital circuit is the semiconductor memory, and billions are manufactured each year and are associated with almost every type of digital system. Memories are often integrated onto the same chip containing many other digital processes, as in a microcomputer, and they are also built as discrete chips. The basic semiconductor memory is an array of transistors interconnected by a rectangular grid of wires used to "address" each memory cell, as illustrated in Fig. 20.32. In this figure, a cell consists of four transistors. Each cell stores 1 bit. One of the big "games" among semiconductor manufacturers in recent years has been to try to pack the maximum number of memory cells into a unit area of silicon, and the results have been fantastic, as noted below.

Two basic types of memory are used in digital systems:

1 *Random-access memory (RAM).* This is a volatile memory used in time-function operations of a digital system, such as the process of arithmetic operations in a calculator or computer.

2 *Read-only memory (ROM).* This is the nonvolatile memory of a digital system that contains instructions, operating routines, and mathematical routines or stores output data, etc.

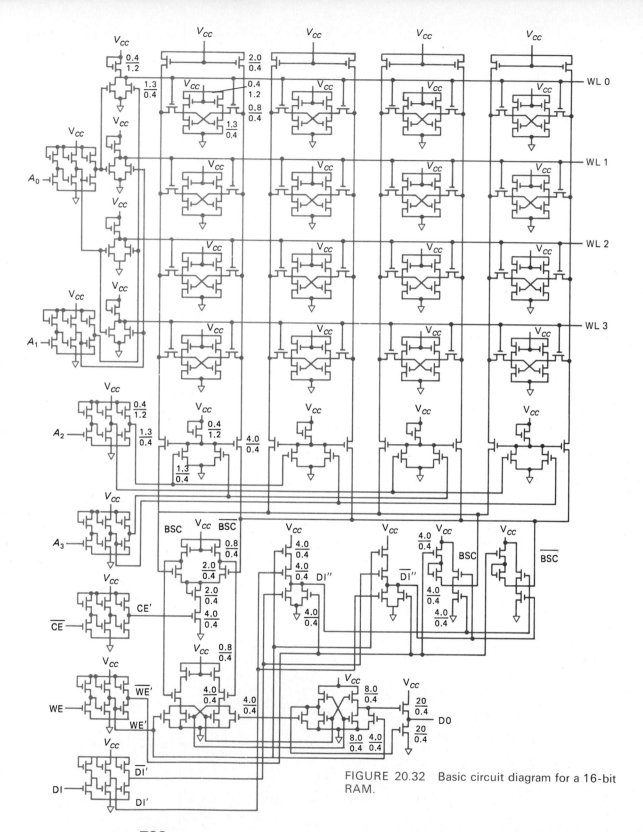

FIGURE 20.32 Basic circuit diagram for a 16-bit RAM.

732

There are further divisions among these two basic memory types:

1   *Static RAMs (SRAMs)*. This is the highest-quality memory chip; it is used for long-term random-access memory in all computer systems, microprocessors, and process controllers.

2   *Dynamic RAM (DRAM)*. The DRAM uses a capacitor to store 1 bit and basically requires only one transistor per bit.

3   ROMs are created by the semiconductor manufacturer and are unalterable. They consist of a vast array of diodes or transistors and have the highest packing density of all memories.

4   PROMs are programmable ROMs using burnable fuses.

5   EPROMs are "erasable" PROMs, where the erasing involves ultraviolet light techniques.

6   *EEPROMs* or *E2PROMs* are "electrically erasable" PROMs. EEPROMs are intended to be rewritten repeatedly, usually by the computer itself. Up to 10,000 read/write cycles are guaranteed by some manufacturers, but at a sacrifice of reduced packing density and added cost.

7   Flash EEPROM. This is a fairly recent erasable chip in which the entire chip is erased by an electrical pulse. It is cheaper, has higher packing density (up to 4096 kilobits per chip in 1990), but fewer rewrite cycles than the EEPROM.

Table 20.4 lists the trends in memory packing density, i.e., the bits of memory that can be built into one chip. The design rule refers to wire widths of CMOS technology.

There is a third category of computer memory known as *cache memory*. This is a mechanism in the memory architecture of a computing system between the main memory and the CPU (central processing unit) used to improve effective memory transfer rates and raise processing speeds. The name is based on the fact that this mechanism is essentially "hidden" and appears transparent to the user. The cache concept anticipates the likely reuse by the CPU of data in main storage by organizing a copy of it in cache memory. The cache memory is usually implemented by an SRAM chip, which is a very high-speed chip, and the main memory utilizes a less costly

TABLE 20.4 CMOS Memory Density Trends

Type of Memory	Design Rules	
	1 μm	0.7 μm
6-transistor SRAMs	64 kilobits (1984)	256 kilobits (1986/1987)
4-transistor SRAMs	256 kilobits (1986/1987)	1 megabit (1988/1989)
DRAMs	1 megabit (1985)	4 megabits (1988/1989)
EPROMs	2–4 megabits (1986)	8–16 megabits (1988)
EEPROMs	256 kilobits–1 megabit (1987/1988)	1–4 megabits (1988/1990)
ROMs	1 megabit (1984)	4–16 megabits (1986/1987)

chip. A 1990 cache SRAM with 16 kilobits per chip is available at 25-, 33-, and 40-MHz operating frequency and has a minimum cost of $65.00.

Still other types of memory chips are known, such as FIFO (first in, first out) and LIFO (last in, first out). These are short-time storage chips and are based on the simple concept of people (or assembly-line products) waiting in a line. In FIFO, the first data stored temporarily in its memory are serviced first; in LIFO, the last in is serviced first.

20.5.2 Programmable Logic Devices

PLDs are chips on which a number of AND and/or OR gates have been etched, along with other functions such as registers and buffers. The devices and input-output terminals are interconnected with networks of fusible wiring which can be broken or interconnected by light or electrical signals to achieve the desired interconnection of the logic devices and input-output ports. Figure 20.33 shows the circuit diagram of a simple PLD chip. Figure 20.34 illustrates a PLD sequencer containing 45 AND and 21 OR gates and which includes several *JK* flip-flops and six bidirectional *I/O* lines. Methods for designing digital systems with the desired characteristics using PLDs are available from the manufacturer, such as the computer design program AMAZE supplied by Signetics. AMAZE is a software development program that has

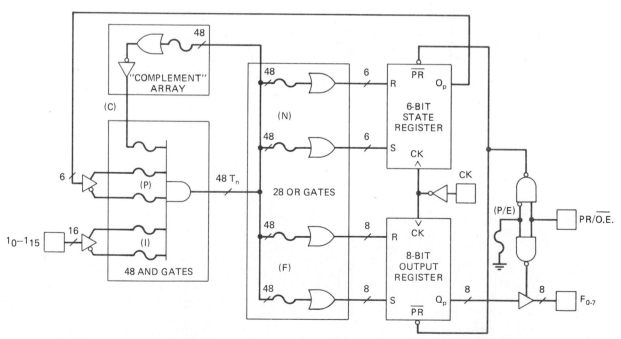

FIGURE 20.33 Circuit diagram for a basic PLD. (Courtesy Signetics Co., Sunnyvale, CA.).

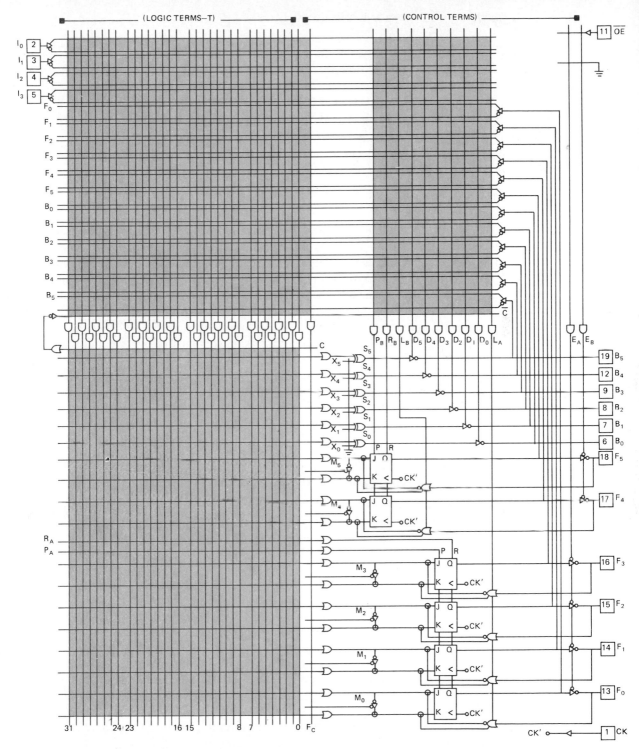

FIGURE 20.34 Circuit diagram for the Signetics PLS-157-field programmable logic sequencer. (Courtesy of Signetics Co., Sunnyvale, CA.).

been developed for both mainframe and microcomputer environments, and it allows data entry in a simple format that can include both boolean algebra and state-variable equations to express the desired logic function.

There are several similar types of programmable logic available, such as *programmable array logic* (PAL), which is a fixed OR gate array with fused interconnections.

20.5.3 IC Parameters

The IC is without doubt the most significant electronic circuit structure in modern technology; therefore, a few of its characteristics merit comment here. As noted above, the technology for designing and manufacturing custom ICs is far beyond the scope of this text, so the standard cell technology is used as the basis for our discussion.

Again we use a simple (2-bit) adder (Fig. 20.35) as the basis for describing some IC characteristics. In this figure, the logic circuit diagram, truth table, and circuit capacitance are shown. Also note that it is an inverting adder and that the nominal temperature is 25°C. The design of an IC using standard cell techniques starts with a review of the library, to ensure that the required logic functions are available; then the physical layout of the logic devices and the I/O buffers and pins, using a more detailed version of Fig. 20.31, is made; the logic of the digital circuit being designed should be checked out with a logic analyzer to ensure that the circuit is capable of achieving its design goals in terms of logic; then a *net list*—the interconnection among devices and with I/O—is made. Most of these operations are performed on a work station which includes graphical facilities for viewing the actual layout and for facilitating changes in device location and wiring. There is an interactive process involved, generally called *optimization*, between the circuit designer and vendor using the work station to achieve the best possible design for a given application. This usually requires some prototyping, possibly using PROMs or PLDs, since the standard cell method is usually associated with fairly large production of a given design; this is even more true of custom IC design. Another degree of optimization available from some vendors is the optimization of the cell itself; the cells of Fig. 20.35 are available optimized for minimum chip area (area) or for performance (perf), which generally refers to the response time. An important circuit parameter in IC performance is the circuit capacitance, which is shown in Fig. 20.35 for the 2-bit adder. Capacitance affects the circuit response time and the power required by the chip.

To evaluate the performance of an IC, computer modeling is most useful. Modeling means are usually supplied with most IC design work stations and are also available from many computer manufacturers and IC vendors. For example, the program AMAZE discussed above for PLD design contains models for ICs using PLDs and includes estimated capacitances within the circuits. The model can predict circuit performance and power losses within the circuit. Losses in resistors on the chip, such as losses in the weighted resistors in a DAC, are calculated in the usual manner. Other IC losses include power required to charge up transistor gates and

Full Adder

This cell is a 2-bit full adder which provides inverted sum and carry outputs. FA cells can be combined to make arbitrary length adders.

Grids 15, **Transistors** 24

Inputs
A,B,C

Outputs
ZCN,ZSN

Capacitances

	A	B	C
Area	0.114pF	0.115pF	0.089pF
Perf	0.415pF	0.416pF	0.315pF

Truth Table

INPUTS			OUTPUTS	
A	B	C	ZSN	ZCN
0	0	0	1	1
0	0	1	0	1
0	1	0	0	1
0	1	1	1	0
1	0	0	0	1
1	0	1	1	0
1	1	0	1	0
1	1	1	0	0

Delay Information

From Input	To Output	Propagation Delay			
		Area		Performance	
		Extrinsic	Intrinsic	Extrinsic	Intrinsic
A ↓	ZCN ↑	8.06ns/pF	0.73ns	1.68ns/pF	0.49ns
A ↓	ZSN ↓	16.45ns/pF	1.23ns	3.57ns/pF	0.85ns
A ↓	ZSN ↑	10.77ns/pF	0.70ns	2.30ns/pF	0.40ns
A ↑	ZCN ↓	4.03ns/pF	1.09ns	1.07ns/pF	0.45ns
A ↑	ZSN ↓	5.95ns/pF	1.01ns	1.56ns/pF	0.45ns
A ↑	ZSN ↑	11.10ns/pF	1.50ns	2.59ns/pF	0.81ns
B ↓	ZCN ↑	8.06ns/pF	0.73ns	1.68ns/pF	0.49ns
B ↓	ZSN ↓	16.45ns/pF	1.23ns	3.57ns/pF	0.85ns
B ↓	ZSN ↑	10.77ns/pF	0.70ns	2.30ns/pF	0.40ns
B ↑	ZCN ↓	4.03ns/pF	1.09ns	1.07ns/pF	0.45ns
B ↓	ZSN ↓	5.95ns/pF	1.01ns	1.56ns/pF	0.45ns
B ↑	ZSN ↑	11.10ns/pF	1.50ns	2.59ns/pF	0.81ns
C ↓	ZCN ↑	6.47ns/pF	0.58ns	1.36ns/pF	0.40ns
C ↓	ZSN ↓	14.14ns/pF	1.11ns	3.12ns/pF	0.72ns
C ↓	ZSN ↑	10.77ns/pF	0.70ns	2.30ns/pF	0.40ns
C ↑	ZCN ↓	4.03ns/pF	1.09ns	1.07ns/pF	0.45ns
C ↑	ZSN ↓	5.88ns/pF	1.03ns	1.56ns/pF	0.45ns
C ↑	ZSN ↑	11.10ns/pF	1.50ns	2.59ns/pF	0.81ns

VDD=5V, T=25°C, Nominal Process.

Motis Model

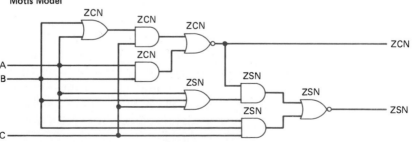

FIGURE 20.35
A full adder
(Courtesy AT&T).

internal transistor losses:

$$P_c = CV^2 f = \text{ac dissipation in charging capacitance of internal gates} \qquad (20.12)$$

where C = total capacitance being charged, F

$\qquad V$ = power-supply maximum voltage, V

$\qquad f$ = charging frequency, Hz

$$P_{\text{int}} = P_g N f_c A = \text{ac power for logic gates} \qquad (20.13)$$

where P_g = power per gate (typically 5 to 10 μW/gate/MHz)

$\qquad N$ = total number of gates

$\qquad f_c$ = clock frequency, MHz

$\qquad A$ = activity factor (typically 20 to 50 percent), which is an
acknowledgment that not all ICs are working at one time

$$P_{\text{ext}} = \tfrac{1}{2} V^2 C_0 f_c A$$
$$= \text{ac dissipation in charging output or buffer capacitance} \qquad (20.14)$$

where C_0 = sum of capacitance on all chip outputs.

Example 20.9   Determine the internal dissipation for an IC with 5000 gates in a synchronous design with a clock frequency of 8 MHz and an activity factor of 20 percent.

Solution   Assume the minimum power per gate of 5 μW; the dissipation is then

$$P_{\text{int}} = 5 \times 10^{-6} \times 5000 \times 8 \times 0.2 = 40 \text{ mW}$$

In general, the total chip power is

$$P_{\text{chip}} = P_c + P_{\text{int}} + P_{\text{ext}} \qquad (20.15)$$

Chip power dissipation is not negligible, and the removal of heat from the chip must be considered in its design. Silicon chips can generally be operated at chip temperatures up to 125°C; gallium arsenide chips can operate at much higher temperatures.

A final parameter of interest is the size of a chip. From much of the glamor and media coverage associated with ICs, one might think that a chip is of negligible size. But all ICs have a finite size, and the silicon or gallium arsenide that is required for a million chips can represent a sizable cost item. Reference 3 gives a means of estimating the size of a chip designed from standard cell technology

$$S = \sqrt{k_1 N_g^2 + k_2 N_g + 1.2 A_{\text{sp}}} + H_b \qquad (20.16)$$

where S = length of one chip side, mils (thousandths of an inch)

N_g = number of grids on chip; grid is a measure of width in standard cell design. For example, a standard cell AND or OR gate is 2.6 grids wide (it contains four transistors, in general; therefore, a transistor is 0.67 grid wide).

k_1, k_2 = constants to account for routing; in general, $k_1 = 5 \times 10^{-6}$ and $k_2 = 0.45$.

A_{sp} = area (in square mils) of special blocks, such as memory or PLDs, used on chip

H_b = height of buffer ring, $\approx$ 30 to 50 mils

Formula (20.16) is said to give size estimates within ± 10 percent; accurate size can be determined from the graphical layout on the work station.

20.6 MICROPROCESSORS

A microprocessor, also known as a *microcontroller*, is an assemblage of digital circuits used to perform a specific function. (Most of these circuits have been described in previous sections of this text.) A microprocessor consists of three basic elements: an arithmetic logic unit (ALU), short-term memory registers and accumulators, and control block. A *microcomputer* is, in the simplest form, a microprocessor to which have been added on-chip (or on-board) memory location, input-output (I/O) ports (or terminals), interrupt functions, and additional ROM (read-only memory). There are also secondary elements in microprocessors: buffers, used to accept input data and control signals; buses, which may be a common set of physical connections or a channel along which data can be sent; a small amount of memory or long-term data storage, usually for the purpose of storing the processor program or set of instructions; and a clock or timing control device.

The microprocessor is actually a miniaturization of the basic computing element of a computer, often called the *central processing unit* (CPU). With the addition of permanent data storage (memory) and facilities for communication with peripherals such as printers, video terminals, disk storage, and tape decks, the microprocessor becomes a minicomputer or microcomputer. A microprocessor is often called the "guts" or the "heart" of a computer system. Each microprocessor is supplied with an instruction set—the software containing the total list of instructions that can be executed by the microprocessor.

However, a microprocessor is also a stand-alone system and has found widespread use in almost every area of system control. A great many automobiles today have engines controlled by microprocessors. These microprocessors accept inputs from sensors which sense such parameters as vehicle speed, engine rotation speed, engine torque (or manifold air pressure), exhaust vacuum, and exhaust temperature and determine the best mode of engine (and sometimes transmission) operation to minimize exhaust pollutants and maintain good vehicle efficiency at each condition of vehicle speed and torque. In the chemical and petroleum industries, as well as in similar process control industries, microprocessors control many refining, mixing, and other processes. Microprocessors are also widely used in operations as diverse as the control of steel or aluminum rolling processes, robotic control (as seen in Fig. 20.19), food processing, and many, many others.

One example of a block-diagram representation of a microprocessor is shown in Fig. 20.36. There are virtually countless variations to microprocessor circuit configurations (often called the system *architecture*), which is an indication of the great diversity of control and computing functions that can be performed by microprocessors. Most of the blocks in Fig. 20.36 should be familiar to the reader by now; the remainder may be described as follows:

The *arithmetic logic unit* (ALU) is basically an assemblage of adder/subtractor circuits plus some logic circuitry. The ALU carries out the fundamental arithmetic operations of adding and subtracting and the logic operations of AND, OR, or their complements. In some microprocessors, multiplication is also included in the ALU. The ALU accepts data from the data bus, processes the data according to fixed instructions (from program storage) and/or to external control signals, and feeds the results into temporary storage, from which various external control and actuation functions are performed.

The *accumulators* are parallel storage registers (see Sec. 19.4.2) used for processing the work in progress. Addresses and data may be stored in them temporarily, and some are used for housekeeping functions.

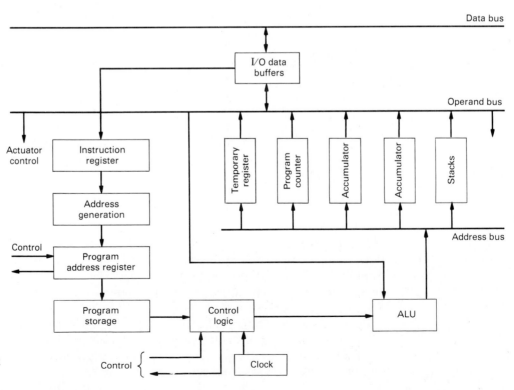

FIGURE 20.36 Microprocessor generalized block diagram.

The *stacks* provide temporary data storage in a sequential order—the last-in first-out (LIFO) order. The stacks are used mainly during the execution of subroutines and eliminate the need to read and write temporary data into memory.

The *program counter* is a register/counter which holds the *next* instruction address. It is also used in program control.

Microprocessors have instruction sets ranging from around 20 to several hundred instructions. These instructions—called *microprograms*—consist of a series of operations of both the arithmetic and logical types, and they include instructions for fetching and transferring data. The microprograms are stored in ROM and initiate the routines of the microprocessor.

Microprocessors are categorized by the *word size* in bits; 1-, 4-, 8-, and 16-bit microprocessors are common. Usually, the larger the word size, the more powerful the processor. Another term related to word size is the *byte*, which is defined as 8 bits.

20.6.1 Microprocessor Buses and Internal Connections

The term *bus* has essentially the same meaning in microprocessing as in electric power systems—namely, a location where common electrical connections are made. In power systems, the word also connotates a point of fixed electric potential, which is not necessarily true of microprocessor buses. In power systems, buses are frequently constructed of long, parallel electrical conductors, often called *busbars*. A microprocessor bus is also constructed of parallel electrical conductors (although of vastly different current-carrying ability), with each conductor representing a digit in the word length. A bus is used to facilitate communication between more than two devices. A microprocessor system bus consists of three physical buses: the address bus, the data bus, and the control bus. In a power system, all incoming circuits to a bus are generally energized simultaneously. The types of circuits connected to microprocessor buses are registers, accumulators, or buffer circuits between the bus and the external memory or input-output circuits. It is required that these many incoming circuits to a bus be electrically connected to the bus individually, one at a time—a process called READ if the signal is incoming to the bus and WRITE if it is outgoing from the bus. Every time that the system clock generates a READ or WRITE pulse, this causes the registers to feed signals into or out of a bus; at all other times, the registers normally read a binary 1 into all conductors of the bus. Therefore, if the selected register is to read in some binary 0s, there will be a conflict on the lines to which 0s are being read:

1 *Wired-OR connection*: In this technique, the registers and buses are designed such that a binary 0 always overrides a binary 1. This is illustrated in Fig. 20.37.

2 *Tristate output.* This method adds a third operating state to the register. This is a state of very high impedance, which more or less disconnects the register from the bus when it is not being read.

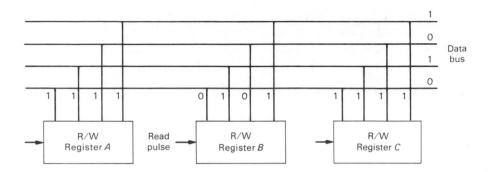

3 *Multiplexing.* This is probably the most obvious solution, which is to connect each register to the bus on a time-shared basis only when it is being read.

There are several additional classes of buses used in computer systems, including the *operating bus*, used to transfer various internal operations and commands, and *interface buses.* The latter are means of communicating between the computer and the "outside world." By far the most important interface bus is the IEEE-488 or GPIB (general-purpose interface bus). IEEE stands for the Institute of Electrical and Electronic Engineers by whom this bus was developed in 1975. This bus permits communications with external devices, such as oscilloscopes and other types of instruments, data collection devices, display devices, etc., as long as such devices are equipped with the IEEE-488 bus. Most modern instruments are equipped with this bus interface.

20.6.2 Addressing

Addressing is the process of locating bits of information in memory. Large microprocessors generally have an address bus of 16 lines or 16 bits. As we have seen in Sec. 20.4.3, n binary bits can represent up to 2^n decimal numbers; therefore, a 16-bit address bus can address up to $2^{16} = 65,536$ memory locations, an 8-bit bus can address 256, and so on.

In many microprocessors and computers, addressing is performed with the hexadecimal code to avoid long strings of binary numbers. The relationship between binary and hexademical is given in Table 20.3. Groups of hexadecimal (or binary) numbers which represent alphanumeric characters (numerals, letters, and other symbols) are called a *character code*. The most common code in microprocessors is the American Standard Code for Information Interchange (ASCII), which is a 7-bit code and is shown in Table 20.5. The relationship between the bit location, binary values, and hexadecimal values for the ASCII character ESC is

	Left	Right
Binary	001	1011
Hexadecimal	1	B

TABLE 20.5 ASCII 7-Bit Character Code*

| Right | | Left | | | | | | | |
| | | Binary | 000 | 001 | 010 | 011 | 100 | 101 | 110 | 111 |
Binary	Hexadecimal	Hexademical	0	1	2	3	4	5	6	7
0000	0		NUL	DLE	SP	0	@	P	/	P
0001	1		SOH	DC1	!	1	A	Q	a	q
0010	2		STX	DC2	"	2	B	R	b	r
0011	3		EXT	DC3	#	3	C	S	c	s
0100	4		EOT	DC4	S	4	D	T	d	t
0101	5		ENQ	NAK	%	5	E	U	e	u
0110	6		ACK	SYN	&	6	F	V	f	v
0111	7		BEL	ETB	.	7	G	W	g	w
1000	8		BS	CAN	(	8	H	X	h	x
1001	9		HT	EM	)	9	I	Y	i	y
1010	A		LF	SUB	*	:	J	Z	j	z
1011	B		VT	ESC	+	;	K	[	k	{
1100	C		FF	FS	,	<	L	/	l	/
1101	D		CR	GS	−	=	M	]	m	}
1110	E		SO	RS	0	>	N	↑	n	~
1111	F		SI	US	/	?	O	←	o	DEL

* *Right* refers to the four rightmost, or least significant, digits; *left* to the left three, or most significant, digits.

ASCII words are frequently stored as 8-bit words. It is common to use a 1 or 0 for the leftmost bit (MSB). Another practice is to use this extra bit as a means of checking that a word is unaltered as it is moved around the processor, in which case the extra bit is called the *parity bit*.

20.6.3 Microprocessor Instructions

Instructions generally consist of 1, 2, or 3 bytes. The first byte contains the type of operation to be performed and is known as the *operating code*. The second and third bytes contain an address and possibly some data. The three main types of instructions in a microprocessor are

1 Data transfer—moving data in and out of registers from memory or inputs

2 Arithmetic and logic—the normal mathematical and logic functions that we associate with calculators and computers

3 Test and branch—the word or data are tested to verify correctness and also to determine the next step in the program

TABLE 20.6 Symbols in Common Use	
Symbol	Meaning
←	Is transferred to
( )[ ]	Contents of register or memory location named in the parentheses
↔	Is exchanged with
$[A]'$, $(A)'$	Contents of register A are complemented

Symbols in common use for describing some of the instruction processes are shown in Table 20.6. Instructions are designated by a code name called a *mnemonic*. This code name is not an abbreviation; rather, it is a group of letters used to describe the instruction when programming is done in a microprocessor in *assembler language*, which is a level of programming between the higher-level languages, such as BASIC (beginner's all-purpose symbolic instruction code) or FORTRAN (formula translation), and the machine code itself. For example, in the Motorola 6800 microprocessor an instruction is described as follows:

ABA: Add accumulator A to accumulator B
$A \leftarrow (A) + (B)$
Binary code: 00011011
Hexadecimal code: $1B$

The upper line is the instructions mnemonic and its description, the second line is the symbolic representation of what the instruction will do, and the third and fourth

TABLE 20.7 Characteristics of Selected Microprocessors

Designation	Manufacturer	Technology	Maximum Power Dissipation, W	Minimum Instruction Time, μs	Word Length Data/Instruction, Bits	On-Board Clock/Pulse Rate, MHz	Direct Address Range, Words
Mc14500	Motorola	CMOS		1.0	1/4	Yes/1.0	0
COP 402	National	NMOS	0.75	4.0	4/8	Yes/1.0	1K
8 × 305	Signetics	Bipolar	1.65	0.25	8/8	No/10.0	8K
8080A	Intel	NMOS		1.5	8/8	No/2.0	64K
80188	Intel	NMOS	3.0	0.2	8/16	Yes/8.0	1M
80286	Intel	NMOS	3.0	0.25	16/16	No/8.0	1G (Virtual)
MC68020	Motorolo	HCMOS	1.5	0.13	32/32	16.67	16M
Micro/J-11	D.E.C.	CMOS	1.0	0.2	32/16	Yes/20	4M

lines are the binary and hexadecimal codes for the instruction. The complete meaning of this particular instruction *ABA* is, "Add the contents of accumulator *A* to the contents of accumulator *B*, placing the sum in *A* and leaving the contents of *B* unchanged." For example, if the contents of *A* were 01100010 and the contents of *B* were 00001111 when this instruction command was given, the following changes in the two accumulators would occur:

	A	*B*
Before	01100010	00001111
After	01110001	00001111

Each instruction of a microprocessor causes the microprocessor to go through a number of routines and operations, i.e., through what we have described as a microprogram. A microprogram has two basic sets of routines, known as *fetch* and *execute*. In the fetch portion of a microprogram, the instructions to be implemented are retrieved from memory and stored in registers of the microprocessor; the operating code of the instruction is always stored in a register known as the *index register*. In the execute portion, the mathematical, logical, transfer, or cycling processes called for in the instructions are executed to completion. A collection of microprograms arranged to perform one or more functions is known as a microprocessor program. For further information on microprocessor programming, consult Refs. 8 to 11.

Table 20.7 summarizes the data on the MC14500B and several other microprocessors in general use.

No. Basic Instructions	No. General- Purpose Registers	Floating-Point Arithmetic?	Interrupt Levels	Package Type/Pins	High-Level Language?	No. Stack Registers	RAM, Bits
16	1	No	1	DIP/16	No	0	—
49	64	Yes	3	DIP/40	No	RAM	64 × 4
8	8	No	—	DIP/50	No	0	
78	8	Yes	1	DIP/40	Yes	RAM	128 × 8
88	8	Yes	4	SMT/68	Yes	4 × 16	
92	8	Yes	1	SMT/68	Yes	4 × 16	
61	16	Yes	7		Yes	3 × 32	
112	12	Yes	4	DIP/60	Yes		

20.7 MICROCOMPUTERS

As we have noted, a microprocessor is actually the heart of a full computing system. It is essentially what gives a computer its intelligence. Any computer, whether a microcomputer, a minicomputer, or a mainframe computer, consists of four principal parts: the central processing unit (CPU), which is basically a microprocessor unit (MPU); storage (memory); input components; and output components. Whereas an MPU is generally dedicated to performing fairly specific tasks, the CPU of a computer must have much more general and diverse capabilities. Therefore, it generally has an expanded control section, many more instructions, and additional buses.

In addition to the ROM used to contain instructions and other basic computer housekeeping functions, computer storage is generally of two types: main memory and off-line memory. Main memory is the working memory of a computer. Programs permanently stored on disks (or other types of off-line memory, such as drums, tape decks, and optical storage) are brought into the working memory and partitioned and stored temporarily in such a way as to make the operation of the program efficient. Both the working and off-line memories are of the type known as random-access memory (RAM); as its name implies, any bit of information stored in RAM can be accessed within a very short time.

The working memory is constructed of semiconductor memory elements, although magnetic bubble memories are becoming more common. The off-line memory most often consists of rotating disks, although rotating drums are still used in some large storage requirements. Disks and drums consist of surfaces coated with a thin magnetic oxide film upon which small regions of magnetization represent a binary 1 or 0. The packing density (bits per square inch) is very high, and access by means of a high-speed pickup head is very rapid. Memory size is one means of quantizing the size of a computer. Memory size is designated by rounding off the memory locations discussed in Sec. 20.5.2. A 16-bit memory size which can store 2^{16} (or 65,536) bits is designated as a 64K memory; likewise, a 2^{10}-bit (or 1024-bit) memory is designated as a 1K memory. A computer with a core memory of 64K and above is generally classified as a minicomputer. Disk memory sizes vary from a few K bits in the very small floppy-disk memories to many megabytes in the large steel disks used in minicomputers and mainframe computers.

Disks, tape decks, and printers are the principal input and output devices associated with microcomputers and minicomputers and are jointly termed *peripherals*. Figure 20.38 illustrates the basic organization of a microcomputer. A microprocessor is the arithmetic and control section of the microcomputer, to which have been added memory, input-output (I/O) ports, additional clocks and/or timing circuits, and interrupts. The function of an interrupt is to halt the normal computer program in response to certain conditions, such as a logic decision in the computer program, a flag (or warning) about a program system malfunction, or a request for higher-priority operations. After the interrupt, the normal program may or may not be resumed, depending on the reason for the interrupt and the program philosophy. The power supply and interfaces to peripheral devices are external to the microcomputer.

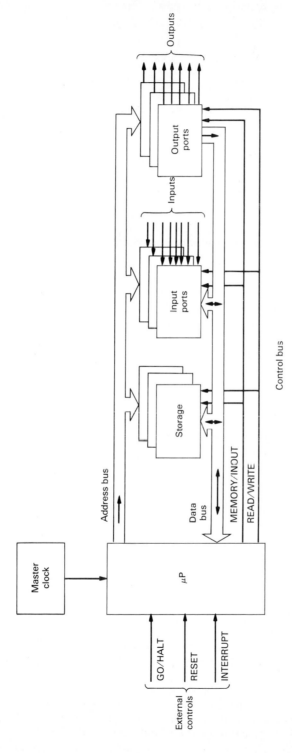

FIGURE 20.38   Generalized block diagram of a microcomputer. (Courtesy of Integrated Computer Systems)

TABLE 20.8 Characteristics of Selected Microcomputers

Designation	Manufacturer	Technology	Maximum Power Dissipation, W	Word Length Data/Instruction, Bits	On-Board Clock/Pulse Rate, MHz	Minimum Instruction Time, µs	Number of Basic Instructions	On-Board RAM, bits
COP420	National	NMOS	0.75	4/8	Yes/2.0	4.0	26	64 × 4
PPS–4/1	Rockwell	PMOS	1.0	4/8	Yes/0.2	5.0	50	48 × 4
Z8	Zilog	NMOS		8/8	Yes/4.0	1.5	47	144 × 8
MC6801	Motorola	NMOS	1.2	8/8	Yes/4.0	2.0	82	128 × 8
MC6801U4	Motorola	HNMOS	1.2	8/8	Yes/4.0		98	192 × 8
MC6805	Motorola	HNMOS	0.7	8/8	Yes/4.0	2.0	61	96 × 8
MC68HC11	Motorola	HCMOS	0.11	8/8	Yes/8A		139	256 × 8
8022	Intel	NMOS		8/8	Yes/3.6	8.4	70	64 × 8
8048AH	Intel	HNMOS	1.5	8/8	Yes/6.0	2.5	96	64 × 8
8748	Intel	CMOS	1.0	8/8	Yes/6.0	1.36	96	64 × 8
80C51	Intel	CHMOS	20	8/8	Yes/12	1.0	111	128 × 8
8751	Intel							
8396	Intel	NMOS	1.5	16/8	Yes/12	1.0	83	232 × 8
68200	MOSTEK	NMOS	1.0	16/16	Yes/6.0	1.0		128 × 16

Table 20.8 summarizes some of the characteristics of several microcomputers in general use. The reader is probably well aware that the computing systems designated by the prefixes "mini" and "micro" are often more powerful and capable than many of the large mainframe computers of just a few years ago. Figure 20.39 illustrates the architecture of a recent "lap-top" computer using the Intel 32-bit 386SL chip.

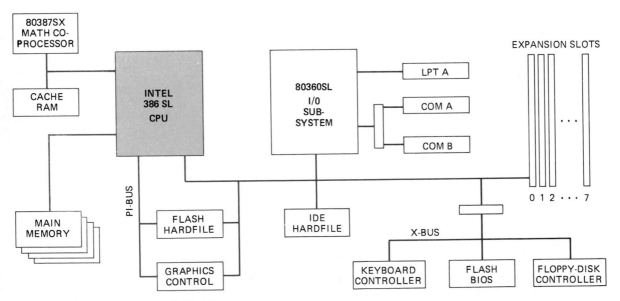

FIGURE 20.39
Intel's SL SuperSet combines in two VLSI components—the 386 SL microprocessor and 82360SL peripheral subsystem chip—virtually all the critical elements needed for a range of 386 CPU-based products. (Courtesy of Intel Corp.).

On-Board ROM, Kbits	On-Board EPROM, Bhts	I/O Ports		Interrupt Levels	Package/Pins	Floating-Point Arithmetic?	Number of General-Purpose Registers	On-Board Timers	On-Board ADC	Maximum Temperature °C	
		Ser	Par							Op.	Ston
1×8	0	3	20	1	DIP/28	No	4		—	70	150
2×8	0	3	28	2	DIP/40	Yes	1		—		
$\frac{2}{4} \times 8$	0	4	28	6	DIP/40	Yes	124		—		
2×8	0	1	29	2	DIP/40	Yes		1	—		
4×8	0	1	29	2	DIP/40	Yes	2	1	—		
3.6×8	128	1	32	1	DIP/40	Yes		2	—		
8×8	512	2	5	2	SMT/52	Yes	7	1	Yes	85	150
2×8	0	3	24	1	DIP/28	Yes	8	1	Yes	85	150
1×8	0	3	24	1	DIP/40	Yes	7	1	—	70	150
—	1024	3	24	1	DIP/40	Yes	7	1	—	70	150
4×8	0	2	32	2	DIP/40	Yes	32	2	—		
8×8	0	2	5	8	SMT/68	Yes		4	Yes	70	150
2×16		1	6		DIP/48	Yes	8	3	—		

PROBLEMS

20.1 Determine the first three terms of the Fourier series for a square wave with a period of 1 ms. Assuming these three terms adequately represent the square wave, what is the minimum sampling frequency, or Nyquist rate?

20.2 In the PWM scheme illustrated in Fig. 20.9, it is desired to cover a range of numerical values from 0 to 20 by means of the square pulse widths shown in Fig. 20.9c. If the half period of the analog function $f(t)$ is 2 ms, suggest a magnitude and frequency for the sawtooth wave (Fig. 20.90b) to accomplish this, and sketch the resulting waveforms. What is the average value (over a full period) of the modulated square pulse when representing a magnitude of 2? of 5? of 20?

20.3 A certain signal is represented by the equation $f(t) = 4 \cos \omega t = \cos 3 \omega t$, where $\omega = 377$. A PAM scheme is to sample at 4 times the Nyquist rate. Sketch the resulting pulses for (a) bipole square-topped PAM and (b) unipolar conformal PAM. (c) What is the pulse period T?

20.4 What is the pulse capacity C of a binary PCM scheme with five pulses per group? How many values of the analog signal can be transmitted?

20.5 Prepare a weighted network diagram (Fig. 20.12) and determine the levels of analog voltage output for an 8-bit DAC using a 12-V logic family.

20.6 A DAC has a maximum output of 10 V and accepts 6 binary bits as inputs. How many volts does each analog step represent?

20.7 Repeat Example 20.2 for a 3-bit ladder network.

20.8 Prove that the ladder networks of the type shown in Figs. 20.14 and 20.15 have constant output impedance.

20.9 Repeat Example 20.3 when the analog signal has a value of 10.

20.10 Develop an adder system to add the numbers 18 and 20, and show the system block diagram and the readout.

20.11 Repeat Prob. 20.10 for subtracting 18 from 20.

20.12 Find the 2's complement representation of -25. Subtract 25 from 10, using 2's complement notation. Assume an 8-bit word length.

REFERENCES

1 C. E. Shannon, "Communication in the Presence of Noise," *Proceedings IRE*, vol. 37, no. 1, January 1979.

2 Mischa Schwartz, *Information Transmission, Modulation, and Noise: A Unified Approach*, 3d ed., McGraw-Hill, New York, 1980.

3 "AT&T Standard Cells and Function Blocks." AT&T Manual MR89-011CMOS, January 1990.

4 *Analog Switches and Their Applications*, Siliconix, Inc., Santa Clara, California, 1980.

5 D. F. Hoeschele, Jr., *Analog-to-Digital, Digital-to-Analog Conversion Techniques*, Wiley, New York, 1968.

6 H. Schmid, "Electronic Design Practical Guide to D/A and A/D Conversion," *Electronic Design*, October 1968.

7 J. D. Greenfield, *Practical Digital Design Using ICs*, 2d ed., Wiley, New York, 1983.

8 D. E. Heffer, G. A. King, and D. Keith, *Basic Principles and Practices of Microprocessors*, Halsted Press, New York, 1981.

9 E. A. Lamagna, "Fast Computer Algebra," *Computer* (IEEE), vol. 15, no. 9, September 1982.

10 Ron Bishop, *Basic Microprocessors and the Sixty-Eight Hundred*, Hayden Book, Rochelle Park, New Jersey, 1979.

11 I. Flores and C. Terry, *Microcomputer Systems*, Van Nostrand Reinhold, New York, 1982.

12 M. D. Freedman and L. B. Evans, *Designing Systems with Microcomputers: A Systematic Approach,*, Prentice-Hall, Englewood Cliffs, New Jersey, 1983.

13 D. F. Stout, *Microprocessor Application Handbook*, McGraw-Hill, New York, 1982.

14 P. Lister (ed), *Single-Chip Microcomputers*, McGraw-Hill, New York, 1984.

SELECTED BIBLIOGRAPHY

R. C. Genn, Jr., and E. L. Genn, *Microcomputer Handbook with Tested BASIC Programs*, Prentice-Hall, Englewood Cliffs, N.J., 1984.

A. B. Williams, *Designer's Handbook of Integrated Circuits*, McGraw-Hill, New York, 1984.

S. Cowan, *Handbook of Digital Logic*, Prentice-Hall, Englewood Cliffs, N. J., 1985.

A. D. Freedman, *Fundamentals of Logic Design and Switching Theory*, Computer Science Press, 1986.

D. E. Heffer, G. A. King, and D. Keith, *Basic Principles and Practices of Microprocessors*, Halsted Press, New York, 1981.

C. K. Adams. *Master Handbook of Microprocessor Chips*, Tab Books, Inc., Blue Ridge Summit, Pa., 1981.

M. F. Hordeski, *The Illustrated Dictionary of Microcomputers*, 3d ed. Tab Books, Inc., Blue Ridge Summit, Pa., 1990

Appendix

Symbol	Description	One (SI Unit)	Is Equal to (USGS Unit)
		Unit Conversion	
B	Magnetic flux density	tesla (T) (= 1 Wb/m^2)	6.452×10^4 lines/in^2
H	Magnetic field intensity	ampere per meter (A/m)	0.0254 A/in
ϕ	Magnetic flux	weber (Wb)	10^8 lines
D	Viscous damping coefficient	newton meter-second (N · m · s)	0.73756 lb · ft · s
F	Force	newton (N)	0.2248 lb
J	Inertia	kilogram-square meter (kg · m^2)	23.73 lb · ft^2
T	Torque	newton-meter (N · m)	0.73756 ft · lb
W	Energy	joule (J)	1 W · s

Table of Certain Constants

Permeability of free space $= 4\pi \times 10^{-7}$ H/m
Permittivity of free space $= 8.85 \times 10^{-12}$ F/m
Resistivity of copper $= 1.724 \times 10^{-8}\ \Omega \cdot$ m
Resistivity of aluminum $= 2.83 \times 10^{-8}\ \Omega \cdot$ m
Charge of an electron $= -1.6 \times 10^{-19}$ C
Electron mass $= 9.11 \times 10^{31}$ kg
Acceleration of gravity $= 9.807$ m/s^2

Fourier Series

The fourier series is a representation of periodic functions that fulfill certain conditions in terms of their harmonics. A function is said to be periodic if it repeats at certain intervals. Explicitly, for periodicity,

$$F(t) = F(t + T) \tag{1}$$

where T is the period.

Stated formally, if $F(t)$ is a single-valued periodic function which is finite, has a finite number of discontinuities, and has a finite number of maxima and minima over a period T, then $F(t)$ may be represented over a complete period by a series of simple harmonic functions, the frequencies of which are integral multiples of the fundamental frequency. This series is known as *Fourier series*. Expressed mathematically, we have

$$F(t) = \frac{A_0}{2} + \sum_{n=1}^{\infty} (A_n \cos n\omega t + B_n \sin n\omega t) \tag{2}$$

where the coefficients A_n and B_n are given by

$$A_n = \frac{2}{T} \int_0^T F(t) \cos n\omega t \, dt \tag{3}$$

and

$$B_n = \frac{2}{T} \int_0^T F(t) \sin n\omega t \, dt$$

The term $A_0/2$ in Eq. (2) is introduced so that the general result of Eq. (3) is applicable when $n = 0$, that is, for A_0 as well. It may be seen from (2) that $A_0/2$ is the average value of the function over the period. Using (3) for A_0, we have

$$A_0 = \frac{2}{T} \int_0^T F(t) \, dt \tag{5}$$

753

Laplace Transforms

We define the *Laplace transform* of a function $f(t)$, where $f(t) \equiv 0$ for $t \leq 0$, as

$$\mathscr{L}\{f(t)\} \equiv F(s) = \int_0^\infty e^{-st} f(t)\, dt \tag{1}$$

In Eq. (1), s is a complex variable, $s = \sigma + j\omega$, with σ chosen large enough to make the infinite integral converge. Given $f(t) = e^{-at}(a > 0)$, we find $F(s)$.
   Substituting in Eq. (1) yields

$$F(s) = \int_0^\infty e^{-st} e^{-at}\, dt = \frac{e^{-(s+a)t}}{-(s+a)}\bigg|_0^\infty = \frac{1}{s+a}$$

where, to make the exponential vanish at the upper limit, it has been assumed that $\sigma > -a$. However, this restriction may be dropped, and the result

$$F(s) = \frac{1}{s+a}$$

may be considered valid over the entire complex S plane.
   Hence, the Laplace transform pairs in Table D.1 are developed.

TABLE D.1 Laplace Transform Pairs

	$f(t)$	$F(s)$
1	$\delta(t)$	1
2	a	$\dfrac{a}{s}$
3	t	$\dfrac{1}{s^2}$
4	e^{-at}	$\dfrac{1}{s+a}$
5	te^{-at}	$\dfrac{1}{(s+a)^2}$
6	$\sin \omega t$	$\dfrac{\omega}{s^2+\omega^2}$
7	$\cos \omega t$	$\dfrac{s}{s^2+\omega^2}$
8	$e^{-at}\sin \omega t$	$\dfrac{\omega}{(s+a)^2+\omega^2}$
9	$e^{-at}\cos \omega t$	$\dfrac{s+u}{(s+a)^2+\omega^2}$
10	$\dfrac{d}{dt}[f(t)]$	$sF(s) - f(0+)$
11	$\dfrac{d^2}{dt^2}[f(t)]$	$s^2F(s) - sf(0+) - f'(0+)$
12	$\dfrac{d^n}{dt^n}[f(t)]$	$s^nF(s) - s^{n-1}f(0+) - s^{n-2}f'(0+) - \cdots - f^{n-1}(0+)$
13	$\displaystyle\int_0^t f(\tau)\,d\tau$	$\dfrac{1}{s}F(s)$
14	$af(t) + bg(t)$	$aF(s) + bG(s)$
15	$\displaystyle\int_0^t f(\tau)g(t-\tau)\,d\tau$	$F(s)G(s)$
16	$f(\infty)$ final value	$\displaystyle\lim_{s\to 0} sF(s)$
17	$f(0+)$ initial value	$\displaystyle\lim_{\substack{s\to\infty\\ s\,\text{real}}} sF(s)$

Solution of Simultaneous Equations

Since the node voltage and mesh current methods of Chapter 3 involve writing (and solving) linear simultaneous algebraic equations, it is appropriate to examine methods for solving these types of equations.

A set of two linear simultaneous algebraic equations in two unknowns is written in the form

$$a_{11}x_1 + a_{12}x_2 = b_1$$
$$a_{21}x_1 + a_{22}x_2 = b_2$$

(1)

Here x_1 and x_2 are the two unknowns to be solved for. The coefficients a_{11}, a_{12}, a_{21}, and a_{22} are known quantities. The first subscript on these quantities denotes the equation and the second subscript denotes the unknown which it multiplies. The two quantities on the right-hand side, b_1 and b_2, are also known quantities. A set of three linear simultaneous algebraic equations in the unknowns x_1, x_2, x_3, would appear as

$$a_{11}x_1 + a_{12}x_2 + a_{13}x_3 = b_1$$
$$a_{21}x_1 + a_{22}x_2 + a_{23}x_3 = b_2$$
$$a_{31}x_1 + a_{32}x_2 + a_{33}x_3 = b_3$$

(2)

These equations are said to be linear since none of the unknowns x_1, x_2, x_3 appears in the equations as products of each other or raised to a power other than unity. If any equation contained, for example, $x_1 x_3$ or x_2^3, then the entire set would be classified as a nonlinear set. Linear equations are considerably easier to solve than nonlinear ones. The equations are said to be algebraic since they do not involve derivatives of the unknowns, such as $dx_2(t)/dt$. Algebraic equations are also easier to solve than differential equations, which involve derivatives of the unknowns.

756

Probably the simplest technique for solving linear simultaneous algebraic equations is the method of Gauss elimination. Gauss elimination is also the technique which is used exclusively to solve these equations with a high-speed digital computer. The basic idea is to reduce the equations to an equivalent form which is triangular. For example, we would try to reduce the set of three equations in Eq. (2) to the equivalent, triangular form:

$$c_{11}x_1 + c_{12}x_2 + c_{13}x_3 = d_1$$
$$c_{22}x_2 + c_{23}x_3 = d_2 \tag{3}$$
$$c_{33}x_3 = d_3$$

If the solutions for x_1, x_2, x_3, of Eqs. (3) and (2) are the same, the equations are said to be equivalent. However, Eq. (3) is much easier to solve than is Eq. (2). Once the set is in triangular form, as in Eq. (3), we may use the "back substitution" technique. First solve the last equation for x_3:

$$x_3 = \frac{d_3}{c_{33}} \tag{4}$$

Then substitute into the second equation

$$c_{22}x_2 + c_{23}\frac{d_3}{c_{33}} = d_2 \tag{5}$$

and solve:

$$x_2 = \frac{1}{c_{22}}\left[-c_{23}\left(\frac{d_3}{c_{33}}\right) + d_2\right] \tag{6}$$

Now substitute x_3 from Eq. (4) and x_2 from Eq. (6) into the first equation of Eq. (3) and solve for x_1.

The final point we have to address is how to reduce the set in Eq. (2) to the equivalent triangular form in Eq. (3). We can multiply any equation in the set by a nonzero number without changing the solution; we can also add or subtract two equations and replace one of the two with the result. These are the only two techniques needed to reduce a set to equivalent triangular form. Once triangular form is achieved, we can solve for the unknowns by back substitution.

Example 1   Consider the set of linear simultaneous algebraic equations

$$x + 3y - z = 4$$
$$2x - 4y + 2z = 3$$
$$3x + y + z = 2$$

Multiply the first equation by -2 and add the result to the second equation;

$$x + 3y - z = 4$$
$$0 - 10y + 4z = -5$$
$$3x + y + z = 2$$

Note that this has removed x from the second equation. Now multiply the first equation by -3 and add the result to the third equation:

$$x + 3y - z = 4$$
$$0 - 10y + 4z = -5$$
$$0 - 8y + 4z = -10$$

Thus, we have eliminated x from the third equation. Now let us eliminate y from the third equation and we will have achieved the triangular form. Multiply the second equation by $-\frac{8}{10}$ and add the result to the third equation:

$$x + 3y - z = 4$$
$$0 - 10y + 4z = -5$$
$$0 + 0 + \tfrac{4}{5}z = -6$$

Now solve, by back substitution, for the variables. The third equation yields

$$z = \frac{-6}{\frac{4}{5}}$$
$$= -\tfrac{15}{2}$$

Substituting this result into the second equation gives

$$-10y + 4(-\tfrac{15}{2}) = -5$$

from which we obtain

$$y = -\tfrac{5}{2}$$

Substituting these results into the first equation gives

$$x + 3(-\tfrac{5}{2}) - (-\tfrac{15}{2}) = 4$$

or

$$x = 4$$

We should substitute these supposed solutions into the original set to see if any numerical errors were made:

$$(4) + 3(-\tfrac{5}{2}) - (-\tfrac{15}{2}) \overset{?}{=} 4$$
$$2(4) - 4(-\tfrac{5}{2}) + 2(-\tfrac{15}{2}) \overset{?}{=} 3$$
$$3(4) + (-\tfrac{5}{2}) + (-\tfrac{15}{2}) \overset{?}{=} 2$$

This shows that our "solutions" do indeed satisfy the original equations. It is a good idea always to substitute back into the original equations to see whether we have obtained the solution.

A second technique which is useful for hand calculations is Cramer's rule. It is generally not used for digital computer calculations, since Gauss elimination is much faster. Cramer's rule requires the use of the concept of a determinant. To illustrate why this occurs, let us solve the set of two equations in Eq. (1) in general form by Gauss elimination. We obtain (the reader should verify this)

$$x_1 = \frac{a_{22}b_1 - a_{12}b_2}{a_{11}a_{22} - a_{12}a_{21}}$$

$$x_2 = \frac{a_{11}b_2 - a_{21}b_1}{a_{11}a_{22} - a_{12}a_{21}} \tag{7}$$

Note that the denominator of each solution involves

$$a_{11}a_{22} - a_{12}a_{21} \tag{8}$$

If we arrange the coefficients of Eq. (1) in the following array

$$\begin{bmatrix} a_{11} & a_{12} \\ a_{21} & a_{22} \end{bmatrix} \tag{9}$$

then Eq. (8) is said to be the determinant of the 2×2 array. The terms a_{11} and a_{22} are said to be the main diagonal terms of the array, whereas the terms a_{12} and a_{21} are said to be the off-diagonal terms. The determinant of the 2×2 array is denoted with vertical bars and is the product of the main diagonal terms minus the product of the off-diagonal terms:

$$\begin{vmatrix} a_{11} & a_{12} \\ a_{21} & a_{22} \end{vmatrix} = a_{11}a_{22} - a_{12}a_{21} \tag{10}$$

Now we see a simple rule for computing the solutions in Eq. (7):

$$x_1 = \frac{\begin{vmatrix} b_1 & a_{12} \\ b_2 & a_{22} \end{vmatrix}}{\begin{vmatrix} a_{11} & a_{12} \\ a_{21} & a_{22} \end{vmatrix}}$$

$$x_2 = \frac{\begin{vmatrix} a_{11} & b_1 \\ a_{21} & b_2 \end{vmatrix}}{\begin{vmatrix} a_{11} & a_{12} \\ a_{21} & a_{22} \end{vmatrix}}$$

$$(11)$$

This result is called Cramer's rule and we will generalize it to larger numbers of equations. Note that the solution for an unknown is the ratio of two determinants. The denominator determinant is the determinant of the array. The numerator determinant is formed from the coefficient array by replacing the column of the array corresponding to the unknown to be solved for with the right-hand side known quantities of the equations.

The extension of Cramer's rule to more than two equations is identical to the results for two equations but is slightly more involved in the evaluation of the resulting determinants. For example, Cramer's rule for the set of three equations in Eq. (2) gives the solutions as

$$x_1 = \frac{\begin{vmatrix} b_1 & a_{12} & a_{13} \\ b_2 & a_{22} & a_{23} \\ b_3 & a_{32} & a_{33} \end{vmatrix}}{\Delta}$$

$$x_2 = \frac{\begin{vmatrix} a_{11} & b_1 & a_{13} \\ a_{21} & b_2 & a_{23} \\ a_{31} & b_3 & a_{33} \end{vmatrix}}{\Delta}$$

$$x_3 = \frac{\begin{vmatrix} a_{11} & a_{12} & b_1 \\ a_{21} & a_{22} & b_2 \\ a_{31} & a_{32} & b_3 \end{vmatrix}}{\Delta}$$

$$(12)$$

where Δ is the 3×3 determinant of the array of coefficients:

$$\Delta = \begin{vmatrix} a_{11} & a_{12} & a_{13} \\ a_{21} & a_{22} & a_{23} \\ a_{31} & a_{32} & a_{33} \end{vmatrix}$$

$$(13)$$

Evaluation of each of these determinants is done in terms of the 2×2 determinants each contains. Select any row or column and multiply each element in that selected row or column by its cofactor. The sum of these products is the value of the determinant of the 3×3 array. A cofactor of a particular element appearing in the ith row and jth column is the product of $(-1)^{i+j}$ and the determinant formed by removing the ith row and jth column. For example, to evaluate the 3×3 determinant in Eq. (13) we may (arbitrarily) select the second column and the determinant is

$$\Delta = a_{12}(-1)^{1+2} \begin{vmatrix} a_{21} & a_{23} \\ a_{31} & a_{33} \end{vmatrix}$$

$$+ a_{22}(-1)^{2+2} \begin{vmatrix} a_{11} & a_{13} \\ a_{31} & a_{33} \end{vmatrix}$$

$$+ a_{32}(-1)^{3+2} \begin{vmatrix} a_{11} & a_{13} \\ a_{21} & a_{23} \end{vmatrix} \tag{14}$$

The remaining 2×2 determinants are evaluated as before. We will consider determinants of no higher order than 3×3, so the above discussion suffices for Cramer's rule.

The sign of each cofactor can be found by using the simple "checkerboard sign pattern"

$$\begin{vmatrix} + & - & + \\ - & + & - \\ + & - & + \end{vmatrix} \tag{15}$$

Example 2   Consider the set of three equations in Example 1:

$$x + 3y - z = 6$$

$$2x - 4y + 2z = 3$$

$$3x + y + z = 2$$

Solve for x, y, and z using Cramer's rule.

Solution   The array of coefficients is

$$\begin{bmatrix} 1 & 3 & -1 \\ 2 & -4 & 2 \\ 3 & 1 & 1 \end{bmatrix}$$

The determinant of this array is found by expanding along the (arbitrarily selected) third row:

$$\Delta = 3(-1)^{3+1} \begin{vmatrix} 3 & -1 \\ -4 & 2 \end{vmatrix} + 1(-1)^{3+2} \begin{vmatrix} 1 & -1 \\ 2 & 2 \end{vmatrix} + 1(-1)^{3+3} \begin{vmatrix} 1 & 3 \\ 2 & -4 \end{vmatrix}$$

$$= 3(6 - 4) - 1(2 + 2) + 1(-4 - 6)$$

$$= -8$$

The solutions are therefore

$$x = \frac{\begin{vmatrix} 4 & 3 & -1 \\ 3 & -4 & 2 \\ 2 & 1 & 1 \end{vmatrix}}{-8}$$

$$= \frac{-32}{-8}$$

$$= 4$$

$$y = \frac{\begin{vmatrix} 1 & 4 & -1 \\ 2 & 3 & 2 \\ 3 & 2 & 1 \end{vmatrix}}{-8}$$

$$= \frac{20}{-8}$$

$$= -\frac{5}{2}$$

$$z = \frac{\begin{vmatrix} 1 & 3 & 4 \\ 2 & -4 & 3 \\ 3 & 1 & 2 \end{vmatrix}}{-8}$$

$$= \frac{60}{-8}$$

$$= -\frac{15}{2}$$

as were obtained by Gauss elimination in Example 1.

Answers to Selected Problems

CHAPTER 2

2.1 1.012×10^6 tons.   **2.3** 7.2×10^{-9} N.   **2.5** 24×10^{-6} J.   **2.7** -5.25 V.
2.9 2A; 3A; 0 A; -3 A; 0 A.
2.13 (a) -6 W; (b) -12 W; (c) -6 W; (d) -4 W; (e) -6 W; (f) 12 W.
2.15 $i_y = 5$ A; $i_z = 2$ A.   **2.17** $i_1 = -3$ A; $i_2 = -1$ A.
2.19 $V_x = -6$ V; $V_y = -4$ V; $V_{ba} = 2$ V.   **2.21** $V_x = -1$ V; $V_y = 4$ V; $V_z = -4$ V.
2.23 $V_x = -2$ V; $V_y = -1$ V; $V_z = -4$ V.   **2.25** -6 V
2.27 -4 W; -4 W; 12 W: -4 W.   **2.29** 2 W; 3 W; -5 W.
2.31 $v_x = -10/3$ V; $i_x = -5/3$ A.   **2.33** $v_x = -26/7$ V; $i_x = -8/7$ A.
2.35 $v_x = 5/2$ V; $i_x = 5/4$ A.   **2.37** $v_x = 39/7$ V; $i_x = 1/7$ A.
2.39 $v_x = -51/5$ V; $i_x = 13/5$ A.

CHAPTER 3

3.1 $8/3$ Ω.   **3.3** 2 Ω.   **3.5** 5 Ω.   **3.7** $5/3$ V.   **3.9** $10/3$ V.   **3.11** 2 V
3.13 $5/3$ A.   **3.15** $5/2$ A.   **3.17** $-9/8$ A.   **3.19** $v_x = 4$ V; $i_x = 1$ A.
3.21 $v_x = -13/4$ V; $i_x = -3/4$ A.   **3.23** $v_x = 11/2$ V; $i_x = 1/4$ A
3.25 $v_x = -14/3$ V; $i_x = -19/3$ A.   **3.27** $11/2$ V; $11/2$ A.
3.29 $V_{OC} = 4$ V; $I_{SC} = 2$ A; $R_{TH} = 2$ Ω.   **3.31** $V_{OC} = 15/2$ V; $I_{SC} = 15/4$ A.
3.33 $7/10$ A.   **3.35** $7/4$ A.   **3.37** 3 V; $7/2$ A.   **3.39** 1 V; $-1/4$ A.
3.41 $-3/2$ V; $-1/4$ A.   **3.43** $10/31$ V; $-34/31$ A.   **3.45** 1 V; -2 A.
3.47 1 Ω; $49/16$ W.   **3.49** $1/12$ W.

CHAPTER 4

4.1 $10\ \mu$A; 0; $-5\ \mu$A; 0.   **4.3** $v(t) = \frac{1}{2}\int_{-\infty}^{t} i_s(t)\,dt$; $w(t) = V^2$.   **4.5** $w(t) = \frac{1}{10}V^2$.
4.9 $28/15$ F.   **4.11** ± 0.925 V; ∓ 0.925 A.   **4.13** -20 kA; 20 kA; 40 kA; 20 kA; 40 kA.
4.14 $w(t) = i^2$ **4.19** $16/5$ H.   **4.21** $22/5$ H.   **4.25** $-0.81 \sin 3t$ V.
4.27 $(p^2 + \frac{1}{6}p + \frac{1}{12})i_{out} = \frac{1}{12}t$.   **4.29** $(p + \frac{4}{3})v_{out} = 4 \cos 4t$.
4.31 $(p + 15)i_{out} = 75t + 30 \cos 3t$.   **4.33** $(p^2 + p + 2)i_{out} = 6 + 10t$.   **4.35** $-7/5$ A.
4.37 1 V.

CHAPTER 5

5.1 (a) $3\cos(2t - 120°)$; (c) $\sin(t - 45°)$; (e) $3\sin(3t + 25°)$; (g) $2\cos(3t - 120°)$.
5.3 (a) $4.47\sin(2t - 26.57°)$ A; (b) $3.84\cos(3t - 9.8°)$ A.   **5.5** $1.665\sin(3t - 116.319°)$ A.
5.7 (a) $4.844 \ \angle -55.67°$; (b) $1.0 \ \angle 150°$; (c) $6.324 \ \angle -101.57°$; (e) $1.581 \ \angle -41.56°$;
(g) $0.316 \ \angle 71.57°$; (i) $4.937 \ \angle 28.23°$.
5.9 (a) $4.472\sin(2t - 26.57°)$; (b) $3.841\cos(3t - 9.806°)$.   **5.11** $1.664\sin(3t - 116.31°)$.
5.13 $2.828\cos(2t - 105°)$.   **5.15** $2\cos(2t + 66.87°)$.   **5.17** $4.808\cos(2t - 116.3°)$.
5.19 $8.32\sin(3t + 3.69°)$.   **5.21** $5\cos(2t - 90°)$.
5.23 $7.071\sin(3t + 90°) + 3.328\cos(2t - 3.69°)$.   **5.25** $5.475\cos(3t + 63.833°)$.
5.27 $1.487\cos(2t - 51.09°)$.   **5.29** $0.754\cos(4t - 145.65°)$.   **5.31** 9.62 W; 9.61 W.
5.33 6.16 W; 1.69 W; 4.44 W; 6.13 W.   **5.35** $2/25$ F.   **5.41** 0.796 mH.
5.43 10.6 pF; 100.   **5.45** $0.116\ \mu$H.

CHAPTER 6

6.1 $2.5(1 - e^{-4t/3})$.   **6.3** $-3e^{-2t}$.   **6.5** $-5/2$ A.   **6.7** $5/3$ A.   **6.9** $5e^{-6t/25} - 1$.
6.11 $10(1 - e^{-3t/2})$.   **6.13** $4e^{-15t/4} + 8$.   **6.15** $10(1 - e^{-2t})$.   **6.17** $3e^{-5t/3}$.
6.19 $5(e^{2t} + 1)$.   **6.21** $i_L(t) = 5(e^{-t} - e^{-3t})$; $v_c(t) = 5(-3e^{-t} + e^{-3t} + 2)$.
6.23 $i_L(t) = \frac{5}{3}e^{-2t}\sin 3t$; $v_c(t) = e^{-2t}(-10\cos 3t - \frac{20}{3}\sin 3t) + 10$.
6.25 $i_L(t) = -5e^{-5t} - 25te^{-5t} + 5$; $v_c(t) = 100te^{-5t}$.   **6.27** $i_L(t) = -\frac{4}{3}e^{-t} + \frac{4}{3}e^{-3t}$,
$v_c(t) = 4e^{-t} - \frac{4}{3}e^{-3t} + 6$.   **6.29** $i_L(t) = -\frac{5}{2}e^{-5t} - \frac{25}{2}te^{-5t} + \frac{15}{4}$; $v_c(t) = 50te^{-5t}$.
6.31 $i_L(t) = 5e^{-2t}(\frac{1}{2}\cos 3t - \frac{1}{3}\sin 3t)$; $v_c(t) = -\frac{65}{3}e^{-3t}\sin 3t$.
6.33 $i_L(t) = -9.656e^{-t} + 10.867e^{-2t} + 3.947\sin(3t - 17.87°)$;
$v_c(t) = 6.477e^{-t} - 3.642e^{-2t} + 2 + 0.877\sin(3t - 107.87°)$.   **6.35** $i_L(t) = \frac{10}{3}te^{-3t}$;
$v_c(t) = -10e^{-3t} - 30te^{-3t} + 10$.

CHAPTER 7

7.3 $0.226\ \mu$A; $2.27\ \mu$A.   **7.5** 1.8 mA; 0.99 mW.   **7.7** 2.5 mA; 0.6 V; 132 mW.
7.9 1.8 mA; 0.55 V; 10.2 mW.   **7.11** 2 mA; 0.275 mA.   **7.19** $500\ \Omega$

CHAPTER 8

8.1 200; 1000.   **8.5** $I_D = 6$ mA; $R_D = 1.333$ kΩ; $A_V = -2$.   **8.7** $R_S = R_D = 1$ kΩ;
$R_{G1} = 78947\ \Omega$; $R_{G2} = 1.5$ MΩ; $A_V = -1$.   **8.9** $A_V = -3.182$; $A_V = -0.780$.
8.11 $A_0 = 0.0833$ **8.13** $A_V = -3.03$; $A_1 = -20.2$; $Z_i = 667.7$ kΩ; $Z_0 = 4.325$ kΩ;
neglecting $1/y_{os}$; $A_V = -3.33$; $A_I = -22.2$; $Z_i = 66.7$ kΩ; $Z_0 = 5$ kΩ.
8.15 $R_D = 10$ kΩ: $A_V = -6$; $A_I = -15$; $A_p = 90$; $R_D = 5$ kΩ: $A_V = -4$; $A_I = -10$;
$A_p = 40$.   **8.17** $A_V = -0.1174 \ \angle 76°$.   **8.19** $A_V = -15$; $Z_i = 323\ \Omega$; $Z_0 = 10$ kΩ;
$A_I = 0.485$.   **8.21** $R_B = 158.33$ kΩ.   **8.23** $R_B = 162.857$ kΩ.   **8.25** $I_B = 58.25\ \mu$A;
$I_C = 3.67$ mA.   **8.27** $A_V = -171.43$; $Z_i = 250\ \Omega$; $A_I = -42.86$; $Z_o = 909\ \Omega$;
$A_p = 7347.49$.   **8.29** $A_V = -2.19$; $Z_i = 2236\ \Omega$; $Z_o = 2$ kΩ; $A_I = -4.89684$; $A_p = 10.72$.
8.31 $Z_i = 23.9\ \Omega$; $Z_o = 10$ kΩ; $A_I = 0.478$; $A_p = 95.9$.
8.33 $A_V = -30.6$; $A_I = -24.4$; $Z_i = 796\ \Omega$; $A_p = 745$.   **8.35** $A_V = 2 \times 10^{-3}$.

CHAPTER 9

9.1 -2 V.   **9.3** 2 mV.   **9.5** $v_0 = (v_2 - v_1)R_2/R_1$.   **9.7** $(R_f + R_i)/R_i$.   **9.9** -10 kΩ.
9.11 $R_i = 1592$ Ω; $R_f = 3182$ Ω.   **9.15** $-25/2$ A.   **9.17** 50/3 A; 20/3 V; 100/9 W.
9.19 $R_1 = R_f = 1$ kΩ; $C = 100$ pF; $R = 1592$ Ω

CHAPTER 10

10.7 $R_C = 2400$ Ω; $R_B = 43$ kΩ.   **10.11** $V_i = 1.5$ V.   **10.13** $R_B = 763$ Ω.

CHAPTER 11

11.1 $I_l = 14.43 \angle 45°$ A; $I_p = 8.33 \angle 15°$ A; 2500 VA.   **11.3** $14.5 \angle -159.9°$ A;
$10.77 \angle 51.8°$ A; $7.81 \angle -26.3°$ A.   **11.5** $14.33 \angle -81°$ A; $10.77 \angle 111.8°$ A; $6.07 \angle 45.5°$ A.
11.7 4.8 kvar.   **11.9** (a) 25.98 kvar; (b) $V_l = 346.42$ V.   **11.11** $2.05 \angle 79°$ A.
11.13 92.38; 64 kW.   **11.15** 2734 VA.   **11.17** Wye-connected load: 19.95 kW;
11.5 kvar; delta-connected load: 12.83 kW; 15.28 kvar.   **11.19** $189.6 \angle -79.65°$ A.

CHAPTER 12

12.1 8800; 500.   **12.3** 2.46×10^6 At.   **12.5** 6.0 A.   **12.7** 47 turns.   **12.9** 48 V.
12.11 2; 25 A; 50 A.   **12.13** 0.446.   **12.15** 120 V; 110 V; 100 A; 9.09 A.
12.17 (a) 0.8 Ω; 80 Ω.   **12.19** Maximum efficiency when core loss = I^2R-loss.
12.21 96.4 percent.

CHAPTER 13

13.1 (a) 36 V; (b) 67.5 V.   **13.3** 114.6 Nm.   **13.5** (a) 96.875 kW; (b) 98 kW.
13.7 0.1215 Ω   **13.9** 11.428 kW.   **13.13** (a) 254 V; (b) 226 V.   **13.15** 90.6 percent.
13.17 802.8 r/min.   **13.19** 692.8 r/min.   **13.21** 1297.6 r/min   **13.23** 226.52 Ω.
13.25 (a) 0.704 Ω; (b) 479.5 Nm.   **13.27** 43.2 V.   **13.29** 4 A.   **13.31** 18.52 A.

CHAPTER 14

14.1 1000 r/min.   **14.3** 52.72 percent.   **14.5** 11.54 kV (line-to-line).
14.7 82.17 percent.   **14.9** 8077 V/phase; 8.74°.   **14.11** 274 V/phase; 11.75°.
14.13 0.83 leading.   **14.15** (a) 10.48 kV; (b) 11.04 kV; (c) 11.56 kV (all values line-to-line).
14.17 0.53 leading; 196.9 A.   **14.19** 107.67 kVA.

CHAPTER 15

15.1 1200 r/min: 4 percent.   **15.3** (a) 5 percent; (b) 3 Hz; (c) (i) 1800 r/min; 188.5 r/s.
(ii) same as (i).   **15.5** 63.44 Nm.   **15.7** 0.4 Ω.   **15.9** 3341 r/min.
15.11 76.95 percent.   **15.13** (a) 24.08 kW; (b) 124.98 Nm; (c) 127.8 Nm.   **15.15** 4.06.
15.17 3.2 Ω.   **15.19** (a) 14.61 A; (b) 0.839 lagging.   **15.21** 0.6 Ω.
15.23 (a) 298.14 V; (b) 111.8 A.   **15.25** 25 percent.

CHAPTER 16

16.1 (*a*) 16.67 percent; (*b*) 183.33 percent.   **16.3** 2.246 A; 0.5 lagging; 73.1 W.
16.5 6.92.   **16.7** 0.993 Nm.   **16.9** 0.38 Nm.

CHAPTER 17

17.1 (*a*) $s\tau/(1 + s\tau)$; (*b*) $(1 + \tau_1 s)/(1 + \tau_2 s)$; (*c*) $R_2[(1 + s\tau_1)/(1 + s\tau_1 + s^2\tau_1\tau_2)]$,
$\tau_1 = 2R_1 C$, $\tau_2 = R_2 C/2$; (*d*) $(1 + s\tau_1 + s^2 LC)/(1 + s\tau_2 + s^2 LC)$, $\tau_1 = R_2 C$,
$\tau_2 = (R_1 + R_2)C$.   **17.3** $-R(s^4 + \omega_0^2 s^2 + \omega_0^4)/(s^4 + 2\omega_0^2 L s^3 + 3\omega_0^2 R s^2 + 2\omega_0^2 c + \omega_0^4 R)$,
with $\omega_0^2 = 1/LC$.   **17.5** V_{DC}/V_{CM}.   **17.11** 11.7 cm^2/s.   **17.13** 11,307 cm^2/s.

CHAPTER 18

18.1 0.5 V; ± 0.67 percent.   **18.3** Error in calculation $= 2.42$ percent.
18.5 Error in reading $= \pm 2$ percent.

CHAPTER 19

19.1 (*a*) 1001100; (*b*) 10100000001; (*c*) 1011101110; (*d*) 0.0000111110001;
(*e*) 0.1110001; (*f*) 0.001011001.   **19.7** Yes   **19.9** (*a*) $\sum (7, 5, 4, 3, 1)$;
(*b*) $\sum (15, 14, 13, 12, 11, 7, 6, 5, 4, 1, 0)$; (*c*) $\sum (15, 11, 10, 9, 8, 7, 5, 1)$.
19.11 (*a*) $\pi(6, 2, 0)$; (*b*) $\pi(10, 9, 8, 3, 2)$; (*c*) $\pi(14, 13, 12, 6, 4, 3, 2)$.
19.13 $f(A, B, C, D) = A\bar{B} + ACD + \bar{A}\bar{C}\bar{D} + \bar{A}\bar{B}\bar{C}$.
19.15 $f(W, X, Y, Z) = (X + \bar{Y})(X + Y + Z)(W + \bar{Y} + \bar{Z})$.

CHAPTER 20

20.1 $I_1 = 1$ ms; $f_1 = 1000$ *Hz*; $f_5 = 5000$ Hz.   **20.3** Nyquist rate $= 360$ Hz;
Sampling rate $= 1440$ Hz. (*c*) $T = 0.694$ ms.   **20.5** 0, 0.09375..., 11.90625, 12.0 V.
20.9 Time $= 57\ \mu$s.

INDEX

Ac circuits, 145
Ac commutator motor, 582
Ac tachometer, 584
Admittance, 167
Algebraic sum, 23
Ammeter, clip-on, 641
Ampere's law, 73
Amplifier, 308
    BJT, 342, 347, 356, 370
    current gain, 310, 327, 333, 335,
      363, 365, and 368
    JFET, 311
    linear, 308, 309
    power gain, 311
    small-signal analysis of, 327, 356
    voltage gain, 310, 319, 326, 331,
      362
Amplitude modulation, 36
Analog computers, 405–408
Analog meters, 633, 638
    characteristics of, 638
    dynamometer, 635
    frequency range, 635
    iron vane, 635
    thermocouple, 638
Arithmetic digital system, 716–729
Average power, 37, 172, 183
Average value, 36

Band-pass filter, 193
Band-reject filter, 196
Bandwidth, 195, 649
B-H curve, 471, 474
Binary system, 653
    base 2 system, 653
    decimal-to binary conversion,
      654
Bit, 653
BJT (bipolar junction transistor),
    342
    npn, 342
    pnp, 342
BJT common-base amplifier, 369,
    386
BJT common-collector amplifier,
    369–387
BJT common-emitter amplifier, 369
    ac circuit, 361, 367
    dc circuit, 361
BJT switch, 417
Block diagram 592
Boltzmann constant, 271
Boolean algebra, 656–671
    AND, 656
    connective, 658
    DeMorgan's theorem, 657
    laws, 656–658

    literals, 662
    NAND, 658
    NOT, 657
    OR, 656
    SOP, 664
    truth table, 658
Boolean logic, implementation of,
    661–671
    don't care logic, 670
    gates, 661
    Karnaugh map, 665
    minimizing logic functions, 667
Branch, 21
Brush-slip ring, 498, 548

Capacitance, 104
Capacitor, 104
    energy stored in, 107
    nonlinear, 106
    parallel connection, 110
    series connection, 112
    terminal relations, 106, 107
Carrier frequency, 25
Characteristic equation, 239
Chopper, 515–519
Circuit equations, 128

CMOS (complementary metal-oxide semiconductor), 431
Combinational logic, 437
Comparator, 440
Complex, exponentials, 161
Complex numbers, 151
  conjugate of, 155
  imaginary part of, 151
  in polar form, 152
  real part of, 151
  in rectangular form, 151
Conductance, 28
Conduction band, 268
Control of ac motors, 561
Control of dc motors, 515
Converters, 519
Coulomb's law, 12
Coupled coils dot convention, 119–121
Covalent bond, 269
Cramer's rule, 759
CSMOS (complementary-symmetry MOS), gate, 431
Current division, 67
Current gain, 310
Current source, 32
Current transformation, 478
Current transformer, 642

Damping ratio, 613
Data acquisition, 713
Dc machines, 492–528
  armature, 499
  armature windings, 499–502
  back emf, 506
  characteristics of, 510–513
  classification of, 503
  commutator, 498, 501
  compound, 503, 511
  construction of, 499
  critical resistance of, 510
  efficiency of, 516
  emf equation, 504
  field, 499
  generator, 492
  heteropolar, 493, 495
  homopolar, 493
  losses, 516
  motor, 492

  series, 503
  shunt, 503, 509
  speed control, 515–523
  speed equation, 506
  torque equation, 505
Demodulation, 703
DeMorgan's theorem, 657
Depletion region, 270
  Digital electronic circuits, 411
Digital logic circuits, 653–689
Digital meters, 680–682, 687
Digital systems, 691–747
  chip complexity, 693
  circuit complexity, 693
  data acquisition, 713
  encoder, 699
  signal conversion, 705
Diode:
  actual, 272
  breakdown of, 298
  characteristics of, 272, 273
  equation, 271
  equivalent circuit, 289
  formation of, 267
  forward-biased, 271
  freewheeling, 582, 589
  ideal, 283
  reverse-biased, 271
  Shockley equation, 271
  small-signal analysis of, 278
  symbol for, 272
  zener, 298
Diode gates, 412
  AND gate, 412
  NAND gate, 415
  NOR gate, 415
  NOT gate, 414
  OR gate, 412
DIP (dual in-line package), 373
Donor, 269
Drain, 312
DTL (diode-transistor logic), 425

ECL (emitter-coupled logic), 431
Eddy-current loss, 473
Electric charge, 11
  force on, 12
Electric circuits, 9

Electric current, 16
Electric field, 14
Electric force,14
Electric generator, 492, 493, 496, 537
Electric machine(s), 2
  ideal, 492
Electric motor, 492
  control of, 515–561
Electric potential, 13
Electrical engineering, 1
Electrical instrumentation, 630
  accuracy, 631
  calibration, 633
  current measurement, 634
  error, 631
  power measurement, 634, 642
  precision, 631
  voltage measurement, 634
Electronic circuits, 3
Encoder, 699
  Energy band, 268
Energy conversion, 492
Energy storage element, 104
Euler's identity, 152

Farad, 105
Faraday disk, 493
Faraday's law, 114, 494, 537
Feedback, 591
Feedback control systems, 591–625
  analysis of, 596
  classification, 617, 619
  closed loop, 594
  continuous, 596
  definitions relating to, 591, 592
  digital, 596
  frequency response, 612
  general representation of, 592
  multiple-input, 600
  multiple-loop, 600
  on-off, 598
  open-loop, 594
  open-loop response, 613
  return difference, 595
  stability of, 616
  steady-state gain, 614
  types, 617
FET (field-effect transistor), 311

FET switch, 431
Filters, 185
    active bandpass, 399
    band-reject, 196
    band-pass, 193
    high-pass, 192
    low-pass, 190
Flip-flop, 437
Flux-cutting rule, 494
Flux linkage, 113
Forbidden band, 268
Force equation, 472
Fourier series, 185, 753
Frequency, 35, 145
Frequency modulation, 36, 516
Fringing flux, 472

Gravitational field, 13

Half-power point, 192
Harmonic factor, 520
Harmonics, 185
Hay bridge, 647
High-pass filter, 192
Hybrid parameters, 360
Hysteresis loop, 473
Hysteresis loss, 473
Hysteresis motor, 581

IC (integrated-circuit) logic family,
    428, 680, 692
    CMOS logic, 431, 682
    emitter-coupled logic, 431, 686
    transistor-transistor logic, 428,
    686
IGFET (insulated-gate FET), 378
Ignition coil, 119
Impedance, 166
    measurement, 646
    transformation, 477
Induced voltage, 476, 494, 537
Inductance, 113
    mutual, 119
    self, 119
Induction machines, 548–572
    brushless, 548

deep-bar, 590
double-cage, 570
efficiency of, 556
equivalent circuits, 552–554
generator operation, 548
motor operation, 549–570
performance of, 555–561
slip, 550
slip frequency, 551
slip ring, 548
speed control, 561–567
starting, 568–571
torque, 556
wound-rotor, 548
Induction motor (*See also*
    Induction machines)
single phase, 573–580
Induction motor, variable-
    frequency operation, 565
Inductor, 113
    current through, 114
    energy stored in, 115
    nonlinear, 114
    parallel connection, 116
    series connection, 116
    terminal relations, 114, 115
Insulator, 269
Integrated circuit (IC), 373
    bipolar junction transistor, 376
    fabrication techniques, 375
    field-effect transistor, 378
Integrated circuit logic family (*see*
    IC logic family)
Integrated-circuit technology, 373
Inverter, 562
    ACSI (ac current source
        inverter), 563
    AVI (adjustable voltage inverter),
        564
    PFM (pulse-frequency-
        modulated), 565
    PWM (pulse-width-modulated),
        565
Iron loss, 473

j operator, 151
JFET (junction field-effect
    transistor), 311

JFET amplifier, 316
    common-drain, 337, 385
    common-gate, 337, 385
    frequency response of, 338
    high-frequency model, 339
    small-signal equivalent circuit,
        330

Karnaugh map, 665
Kelvin bridge, 647
Kirchhoff's current law (KCL), 21
Kirchhoff's voltage law (KVL), 22

Ladder network, 708
Lamination, 474
Laplace transform, 592, 754
Leakage flux, 472
Left-hand rule, 499
Lenz's law, 476
    Limiter, 298
    ideal, 298
    synthesis of, 298
Linear amplifier, 308
Linear system, 611
Load line, 275, 318
Logic circuits, hardware
    implementation of, 679
Logic families, 428, 680
Logic gates, 412, 661
Low-pass filter, 190
LSI (large-scale integration), 411
LTC (load tap-changer), 609

Magnetic circuit, 469
Magnetic field, 16
Magnetic flux, 113
Magnetic force, 16
Majority carrier, 270
Matched impedance, 184
Maximum power transfer, 184
Maxwell bridge, 647
Mesh current, 88
Mesh current analysis, 88
Microcomputers, 746–749
Microprocessors, 739–745
Miller effect, 341, 372
Mmf (magnetomotive force), **470**

Modulation, 695
MOS (metal-oxide semiconductor), 379, 431
MOSFET (metal-oxide-semiconductor field-effect transistor), 379, 431
MSI (medium-scale integration), 441
Multivibrator, 437

n-type semiconductor, 267
Node, 21
  reference, 81
Node voltage, 81
Node voltage analysis, 81
Noise, 747
Noise margins, 423
Nonlinear resistor, 28, 267
Norton equivalent circuit, 76, 80

Ohmmeter, 646
Ohm's law, 28
Open circuit, 69
Operating point, 278
  selection of, 319
  stability of, 351
Operational amplifier, 389
  active filter, 399
  frequency response of, 398
  gains, 395
  ideal, 390
  integrator, 397
  inverting, 394, 400
  negative impedance converter, 398
  noninverting, 395
  practical, 398
  summer, 396
Oscilloscope, 648–650

p-type semiconductor, 270
Parallel connection, 57
PAM (pulse-amplitude modulation), 695, 698
PCM (pulse-code modulation), 695, 701
Period, 35, 146

Permeability, 471
  relative, 472
PFM (pulse-frequency modulation), 696
Phasor circuit, 164
  circuit components, in, 131
Phasor notation, 164
pn junction, 267
Polyphase circuits, 455
Potential difference, 14
Power, 18
  average, 37, 174
  complex, 175
  conservation of, 27
  instantaneous, 37, 172
  maximum, 183
  measurement of, 643, 645
  reactive, 176
Power angle, 541
Power factor, 180, 644
Power semiconductors, 515
PPM (pulse-position modulation), 696
PTM (pulse-time modulation), 696
PWM (pulse-width, modulation), 516, 695, 700

Reactive power, 176
Rectifier, 288
  bridge, 297
  full-wave, 294
  half-wave, 288
Resistance, 28
Resistance measurement, 647
Resistivity, 29
Resistor, 28
Resonance, 193
Right-hand rule, 494
Rms (root-mean-square) value, 37
Rotating magnetic field, 534
RTL (resistor-transistor logic), 425

S plane, 614
  poles, 614
  zeros, 614
Sampling, 695
  natural, 700
  Nyquist rate, 696

sample-and-hold circuit, 697
sampling rate, 696
sampling theorem, 696
uniform, 700
Semiconductor, 267
  diode, 267
  extrinsic, 270
  intrinsic, 270
  n-type, 267
  p-type, 270
Sequential logic, 671–679
  clock, 674
  counter, 677
  flip-flop, 672
  register, 674
  shift register, 677
Series connection, 57
Series-parallel reduction, 36–39
Servomotor, 585
Short circuit, 69
Sign convention, 18
Signal conversion, 705
  analog-to-digital, 710
  digital-to-analog, 705
Silicon-controlled rectifier (SCR) (*See* Thyristor)
Sinusoidal source, 36
Skin effect, 590
Small-signal analysis, 280, 327, 356
Square-law resistor, 284
SSI (small-scale integration), 411
Stability of operating point, 351
Stacking factor, 474
Standard meter, 634
Steady-state solution, 210
Stepper motor, 573, 586
Strain gauge, 31
Superposition, principle of, 69, 168
Switching operation, 210
Synchronous machines, 529–547
  armature mmf, 533–536
  construction of, 530
  damper windings on, 533
  emf, 538
  frequency, 537
  generator, 537
  power angle, 541
  reactance, 539
  reluctance torque, 542
  round-rotor, 530, 537–539, 543

salient-pole, 530, 542
speed of, 530, 537
synchronous impedance, 539
synchronous reactance, 539
V curves, 544
voltage regulation, 539
Synchronous motor, 542–544
small, 580–582
Synchronous speed, 536, 537
Thévenin equivalent circuit, 76
Thévenin impedance, 169
Three-phase systems, 455
balanced system, 456
delta connection, 457
four-wire system, 458
line current, 458
line voltage, 458
phase current, 458
phase voltage, 458
power in, 460
wye connection, 457
wye-delta equivalence, 464
Thyristor, 515
commutation, 515
ideal, 515
line commutation, 519
Time constant, 216
Time-domain circuit, 165

Transfer function, 223, 231, 592
Transformers, 469, 476–489
auto-, 488
efficiency of, 483, 484
equivalent circuit, 481–483
ideal, 477–479
nonideal, 481
operation of, 477
polarity of, 486–487
turns ratio, 487
two-winding, 478
voltage regulation, 482
Transient solution, 210
Transients, 210
in LC circuit, 235
in RC circuit, 222
in RL circuit, 212
Transistor, 308
Truth table, 658
TTL (transistor-transistor logic), 428
Turns ratio, 480
Two-phase motors, 584

Unit conversion, 751
Universal bridge, 647

Valence band, 268
Vector addition, 153
Vector subtraction, 153
VLSI (very large-scale integration), 441, 693
Voltage, 13
Voltage division, 65
Voltage drop, 23
Voltage gain, 310
Voltage regulation, 482, 539
Voltage regulator, 609
Voltage rise, 22
Voltage source, ideal, 31
Voltage transformation, 477

Ward-Leonard system, 523, 608
Wattmeter, 643
connections, 645
electronic, 643
Hall-type, 643
Waveforms, 24
Wein bridge, 606
Wheatstone bridge, 647

Zener diode, 298